CAMBRIDGE STUDIES IN
ADVANCED MATHEMATICS 69

MODULAR FORMS AND GALOIS COHOMOLOGY

Books in this series

1 W.M.L. Holcombe *Algebraic automata theory*
2 K. Petersen *Ergodic theory*
3 P.T. Johnstone *Stone spaces*
4 W.H. Schikhof *Ultrametric calculus*
5 J.-P. Kahane *Some random series of functions, 2nd edition*
6 H. Cohn *Introduction to the construction of class fields*
7 J. Lambek & P.J. Scott *Introduction to higher-order categorical logic*
8 H. Matsumura *Commutative ring theory*
9 C.B. Thomas *Characteristic classes and the cohomology of finite groups*
10 M. Aschbacher *Finite group theory*
11 J.L. Alperin *Local representation theory*
12 P. Koosis *The logarithmic integral I*
13 A. Pietsch *Eigenvalues and S-numbers*
14 S.J. Patterson *An introduction to the theory of the Riemann zeta-function*
15 H.J. Baues *Algebraic homotopy*
16 V.S. Varadarajan *Introduction to harmonic analysis on semisimple Lie groups*
17 W. Dicks & M. Dunwoody *Groups acting on graphs*
18 L.J. Corwin & F.P. Greenleaf *Representations of nilpotent Lie groups and their applications*
19 R. Fritsch & R. Piccinini *Cellular structures in topology*
20 H Klingen *Introductory lectures on Siegel modular forms*
21 P. Koosis *The logarithmic integral II*
22 M.J. Collins *Representations and characters of finite groups*
24 H. Kunita *Stochastic flows and stochastic differential equations*
25 P. Wojtaszczyk *Banach spaces for analysts*
26 J.E. Gilbert & M.A.M. Murray *Clifford algebras and Dirac operators in harmonic analysis*
27 A. Frohlich & M.J. Taylor *Algebraic number theory*
28 K. Goebel & W.A. Kirk *Topics in metric fixed point theory*
29 J.F. Humphreys *Reflection groups and Coxeter groups*
30 D.J. Benson *Representations and cohomology I*
31 D.J. Benson *Representations and cohomology II*
32 C. Allday & V. Puppe *Cohomological methods in transformation groups*
33 C. Soulé et al *Lectures on Arakelov geometry*
34 A. Ambrosetti & G. Prodi *A primer of nonlinear analysis*
35 J. Palis & F. Takens *Hyperbolicity and sensitive chaotic dynamics at homoclinic bifurcations*
36 M. Auslander, I. Reiten & S. Smalo *Representation theory of Artin algebras*
37 Y. Meyer *Wavelets and operators*
38 C. Weibel *An introduction to homological algebra*
39 W. Bruns & J. Herzog *Cohen-Macaulay rings*
40 V. Snaith *Explicit Brauer induction*
41 G. Laumon *Cohomology of Drinfeld modular varieties I*
42 E.B. Davies *Spectral theory and differential operators*
43 J. Diestel, H. Jarchow & A. Tonge *Absolutely summing operators*
44 P. Mattila *Geometry of sets and measures in Euclidean spaces*
45 R. Pinsky *Positive harmonic functions and diffusion*
46 G. Tenenbaum *Introduction to analytic and probabilistic number theory*
47 C. Peskine *An algebraic introduction to complex projective geometry I*
48 Y. Meyer & R. Coifman *Wavelets and operators II*
49 R. Stanley *Enumerative combinatories*
50 I. Porteous *Clifford algebras and the classical groups*
51 M. Audin *Spinning tops*
52 V. Jurdjevic *Geometric control theory*
53 H. Voelklein *Groups as Galois groups*
54 J. Le Potier *Lectures on vector bundles*
55 D. Bump *Automorphic forms*
56 G. Laumon *Cohomology of Drinfeld modular varieties II*
59 P. Taylor *Practical foundations of mathematics*
60 M. Brodmann & R. Sharp *Local cohomology*
64 J. Jost & X. Li-Jost *Calculus of variations*
66 S. Morosawa et al. *Holomorphic dynamics*
68 Ken-iti Sato *Lévy processes and infinitely divisible distributions*
69 H. Hida *Modular forms and Galois cohomology*

Modular Forms and Galois Cohomology

Haruzo Hida
UCLA

CAMBRIDGE UNIVERSITY PRESS
Cambridge, New York, Melbourne, Madrid, Cape Town, Singapore, São Paulo

Cambridge University Press
The Edinburgh Building, Cambridge CB2 8RU, UK

Published in the United States of America by Cambridge University Press, New York

www.cambridge.org
Information on this title: www.cambridge.org/9780521770361

First published 2000
This digitally printed version 2008

A catalogue record for this publication is available from the British Library

ISBN 978-0-521-77036-1 hardback
ISBN 978-0-521-07208-3 paperback

Contents

Preface

In the past few years (1995–98), I have given several advanced graduate courses at UCLA in order to provide a comprehensive account of the proof by Wiles (and Taylor) of the identification of certain Hecke algebras with universal deformation rings of Galois representations. Assuming a good knowledge of Class field theory, I started with an overview of the theory of automorphic forms on linear algebraic groups, specifically, $GL(n)$ over number fields. Since second year graduate students often lack knowledge of representation theory of profinite groups, necessary to carry out the task, I went on to describe basic representation theory, the theory of pseudo-representations and their deformation. To reach this point, I had already covered almost a one-year course. Then I continued to give a sketch of the rationality and the control theorems of the space of elliptic modular forms, which is the basis of the definition of the Hecke algebra. In the meantime, K. Fujiwara and F. Diamond independently gave, in 1996, a substantial simplification of the proof of Wiles, which I incorporated in my course. After having proved the theorem, assuming many things, I came back to the material I used in the proof, in particular the duality theorems (due to Poitou and Tate) of Galois cohomology groups. Thus the first chapters follow faithfully my series of courses; so, logically the reader might have to jump around between chapters. However, except for the construction of modular Galois representations, which has been described in some literature already, basically all ingredients of the proof of Wiles are at least covered to some extent. The final chapter (Chapter V) is added after finishing the series of courses, in order to give some indication of further study, and this part contains some new results of mine. An outline of the book can be found in Subsection 2.1 in Chapter 1.

Although I have not covered the proof by Wiles of the Shimura–Taniyama conjecture and Fermat's last theorem, I hope that graduate students can, after finishing this book, thoroughly understand Wiles' original paper treating these two profound results. I hope to return to the theory of elliptic curves and modular Galois representations in a book in the near future.

While I was preparing this book, I received help from many people (including my present and former students) who read the manuscript and provided useful advice and corrections on mathematical, linguistic and historical matters. I wish to thank all of them. I would also like to acknowledge partial support from the national science foundation while I was preparing this book.

May 25th, 1999 at Los Angeles,
Haruzo Hida

1
Overview of Modular Forms

It is difficult to provide a brief summary of techniques used in modern number theory. Traditionally, mathematical research has been classified by the method mathematicians exploit to study their research areas, except possibly for number theory. For example, algebraists study mathematical questions related to abstract algebraic systems in a purely algebraic way (only allowing axioms defining their algebraic systems), differential geometers study manifolds via infinitesimal analysis, and algebraic geometers study geometry of algebraic varieties (and its siblings) via commutative algebras and category theory. There are no central techniques which distinguish number theory from other subjects, or rather, number theorists exploit any techniques available to hand to solve problems specific to number theory. In this sense, number theory is a discipline in mathematics which cannot be classified by methodology from the above traditional viewpoint but is just a web of rather specific problems (or conjectures) tightly and subtly knit to each other. We just study numbers, those simple ones, like integers, rational numbers, algebraic numbers, real and complex numbers and p-adic numbers, and that is it.

What has emerged from our rather long history is that we continue to study at least two aspects of these numbers: the numbers of the base field and the numbers of its extensions. For example, the *quadratic reciprocity law* describes in a simple way how rational primes decompose as a product of prime ideals in a quadratic extension only using data from rational integers. More generally, by class field theory, we know how rational primes decompose in an abelian extension out of the datum from rational numbers. Thus we have two sets of numbers, the first is the numbers of the base field and the other from an extension of the base field. Nowadays, class field theory is often described using transcendental numbers from all possible completions of the base fields, involving complex, real and p-adic numbers. The *adele ring* $\mathbb{A}$ is just a subring of $(\prod_p \mathbb{Q}_p) \times \mathbb{R}$ generated by p-adic integers (for all primes p), real numbers and rational numbers (even additively):

$$\mathbb{A} = ((\widehat{\mathbb{Z}} \times \mathbb{R}) + \mathbb{Q}) \subset \left(\prod_p \mathbb{Q}_p\right) \times \mathbb{R} \quad \left(\widehat{\mathbb{Z}} = \prod_p \mathbb{Z}_p\right),$$

where we regard $\mathbb{Q} \subset \mathbb{A}$ by the diagonal embedding $\xi \mapsto (\xi, \xi, \dots, \xi, \dots) \in \prod_p \mathbb{Q}_p \times \mathbb{R}$. Thus for a given *number field* F (that is, a finite extension of the rational numbers $\mathbb{Q}$), the adele ring $F_{\mathbb{A}} = F \otimes_{\mathbb{Q}} \mathbb{A}$ of F represents all data from the base field. For a given algebraic group G defined over F, which we may think of as just a coherent rule assigning a group $G(A)$ to any F-algebra A, $G(F_{\mathbb{A}})$ is an immediate source of information. For example, $A \mapsto GL_n(A)$, the group of invertible $n \times n$ matrices with coefficients in A, is an algebraic group. Global class field theory is typically described as a canonical exact sequence:

$$1 \to \overline{GL_1(F)C} \to GL_1(F_{\mathbb{A}}) \to \mathrm{Gal}(F^{ab}/F) \to 1$$

for the identity connected component C of $GL_1(F_{\mathbb{A}})$, where '$\overline{X}$' indicates the topological closure of X, and F^{ab}/F is the composite of all Galois extensions M/F (inside an algebraic closure $\overline{F}$ of F) with $\mathrm{Gal}(M/F)$ abelian (such an extension is called an *abelian extension* of F). Thus we have the second set of numbers F^{ab}: those numbers in a Galois extension specific to our choice of the algebraic group $G = GL_1$. In this first example, $G = GL_1$, which is the simplest (and most important) of all abelian algebraic groups. Thus we might call the study of extensions of a base field the *Galois side* of number theory.

The above example tells us that it is important to study the geometry of the homogeneous space $G(F)\backslash G(F_{\mathbb{A}})$. Most geometers, if they are given a topological space, start studying functions on the space, because they know by experience that functions are easier to manipulate and eventually determine the space. We call functions on $G(F)\backslash G(F_{\mathbb{A}})$ *modular forms.* The homogeneous space $G(F)\backslash G(F_{\mathbb{A}})$ often classifies geometric objects, like abelian varieties and motives (as is often the case for a quotient of a big group by a discrete subgroup, because the big group is somehow a (local) transformation group of a collection of geometric objects, and elements of the discrete subgroup give rise to (global) isomorphisms between the objects). For example, when $G = GL_2$, $X = G(\mathbb{Q})\backslash G(\mathbb{A})/G(\widehat{\mathbb{Z}})Z(\mathbb{R})SO_2(\mathbb{R})$ for the maximal connected compact subgroup $SO_2(\mathbb{R}) \subset GL_2(\mathbb{R})$ and the center $Z(\mathbb{R}) \subset G(\mathbb{R})$ classifies isomorphism classes of elliptic curves over

$\mathbb{C}$, and therefore, gives rise to the set of complex points of the (coarse) moduli scheme $\mathbf{P}^1(j)$ (defined over $\mathbb{Q}$) classifying elliptic curves over $\mathbb{Q}$. Because of the classification property of X, we have a canonical algebraic variety $\mathbf{P}^1(j)$ defined over $\mathbb{Q}$ (and actually defined over $\mathbb{Z}$) which gives rise to X. The scheme $\mathbf{P}^1(j)$ is called a *canonical model* of X. This phenomenon that the homogeneous space $G(F)\backslash G(F_{\mathbb{A}})$ classifies some algebro-geometric objects is prevalent in many other cases of different algebraic groups (like symplectic groups $G = Sp(2g)$ and unitary groups $U(m,n)$), and the resulting canonical models are called *Shimura varieties* of PEL-type. In any case, a general homogeneous space $X(U) = GL_2(\mathbb{Q})\backslash GL_2(\mathbb{A})/U \cdot Z(\mathbb{R})SO_2(\mathbb{R})$ for an open subgroup $U \subset GL_2(\widehat{\mathbb{Z}})$ classifies elliptic curves with some additional structure (such as a given point of order N) over $\mathbb{Z}$ (see [AME] and [GMF]). Then the canonical model $X(U)$ is called a *modular curve*, because it is a finite covering of $\mathbf{P}^1(j)$ and hence is an algebraic curve. Thus finding an elliptic curve (with a given additional structure) defined over $\mathbb{Q}$ (or $\mathbb{Z}$) is equivalent to finding a rational (or integral) solution to the defining equations of a specific modular curve $X(U)$. In this way, our effort in understanding the homogeneous space $X(U)$ provides us with another number theoretic question: a *Diophantine problem* of the equations of modular curves. This is a typical example in Number theory of where a serious study of one good problem yields another interesting question, making the life of the theory virtually inexhaustible.

An elliptic curve E defined over a number field $\mathbb{Q}$ is a natural source of a Galois representation $\rho_{E,p} : \mathrm{Gal}(\overline{\mathbb{Q}}/\mathbb{Q}) \to GL_2(\mathbb{Z}_p)$ ramifying at p and a finite set S of primes (independent of p). This comes from the fact that the group $E[p^r]$ of p^r-torsion points of an elliptic curve $E_{/\mathbb{Q}}$ is isomorphic to $(\mathbb{Z}/p^r\mathbb{Z})^2$ and that the Galois action on $E[p^r]$ therefore gives rise to a Galois representation $\rho_E \mod p^r : \mathrm{Gal}(\overline{F}/F) \to GL_2(\mathbb{Z}/p^r\mathbb{Z}) \cong Aut(E[p^r])$. This Galois representation has a remarkable property, found by Hasse, that $L_p(X) = \det(1 - \rho_{E,p}(Frob_\ell)X) = 1 - a(\ell)X + \ell X^2$ has rational integral coefficients $a(\ell)$ independent of p for primes $\ell \notin S \cup \{p\}$ (see, for example, [AME] or [GMF]). Here $Frob_\ell$ is the Frobenius element in the Galois group. Then it is traditional to make an Euler product:

$$L(s,E) = \prod_p L_p(p^{-s})^{-1}.$$

This *Hasse–Weil L-function* is absolutely convergent if $\mathrm{Re}(s) > \frac{3}{2}$, and Hasse and Weil conjectured that it should have an analytic continuation to the whole s-plane with a functional equation relating $L(s,E)$ to $L(2-$

$s, E)$. This is a hard question, because $L(s, E)$ is defined in a purely algebraic way, while the conjecture predicts a purely analytic property (typical for Number theoretic questions, as number theory belongs neither to algebra nor to analysis).

Since a modular form f is a function on a topological group $GL_2(\mathbb{A})$, it is natural to make a convolution product with a compactly supported function ϕ on $GL_2(\mathbb{A})$. This operator $f \mapsto \phi * f$ is called a Hecke operator. Sometime in the 1930's, Hecke discovered that the space of holomorphic modular forms on $X(U)$ has a base made of common eigenforms of standard Hecke operators $T(n)$ indexed by positive integers n (see Section 1.2 for a description of $T(n)$). Pick a common eigenform f, and write the eigenvalues for $T(n)$ as $\lambda(T(n))$. Hecke made an L-function: $L(s, \lambda) = \sum_{n=1}^{\infty} \lambda(T(n))n^{-s}$. This is a (modular) Hecke L-function, which satisfies a functional equation relating $L(s, \lambda)$ to $L(k - s, \lambda)$ for a positive integer k called the *weight* of f. A remarkable fact is that the eigenvalues are algebraic integers in a number field $\mathbb{Q}(\lambda)$ (as implied by Theorem 3.13 in Chapter 3) independent of n. It is not very often but not rare either that $\lambda(T(n)) \in \mathbb{Z}$ for all n when the weight k is 2 (although $\mathbb{Q}$-rational eigenforms become sporadic as k grows); thus, $\mathbb{Q}(\lambda) = \mathbb{Q}$ in such cases. Another remarkable fact is that this L-function has an Euler product: $L(s, f) = \prod_p H_p(p^{-s})^{-1}$ with an Euler factor $H_p(X) = 1 - a(p)X + \psi(p)p^{k-1}X^2$ for the weight $k \geq 1$ and a Dirichlet character ψ, which is called the *'Neben'* character of f by Hecke. Thus when $k = 2$ and $\psi = 1$, the case Hecke called 'Haupt typus' (principal type), the L-function looks like a Hasse–Weil L-function. Since Hecke initiated the study of the modular side (in the non-abelian case), it would be appropriate to call the study of modular forms (or the numbers of the base field) the *Hecke side* of Number theory.

The *Shimura–Taniyama conjecture* states that the Hasse–Weil L-function of every elliptic curve rational over $\mathbb{Q}$ appears as a Hecke L-function of a rational Hecke eigen cusp form, or equivalently, (and more geometrically) that every $\mathbb{Q}$-rational elliptic curve appears as a factor of the jacobian of a modular curve (see [Lg] and [Sh3] for the history of the conjecture, and see also [Sh4] for an account of Shimura's work in the 50's and 60's). As was shown by Shimura ([IAT] Chapter 7), to each Hecke eigen cusp form of weight 2 defined on a modular curve $X(U)$, one can attach a canonical subabelian variety A (or a quotient) of the jacobian of $X(U)$ so that the L-function of A coincides with the Hecke L-function of the cusp form. This fact implies that a Hecke eigen cusp form with eigenvalue $\lambda(T(\ell))$ and with 'Neben' character ψ has a unique two-dimensional p-adic Galois representation ρ_λ whose characteristic polynomial of the Frobenius element is given by $X^2-\lambda(T(\ell))X+\psi(\ell)\ell$ for almost all primes ℓ. Later, the association of such a Galois representation to a cusp form was generalized to all weights (≥ 1) by Deligne, Shimura and Deligne–Serre (see Theorem 3.26 in Chapter 3). If one varies p, these p-adic Galois representations indexed by p have a peculier property that the characteristic polynomials of the Frobenius elements at (unramified) primes ℓ different from p are independent of p. This type of system of Galois representations is called a compatible system. One might then ask, in the spirit of Shimura and Langlands, whether every such compatible system of two-dimensional Galois representations is associated to an elliptic cusp form. This is a typical example of inter-related problems, which in aggregate form a grand program, initiated by Shimura and later developped by Langlands, connecting intricately arithmetic of the Galois side and the Hecke side.

We assume that the p-adic member of a given compatible system is p-adically close to a p-adic Galois representation associated to a cusp form, to ease further the difficulty when we study the above question of modularity of the system. Then one could approach this problem directly from the theory of p-adic Galois representations. For a given p-adic Galois representation $\varphi : \mathrm{Gal}(\overline{\mathbb{Q}}/\mathbb{Q}) \to GL_2(\mathbb{Z}_p)$, we take its reduction modulo p: $\overline{\rho} = (\varphi \mod p) : \mathrm{Gal}(\overline{\mathbb{Q}}/\mathbb{Q}) \to GL_2(\mathbb{F}_p)$. Then we consider all Galois representations $\rho : \mathrm{Gal}(\overline{\mathbb{Q}}/\mathbb{Q}) \to GL_2(A)$ for p-adic local rings A with residue field $\mathbb{F}_p$ which give rise to $\overline{\rho}$ after reducing modulo the maximal ideal of A. A representation ρ with the above property is called a *deformation* of $\overline{\rho}$. As pointed out by Mazur, the totality of such ρ unramified outside $S \cup \{p\}$ for a fixed finite set S is indexed by an affine local ring R; in other words, any such ρ is induced by a universal representation $\varrho : \mathrm{Gal}(\overline{\mathbb{Q}}/\mathbb{Q}) \to GL_2(R)$ composed by a unique algebra homomorphism $R \to A$. If $\overline{\rho}$ is, for example, given by a modular form as $\overline{\rho} = \rho_\lambda \mod p$, one could ask if all deformations are associated to modular forms; in other words, if the algebra R is isomorphic to a factor of the algebra generated by Hecke operators $T(n)$ over $\mathbb{Z}_p$. Mazur made this type of conjecture which asserts the identity of the universal ring R with a Hecke algebra (see [MT] and [FM]). The conjecture is, of course, interesting on its own, but it gives at least some meaning to the rather random decomposition of the p-adic Hecke algebra into local pieces. A question of Taniyama (see [Sh3]) to find a way of decomposing the jacobian of a given modular curve into rational simple factors is, after the work of Shimura ([IAT] Chapter 7), obviously related to decomposing the Hecke algebra over $\mathbb{Q}$ into simple factors. Computed examples of such a decomposition over $\mathbb{Q}$ look more random at this moment than p-adic decomposition (although Maeda conjectured that the Hecke algebra for $SL_2(\mathbb{Z})$ at each weight is simple; see, [HM]). Deciding theoretically the number fields (the Hecke fields) appearing as a simple piece of the Hecke algebra seems difficult at this moment (although it may not be out of reach).

Another development in knitting conjectures in Number theory was given by Frey. In 1986, Frey constructed a semi-stable elliptic curve: $y^2 = x(x+u^p)(x-v^p)$ out of a (hypothetical) rational solution of Fermat's equation: $u^p + v^p + w^p = 0$ with $uvw \neq 0$ and suggested that this curve could not be modular (that is, cannot be identified with a Hecke eigenform). A close study of the ramification of the p-adic Galois representation of the elliptic curve initiated by Serre and studied by Ribet tells us that the elliptic curve has to be associated to a weight 2 modular forms of 'Haupt' type of level 2. By computation, such a modular form does not exist; so, the Shimura–Taniyama conjecture implies Fermat's last theorem (see [Se1] and [R2]).

Wiles' strategy to prove the Shimura-Taniyama conjecture for semi-stable elliptic curves (and hence Fermat's last theorem) is separated into three steps: First, starting from a modular irreducible $\overline{\rho}$ with minimal ramification, prove that the universal ring with minimal ramification and fixed determinant is actually isomorphic to a Hecke algebra. Then, using congruences between minimal (primitive) cusp forms and non-minimal ones (studied principally by Ribet), extend the identification (of the universal Galois deformation ring with a Hecke algebra) to non-minimal Galois representations. Thus for a given semi-stable elliptic curve $E_{/\mathbb{Q}}$, if the Galois module $E[p]$ is modular irreducible, E is modular (basically, forgetting the conditions of ramification), because its p-adic representation is a deformation of $E[p]$. Second, look at $E[3]$. Since $GL_2(\mathbb{F}_3)$ is soluble, the representation on $E[3]$ is modular by a result of Langlands and Tunnell if it is absolutely irreducible. In the two-dimensional soluble cases, Langlands and Tunnell have proved Artin's conjecture identifying the Artin L-function with a modular Hecke L-function of weight 1 (see [Ro], for example). Since a Galois representation into $GL_2(\mathbb{C})$ has finite image (see Proposition 2.2 in Chapter 2), it has values in $GL_2(\mathbb{Z}[\mu_N])$ for the group μ_N of appropriate Nth roots of unity. After reducing modulo a prime ideal of $\mathbb{Z}[\mu_N]$, one may consider the representation as having values in a finite field; in particular, an irreducible Galois representation into $GL_2(\mathbb{F}_3)$ is modular. Third, even if

$E[3]$ is not irreducible, look at $E[5]$ and write its Galois representation as $\overline{\rho}$. By sheer luck, the modular curve $X(\overline{\rho})$ classifying elliptic curves $\mathcal{E}$ with specified Galois module structure $\mathcal{E}[5] \cong \overline{\rho}$ is of genus 0. Since $(E, E[5])$ is rational over $\mathbb{Q}$, $X(\overline{\rho})$ has infinitely many rational points (including the one corresponding to $(E, E[5])$). Out of elliptic curves sitting on the rational points of $X(\overline{\rho})$, Wiles found (basically by Hilbert's irreducibility theorem) a particular one E' with absolutely irreducible $E'[3]$. Thus E' is modular by the first step; hence, $\overline{\rho} \cong E'[5]$ is modular. Applying deformation theory in the 5-adic setting of $\overline{\rho}$, E itself is known to be modular. We refer readers to the details of this argument in the original paper of Wiles [W2] Chapter 5.

In this book, I shall give a detailed exposition of the deformation theory of Galois representation (Mazur's approach) and the identification of the universal deformation ring R with a Hecke algebra (a result due to Wiles and Taylor), restricting ourselves to the case where ramification is limited to a single prime $p > 3$ and ∞. At the end (Chapter 5), I shall briefly give an application of the theory to the special values of the adjoint L-functions of modular forms and their Selmer groups, recalling some of my old and new results ([H81a] and [H99b]).

In this chapter, I shall give an overview of the theory of modular forms, expanding a bit the above description, as an introduction to the subject of the book, starting from Hecke characters, which can be regarded as abelian modular forms. Since this is the introductory part, proofs of some results may not be given here, putting them off until later chapters (or to somewhere else as indicated). The reader can find an outline of the chapters of this book in Subsection 1.2.1.

1.1 Hecke Characters

Let F be a number field, that is, a finite extension of $\mathbb{Q}$. A continuous character $\varphi : F_{\mathbb{A}}^{\times}/F^{\times} \to \mathbb{C}^{\times}$ is called a Hecke character. A Hecke character is a function on the homogeneous space $GL_1(F)\backslash GL_1(F_{\mathbb{A}})$, and hence, the simplest among modular forms on algebraic groups. We study in this section Hecke characters in terms of class field theory.

1.1.1 Hecke characters of finite order

Let $\mathbb{A}$ be the adele ring of $\mathbb{Q}$, and put $F_{\mathbb{A}} = F \otimes_{\mathbb{Q}} \mathbb{A}$, which is the adele ring of F. We write $O = O_F$ for the integer ring of F. We write $\mathbb{A}^{(\infty)}$ for the finite part of the adele ring. Thus $\mathbb{A} = \mathbb{A}^{(\infty)} \times \mathbb{R}$, $F_{\mathbb{A}^{(\infty)}} = F \otimes_{\mathbb{Q}} \mathbb{A}^{(\infty)}$

and $F_{\mathbb{A}} = F_{\mathbb{A}^{(\infty)}} \times F_{\mathbb{R}}$ for $F_{\mathbb{R}} = F \otimes_{\mathbb{Q}} \mathbb{R}$. We write $F^{\times}_{\mathbb{R}+}$ for the connected component of $F^{\times}_{\mathbb{R}}$ containing the identity (the identity component). Then for each prime ideal $\mathfrak{p}$ of O, we consider the $\mathfrak{p}$-adic integer ring $O_{\mathfrak{p}} = \varprojlim_n O/\mathfrak{p}^n$, which is a valuation ring free of finite rank over $\mathbb{Z}_p$. Here p is the prime generating $\mathbb{Z} \cap \mathfrak{p}$. By class field theory, we have an exact sequence

$$1 \to \overline{F^{\times}F^{\times}_{\mathbb{R}+}} \to F^{\times}_{\mathbb{A}} \xrightarrow{[\ ,F]} \mathrm{Gal}(F_{ab}/F) \to 1, \qquad \text{(CFT)}$$

where F_{ab} is the maximal abelian extension of F, $\overline{F^{\times}F^{\times}_{\mathbb{R}+}}$ is the topological closure of $F^{\times}F^{\times}_{\mathbb{R}+}$ in $F^{\times}_{\mathbb{A}}$ and $[\ ,F]$ is the Artin reciprocity law map (see [CFN], Chapter III). Let $\varphi : F^{\times}_{\mathbb{A}}/F^{\times} \to \mathbb{C}^{\times}$ be a Hecke character of finite order. We look at the restriction φ_{∞} of φ to $F^{\times}_{\mathbb{R}}$. Since the inclusion $F^{\times}_{\mathbb{R}} \hookrightarrow F^{\times}_{\mathbb{A}}$ is continuous, $\varphi_{\infty} : F^{\times}_{\mathbb{R}} \to \mathbb{C}^{\times}$ is a continuous character. If φ is of order N, φ_{∞} also has values in the discrete group $\mu_N(\mathbb{C})$ of Nth roots of unity. Thus φ_{∞} has to be trivial on the connected component $F^{\times}_{\mathbb{R}+}$, and φ factors through $\mathrm{Gal}(F_{ab}/F)$. This shows the following one-to-one onto correspondence:

$$\{\textit{Hecke characters of finite order}\} \leftrightarrow \{\textit{Galois characters of finite order}\}.$$

Let $\varphi : \mathrm{Gal}(F_{ab}/F) \to \mathbb{C}^{\times}$ be a continuous character, and suppose that $|\varphi(\sigma) - 1| < 1/2$. If $\varphi(\sigma) \neq 1$, we see that $|\varphi(\sigma)^N - 1| > 1/2$ for some integer N. Since $\mathrm{Gal}(F_{ab}/F)$ is a profinite group (see Chapter 2 2.1.2 for a brief description of profinite groups), its image under φ is a compact set in $\mathbb{C}$, and for a sufficiently small open normal subgroup U of $\mathrm{Gal}(F_{ab}/F)$, $\varphi(U)$ has values in an open disk of radius $\frac{1}{2}$ centered at 1. Thus the above argument for $\sigma \in U$ shows that $\varphi(U) = 1$. Therefore φ factors through the finite group $\mathrm{Gal}(F_{ab}/F)/U$, and φ is of finite order. This argument tells us that any continuous character from a profinite group into $\mathbb{C}^{\times}$ has a finite image.

Here are some examples. Let $F = \mathbb{Q}$. Then for $\widehat{\mathbb{Z}}^{\times} = \prod_{p:primes} \mathbb{Z}_p^{\times}$, we have

$$\mathbb{A}^{\times} = \mathbb{Q}^{\times}\widehat{\mathbb{Z}}^{\times}\mathbb{R}^{\times}_{+} \cong \mathbb{Q}^{\times} \times \widehat{\mathbb{Z}}^{\times} \times \mathbb{R}^{\times}_{+}$$

in the following way. For each $x = (x_v) \in \mathbb{A}^{\times}$, we consider the rational number $rat(x) = (x_{\infty}/|x_{\infty}|)\prod_p p^{v_p(x_p)}$ for the valuation v_p at p with $v_p(p) = 1$. Then $rat(x)^{-1}x^{(\infty)} \in \widehat{\mathbb{Z}}^{\times}$ and $rat(x)^{-1}x_{\infty} \in \mathbb{R}^{\times}_{+}$. Thus the above isomorphism is induced by

$$x \mapsto (rat(x), rat(x)^{-1}x^{(\infty)}, rat(x)^{-1}x_{\infty}).$$

Let N be a positive integer, and write $\mu_N(A)$ for the multiplicative group of N-th roots of unity for any ring A. Let $\overline{\mathbb{Q}}$ be the totality of algebraic numbers in $\mathbb{C}$. We consider the cyclotomic field $\mathbb{Q}(\mu_N)$. We can define the cyclotomic character $\chi_N : \mathrm{Gal}(\mathbb{Q}(\mu_N)/\mathbb{Q}) \to (\mathbb{Z}/N\mathbb{Z})^\times$ by $\zeta^\sigma = \zeta^{\chi_N(\sigma)}$ for $\zeta \in \mu_N(\overline{\mathbb{Q}})$. By the compatibility of local and global class field theory, $[x, \mathbb{Q}] = \prod_v [x_v, \mathbb{Q}_v]$, where v runs over places of $\mathbb{Q}$ including ∞. If x_∞ is negative, $[x_\infty, \mathbb{R}]$ is the complex conjugation (unique in $\mathrm{Gal}(\mathbb{Q}_{ab}/\mathbb{Q})$), and $[x_\infty, \mathbb{R}] = 1$ if $x_\infty > 0$. For a prime ℓ outside N, ℓ is unramified in $\mathbb{Q}(\mu_N)$. Thus $[\ell, \mathbb{Q}_\ell]|_{\mathbb{Q}(\mu_N)}$ is the Frobenius $Frob_\ell$ at ℓ, and $\chi_N([\ell, \mathbb{Q}_\ell]) = \ell \mod N$ for $\ell \nmid N$. Writing ℓ_N for an idele such that its p-component for $p|N$ is equal to ℓ and it is equal to 1 outside N, we define $\ell^{(N)} \in \mathbb{A}^\times$ by $\ell_N \ell^{(N)} = \ell$. Note that $[\ell_p^{(N)}, \mathbb{Q}_p]|_{\mathbb{Q}(\mu_N)} = 1$ unless $p = \ell$ because p is unramified in $\mathbb{Q}(\mu_N)$ if $p \nmid N$ and $\ell_p^{(N)} \in \mathbb{Z}_p^\times$ if $\ell \neq p$. Then we see, for $\ell \nmid N$,

$$\chi_N([\ell^{(N)}, \mathbb{Q}]) = \chi_N(\prod_{p \nmid N} [\ell_p, \mathbb{Q}_p]) = \chi_N([\ell_\ell, \mathbb{Q}_\ell]) = \ell \mod N.$$

Since $[\ell, \mathbb{Q}] = 1$,

$$\chi_N([\ell_N, \mathbb{Q}]) = \chi_N([\ell^{(N)}, \mathbb{Q}])^{-1} = (\ell \mod N)^{-1}.$$

Let $\varphi : (\mathbb{Z}/N\mathbb{Z})^\times \to \mathbb{C}^\times$ be a Dirichlet character. We can regard φ as an idele character in two ways: Since we have a natural projection $\pi : \widehat{\mathbb{Z}}^\times \to (\mathbb{Z}/N\mathbb{Z})^\times$, we just define $\varphi_* : \mathbb{A}^\times/\mathbb{Q}^\times \to \mathbb{C}^\times$ by $\varphi_*(x) = \varphi(\pi(rat(x)^{-1} x^{(\infty)}))$. The second way is to define $\varphi^* : \mathbb{A}^\times/\mathbb{Q}^\times$ by $\varphi^*(x) = \varphi(\chi_N([x, \mathbb{Q}]))$. By the above computation, for primes $\ell \nmid N$,

$$\varphi^*(\ell_\ell) = \varphi(\ell) \quad \text{and} \quad \varphi_*(\ell^{(\ell)}) = \varphi(\ell).$$

We have in either way an onto correspondence:

$$\{\textit{Dirichlet characters}\} \twoheadrightarrow \{\textit{finite order Hecke characters of } \mathbb{A}^\times/\mathbb{Q}^\times\}.$$

To make this one to one, we need to impose primitivity on Dirichlet characters:

$$\{\textit{primitive Dirichlet characters}\} \leftrightarrow \{\textit{finite order Hecke characters of } \mathbb{A}^\times/\mathbb{Q}^\times\}.$$

1.1.2 Arithmetic Hecke characters

Let p be a prime. We simply write χ_n for χ_{p^n}. By taking the projective limit, we get the p-adic *cyclotomic* character

$$\chi = \varprojlim_n \chi_n : \mathrm{Gal}(\mathbb{Q}_{ab}/\mathbb{Q}) \to \varprojlim_n (\mathbb{Z}/p^n\mathbb{Z})^\times = \mathbb{Z}_p^\times.$$

Since $\mathbb{Q}(\mu_{p^n})/\mathbb{Q}$ is unramified at any prime $\ell \neq p$, $[u_\ell, \mathbb{Q}_\ell]$ is trivial on $\mathbb{Q}(\mu_{p^n})$. Thus $\chi([p^{(p)}, \mathbb{Q}]) = 1$. Since $[p, \mathbb{Q}] = 1$, we get $\chi([p_p, \mathbb{Q}]) = 1$. For primes $\ell \neq p$, we see that

$$\chi([\ell_\ell, \mathbb{Q}]) = \chi([\ell, \mathbb{Q}_\ell]) = \chi(Frob_\ell) = \ell.$$

From $[\ell, \mathbb{Q}] = 1$, we get

$$\chi([\ell^{(\ell)}, \mathbb{Q}]) = \chi([\ell_\ell, \mathbb{Q}])^{-1} = \ell^{-1} \in \mathbb{Z}_p^\times.$$

This Galois character is actually associated to an infinite order Hecke character $\varphi(x) = |x|_{\mathbb{A}}^{-1} = (|x_\infty| \prod_p |x_p|_p)^{-1}$, because $\varphi(\ell_\ell) = |\ell|_\ell^{-1} = \ell = \chi(Frob_\ell)$. As a result, it would be legitimate to write $\chi = \varphi^*$. Let $\mathscr{A} = \mathscr{A}_{\mathbb{Q}}$ be the set of Hecke characters $\varphi : \mathbb{A}^\times/\mathbb{Q}^\times \to \mathbb{C}^\times$ such that $x \mapsto \varphi(x)|x|_{\mathbb{A}}^m$ is a finite order character for an integer m. We call an element of $\mathscr{A}$ an arithmetic Hecke character. Since every finite order character of $\mathbb{A}^\times/\mathbb{Q}^\times$ is associated with a Galois character, we get the following correspondence:

$$\mathscr{A}_{\mathbb{Q}} \leftrightarrow \left\{ \begin{array}{l} \text{characters } \psi : \mathrm{Gal}(\overline{\mathbb{Q}}/\mathbb{Q}) \to \overline{\mathbb{Q}}_p^\times \,|\, \psi\chi^{-m} \\ \text{is of finite order for an integer } m \end{array} \right\},$$

where $\overline{\mathbb{Q}}_p$ is an algebraic closure of $\mathbb{Q}_p$.

Is there an intrinsic characterization of characters in $\mathscr{A}$ without using specific characters like $|\ |_{\mathbb{A}}$? There is! If a Hecke character φ restricted to $\mathbb{R}_+^\times$ is just $x \mapsto x^m$, then $\varphi_0 = \varphi|\ |_{\mathbb{A}}^{-m}$ is trivial on $\mathbb{Q}^\times\mathbb{R}_+^\times$ and hence factors through $\mathrm{Gal}(\mathbb{Q}_{ab}/\mathbb{Q})$. As already seen, any complex valued Galois character is of finite order; hence, φ_0 is of finite order. This shows that $\varphi \in \mathscr{A}$. The Lie group $\mathbb{R}_+^\times$ has invariant differential operators. For the additive group, the differential operator $\frac{d}{dx}$ is invariant under the group operation, that is, $\frac{df}{dx}(x+y) = \frac{df(x+y)}{dx}$. Any other invariant differential operator is a polynomial of $\frac{d}{dx}$. For the multiplicative group, the invariant differential operator is given by $\Delta = exp_*(\frac{d}{dx}) = t\frac{d}{dt}$, writing the variable $t = exp(x)$ of $\mathbb{R}_+^\times$. We may regard each Hecke character φ as a function of $t \in \mathbb{R}^\times$ fixing the variable at the finite part. Then we apply Δ. Then

$$\mathscr{A} = \{characters\ \varphi : \mathbb{A}^\times/\mathbb{Q}^\times \to \mathbb{C}^\times | \Delta\varphi = m\varphi \text{ for } m \in \mathbb{Z}\}.$$

We now generalize this fact to an arbitrary number field F. Let $I = I_F$ be the set of all field embeddings of F into $\mathbb{C}$. If $\sigma(F) \subset \mathbb{R}$, we call $\sigma \in I$ real, and otherwise, we call σ complex or imaginary. The Galois group $\mathrm{Gal}(\mathbb{C}/\mathbb{R}) = \langle c \rangle$ acts on I so that $x^{\sigma c} = (x^\sigma)^c$. Then we put $\mathbf{a} = I/\mathrm{Gal}(\mathbb{C}/\mathbb{R})$, which is the set of archimedean places of F. We write F_σ for the σ-completion of F, that is, $\mathbb{R}$ or $\mathbb{C}$ according as σ is real or complex. We then identify

$$F_{\mathbb{R}} = \prod_{\sigma\in\mathbf{a}} F_\sigma = \prod_{\sigma\in\mathbf{a}(\mathbb{R})} \mathbb{R} \times \prod_{\sigma\in\mathbf{a}(\mathbb{C})} \mathbb{C},$$

where $\mathbf{a} = \mathbf{a}(\mathbb{R}) \sqcup \mathbf{a}(\mathbb{C})$ and $\mathbf{a}(\mathbb{R})$ consists of all real archimedean places of F. For each $\sigma \in I$, we put $\Delta_\sigma = t_\sigma \frac{d}{dt_\sigma}$, where t_σ is the variable of the σ-component F_σ. When σ is complex, we agree to put $\Delta_{\sigma c} = \bar{t}_\sigma \frac{d}{d\bar{t}_\sigma}$, where $\bar{t} = t^c$. Then we define the set of arithmetic Hecke characters by

$$\mathscr{A}_F = \{\textit{Hecke characters } \varphi : F_{\mathbb{A}}^\times/F^\times \to \mathbb{C}^\times | \Delta_\sigma\varphi = m_\sigma\varphi \textit{ for integers } m_\sigma\}.$$

We write $\infty(\varphi) = m = \sum_{\sigma\in I} m_\sigma\sigma \in \mathbb{Z}[I]$ and call it the *infinity type* of φ. The infinity type of φ only determines the restriction of φ to the identity component $F_{\mathbb{R}+}^\times$. In order to determine the full infinity part, that is, the restriction to $F_{\mathbb{R}}^\times$, we need to specify the restriction of φ to the maximal compact subgroup of $F_{\mathbb{R}}^\times$, that is $C_\infty = \{\pm 1\}^{\mathbf{a}(\mathbb{R})} \times \mathbb{T}^{\mathbf{a}(\mathbb{C})}$, where $\mathbb{T} = \{z \in \mathbb{C} || z| = 1\}$.

1.1.3 A theorem of Weil

We take $\varphi \in \mathscr{A}_F$ with infinity type $m \in \mathbb{Z}[I]$. We have

$$\varphi(x_\infty) = x_\infty^m = \prod_{\sigma\in I} (x_\infty^\sigma)^{m_\sigma}$$

on $F_{\mathbb{R}+}^\times$. Then for $\xi \in F_+^\times = F_{\mathbb{R}+}^\times \cap F$,

$$\varphi(\xi) = \varphi(\xi^{(\infty)})\varphi(\xi_\infty) = \varphi(\xi^{(\infty)})\xi^m = 1.$$

Thus $\varphi(\xi^{(\infty)}) = \xi^{-m} \in F^{cl}$, where F^{cl} is the composite of all $\sigma(F)$ in $\overline{\mathbb{Q}}$. Let $\widehat{O} = O \otimes_{\mathbb{Z}} \widehat{\mathbb{Z}} = \prod_{\mathfrak{p}} O_{\mathfrak{p}}$. Since $\widehat{O}^\times$ is a profinite group (see Chapter 4 Subsection 2.1.2 for a brief description of profinite groups), as already remarked, the continuity of φ tells us that $\varphi|_{\widehat{O}^\times}$ is a finite order character (see Proposition 2.2 of Chapter 2). Hence $\mathrm{Ker}(\varphi)$ contains

$$U(N) = \left\{x \in \widehat{O}^\times | x \equiv 1 \mod N\widehat{O}\right\}.$$

for a suitable ideal N of O. We also define for a prime p

$$U(Np^\infty) = \bigcap_n U(Np^n) = \{x \in U(N) | x_p = 1\}.$$

Note that $F_{\mathbb{A}}^\times/F^\times U(1) F_{\mathbb{R}+}^\times$ is the strict class group Cl_F^+ of F, which is a finite group. Further $(U(1) : U(N)) = \#(O/N)^\times$ is finite; thus, we see that $x^h \in F^\times U(N)$ for every $x \in F_{\mathbb{A}^{(\infty)}}^\times$ with a suitable positive integer h

independent of x. This shows that $\varphi(x)^h \in F^{cl}$ for every $x \in F^\times_{\mathbb{A}^{(\infty)}}$. In particular, an arithmetic Hecke character φ has values in a finite extension $\Phi/\mathbb{Q}$ on $F^\times_{\mathbb{A}^{(\infty)}}$. We sometimes write $\Phi = \mathbb{Q}(\varphi)$ when Φ is minimal with this property. We now fix an embedding $i_p : \overline{\mathbb{Q}} \hookrightarrow \overline{\mathbb{Q}}_p$ and regard $\overline{\mathbb{Q}}$ as a subfield of $\overline{\mathbb{Q}}_p$. Let $\mathbb{Q}_p(\varphi) = K$ be the topological closure of $\mathbb{Q}(\varphi)$ in $\overline{\mathbb{Q}}_p$. In other words, for the p-adic place $\mathfrak{P}$ of $\mathbb{Q}(\varphi)$ induced by the inclusion: $\mathbb{Q}(\varphi) \hookrightarrow \overline{\mathbb{Q}}_p$, K is the $\mathfrak{P}$-adic completion $\Phi_{\mathfrak{P}}$. Then its p-adic integer ring $\mathcal{O}$ satisfies $\mathcal{O} = \varprojlim_n \mathcal{O}/p^n\mathcal{O}$. We may regard the restriction of φ to the finite part $F^\times_{\mathbb{A}^{(\infty)}}$ as having values in $K^\times$. The character $\varphi^{(\infty)} : F^\times_{\mathbb{A}^{(\infty)}} \to K^\times$ is a continuous character under the discrete topology on $K^\times$. Each $\sigma : O \hookrightarrow \mathcal{O}$ induces a continuous character $\sigma : O_p = \varprojlim_n O/p^nO \to \mathcal{O} = \varprojlim_n \mathcal{O}/p^n\mathcal{O}$. We define $\widehat{\varphi} = \widehat{\varphi}_{\mathfrak{P}} : F^\times_{\mathbb{A}} \to K^\times$ by $\widehat{\varphi}(x) = \varphi(x^{(\infty)})x_p^m$, where $x_p^m = \prod_{\sigma \in I}(x_p^\sigma)^{m_\sigma} \in K^\times$. Then $\widehat{\varphi}(\xi) = \varphi(\xi^{(\infty)})\xi^m = 1$. Thus $\widehat{\varphi}$ is a continuous character (under the p-adic topology on $K^\times$) trivial on $F^\times F^\times_{\mathbb{R}+}$ and hence trivial on $\overline{F^\times F^\times_{\mathbb{R}+}}$. Then by (CFT), φ induces the Galois character $\widehat{\varphi}_{\mathfrak{P}} : \mathrm{Gal}(F_{ab}/F) \to \Phi^\times_{\mathfrak{P}}$ such that $\widehat{\varphi}_{\mathfrak{P}}(Frob_{\mathfrak{l}}) = \varphi(\varpi_{\mathfrak{l}})$ for a prime element $\varpi_{\mathfrak{l}}$ of $O_{\mathfrak{l}}$ for prime ideals $\mathfrak{l} \nmid pN$ of O. Since $\widehat{\varphi}(U(Np^\infty)) = 1$, the abelian extension of F fixed by $\mathrm{Ker}(\widehat{\varphi})$ is unramified for $\mathfrak{l} \nmid pN$. In this case, we call φ *unramified* outside N. We have proved

THEOREM 1.1 (A. WEIL, 1955) *For every arithmetic Hecke character φ, there is a unique system of $\mathfrak{P}$-adic continuous Galois characters $\{\widehat{\varphi}_{\mathfrak{P}}\}$ such that $\widehat{\varphi}_{\mathfrak{P}}$ is unramified outside $N(\mathfrak{P})N$ and $\widehat{\varphi}_{\mathfrak{P}}(Frob_{\mathfrak{l}}) = \varphi(\varpi_{\mathfrak{l}})$ for all prime ideals $\mathfrak{l} \nmid N(\mathfrak{P})N$ of O, where $\mathfrak{P}$ runs over all finite places of $\mathbb{Q}(\varphi)$.*

A remarkable fact is that $\widehat{\varphi}(Frob_{\mathfrak{l}})$ is an algebraic number depending only on φ, although the Galois characters $\widehat{\varphi}_{\mathfrak{P}}$ depend highly on $\mathfrak{P}$. The system $\{\widehat{\varphi}_{\mathfrak{P}}\}$ is called a compatible system of Galois characters associated to an arithmetic character φ. We define the Hecke L-function of φ by

$$L(s,\varphi) = \prod_{\mathfrak{l}\nmid N}(1 - \widehat{\varphi}_{\mathfrak{P}}(Frob_{\mathfrak{l}})N(\mathfrak{l})^{-s})^{-1} = \prod_{\mathfrak{l}\nmid N}(1 - \varphi(\varpi_{\mathfrak{l}})N(\mathfrak{l})^{-s})^{-1}.$$

This L-function converges at $s \in \mathbb{C}$ with sufficiently large real part, is continued to a meromorphic function on the whole complex s-plane and satisfies an appropriate functional equation (a theorem of Hecke, 1920, see [LFE] Sections 2.7 and 8.6).

We can generalize the above definition of a compatible system of Galois characters to a non-abelian situation. Let E be a number field. A system of continuous representations $\rho = \{\rho_{\mathfrak{P}} : \mathrm{Gal}(\overline{F}/F) \to GL_n(E_{\mathfrak{P}})\}$

is called a compatible system with coefficients in E if the following two conditions are satisfied:

(unr) $\rho_{\mathfrak{P}}$ is unramified outside $N(\mathfrak{P})N$ for an ideal N of O independent of $\mathfrak{P}$. Here $\rho_{\mathfrak{P}}$ is unramified at a prime ideal $\mathfrak{l}$ if the image of the inertia group for any prime over $\mathfrak{l}$ under $\rho_{\mathfrak{P}}$ is trivial;

(chr) $\det(1_n - \rho_{\mathfrak{P}}(Frob_{\mathfrak{l}})X)$ is contained in $E[X]$ (not just $E_{\mathfrak{P}}[X]$) and is independent of $\mathfrak{P}$ as long as $\mathfrak{l} \nmid N(\mathfrak{P})N$.

If such a system ρ is given, we can define a non-abelian L-function by

$$L(s,\rho) = \prod_{\mathfrak{l}\nmid N} \det(1_n - \rho_{\mathfrak{P}}(Frob_{\mathfrak{l}})N(\mathfrak{l})^{-s})^{-1}.$$

Here $\mathfrak{P}$ is always chosen so that $\mathfrak{l} \nmid N(\mathfrak{P})$. Choose an arithmetic Hecke character φ of F. For each finite extension F' of F, we consider the induced representation (see Chapter 2 Subsection 2.1.6 in for the definition of induced representations):

$$\rho_{\mathfrak{P}} = \mathrm{Ind}_{\mathrm{Gal}(\overline{F}/F')}^{\mathrm{Gal}(\overline{F}/F)} \widehat{\varphi}_{\mathfrak{P}} : \mathrm{Gal}(\overline{F}/F) \to GL_n(E_{\mathfrak{P}}),$$

where $E = \mathbb{Q}(\varphi)$ and $n = [F' : F]$. Then $\rho = \{\rho_{\mathfrak{P}}\} = \mathrm{Ind}_{F'}^{F} \widehat{\varphi}$ is an example of a compatible system. We claim

$$L(s, \mathrm{Ind}_{F'}^{F} \widehat{\varphi}) = L(s,\varphi). \tag{1.1}$$

Proof Write ρ for $\mathrm{Ind}_{F'}^{F} \widehat{\varphi}$. Let ϕ be the Frobenius for a prime $\mathfrak{l}$ of F. Then writing the eigenvalues of $\rho(\phi)$ as $\alpha_1, \ldots, \alpha_d$ for $d = [F' : F]$, we have

$$\sum_{n=0}^{\infty} \mathrm{Tr}(\rho)(\phi^n)X^n = \sum_{n=0}^{\infty}\sum_{i=1}^{d} \alpha_i^n X^n = \prod_i (1-\alpha_i X)^{-1} = \det(1_n - \rho(\phi)X)^{-1}. \tag{$*$}$$

Let $G = \mathrm{Gal}(\overline{\mathbb{Q}}/F)$ and $H = \mathrm{Gal}(\overline{\mathbb{Q}}/F')$. We fix a representative set $\Sigma = \{\sigma_j | j = 1, \ldots, h\}$ so that $G = \bigsqcup_{j=1}^{h} \sigma_j H$. Then we can realize ρ in a matrix form as follows: $\rho(\tau) = (\widehat{\varphi}(\sigma_i^{-1}\tau\sigma_j))_{i,j}$, where we agree to put $\widehat{\varphi}(g) = 0$ if $g \notin H$. Thus $\mathrm{Tr}(\rho)(\tau) = \sum_{j=1}^{h} \widehat{\varphi}(\sigma_j^{-1}\tau\sigma_j)$. Since the isomorphism class of ρ only depends on $\widehat{\varphi}$ but not on the choice of Σ, $\mathrm{Tr}(\rho)$ is independent of the choice of Σ. We fix one prime $\mathfrak{L}$ over $\mathfrak{l}$ in $\overline{\mathbb{Q}}$. Let D be the decomposition subgroup for $\mathfrak{L}$ in G, and we assume that $\mathfrak{l}$ is unramified in F'/F. We decompose $G = \bigsqcup_{i=1}^{g} D\delta_i H$. Then $\{\mathfrak{L}_i = \mathfrak{L}^{\delta_i} \cap F' | i = 1, \ldots, g\}$ is the set of all primes of F' over $\mathfrak{l}$, and $f_i = \#(D\delta_i H/H)$ is the relative degree of $\mathfrak{L}_i$, that is, $N(\mathfrak{L}_i) = N(\mathfrak{l})^{f_i}$.

Moreover $\Phi_i = \delta_i^{-1}\phi^{f_i}\delta_i$ is the Frobenius element of $\mathfrak{L}_i$. For each integer $m \geq 0$, we write $\Phi_i^{m/f_i} = \delta_i^{-1}\phi^m\delta_i$. Thus choosing $\Sigma \subset \sqcup_i D\delta_i$, we see $\mathrm{Tr}(\rho)(\phi^m) = \sum_i f_i\widehat{\varphi}(\Phi_i^{m/f_i})$. Note that $\varphi(\Phi_i^{m/f_i}) = 0$ unless $f_i|m$ by our convention. Thus, writing μ_{f_i} for the group of f_i-th roots of unity and knowing $\sum_{\zeta\in\mu_{f_i}}\zeta^m = f_i$ or 0 according as $f_i|m$ or not, we have

$$\sum_i f_i\widehat{\varphi}(\Phi_i^{m/f_i}) = \sum_i\sum_{\zeta\in\mu_{f_i}}\zeta\widehat{\varphi}(\Phi_i^m)^{1/f_i}.$$

Replacing $\mathrm{Tr}(\rho)(\phi^n)$ in (∗) by the above formula, we get

$$\sum_{n=0}^{\infty}\mathrm{Tr}(\rho)(\phi^n)X^n = \sum_{n=0}^{\infty}\sum_i\sum_{\zeta\in\mu_{f_i}}\zeta\widehat{\varphi}(\Phi_i^n)^{1/f_i}X^n = \prod_i(1-\widehat{\varphi}(\Phi_i)X^{f_i})^{-1}.$$

Thus we get

$$\det(1_n - \rho(\phi)X) = \prod_i(1-\widehat{\varphi}(\Phi_i)X^{f_i}).$$

Replacing X by $N(\mathfrak{l})$ and noting $N(\mathfrak{L}_i) = N(\mathfrak{l})^{f_i}$, we get the desired identity Euler factor by Euler factor. □

By étale cohomology theory due principally to Grothendieck (cf. [ECH]), to any projective smooth algebraic variety $V_{/F}$, we can attach a compatible system of Galois representations. If we take a projective space $\mathbf{P}^1$, we find in this way the p-adic cyclotomic characters. If we take an abelian variety of CM type, then we get more general $\widehat{\varphi}$ as above. For more generic V, the compatible system thus obtained is expected to be highly non-abelian and not of the form $\mathrm{Ind}_F^{F'}\widehat{\varphi}$. Thus we have at least an abundance of compatible systems to study.

1.2 Introduction to Modular Forms

In this section, we briefly sketch the definition of modular forms on $GL(n)$. We will study in detail the theory of holomorphic modular forms for $n = 2$ later in Chapter 3.

1.2.1 Modular forms

We consider the topological group $GL_n(F_{\mathbb{A}})$. Then

$$GL_n(F_{\mathbb{A}}) = GL_n(F_{\mathbb{A}}^{(\infty)}) \times GL_n(F_{\mathbb{R}}).$$

The system of neighborhoods of 1 in the finite part $GL_n(F_{\mathbb{A}}^{(\infty)})$ is given by

$$S(N) = \left\{u \in GL_n(\widehat{O}) | u - 1_n \in M_n(\widehat{N})\right\} \quad \text{for ideals } N \text{ of } O,$$

where $\widehat{N} = N\widehat{O}$. Thus a subgroup U of $GL_n(F_{\mathbb{A}}^{(\infty)})$ containing $S(N)$ for some N is open. Since $GL_n(\widehat{O})$ is profinite (and hence compact), $S(N)$ is also profinite and compact. A modular form on $GL(n)_{/F}$ is a function $f : GL_n(F)\backslash GL_n(F_{\mathbb{A}}) \to \mathbb{C}$ satisfying the following four conditions:

(LV) $f(xu) = f(x)$ for all u in an open compact subgroup $U \subset GL_n(F_{\mathbb{A}}^{(\infty)})$;

(∞) $x_\infty \mapsto f(x^{(\infty)}x_\infty)$ for a fixed $x^{(\infty)} \in GL_n(F_{\mathbb{A}}^{(\infty)})$ is an eigenfunction of invariant differential operators on $GL_n(F_{\mathbb{R}})_+$, where $GL_n(F_{\mathbb{R}})_+$ is the identity component of the Lie group $GL_n(F_{\mathbb{R}})$;

(Cpt) On the standard maximal compact subgroup

$$C_\infty = \prod_{\sigma\in\mathbf{a}(\mathbb{R})} O_n(\mathbb{R}) \times \prod_{\sigma\in\mathbf{a}(\mathbb{C})} U_n(\mathbb{C})$$

of $GL_n(F_{\mathbb{R}})$, $f(x) \mapsto f(xu)$ is a given irreducible representation of C_∞;

(Gr) $|f(\delta(a_1,\dots,a_n)x^{(\infty)})|$ for each fixed $x^{(\infty)} \in GL_n(F_{\mathbb{A}}^{(\infty)})$ grows moderately as $|a_i/a_{i+1}|_{\mathbb{A}}$ goes to infinity (that is, at most a polynomial growth), where $\delta(a_1,\dots,a_n)$ is the diagonal matrix with diagonal entries $(a_1,\dots,a_n)$ in this order.

Under these conditions (appropriately chosen), the space of modular forms $\mathcal{M}(U)$ is of finite dimension. We will specify the conditions above in many cases later. We now look at a double coset UxU in $GL_n(F_{\mathbb{A}}^{(\infty)})$, which is a compact set. Since U is also open, UxU/U is compact and discrete and hence is finite. Choose a complete set of representatives R for the quotient UxU/U, that is, $UxU = \sqcup_{r\in R} rU$. For each $f \in \mathcal{M}(U)$, we define $[UxU](f)(y) = \sum_{r\in R} f(yr)$. Note that the characteristic function χ_{UxU} is locally constant compactly supported on $GL_n(F_{\mathbb{A}}^{(\infty)})$ and that the operator $[UxU]$ is just the convolution product $f \mapsto \chi_{UxU} * f$ if the Haar measure on $GL_n(F_{\mathbb{A}}^{(\infty)})$ is chosen in an appropriate way. By (LV), the function $[UxU](f)$ is well defined and independent of the choice of R. Since $r_\infty = 1$, the right shift $f(g) \mapsto f(gr)$ does not affect the conditions (∞) and (Cpt). Since $[UxU](f)$ is a finite sum of $f(gr)$, it still grows moderately. Thus $[UxU](f) \in \mathcal{M}(U)$. Since U is open, it contains a subgroup of the form $S(N)$. The ideal N maximal among ideals M such that $U \supset S(M)$ is called the *level* of U. We hereafter write N for the level of U. Here is a fundamental fact (cf. Shimura's book, [IAT] Chapter 3):

(Hecke) *The subalgebra of* $End_{\mathbb{C}}(\mathcal{M}(U))$ *generated (over* $\mathbb{C}$*) by the Hecke operators* $[UxU]$ *for* x *with* $x_{\mathfrak{l}} = 1$ *for all* $\mathfrak{l}|N$ *is a semi-simple commutative algebra.*

This is a great discovery of Hecke (1938) in the case of $n = 2$, and the above algebra is called the Hecke algebra of $\mathcal{M}(U)$ with coefficients in $\mathbb{C}$. We write $\mathcal{H}(U) = \mathcal{H}(U;\mathbb{C})$ for the Hecke algebra of $\mathcal{M}(U)$ with coefficients in $\mathbb{C}$. We have a little more information on the Hecke algebra. Write $\varpi = \varpi_{\mathfrak{p}}$ for a generator of $\mathfrak{p}$ in $O_{\mathfrak{p}}$ and put

$$T_i(\mathfrak{p}) = U\delta(\overbrace{\varpi,\dots,\varpi}^{i},\overbrace{1,\dots,1}^{n-i})U.$$

Then $\mathcal{H}(U)$ is generated over $\mathbb{C}$ by $T_i(\mathfrak{p})$ for primes $\mathfrak{p} \nmid N$. Since a semi-simple commutative algebra $\mathcal{H}(U)$ acts on a finite dimensional vector space $\mathcal{M}(U)$, $\mathcal{M}(U)$ is spanned by common-eigenvectors of $\mathcal{H}(U)$. Let f be one of the common-eigenvectors. Thus for $T \in \mathcal{H}(U)$, $T(f) = \lambda(T)f$ for the eigenvalue $\lambda(T)$. Then $\lambda : \mathcal{H}(U) \to \mathbb{C}$ is a $\mathbb{C}$-algebra homomorphism. The L-function of λ is defined by

$$L(s,\lambda) = \prod_{\mathfrak{p}\nmid N}\left\{\sum_{i=0}^{n}(-1)^i\lambda(T_i(\mathfrak{p}))N(\mathfrak{p})^{\frac{i(i-1)}{2}-is}\right\}^{-1}. \tag{L}$$

This is the standard L-function of λ for $GL(n)_{/F}$. It converges for $s \in \mathbb{C}$ with sufficiently large real part, is continued to a meromorphic function on the whole complex s-plane and satisfies an appropriate functional equation (by Hecke 1938 for $n = 2$ and holomorphic modular forms on $GL_2(\mathbb{Q})$ and by Godement–Jacquet 1972 [ZSA] for general n).

We consider the totality $\mathcal{A}_{n,F}$ of all such systems of Hecke eigenvalues λ for $GL(n)_{/F}$. A special case of Langlands' version of non-abelian class field theory is the (conjectural) existence of a natural map:

$$\left\{\begin{matrix}\textit{compatible systems } \rho \textit{ of representations}\\ \textit{of } \mathrm{Gal}(\overline{F}/F) \textit{ into } GL(n)\end{matrix}\right\} \xrightarrow{\iota} \mathcal{A}_{n,F}$$

such that $L(s,\rho) = L(s,\iota(\rho))$ (see [ARL] Part 2). We are far from establishing ι, but there is some evidence for the existence of ι. Another intriguing point is that the Hecke side $\mathcal{A}_{n,F}$ is bigger than the Galois side, although one could propose some conjectural characterization of the image of ι. This point suggests to us that the prevalent way of thinking that the Galois side and the Hecke side are somehow equivalent might not actually be true and that the Hecke side actually has more information than the Galois side. In other words, the reverse direction of the arrow ι is more difficult to prove, although, historically, the reverse direction has been studied more extensively (for example, the direction of the Artin reciprocity map is the reverse, but its proof is usually given in the above direction and then one defines the reciprocity map by its inverse; see Theorem 4.40 in Chapter 4), because there were no apparent ways to start the attack on the problem in our direction.

One of the purposes of this book is to give an exposition of an idea of A. Wiles to prove some cases of the map ι for $n = 2$ in a p-adic sense (not in the archimedean sense as formulated above). This is the first example of the attack on the problem in the direction of our formulation of ι. We have used the word 'p-adic', because the modular forms we study are p-adic ones, which include some classical ones as above, but are slightly different, since we take p-adic rings as coefficient rings, in contrast with the archimedean complex field taken in the above formulation of ι. Out of p-adic modular forms, we construct p-adic Hecke algebras. In the late 80's, Mazur conjectured that for $n = 2$, these p-adic Hecke algebras are universal deformation rings of a modular Galois representation, under some precise conditions. Many cases of Mazur's conjecture have been proved by Wiles, and we shall give an exposition of the result restricting ourselves to p-ordinary level p cases (see Theorem 3.31 for the theorem of Wiles restricted to level p, Theorem 3.26 for modular Galois representations, and 3.2.4 for the deformation rings, all in Chapter 3). Wiles then went on to prove the Shimura-Taniyama conjecture for many elliptic curves over $\mathbb{Q}$ including semi-stable ones (which of course implies the existence of the image of the compatible systems associated to such elliptic curves under the global correspondence ι), and finally, Wiles pulled out of the conjecture Fermat's last theorem. The fact that the Simura-Taniyama conjecture implies Fermat's last theorem has been known before Wiles by Frey and Ribet (see [Se1] and [R2]). This shows how important it is to establish even a part of the correspondence ι.

We do not touch on these last two points (the Shimura-Taniyama conjecture and Fermat's last theorem), simply because the description of many ingredients in the proof of Theorem 3.31 (Chapter 3) consumes more than three-quarters of the book. The reader should proceed to Wiles' original paper [W2] for these final results. I hope that this book gives enough Galois cohomological theory (Chapter 4) and representation theoretic foundation (Chapter 2) for graduate students to understand the paper of Wiles. There are two more essential ingredients of the proof not covered in this book – a detailed description of arithmetic moduli of elliptic curves and the construction of modular Galois representations. The reader can consult Shimura's book [IAT] for Galois representations attached to weight 2 cusp forms, [LFE] 7.5 for a brief discussion of how to extend the weight 2 result to higher weight and the book of Katz–Mazur [AME] for moduli of elliptic curves. All these ingredients missing from this book can be also found in my forthcoming book [GMF].

Having written about many ingredients of the proof, I would still like to say that the idea of Wiles itself is fairly simple compared to its amazing outcome. This is the use of the marvelous interaction of local Frobenius elements (or local Galois groups) inside the absolute Galois group to show that the deformation rings are actually small enough to be isomorphic to the Hecke algebras (Subsections 3.2.6, 3.2.7 and 3.2.8, all in Chapter 3). This simplicity of the proof, I believe, is once again evidence of the naturalness of our direction of the map ι.

Although the map ι is far from being established, we may study the consequence of the existence of ι, and if the result is positive, we may count as evidence in support of the existence (and the naturalness) of the map ι. For example, if ι exists, then the following should be true:

(1) For n arithmetic Hecke characters $\varphi_1, \dots, \varphi_n$ of F, there should be $\lambda \in \mathscr{A}_{n,F}$ such that $L(s, \widehat{\varphi}_1 \oplus \cdots \oplus \widehat{\varphi}_n) = L(s, \lambda)$ (Eisenstein series);
(2) For each arithmetic Hecke character φ of F and a subfield F' of F, there should be $\lambda \in \mathscr{A}_{n,F'}$ for $n = [F : F']$ such that $L(s, \lambda) = L(s, \mathrm{Ind}_F^{F'} \widehat{\varphi}) = L(s, \varphi)$ (Automorphic induction; [Cl]);
(3) For a compatible system $\rho_{F'}$ over F' corresponding to $\lambda_{F'} \in \mathscr{A}_{n,F'}$, the restriction ρ_F of $\rho_{F'}$ to $\mathrm{Gal}(\overline{F}/F)$ should correspond to $\lambda_F \in \mathscr{A}_{n,F}$ such that $L(s, \rho_F) = L(s, \lambda_F)$ (Base-change; [ARL] Chapter 2).

All the problems involving the fixed value: $n = 1$ are the consequence of class field theory (an exercise). For $n > 1$, a brief historical remark and a description of the present situation of the progress are as follows: As for Problem (1), the Hecke eigenvalue system is constructed via Eisenstein series, and the serious study of the problem for $n = 2$ goes back to Hecke (1927). It is basically solved by Langlands in 1976 and later by Mœglin and Waldspurger [MoW], although there are many finer problems (such as pin-pointing the exact rationality of Eisenstein series and their residues) which remain open in general. As for Problem

(2), it is known from the time of Hecke (1926) when $[F : F'] = 2$ (theta series associated to the norm forms of quadratic extensions), and Hecke's result is generalized to prime-cyclic extensions F/F' by Kazhdan [Kz] and Flicker [Fk] in the 80's. One could think about the problem (2) for a general system ρ_F associated to $\lambda \in \mathscr{A}_{n,F}$. The question is solved for prime-cyclic cases by Arthur–Clozel in 1989. As for (3), the research was started by K. Doi and H. Naganuma [DN] in 1969 for real quadratic fields $F/F' = F/\mathbb{Q}$, and after the work of some other Japanese mathematicians, in prime-cyclic cases, Langlands proved the existence in 1980 for $n = 2$, and Arthur–Clozel generalized it to general n in prime-cyclic cases. Recently I proved with Maeda the existence of infinitely many base-changes for totally real F and $F' = \mathbb{Q}$ even in non-Galois and non-solvable cases ([HM]). The method in [HM] is formulated through Mazur's approach and uses the theorem of Wiles and its Hilbert modular version due to Fujiwara [Fu], which we will describe later in more detail (Chapter 5).

In the final chapter (5), we shall give a brief sketch of an application of the theorem to a non-abelian class number formula, the base-change of deformation rings, p-adic Hecke algebras and computing the Selmer groups attached to adjoint modular Galois representations. This part is added to give the reader a glimpse of, firstly, the relation between the structure of the Hecke algebra and the values of the adjoint L-function $L(s, Ad(\lambda))$ (associated with Hecke eigenvalues in a way that is different from $L(s, \lambda)$), secondly, a generalization of the definition of p-adic Hecke algebras to cohomological modular forms on $GL(2)$ over an arbitrary number field (including Hilbert modular forms) and thirdly, a pleasing behavior of the universal deformation rings and Hecke algebras under base-change. I have not given detailed proofs towards the end of the final chapter, and we have to assume, at the end of the book, a good knowledge of arithmetic of Hilbert and cohomological modular forms, on which the author hopes to write in the near future.

1.2.2 Abelian modular forms and abelian deformation

Here we shall describe, as a concrete example of the theory briefly described in the previous subsection, archimedean and p-adic modular forms in the simplest case where $n = 1$. We fix an infinity type $m \in \mathbb{Z}[I]$. We extend m to a character of $F_{\mathbb{R}}^{\times}$ just by $x \mapsto x^m$. We fix an ideal $N \subset O$. Then we look into the space $\mathscr{M}_m(N)$ of functions f on $GL_1(F_{\mathbb{A}}) = F_{\mathbb{A}}^{\times}$

satisfying the following three conditions:

(LV) f factors through $F^\times\backslash F_{\mathbb{A}}^\times/U(N) = F_{\mathbb{A}}^\times/F^\times U(N)$;

(∞) $\Delta_\sigma f = m_\sigma f$ for $m = \sum_\sigma m_\sigma \sigma \in \mathbb{Z}[I]$;

(Cpt) $f(xu) = u^m f(x)$ for $u \in C_\infty$.

We write $\mathscr{H}_m(N;\mathbb{C})$ for the Hecke algebra of $\mathscr{M}_m(N)$. First suppose $m = 0$. Then f factors through $Cl(N) = F_{\mathbb{A}}^\times/F^\times U(N)F_{\mathbb{R}}^\times$. This group fits into the following exact sequence:

$$U(1)/U(N) \to Cl(N) \to Cl(1) \to 1.$$

Since $Cl(1)$ is the class group, it is a finite group. By class field theory, the Artin symbol $[\ ,F]$ induces $Cl(1) \cong \mathrm{Gal}(H/F)$ for the Hilbert class field H/F. We see $U(1)/U(N) \cong (O/N)^\times$, which is again a finite group. Thus f is a function on the finite group $Cl(N)$, and

$$\mathscr{M}_0(N) = Maps(Cl(N), \mathbb{C}) \cong \mathbb{C}[Cl(N)],$$

which is a finite dimensional space. Then for $\mathfrak{l} \nmid N$, $T_1(\mathfrak{l})(f)(x) = f(x\varpi_{\mathfrak{l}})$. Therefore the action of $T_1(\mathfrak{l})$ is just the right shift by the class $[\mathfrak{l}]$ of $\mathfrak{l}$ in $Cl(N)$. This shows that $\mathscr{H}_0(N;\mathbb{C}) \cong \mathbb{C}[Cl(N)]$ via $T_1(\mathfrak{l}) \mapsto [\mathfrak{l}]$. In this case, we can think of the subalgebra of $\mathscr{H}_0(N;\mathbb{C})$ generated by the $T_1(\mathfrak{l})$'s over $\mathbb{Z}$, which we write $\mathscr{H}_0(N;\mathbb{Z})$. Obviously we get

$$\mathscr{H}_0(N;\mathbb{Z}) \cong \mathbb{Z}[Cl(N)] \quad \text{and} \quad \mathscr{H}_0(N;\mathbb{C}) = \mathscr{H}_0(N;\mathbb{Z}) \otimes_{\mathbb{Z}} \mathbb{C}.$$

This is a non-trivial fact, because in general the subalgebra $\mathscr{H}(U;\mathbb{Z})$ of $\mathscr{H}(U;\mathbb{C})$ generated over $\mathbb{Z}$ by the $T_i(\mathfrak{l})$'s may even be of infinite rank over $\mathbb{Z}$. For example, if one eigenvalue of $T_i(\mathfrak{l})$ is transcendental over $\mathbb{Q}$, this happens. The point here is that the eigenvalue of $T_1(\mathfrak{l})$ in the $GL(1)$-case is algebraic (or even integral) over $\mathbb{Z}$.

Now suppose that $m \neq 0$. Let R be a complete set of representatives for $Cl(N)$. Since every $x \in F_{\mathbb{A}}^\times$ can be written as $x = \xi r u$ for $r \in R$, $u \in U(N)F_{\mathbb{R}}^\times$ and $\xi \in F^\times$, $f(x) = f(r)u_\infty^m$ for $f \in \mathscr{M}_m(N)$. Thus the value of f is determined by the value of f at R, and $\dim_{\mathbb{C}} \mathscr{M}_m(N) \le \#R = |Cl(N)|$. Suppose that $\mathscr{M}_m(N) \neq 0$. Pick a common eigenform f of all Hecke operators $T_1(\mathfrak{l})$. We write $f(gx) = \varphi(x)f(g)$ for $x \in F_{\mathbb{A}}^\times$ with $x_\infty = 1$. Since $f \neq 0$ and $f(x) = \varphi(x)f(1)$, $f(1) \neq 0$. Dividing f by $f(1)$, we may assume that $f(1) = 1$. Then $f(x) = \varphi(x)$ and $\varphi(xy) = \varphi(x)\varphi(y)$. This shows that $f = \varphi$ is an arithmetic Hecke character of infinity type m. Since the value of a Hecke character is non-zero everywhere, we can think of a linear map $\Phi : \mathcal{M}_m(N) \to \mathcal{M}_0(N)$ given by $\Phi(f)(x) = \varphi(x)^{-1}f(x)$. We see easily that

$$\Phi(T_1(\mathfrak{l})(f)) = \varphi(\varpi_\mathfrak{l})(T_1(\mathfrak{l})\Phi(f)).$$

From this we conclude:

PROPOSITION 1.2 *Suppose $\mathcal{M}_m(N) \neq 0$, and choose a Hecke character $\varphi \in \mathcal{M}_m(N)$. Then we have*

$$\mathcal{H}_m(N;\mathbb{Q}(\varphi)) \cong \mathbb{Q}(\varphi)[Cl(N)] \quad \textit{via } T_1(\mathfrak{l}) \mapsto \varphi(\varpi_\mathfrak{l})[\mathfrak{l}],$$

and

$$\mathcal{H}_m(N;\mathbb{C}) \cong \mathcal{H}_m(N;\mathbb{Q}(\varphi)) \otimes_{\mathbb{Q}(\varphi)} \mathbb{C},$$

where $\mathcal{H}_m(N;\mathbb{Q}(\varphi))$ is the subalgebra of $\mathcal{H}_m(N;\mathbb{C})$ generated by the $T_1(\mathfrak{l})$'s over $\mathbb{Q}(\varphi)$. In particular, if $\mathcal{M}_m(N) \neq 0$, then Φ, as above, is a linear isomorphism.

For simplicity, we assume that $N = 1$. By class field theory, the inertia group $I_\mathfrak{p}$ at $\mathfrak{p}$ in $\mathrm{Gal}(F_{ab}/F)$ is isomorphic to the image of $O_\mathfrak{p}^\times$ under $[\ ,F]$. Let $K = \mathbb{Q}_p(\varphi)$ be a $\mathfrak{P}$-adic completion of $\mathbb{Q}(\varphi)$ for a Hecke character $\varphi \in \mathcal{M}_m(1)$. Let $\mathcal{O}$ be the p-adic integer ring of K, and put $\mathcal{O}_0 = \mathcal{O} \cap \mathbb{Q}(\varphi)$. We consider the subalgebra $\mathcal{H}_m(1;\mathcal{O})$ of $\mathcal{H}_m(1;K) = \mathcal{H}_m(1;\mathbb{Q}(\varphi)) \otimes_{\mathbb{Q}(\varphi)} K$ generated over $\mathcal{O}$ by $T_1(\mathfrak{l})$ for all $\mathfrak{l} \nmid N(\mathfrak{P})$. By the same proof as in the proposition, we see

$$\mathcal{H}_m(1;\mathcal{O}) \cong \mathcal{O}[Cl(1)] \quad \text{via } T_1(\mathfrak{l}) \mapsto \widehat{\varphi}_\mathfrak{P}(Frob_\mathfrak{l})[\mathfrak{l}]. \qquad \text{(triv)}$$

We know that $\widehat{\varphi}_\mathfrak{P}([u,F_\mathfrak{p}]) = u^m$ for $u \in O_\mathfrak{p}^\times$. Let A be an $\mathcal{O}$-algebra, and pick a character $\eta : \mathrm{Gal}(\overline{F}/F) \to GL_1(A) = A^\times$ such that (i) η is unramified outside p, and (ii) η coincides with $[u,F_\mathfrak{p}] \mapsto u^m$ on $I_\mathfrak{p}$ for all $\mathfrak{p}|p$. Then $\eta\widehat{\varphi}_\mathfrak{P}^{-1}$ is a finite order character unramified everywhere. By class field theory, $\eta\widehat{\varphi}_\mathfrak{P}^{-1}$ induces a unique character $Cl(1) \to A^\times$ taking $[\mathfrak{l}]$ to $\eta\widehat{\varphi}_\mathfrak{P}^{-1}(Frob_\mathfrak{l})$, which in turn induces a unique $\mathcal{O}$-algebra homomorphism of $\mathcal{O}[Cl(1)]$ into A taking $[\mathfrak{l}]$ to $\eta\widehat{\varphi}_\mathfrak{P}^{-1}(Frob_\mathfrak{l})$. Combining this morphism with the isomorphism (triv), we get a unique $\mathcal{O}$-algebra homomorphism:

$$\iota_\eta : \mathcal{H}_m(1;\mathcal{O}) \to A$$

taking $T_1(\mathfrak{l})$ to $\eta(Frob_\mathfrak{l})$. In other words, defining a character

$$\varrho : \mathrm{Gal}(\overline{F}/F) \to \mathcal{H}_m(1;\mathcal{O}) \text{ by } \varrho(\sigma) = \widehat{\varphi}_\mathfrak{P}(\sigma)[\sigma]$$

for the class $[\sigma] \in Cl(1)$ corresponding to $\sigma|_H$ for the Hilbert class field H/F, we have $\varrho(Frob_\mathfrak{l}) = T_1(\mathfrak{l})$ and $\eta = \iota_\eta \circ \varrho$. We have proved

COROLLARY 1.3 *For each $\mathcal{O}$-algebra A, write $\mathcal{F}_m(A)$ for the set of all characters η : $\mathrm{Gal}(\overline{F}/F) \to GL_1(A)$ such that (i) η is unramified outside p, and (ii) η coincides with $[u, F_\mathfrak{p}] \mapsto u^m$ on $I_\mathfrak{p}$ for all $\mathfrak{p}|p$. Then the pair $(\mathcal{H}_m(1;\mathcal{O}), \varrho)$ satisfies the following universal property: $\mathrm{Hom}_{\mathcal{O}-alg}(\mathcal{H}_m(1;\mathcal{O}), A) \cong \mathcal{F}_m(A)$ via $\phi \mapsto \phi \circ \varrho$.*

In other words, whatever character $\eta \in \mathcal{F}_m(A)$ is given for whatever $\mathcal{O}$-algebra A, there is a unique $\mathcal{O}$-algebra homomorphism $\phi : \mathcal{H}_m(1;\mathcal{O}) \to A$ such that $\phi \circ \varrho = \eta$. The character ϱ is called the universal character (with p-inertia weight m). This is a characterization of the p-adic Hecke algebra $\mathcal{H}_m(1;\mathcal{O})$ for $GL(1)$ as a universal deformation ring of the character φ and is another manifestation of class field theory. Mazur's approach to non-abelian class field theory is to generalize this universal property to n-dimensional Galois representations and Hecke algebras for $GL(n)$, which we describe in the following chapters (Chapter 2 and Chapter 3).

Exercises

(1) Show that $\overline{\mathbb{Q}^\times \mathbb{R}_+^\times} = \mathbb{Q}^\times \mathbb{R}_+^\times$ in $\mathbb{A}^\times$. Is it true that $\overline{F^\times F_{\mathbb{R}_+}^\times} = F^\times F_{\mathbb{R}_+}^\times$ for general number fields F?

(2) Analyze the kernel of the following correspondence:

$$\{\textit{Dirichlet characters}\} \twoheadrightarrow \{\textit{finite order Hecke characters of } \mathbb{A}^\times/\mathbb{Q}^\times\}.$$

(3) If $m \in \mathbb{Z}[I]$ is the infinite type of a Hecke character φ, then $m + mc = [m]t$ for an integer $[m] \in \mathbb{Z}$, where $t = \sum_{\sigma \in I} \sigma$.

(4) Prove rigorously that an arithmetic Hecke character of $F_\mathbb{A}^\times$ has values in a number field Φ on $F_{\mathbb{A}^{(\infty)}}^\times$.

(5) Show that $\varphi^{(\infty)} : F_{\mathbb{A}^{(\infty)}}^\times \to K^\times$ is continuous.

(6) Let $\mathcal{H}$ be a semi-simple commutative $\mathbb{C}$-subalgebra of $\mathrm{End}_\mathbb{C}(V)$ for a finite dimensional complex vector space V. Show that V is spanned by common eigenvectors of $\mathcal{H}$.

(7) Determine $\mathcal{H}_{6\iota}(1;\mathbb{Q})$ for $F = \mathbb{Q}(\sqrt{-3})$ and the identity embedding $\iota : \mathbb{Q}(\sqrt{-3}) \hookrightarrow \mathbb{C}$.

2
Representations of a Group

Here we shall give a detailed exposition of a general theory of Group representations, pseudo-representations and their deformation. These results will be used later. The reader who knows the theory well can skip this chapter.

2.1 Group Representations

A representation of degree n of a group G is a group homomorphism of G into the group of invertible $n \times n$ matrices $GL_n(A)$ with coefficients in a commutative ring A. When the structure of the group is very complicated or when the group is very large, such as the absolute Galois group over $\mathbb{Q}$, it is often easier to study representations rather than the group itself. In this section, we study the basic properties of group representations.

2.1.1 Coefficient rings

Any ring in this section is commutative with the identity $1 = 1_A$. If we refer to an algebra R, then R may not be commutative. A ring A is called local if there is only one maximal ideal $\mathfrak{m}_A$ in A. An A-module M is artinian (resp. noetherian) if the set of A-submodules of M satisfies the descending (resp. ascending) chain condition. If A itself as an A-module is artinian (resp. noetherian), we just call A artinian (resp. noetherian). For an artinian A-module M, if $M \supset M_1 \supset \cdots \supset M_n = \{0\}$ is the maximal descending chain of A-submodules, the number n is called the length of M and is written as $\ell_A(M)$. If A is an artinian ring, A is noetherian (Akizuki's theorem, [CRT] Theorem 1.3.2). This fact is not true for modules. For an artinian local ring A, we equip A with the $\mathfrak{m}_A$-adic topology. Thus the system of neighborhoods of $a \in A$ is given by $\{a+\mathfrak{m}_A^j | 0 \le j \in \mathbb{Z}\}$. In this way, A becomes a topological ring. When A is local artinian with finite residue field $\mathbb{F} = A/\mathfrak{m}_A$, A is a finite (and hence compact) ring. Here are several examples of artinian rings: $\mathbb{Z}/N\mathbb{Z}$, $E[X]/(f(X))$ for a field E. If N is a prime power (resp. $f(X) = P(X)^n$ for an irreducible $P(X)$), then $\mathbb{Z}/N\mathbb{Z}$ (resp. $E[X]/(f(X))$) is local artinian. For a local artinian ring A, $A[[X_1, \dots, X_m]]/(\mathfrak{m}_A + (X_1, \dots, X_m))^j$ is a local artinian ring. The p-adic integer ring $\mathbb{Z}_p$ is not artinian, but it is pro-artinian, that is, a projective limit of artinian rings. In other words, a topological ring A is called pro-artinian if (i) it is complete, (ii) there is a system $\mathfrak{U}$ of neighborhoods made of ideals $\mathfrak{a}$ such that $A/\mathfrak{a}$ is artinian and (iii) it is separated (that is, Hausdorff $\iff \bigcap_{\mathfrak{a}\in\mathfrak{U}} \mathfrak{a} = \{0\}$).

Exercise

(1) Find a non-noetherian but artinian $\mathbb{Z}_p$-module. What is the length of such a $\mathbb{Z}_p$-module?

2.1.2 Topological and profinite groups

A group G is called a *topological group* if it is a group and a topological space at the same time and the group operations are continuous. Thus the product: $(a,b) \mapsto ab$ and the inverse: $a \mapsto a^{-1}$ are continuous maps. Since $a \mapsto a^{-1}$ is an involution (that is, twice the operation is the identity map), it has to be a homeomorphism. If G is a topological group, for a fixed element $a \in G$, $g \mapsto ag$ and $g \mapsto ga$ is a continuous map, because if g_n converges to g, $g_n a$ converges to ga by the continuity of the product. Thus the right and left multiplications are homeomorphisms, because the inverse of $g \mapsto ag$ is given by $g \mapsto a^{-1}g$ which is continuous. In particular, if $\mathcal{U} = \{U\}$ is a system of open neighborhoods of the identity $1 = 1_G$ of G, $a\mathcal{U} = \{aU | U \in \mathcal{U}\}$ and $\mathcal{U}a = \{Ua | U \in \mathcal{U}\}$ are systems of neighborhoods of $a \in G$.

A topological group G is called profinite if the following two conditions are met:

(PF1) G is compact;

(PF2) $1_G \in G$ has a system of neighborhoods made up of open normal subgroups.

Suppose G is a profinite group. Let $\mathcal{U}$ be the system of neighborhoods of $1 \in G$ made up of all open normal subgroups. If $U \in \mathcal{U}$, then $G = \bigcup_{a\in G} aU$ is an open covering. By compactness, we can find a finite subcovering $G = \bigcup_{i=1}^{h} a_i U$. This shows that $U \in \mathscr{U}$ is necessarily of finite index. Since G is compact, it is Hausdorff, and hence $\bigcap_{U \in \mathscr{U}} U = \{1\}$.

We put an order on $\mathscr{U}$ so that $U > V$ if $U \subset V$. Then $U \cap V \in \mathscr{U}$ gives a bound of U and V, that is, $U \cap V > U$ and $U \cap V > V$. An ordering $a > b$ given on a set A is called *inductive* or *filtered* if two elements in A always have a bound. Thus the inclusion ordering on $\mathscr{U}$ is inductive. We write $G_U = G/U$ for $U \in \mathscr{U}$, which is a finite group. If $U > V$, then we have a natural projection map $\pi_{U,V} : G_U \twoheadrightarrow G_V$ sending aU to aV (note here $aU \subset aV$). If $U > V > W$, then $\pi_{V,W} \circ \pi_{U,V} = \pi_{U,W}$. A group $\widehat{G}$ with a group homomorphism $\pi_U : \widehat{G} \to G_U$ is called the projective limit of $\{G_U, \pi_{U,V}\}_{U \in \mathscr{U}}$ if the following condition is satisfied:

(PL1) $(\widehat{G}, \pi_U) = (H, \rho_U)$ makes the following diagram commutative for all $U > V$ in $\mathscr{U}$:

$$\begin{array}{ccc} H & \xrightarrow{\rho_U} & G_U \\ \| \downarrow & & \downarrow \pi_{U,V} \\ H & \xrightarrow{\rho_V} & G_V \end{array}$$

(PL2) If H is a group with a group homomorphism $\rho_U : H \to G_U$ given for all $U \in \mathscr{U}$ making the above diagram commutative, then there is a unique group homomorphism $\rho : H \to \widehat{G}$ such that $\rho_U = \pi_U \circ \rho$ for all $U \in \mathscr{U}$.

If there are two $(\widehat{G}, \pi_U)$ and $(\widehat{G}', \pi'_U)$ satisfying (PL1–2), then we have the unique morphism $\pi' : \widehat{G}' \to \widehat{G}$ and $\pi : \widehat{G} \to \widehat{G}'$ such that $\pi_U = \pi'_U \circ \pi$ and $\pi'_U = \pi_U \circ \pi'$ for all $U \in \mathscr{U}$. Then the following two group homomorphisms $\pi' \circ \pi : \widehat{G} \to \widehat{G}$ and $\mathrm{id}_G : \widehat{G} \to \widehat{G}$ satisfy (PL2) for $H = \widehat{G}$. Thus by the uniqueness, $\pi' \circ \pi$ is the identity map of $\widehat{G}$. Similarly, $\pi \circ \pi'$ is the identity map of $\widehat{G}'$. This shows that $\widehat{G} \cong \widehat{G}'$, and the conditions (PL1–2) uniquely characterize the group $\widehat{G}$ up to isomorphisms. We write $\widehat{G} = \varprojlim_{U \in \mathscr{U}} G_U$.

More generally, a system of groups $\{G_a | a \in I\}$ for an inductively ordered index set I with a group homomorphism $\pi_{a,b} : G_a \to G_b$ for all $a > b$ is called a projective system of groups if $\pi_{b,c} \circ \pi_{a,b} = \pi_{a,c}$ for all triples $a > b > c$. Then we can define the projective limit $\widehat{G} = \varprojlim_{a \in I} G_a$ requiring (PL1–2) for $a \in I$ in place of $U \in \mathscr{U}$. The projective limit always exists; for example, one can easily verify that

$$\widehat{G} = \left\{ x = (x_a) = \prod_{a \in I} G_a \middle| \pi_{a,b}(x_a) = x_b \text{ for all } a > b \right\}$$

with the projection $\pi_a(x_a) = x_a$ satisfies (PL1–2) (see Exercise 1 below).

PROPOSITION 2.1 *If G is a profinite group with the system $\mathscr{U}$ of neighborhoods of $1_G \in G$ made up of all open normal subgroups U, then $G \cong \varprojlim_{U \in \mathscr{U}} G_U$. If $\{G_a, \pi_{a,b}\}_{a \in I}$ is a projective system of finite groups, then $\widehat{G} = \varprojlim_{a \in I} G_a$ is a profinite group whose system of neighborhoods of the identity is given by $\{\mathrm{Ker}(\pi_a)\}_{a \in I}$.*

Proof We first prove the second assertion. As already explained, $\widehat{G}$ is a subgroup of $G' = \prod_{a \in I} G_a$. We equip G_a with a discrete topology and G' with the product topology. Then a base of a system of neighborhoods of $1_{G'}$ is given by $\{G'_a = 1_a \times \prod_{b \neq a} G_b \subset G' | a \in I\}$. So $O_a \cap \widehat{G} = \mathrm{Ker}(\pi_a)$. Thus the topology induced to the subset $\widehat{G}$ is the topology given to $\widehat{G}$. Since the projection $p_a : G' \to G_a$ is continuous, for $a > b$,

$$\begin{aligned} G_{a,b} &= \{(x_c)_{c \in I} \in G' | \pi_{a,b}(x_a) = x_b\} \\ &= ((\pi_{a,b} \circ p_a) \times p_b)^{-1}(\{(x, x) \in G_b \times G_b | x \in G_b\}) \end{aligned}$$

is a closed set. Thus $\widehat{G} = \bigcap_{a > b} G_{a,b}$ is a closed subset in G'. Since G', which is a product of compact sets, is compact ([BTP] Tikhonov's theorem I.9.6), $\widehat{G}$ is compact. Thus $\widehat{G}$ is profinite.

Now we prove the first assertion. Let $\widehat{G} = \varprojlim_{U \in \mathscr{U}} G_U$ with $\pi_U : \widehat{G} \to G_U$. Since the projection map $\rho_U : G \to G_U$ given by $g \mapsto gU$ makes the diagram in (PL1) commutative (replacing H by G), we have a unique morphism $\rho : G \to \widehat{G}$ by (PL2). The map ρ is continuous because $\rho^{-1}(\mathrm{Ker}(\pi_U)) = \mathrm{Ker}(\rho_U) = U$. In particular, $\mathrm{Im}(\rho)$ is a closed set, because G is compact. The map ρ is injective because $\mathrm{Ker}(\rho) = \bigcap_{U \in \mathscr{U}} \rho^{-1}(\mathrm{Ker}(\pi_U)) = \bigcap_{U \in \mathscr{U}} U = \{1\}$. For any $x \in \widehat{G}$ and $U \in \mathscr{U}$, take $y \in G$ such that $yU = \pi_U(x)$. Then $\pi_U(\rho(y)) = \pi_U(x)$ and hence $\rho(y) \in \mathrm{Ker}(\pi_U)x$; namely, we can approximate each $x \in \widehat{G}$ by $\rho(y)$ arbitrarily close. Thus $\mathrm{Im}(\rho)$ is a dense closed subset of $\widehat{G}$. Thus $\mathrm{Im}(\rho) = \widehat{G}$. □

Let $\mathscr{PFG}$ be the category of profinite groups. Thus

$$\mathrm{Hom}_{\mathscr{PFG}}(G, H)$$

is made up of continuous group homomorphisms. If there is no confusion likely, we simply write $\mathrm{Hom}(G, H)$ for $\mathrm{Hom}_{\mathscr{PFG}}(G, H)$.

Let G be a topological group and A be a topological commutative ring. Then the ring $M_n(A)$ of $n \times n$ matrices with coefficients in A is again a topological ring under the product topology on $M_n(A) = A^{n^2}$. Let $GL_n(A) = M_n(A)^\times$, which is a topological group. A continuous homomorphism $\rho : G \to GL_n(A)$ is called a representation of G of degree

n with coefficients in A. Two representations $\rho, \rho' : G \to GL_n(A)$ are equivalent if there exists $x \in GL_n(A)$ such that $\rho(\sigma) = x\rho'(\sigma)x^{-1}$ for all $\sigma \in G$. When ρ and ρ' are equivalent, we write $\rho \sim \rho'$. Let $V(\rho)$ be the column vector space A^n equipped with the G-action $\sigma v = \rho(\sigma)v$. If $\rho(\sigma) = x\rho'(\sigma)x^{-1}$ for all $\sigma \in G$, then $\iota_x : v \mapsto xv$ gives a linear isomorphism $V(\rho') \cong V(\rho)$ commuting with the group action. Thus ι_x is an isomorphism of G-modules, and we have a one-to-one onto correspondence:

$$\{\rho : G \to GL_n(A)\} / \sim \longleftrightarrow \{A\text{-free } G\text{-modules of rank } n\} / \cong .$$

A one-dimensional representation is called a character. Let $\overline{\mathbb{Q}}$ be an algebraic closure of $\mathbb{Q}$. We let the Galois group $\mathrm{Gal}(\overline{\mathbb{Q}}/\mathbb{Q})$ act on p^n-th power roots of unity ζ_n and define $\chi(\sigma) \in \mathbb{Z}_p^\times$ by $\zeta_n^\sigma = \zeta_n^{\chi(\sigma) \bmod p^n}$. In this way, we get the p-adic cyclotomic character $\chi = \chi_p : \mathrm{Gal}(\overline{\mathbb{Q}}/\mathbb{Q}) \to \mathbb{Z}_p^\times = GL_1(\mathbb{Z}_p)$, which is a projective limit of artinian characters: $\chi \bmod p^n : \mathrm{Gal}(\overline{\mathbb{Q}}/\mathbb{Q}) \to GL_1(\mathbb{Z}/p^n\mathbb{Z})$. The fixed field of $\mathrm{Ker}(\chi \bmod p^n)$ is obviously the cyclotomic field of p^n-th roots of unity $\mathbb{Q}(\mu_{p^n})$, where μ_{p^n} is the group of p^n-th roots of unity. Let us compute the degree of $\mathbb{Q}(\mu_{p^n})$ to show the surjectivity of χ. Let $\Phi_1(X) = 1 + X + \cdots + X^{p-1}$ for odd prime p, and put $\Phi_n(X) = \Phi_{n-1}(X^p)$. Then $\Phi_1(1) = p$, which is the constant term of $\Psi_1(X) = \Phi_1(X+1)$. Since $\Phi_1(X)|X^p - 1$ and $X^p - 1 \equiv (X-1)^p \bmod p$, $\Psi_1(X) \bmod p = \Phi_1(X+1) \bmod p$ divides $X^{p-1} \bmod p$. Thus $\Psi_n(X) = \Phi_n(X+1)$ satisfies the Eisenstein criterion for p, and hence is irreducible over $\mathbb{Q}$. In particular, this shows that $\mathbb{Q}(\mu_{p^n})$ has degree $p^{n-1}(p-1) = |GL_1(\mathbb{Z}/p^n\mathbb{Z})|$. Therefore, χ is surjective, and hence $\mathrm{Gal}(\overline{\mathbb{Q}}/\mathbb{Q})$ has a p-adic character with a big image. A similar argument also applies to $p = 2$ (Exercise 6 below).

Proposition 2.2 *Suppose $A = \mathbb{C}$ and that G is a profinite group. Then every continuous representation $\rho : G \to GL_n(\mathbb{C})$ has finite image.*

Proof We equip $V = \mathbb{C}^n$ the Euclidean norm: $|(x_1, \ldots, x_n)| = \sqrt{\sum_j |x_j|^2}$. For each linear endomorphism T of V, we define $\| T \| = Sup_v |Tv|$, where v runs over the $n-1$-dimensional sphere of radius 1. Then we have $\| T \| = 0 \iff T = 0$, $\| TT' \| = \| T \| \cdot \| T' \|$ and $\| T + T' \| \le \| T \| + \| T' \|$. Thus $\| \ \|$ is a well defined norm. One can easily check that the topology of $GL_n(\mathbb{C})$ is given by the norm $\| \ \|$. Choosing a sufficiently small open (and hence compact) subgroup U of G, we may assume that $\rho(U)$ is contained in the open disk of radius $1/2$

centered at 1. Suppose that $T = \rho(u) \neq 1$ for $u \in U$. If all eigenvalues of T are equal to 1, then the Jordan canonical form of T has non-diagonal entry. Thus for some big positive integer N, $\| T^N - 1 \| > \frac{1}{2}$, a contradiction. Then T has an eigenvalue $\zeta \neq 1$. If $|\zeta| \neq 1$, it is obvious that $|\zeta^N - 1| > 1/2$ for a big $|N|$. If $|\zeta| = 1$, the argument of ζ is small. Thus for an appropriate N, $|\zeta^N - 1| > 1/2$, a contradiction. This shows that $\rho(U) = \{1\}$. Since $(G : U) < \infty$, $\rho(G)$ is a finite group. □

This proposition shows that to have a representation with big image of a profinite group, for example, a Galois group, we need to take a profinite topological ring A as a coefficient ring, like the p-adic integer ring $\mathbb{Z}_p$.

A profinite group $G = \varprojlim_{\mathfrak{U}} G/U$ is called *p-profinite* for a prime p if G/U has a p-power order for all $U \in \mathfrak{U}$. For any given finite group G, the sequence of subsets $X_\alpha = \{x^{p^\alpha} | x \in G\}$ stabilizes after a finite number of steps; in other words, $X_N = \cdots = X_\alpha = X_{\alpha+1} =: X_\infty$ if $\alpha \geq N$ for sufficiently large N. Let G' be the subgroup generated by X_∞. Since X_∞ is stable under conjugation, G' is a normal subgroup, and by definition, G/G' has p-power order. It is easy to see that every homomorphism of G into a group of p-power order (p-group) factors through G/G'. The group $G_p = G/G'$ is called the maximal p-group quotient. Then for a profinite group $G = \varprojlim_{\mathfrak{U}} G/U$, its maximal p-profinite quotient is defined by $G_p = \varprojlim_{\mathfrak{U}} (G/U)_p$, where $(G/U)_p$ is the maximal p-group quotient of G/U. Every continuous homomorphism of G into a p-profinite group factors through G_p (Exercise 5). Similarly we define a prime-to-p profinite group to be a profinite group $G = \varprojlim_{\mathfrak{U}} G/U$ with $p \nmid [G : U]$ for all $U \in \mathfrak{U}$.

Exercises

(1) Show that

$$\widehat{G} = \left\{ x = (x_a) = \prod_{a \in I} G_a | \pi_{a,b}(x_a) = x_b \text{ for all } a > b \right\}$$

is a projective limit for the projective system $\{G_a, \pi_{a,b}\}$ as above.

(2) Show that the p-adic integer ring with p-adic topology is a profinite ring.

(3) Let K be a field and $\overline{K}$ be a separable algebraic closure of K. Let $G = \mathrm{Gal}(\overline{K}/K)$, and equip it with the topology whose system of neighborhoods of the identity is made up of subgroups $\mathrm{Gal}(\overline{K}/F)$ for all finite Galois extensions F/K (this topology on $\mathrm{Gal}(\overline{K}/K)$ is called the Krull topology). Then show that G is a profinite group.

(4) On a finite dimensional real vector space V, show that any norm $|\ |$ on V satisfying (i) $|x| = 0 \iff x = 0$, (ii) $|x+y| \le |x|+|y|$ and (iii) $|\lambda x| = |\lambda||x|$ for $\lambda \in \mathbb{R}$ gives rise to the Euclidean topology on V. In particular, the operator norm gives $M_n(\mathbb{C})$ the Euclidean topology.

(5) Show that every homomorphism of a profinite group G into a p-profinite group factors through the maximal p-profinite quotient of G.

(6) Let $p = 2$, $\Phi_1(X) = X^2 + 1$ and $\Phi_n(X) = \Phi_{n-1}(X^2)$. Show that $\Psi_n(X) = \Phi_n(X+1)$ satisfies the Eisenstein criterion for $p = 2$.

2.1.3 Nakayama's lemma

We return to the general set-up. Let R be an algebra (which can be non-commutative). The algebra R is called *simple* if there are no two-sided ideals of R except for $\{0\}$ and R. An R-module M is called *irreducible* or *simple* if $M \neq 0$ and any proper R-submodule of M is trivial. Thus M is irreducible $\iff M \cong R/\mathfrak{m}$ for a maximal left ideal $\mathfrak{m}$ of R. Let M be an R-module of finite type. Then for any given proper R-submodule M_0 of M, we consider the set S of all proper R-submodules of M containing M_0. Here the word 'proper' means that $M_0 \neq M$. If X is an ordered subset of S, then $M_X = \cup_{N \in X} N$ is an R-submodule of M. If $M_X = M$, we find an element N of X such that $M = N$ because M is finitely generated over R. This contradicts our assumption that S is made up of proper submodules. Thus we have $M_X \neq M$ and $M_X \in S$. Namely, any ordered sequence in S has a upper bound in S. Then by Zorn's lemma, S has a maximal element. This shows the existence of a maximal proper R-submodule containing a given M_0. Let $I(M)$ (*radical* of M) be the intersection of all proper maximal R-submodules of M. Then if M is of finite type over R, $I(M) = M$ implies that $M = 0$. Let $I = I(R)$ be the intersection of all maximal left ideals of R, called the *radical* of R. Then $R \neq I$. If M is irreducible, then the annihilator $Ann(M) = \{r \in R | rM = 0\}$ of M is a maximal left ideal of R, because $R/Ann(M) \cong M$ via $r \mapsto rm$ for any $0 \neq m \in M$. Thus $I \subset \bigcap_{M:\text{irreducible}} Ann(M)$. Pick r in the intersection. Then $r(R/\mathfrak{m}) = 0$ for any maximal left ideal $\mathfrak{m}$, since $M = R/\mathfrak{m}$ for a maximal left ideal $\mathfrak{m}$ is irreducible. Thus $r \in rR \subset \mathfrak{m}$, and hence we have

$$I = \bigcap_{M:\text{irreducible}} Ann(M). \tag{2.1}$$

Since $rM = 0$ implies $rxM = 0$ for all $x \in R$, $Ix \subset \bigcap_{M:\text{irreducible}} Ann(M) = I$. Thus I is a two-sided ideal. We claim

$$\textit{For every } r \in I,\ 1 - r \in R^{\times}. \tag{2.2}$$

Proof Since $xr \in I$ for $x \in R$ and $r \in I$, we prove (2.2) for $1 - xr$. Suppose the contrary, that $1-xr$ does not even have a left inverse. Then $R(1-xr) \neq R$, and hence, there exists a maximal left ideal $\mathfrak{m} \supset R(1-xr)$. Thus $1-xr \in \mathfrak{m}$ and $xr \in I \subset \mathfrak{m}$, which shows that $1 = 1-xr+xr \in \mathfrak{m}$, a contradiction. Thus $1-xr$ has a left inverse. In particular, we find $s \in R$ such that $s(1-r) = 1$. Then $s = 1 + sr = 1 - (-s)r$. Applying the above argument for $x = -s$, we find $t \in R$ such that $ts = t(1-(-s)r) = 1$. Then $t = t(s(1-r)) = (ts)(1-r) = 1-r$. This shows that $(1-r)s = ts = 1$ and $1-r \in R^{\times}$. □

Suppose that $\mathfrak{a} \subset R$ is a two-sided ideal with the property that $r \in \mathfrak{a} \Rightarrow 1-r \in R^{\times}$. If $\mathfrak{a} \not\subset I$, then there exists a maximal left ideal $\mathfrak{m}$ such that $\mathfrak{a} \not\subset \mathfrak{m}$. Then $\mathfrak{a} + \mathfrak{m} = R$, and $a + m = 1$ for $a \in \mathfrak{a}$ and $m \in \mathfrak{m}$. Then $\mathfrak{m} \supset Rm = R(1-a) = R$ because $1 - a \in R^{\times}$. This is a contradiction. Thus $\mathfrak{a} \subset I$. We thus have

$$\begin{gathered}\textit{If } a \in \mathfrak{a} \Rightarrow \\ 1 - a \in R^{\times} \textit{for a left (resp. right, two-sided) ideal } \mathfrak{a}\textit{, then } \mathfrak{a} \subset I.\end{gathered} \tag{2.3}$$

By (2.3), we conclude

$$I = \bigcap_{\mathfrak{m}:\text{maximal right ideals}} \mathfrak{m} = \bigcap_{\mathfrak{m}:\text{maximal two-sided ideals}} \mathfrak{m}. \tag{2.4}$$

Lemma 2.3 (Krull–Azumaya, Nakayama)

(1) *Let M be an R-module of finite type. If $M = IM$, then $M = 0$.*

(2) *Let A be a commutative local ring and M be an A-module. If either the A-module M is of finite type or $\mathfrak{m}_A^N M = 0$ for a sufficiently large integer N, then $M = \mathfrak{m}_A M$ implies $M = 0$.*

This follows from the two facts: (i) $I(M) = M \Rightarrow M = 0$ and (ii) $IM \subset I(M)$ under the assumption of the lemma (cf. [CRT] Theorem 1.2.2).

Corollary 2.4 *Let A be a local ring. Let M and N be A-modules and $f : M \to N$ be an A-linear map. Suppose either that $\mathfrak{m}_A$ is nilpotent* *or that N is an A-module of finite type. Then if f induces a surjection $\overline{f} : M/\mathfrak{m}_A M \to N/\mathfrak{m}_A N$, then f itself is surjective.*

Proof Consider $X = N/f(M)$. By assumption, $X/\mathfrak{m}_A X = Coker(\overline{f}) = 0$. Thus $X = \mathfrak{m}_A X$ and hence by Lemma 2.3, $X = 0$. □

2.1.4 Semi-simple algebras

For the time being, we assume that A is a perfect field E. A representation $\rho : G \to GL_n(E)$ is called *reducible* if $V(\rho)$ has a proper non-trivial subspace stable under G. A representation is called *irreducible* if it is not reducible. Let L/E be a field extension of degree n. Choose a base $x_1, \ldots, x_n$ of L over E. Then we define $\rho : L^\times \to GL_n(E)$ by $(\xi x_1, \ldots, \xi x_n) = (x_1, \ldots, x_n)\rho(\xi)$. Then ρ is a representation of the multiplicative group $L^\times$. Note that $\det(X - \rho(\xi))$ is irreducible in $E[X]$ for a generator ξ of L/E. Since E is perfect, L/E is generated by a single element, and hence such a ξ exists. If we have an $L^\times$-stable subspace W of $V = V(\rho)$, then $\det(X - \rho(\xi)) = \det(X - \rho|_W(\xi)) \det(X - \rho|_{V/W}(\xi))$. Thus $V = W$ or $W = \{0\}$, and hence, ρ is irreducible. Let $\overline{E}$ be an algebraic closure of E. We regard ρ as having values in $GL_n(\overline{E})$, and write it as $\rho_{\overline{E}}$. Then $V(\rho_{\overline{E}}) = V(\rho) \otimes_E \overline{E}$. Then obviously ρ is reducible as a representation with coefficients in $\overline{E}$. That is, an irreducible representation with coefficients in E can be reducible for an extension L/E.

Hereafter, we stop insisting that E is perfect and assume only that E is a field. A representation $\rho : G \to GL_n(E)$ is called *absolutely irreducible* if $\rho_{\overline{E}}$ on $V(\rho) \otimes_E \overline{E}$ is irreducible. Let $R = R(\rho)$ be the E-subalgebra of the E-linear endomorphism algebra of $V = V(\rho)$ generated over E by $\rho(\sigma)$ for all $\sigma \in G$. Then $R \subset \mathrm{End}_E(V) \cong M_n(E)$, and hence R is a finite dimensional algebra over E; so, it is artinian and noetherian. Let $I = I(R)$. From Nakayama's lemma, $I^N = \{0\}$ for sufficiently large N. Thus $\bigcap_{\mathfrak{m}:max.two-sided} \mathfrak{m}^N \subset I^N = 0$. For a maximal two-sided ideal $\mathfrak{m}$ and an ideal $\mathfrak{a} \not\subset \mathfrak{m}$, we see that $\mathfrak{m} + \mathfrak{a} = R$, and hence $\mathfrak{m}^j + \mathfrak{a}^j = R$ for all $j > 0$. Applying the Chinese remainder theorem, we have $R = \prod_{\mathfrak{m}} R/\mathfrak{m}^N$ for sufficiently large N, where $\mathfrak{m}$ runs over all maximal two-sided ideals of R. We write $R_\mathfrak{m}$ for $R/\mathfrak{m}^N$. If V is irreducible, then $I = \mathfrak{m}$ for a single maximal two-sided ideal. Then $V \neq \mathfrak{m}V$ again by Lemma 2.3. Therefore $\mathfrak{m}V = 0$. Since R acts faithfully on V, we conclude that $\mathfrak{m} = 0$. Thus R is a simple algebra over E. We have shown that

$$V \text{ is irreducible} \Rightarrow R \text{ is simple.}$$

To study modules over simple algebras, we start a little more generally. Here we only assume R to be an artinian algebra. An R-module V is called *completely reducible* if it is a direct sum of irreducible modules. The algebra R is called *semi-simple* if its radical $I = I(R)$ vanishes. Since maximal left ideals of $R/I(R)$ correspond bijectively to maximal left ideals of R by the homomorphism theorem, we see that $I(R/I(R)) = 0$. This shows that the quotient $R/I(R)$ is semi-simple. Now the following three statements are equivalent:

(SS1) R is semi-simple;
(SS2) The left R-module R is completely reducible;
(SS3) Every R-module V of finite length is completely reducible.

Proof We first prove the implication: (SS1) $\Rightarrow$ (SS2): Let Ω be the set of all maximal left ideals. Then $\mathfrak{m} + \mathfrak{n} = R$ for two distinct elements $\mathfrak{m}, \mathfrak{n} \in \Omega$. Then by the Chinese remainder theorem, $\bigcap_{\mathfrak{m}\in\Omega} \mathfrak{m} = I(R) = 0$ implies that $R \cong \bigoplus_{\mathfrak{m}\in\Omega} R/\mathfrak{m}$. Since R is artinian, Ω is a finite set. By the homomorphism theorem, R-submodules of $R/\mathfrak{m}$ correspond bijectively to left ideals between $\mathfrak{m}$ and R. This shows that $R/\mathfrak{m}$ is irreducible, since $\mathfrak{m}$ is a maximal left ideal.
(SS2) $\Rightarrow$ (SS3): Since $R \cong \bigoplus_{\mathfrak{m}\in\Omega} R/\mathfrak{m}$, we have minimal left ideals $I_\mathfrak{m}$ indexed by $\mathfrak{m} \in \Omega$ such that $R = \bigoplus_{\mathfrak{m}\in\Omega} I_\mathfrak{m}$ with $I_\mathfrak{m} \cong R/\mathfrak{m}$ as left R-modules. Then $1 = \oplus_{\mathfrak{m}\in\Omega} e_\mathfrak{m}$ for $e_\mathfrak{m} \in I_\mathfrak{m}$. Multiplying the left-hand side and the right-hand side by $a \in I_\mathfrak{m}$, we get $I_\mathfrak{m} \ni a = a1 = \oplus_{\mathfrak{m}\in\Omega} a e_\mathfrak{m}$, and therefore $ae_\mathfrak{n} = \delta_{\mathfrak{m},\mathfrak{n}} a$, where $\delta_{\mathfrak{m},\mathfrak{n}} = 1$ or 0 according as $\mathfrak{m} = \mathfrak{n}$ or not. Thus $I_\mathfrak{m} = Re_\mathfrak{m}$. Replacing a by $e_\mathfrak{m}$, we get $e_\mathfrak{m}^2 = e_\mathfrak{m}$ and $e_\mathfrak{m} e_\mathfrak{n} = 0$ if $\mathfrak{m} \neq \mathfrak{n}$. Consider an R-linear map $\varphi : I_\mathfrak{m} \to I_\mathfrak{m} v$ given by $\varphi(i) = iv$ for $v \in V$. Since $\mathrm{Ker}(\varphi)$ is an R-submodule of $I_\mathfrak{m}$, the irreducibility of $I_\mathfrak{m}$ tells us that either $\mathrm{Ker}(\varphi) = 0$ or $\mathrm{Ker}(\varphi) = I_\mathfrak{m}$. Thus either $I_\mathfrak{m} v \cong I_\mathfrak{m}$ or $I_\mathfrak{m} v = 0$. In particular, if $I_\mathfrak{m} v \neq 0$, then $\sum_{\mathfrak{m}\in\Omega} e_\mathfrak{m} v = v \in \sum_{\mathfrak{m}\in\Omega} I_\mathfrak{m} v$. From this, we conclude that $V = \sum_{v\in V-\{0\}} \sum_{\mathfrak{m}\in\Omega} I_\mathfrak{m} v$. Then we can find $\mathfrak{m}_j$ and v_j such that $I_{\mathfrak{m}_j} v_j \neq 0$ and $V = \bigoplus_j I_{\mathfrak{m}_j} v_j \cong \bigoplus_j I_{\mathfrak{m}_j}$ as R-modules, since V is of finite length.
(SS3) $\Rightarrow$ (SS1): By (SS3), R is the direct sum of irreducible R-submodules $I_\mathfrak{m}$. By Lemma 2.3, $I_\mathfrak{m} \neq I(R)I_\mathfrak{m}$. Since $I_\mathfrak{m}$ is irreducible, $I(R)I_\mathfrak{m} = 0$. Then $I(R) = I(R)R = \bigoplus_\mathfrak{m} I(R)I_\mathfrak{m} = 0$. □

We now decompose $\Omega = \Omega_1 \bigsqcup \cdots \bigsqcup \Omega_\lambda$ so that $\mathfrak{m}, \mathfrak{n} \in \Omega_j \iff I_\mathfrak{m} \cong I_\mathfrak{n}$ as left R-modules. Then we write $R_j = \bigoplus_{\mathfrak{m}\in\Omega_j} I_\mathfrak{m}$. If $\mathfrak{m} \in \Omega_j$, $I_\mathfrak{m} a$ for $a \in R$ is either isomorphic to $I_\mathfrak{m}$ or to 0, and therefore it has to be inside R_j. Thus $R_j R \subset R_j$ and hence R_j is a two-sided ideal, and $R = \bigoplus_j R_j$ is an algebra direct sum. Returning to the original setting and applying the above argument to $R/I(R)$, we have

(S) *An artinian algebra R has only one maximal two-sided ideal if and only if there is a unique isomorphism class of irreducible R-modules,*

since the set of maximal two-sided ideals of R naturally corresponds to that of $R/I(R)$ bijectively. We now claim that the following three assertions are equivalent (a theorem of Wedderburn):

(S1) R is a simple algebra;
(S2) R is a direct sum of mutually isomorphic minimal left ideals;
(S3) $R \cong M_n(D)$ for a division algebra D,

where a division algebra D is an algebra such that $D - \{0\} = D^\times$.

Proof (S1) $\Rightarrow$ (S2) follows from (S) because minimal left ideals are all irreducible.
(S2) $\Rightarrow$ (S3): Let V be the minimal left ideal of R. Then $R \cong V^n$ as a left R-module. Then $\mathrm{End}_R(R) \cong M_n(\mathrm{End}_R(V))$. Pick $\phi : V \to V \in \mathrm{End}_R(V)$. Then $\mathrm{Ker}(\phi) = 0$ or V and $\mathrm{Im}(\phi) = V$ or 0 because of the irreducibility of V. This shows ϕ is either bijective or the zero map; hence $D = \mathrm{End}_R(V)$ is a division algebra. On the other hand, it is easy to see that $\phi \mapsto \phi(1)$ induces $\mathrm{End}_R(R) \cong R$, because $\phi_a(x) = xa$ gives an element in $\mathrm{End}_R(R)$ with $\phi_a(1) = a$.
(S3) $\Rightarrow$ (S1): Let $\mathfrak{a}$ be a two-sided ideal of R. If $\mathfrak{a}$ has a non-zero element a, multiplying a by elementary matrices from left and right, we may assume that $a = dE_{ij}$ for $0 \neq d \in D$ with the elementary matrix E_{ij} having non-trivial (i, j) entry. Thus $E_{ij} = d^{-1}a \in \mathfrak{a}$. Then again multiplying E_{ij} by elementary matrices, we find that $E_{ij} \in \mathfrak{a}$ for all (i, j). Thus $\mathfrak{a} = R$. □

Let R be a simple artinian algebra. Let $\overline{E}$ be the center of R. Then by (S3), $\overline{E}$ is a field, which is the center of D. Suppose that $\overline{E}$ is algebraically closed. Then for any $x \in D$, $\overline{E}[x] \subset D$ is a finite field extension of $\overline{E}$. Thus $\overline{E} = \overline{E}[x]$ and $x \in \overline{E}$. This shows that $\overline{E} = D$ and

(S4) *If R is a finite dimensional simple algebra over an algebraically closed field* $\overline{E}$, *then* $R \cong M_n(\overline{E})$.

Let $\rho : G \to M_n(E)$ be an absolutely irreducible representation for a field E. Let R be the E-subalgebra of $M_n(E)$ generated by $\rho(g)$ for all $g \in G$. By absolute irreducibility, $R \otimes_E \overline{E}$ remains simple for an algebraic closure

$\overline{E}$ of E. Since there is no simple algebra except for matrix algebras over an algebraically closed field, $R \otimes_E \overline{E} \cong M_{n'}(\overline{E})$ for some $n' \leq n$ if ρ is absolutely irreducible. Pick $x \neq 0$ in V. We consider a map $f : R \otimes_E \overline{E} \to V \otimes_E \overline{E}$ induced by $r \mapsto rx$. Then $Ker(f)$ is a maximal left ideal of $M_{n'}(\overline{E})$, and hence $V \otimes_E \overline{E} \cong \overline{E}^{n'}$ (see Exercise 1 below). This shows that $n = n'$, and $R \cong M_n(E)$. We record here what we have proved:

PROPOSITION 2.5 *Suppose that $\rho : G \to GL_n(E)$ for a field E is absolutely irreducible. Let R be a E-subalgebra of $M_n(E)$ generated by $\rho(g)$ for all $g \in G$. Then*

(1) (Burnside) $R \cong M_n(E)$ *for* $n = \dim_E V$*;*
(2) (Schur) *If a linear map $f : V \to V$ commutes with ρ, then f is induced by a scalar multiplication.*

This result shows that $V(\rho) \otimes_E X$ (or $\rho : G \to GL_n(X)$) remains irreducible for an arbitrary field extension X/E if ρ is absolutely irreducible.

Exercises

(1) Let $\mathfrak{a}$ be a maximal left ideal of $M_n(E)$ for a field E. Show that $M_n(E)/\mathfrak{a} \cong E^n$ as $M_n(E)$-modules.

2.1.5 Representations of finite groups

For the moment, we suppose that G is a finite group. Then we consider the group algebra $E[G]$, which is the set of formal linear combinations of elements in G with coefficients in E. The product of two such linear combinations is given by

$$\sum_{\sigma \in G} a_\sigma \sigma \sum_{\tau \in G} b_\tau \tau = \sum_{\sigma,\tau} a_\sigma b_\tau \sigma\tau.$$

In this way, $E[G]$ becomes a finite dimensional algebra over E whose multiplicative identity coincides with the identity of G. In particular, writing $G = \{g_1, \ldots, g_m\}$, G forms an E-base of $E[G]$. Thus defining $\varphi(g) \in \mathrm{End}_E(E[G]) \cong M_m(E)$ by $(gg_1, \ldots, gg_m) = (g_1, \ldots, g_m)\varphi(g)$, we get a representation $\varphi : G \to GL_m(E)$, which is called the regular representation of G. Then by definition, $R \cong E[G]$ if we write R for the E-subalgebra of $M_m(E)$ generated by $\mathrm{Im}(\varphi)$. For any R-module M of finite type, we know that M is artinian, and M has a proper maximal R-submodule. Here the word 'proper' means that it is not equal to M.

Let $I(M)$ be the intersection of all maximal R-submodules of M. Thus $M \neq I(M)$. By definition, $IM \subset I(M)$ for $I = I(R)$. If $\rho : G \to GL_n(E)$ is a representation, we can extend ρ to a unique E-algebra homomorphism $\rho : E[G] \to M_n(E)$ so that the following diagram commutes:

$$\begin{array}{ccc} E[G] & \xrightarrow{\rho} & M_n(E) \\ \cup\uparrow & & \uparrow\cup \\ G & \xrightarrow[\rho]{} & GL_n(E). \end{array}$$

Thus $V(\rho)$ is an $E[G]$-module of finite type. Now we suppose that ρ is irreducible. Then we conclude that $IV(\rho) = 0$ from the irreducibility of ρ. Writing $R(\rho)$ for the E-subalgebra of $M_n(E)$ generated by $\mathrm{Im}(\rho)$, we get the surjective E-algebra homomorphism $E[G] \to R(\rho)$. Then the kernel $\mathfrak{m}_\rho$ of this algebra homomorphism is a maximal two-sided ideal of $E[G]$ and we find $E[G] = \prod_\rho E[G]/\mathfrak{m}_\rho^N$ for sufficiently large N, where ρ runs over all irreducible (non-equivalent) representations of G with coefficients in E. In particular, there are finitely many non-isomorphic irreducible representations. We write $\widehat{G}_E$ for the set of isomorphism classes of irreducible representations of G with coefficients in E. We now prove the following result:

THEOREM 2.6 (DENSITY THEOREM) *The natural algebra homomorphism*

$$E[G] \to \bigoplus_{\rho \in \widehat{G}_E} R(\rho)$$

is surjective. If either E is of characteristic 0 *or $|G|$ is prime to the characteristic of E, the above map is an isomorphism. In any case $\{\mathrm{Tr}(\rho)\}_{\rho \in \widehat{G}_E}$ are linearly independent as functions on G.*

Proof The surjectivity follows from the argument given above. Consider the function $\mathrm{Tr}(\rho) : G \to E$ given by the trace of the matrix $\rho(g)$ for each representation ρ. Suppose

$$\sum_{\rho \in \widehat{G}_E} \lambda_\rho \, \mathrm{Tr}(\rho) = 0.$$

Then by the surjectivity, we find $\delta_\rho \in E[G]$ such that $\mathrm{Tr}(\rho'(\delta_\rho)) = \delta_{\rho,\rho'}$. Thus we see that

$$0 = \sum_{\rho \in \widehat{G}_E} \lambda_\rho \, \mathrm{Tr}(\rho)(\delta_{\rho'}) = \lambda_{\rho'},$$

which shows the linear independence of the $\mathrm{Tr}(\rho)$'s as functions on $E[G]$. Since linear forms on $E[G]$ are determined by their values on G, it shows the linear independence of the $\mathrm{Tr}(\rho)$'s as functions on G.

We now give a brief account of the injectivity, since representation theory of finite groups over a field of characteristic 0 is well known. We only need to show that $E[G]$ is semi-simple. First we assume that $E = \mathbb{C}$. Then we define an inner product $(\ ,\ ) : \mathbb{C}[G] \times \mathbb{C}[G] \to \mathbb{C}$ by $(\sum_g a_g g, \sum_h b_h h) = \sum_g a_g \bar{b}_g$. This is a positive definite hermitian pairing, satisfying $(gx, gy) = (x, y)$ for all $x, y \in \mathbb{C}[G]$ and $g \in G$. Thus the left multiplication by g on $\mathbb{C}[G]$ is a unitary representation. Take a minimal left ideal $I \subset \mathbb{C}[G]$. Since $(\ ,\ )$ is positive definite, we can decompose $\mathbb{C}[G] = I \oplus I^\perp$ and $I^\perp$ is stable under G. Now take a minimal $\mathbb{C}[G]$-submodule $I_1 \subset I^\perp$. Again we have $I^\perp = I_1 \oplus I_1^\perp$, where $I_1^\perp$ is the orthogonal complement of I_1 in $I^\perp$. Repeating this, we see that $\mathbb{C}[G]$ is isomorphic to a direct sum of minimal left ideals, and hence the radical vanishes: $I(\mathbb{C}[G]) = 0$. If E is of characteristic 0, $I(E[G]) \otimes_E \mathbb{C} \subset I(\mathbb{C}[G])$ because $I(E[G])$ is also the nilradical of $E[G]$. This shows that $I(E[G]) = 0$, and hence $E[G]$ is semi-simple.

By extending scalars, we may assume that all irreducible representations of G over E are absolutely irreducible. Under the assumption, for each irreducible ρ, we can think about $e_\rho = \frac{1}{|G|}\sum_g \mathrm{Tr}(\rho(g^{-1}))g \in E[G]$. We see that, for all $h \in G$,

$$e_\rho h = \frac{1}{|G|}\sum_g \mathrm{Tr}(\rho(g^{-1}))gh = \frac{1}{|G|}\sum_g \mathrm{Tr}(\rho(h)\rho(g^{-1}))g$$
$$= \frac{1}{|G|}\sum_g \mathrm{Tr}(\rho(g^{-1})\rho(h))g = \frac{1}{|G|}\sum_g \mathrm{Tr}(\rho(g^{-1}))hg = he_\rho.$$

Thus e_ρ is in the center of $E[G]$.

For each representation $\sigma : G \to GL_m(E)$ acting on $W = V(\sigma)$, we consider an operator $e(\phi) = \sum_{g\in G} \rho(g^{-1})\phi \circ \sigma(g)$ acting on $\phi \in \mathrm{Hom}_E(W, V(\rho))$. It is easy to see that $e(\phi)(\sigma(g)w) = \rho(g)e(\phi)(w)$ ($w \in E[G]$). If σ is absolutely irreducible, we see from Schur's lemma (Proposition 2.5 (2)) that $e(\phi)$ is a scalar if $\sigma = \rho$ and $e(\phi) = 0$ if $\sigma \not\cong \rho$. Writing $\sigma(g) = (\sigma_{ij}(g))$ and $\rho(g) = (\rho_{ij}(g))$ for (i, j)-entries σ_{ij} and ρ_{ij}. Then choosing ϕ to be a suitable linear map corresponding to an elementary matrix, it is easy to conclude that

$$\sum_{g\in G} \rho_{ij}(g^{-1})\sigma_{k\ell}(g) = \begin{cases} |G| & \text{if } \sigma = \rho \text{ and } (i, j) = (\ell, k), \\ 0 & \text{if either } \sigma \not\cong \rho \text{ or } (i, j) \neq (\ell, k). \end{cases}$$

This is the orthogonality relation of matrix coefficients (and characters) of representations (see, for example, [LRF] I.2.3 for more explanation).

The above orthogonality relation shows that

$$e_\rho e_{\rho'} = 0 \quad \text{if } \rho \not\cong \rho' \text{ and } \quad e_\rho e_\rho = e_\rho, \tag{2.5}$$

because $\mathrm{Tr}(\rho) = \sum_i \rho_{ii}$. Thus e_ρ gives a system of orthogonal central idempotents of $E[G]$, and hence $E[G] = \bigoplus_{\rho \in \widehat{G}} e_\rho E[G]$. Thus taking $W = E[G]$, $e(\phi)$ for any ϕ kills every minimal left ideal of $E[G]$ on which G acts by an irreducible representation non-isomorphic to ρ. From the orthogonality relation (2.5), we see that e_ρ has the same property. Anyway, if E has characteristic 0, $e_\rho R(\rho') = 0$ if $\rho' \not\cong \rho$, and e_ρ is the identity of $R(\rho)$, since $E[G] = \oplus_\rho R(\rho)$ and G act on $R(\rho)$ by a multiple of ρ.

Suppose now that E is of characteristic p. Since every representation of G is defined over a finite field, we may assume that E is a finite field. Take a finite extension K of $\mathbb{Q}_p$ with integer ring O such that $O/\mathfrak{m}_O = E$. For each representation $\rho : G \to GL_d(K)$, the image is finite and hence compact. Since any maximal compact subgroup of $GL_d(K)$ is a conjugate of $GL_d(O)$ (see Exercise 1 of 3.2.3 in Chapter 3), we may assume that ρ has values in $GL_d(O)$ and that $\mathrm{Tr}(\rho)$ has values in O. We define $\overline{\rho} = \rho \mod \mathfrak{m}_O$, taking reduction modulo $\mathfrak{m}_O$ entry by entry. Then the idempotent e_ρ originally in $K[G]$ actually resides in $O[G]$. We write $R_O(\rho)$ for $e_\rho O[G]$; thus, $O[G] = \oplus_\rho R_O(\rho)$, and therefore $R_O(\rho)$ is generated over O by the image of $g \in G$. We also know that $\mathrm{rank}_O R_O(\rho) = d_\rho^2$ for $d_\rho = \dim_K \rho$. Let Z_A be the center of $A[G]$ for a ring A. For any irreducible representation $\overline{\rho}$ of G over E, we still have the central idempotent $e_{\overline{\rho}} \in Z_E$, because $|G|$ is prime to p. We see that $Z_O \otimes_O E = Z_E$, because Z_A is made up of $\sum_g a_g g$ with a_g depending only on the conjugacy class of g, not on individual g. Take $x \in Z_O$ with $x \mod \mathfrak{m}_O Z_O = e_{\overline{\rho}}$. Then for $q = |E|$, the limit $e = \lim_{n\to\infty} x^{q^n}$ exists (under the $\mathfrak{m}$-adic topology; [LFE] Section 7.2) and gives a central idempotent in Z_O. This fact also follows from Hensel's lemma (see [BCM] III.4.6). In particular, e reduces to $e_{\overline{\rho}}$ modulo $\mathfrak{m}_O$. Thus e is a sum of e_ρ over some irreducible representation ρ of G over K. However, if there are more than one such ρ, say, $\rho_1, \ldots, \rho_m$, $e_{\overline{\rho}} = \sum_{j=1}^m e_{\overline{\rho}_j}$ for $\overline{\rho}_j = \rho_j \mod \mathfrak{m}_O$. This contradicts the simplicity of $R(\overline{\rho})$. Thus there is a unique ρ reducing modulo $\mathfrak{m}_O$ to $\overline{\rho}$. On the other hand, $R_O(\rho) \otimes_O E = R(\overline{\rho}) \cong M_d(E)$ for $d = d_{\overline{\rho}} = d_\rho$. Since irreducible representation over K might be decomposed further after reduction modulo $\mathfrak{m}_O$, the number of irreducible representations over K is greater or equal to that over E. The above argument tells us the reverse inequality. Thus, the reduction map: $\rho \mapsto \overline{\rho}$ is one to one and onto between irreducible representations over K and E. Note that

$$\dim_E(\oplus_{\overline{\rho}} R_E(\overline{\rho})) = \sum_{\overline{\rho}} d_{\overline{\rho}}^2 = \sum_{\rho} d_\rho^2 = \dim_K K[G] = \dim_E E[G] = |G|.$$

From this combined with surjectivity, we get the desired injectivity. □

We record the following fact shown in the above proof of the theorem:

COROLLARY 2.7 *Let K be a finite extension of $\mathbb{Q}_p$ with p-adic integer ring O. Let $E = O/\mathfrak{m}_O$ for the maximal ideal $\mathfrak{m}_O$ of O. Suppose $p \nmid |G|$ and that all irreducible representations of G over K are absolutely irreducible. Then all irreducible representations of G over E are absolutely irreducible, and the reduction map $\rho \mapsto (\rho \mod \mathfrak{m}_O)$ induces a bijection between isomorphism classes of absolutely irreducible representations of G over K and over E, preserving dimension.*

We state one more corollary.

COROLLARY 2.8 (BRAUER–NESBITT THEOREM) *Let G be a finite group. If two representations with coefficients in E have the same trace and one of them is absolutely irreducible, then they are equivalent.*

Proof We may assume that E is algebraically closed. Let ρ and ρ' be two representations of G. Suppose that ρ' is absolutely irreducible. Let $\Omega = \{\rho' = \eta_1, \dots, \eta_m\}$ be the set of all absolutely irreducible representations. Let $V = V(\rho)$. Let V_1 be the minimal non-trivial $E[G]$-stable subspace of V. Such a V_1 exists because V is of finite dimension. Then make the quotient V/V_1 and take the pull back $V_2 \subset V$ of a minimal non-trivial $E[G]$-stable subspace of V/V_1. Repeating this process, we find the Jordan–Hölder composition sequence of $E[G]$-modules: $0 \subset V_1 \subset \dots \subset V_r = V$, that is, $V(\rho_j) = V_j/V_{j-1}$ is absolutely irreducible for all j. Then choosing a base of V_j so that it includes a base of V_{j-1} for all $j = 1, \cdots, r$, the matrix form of ρ is given by

$$\begin{pmatrix} \rho_1 & * & \cdots & * \\ 0 & \rho_2 & \cdots & * \\ 0 & 0 & \ddots & * \\ 0 & \cdots & 0 & \rho_r \end{pmatrix}.$$

Thus $\mathrm{Tr}(\rho) = \sum_{j=1}^r \mathrm{Tr}(\rho_j) = \sum_{\eta \in \Omega} m_\eta \mathrm{Tr}(\eta)$; in other words, among $\rho_1, \cdots, \rho_r$, we find η m_η-times. Since $\mathrm{Tr}(\eta)$ $(\eta \in \Omega)$ are linearly independent, the identity $\mathrm{Tr}(\eta_1) = \mathrm{Tr}(\rho') = \mathrm{Tr}(\rho) = \sum_{\eta\in\Omega} m_\eta \mathrm{Tr}(\eta)$ implies that $\rho \sim \eta_1 \sim \rho'$. □

We call $\rho^{ss} = \oplus_j \rho_j$ the semi-simplification of ρ under the notation of the proof of the above corollary. This representation is semi-simple and has the same trace as the original ρ.

Exercise

(1) Let $\rho : G \to M_n(E)$ be a representation for a field E. Let R be the subalgebra of $\mathrm{End}_E(V(\rho))$ generated by $\mathrm{Im}(\rho)$. Show that $\rho \cong \rho_0^m$ for an irreducible representation ρ_0 if R is simple.

2.1.6 Induced representations

Let G be a profinite group and E be a field. We consider the category $Rep_E(G)$ made up of the following data:

- Objects are finite dimensional E-vector spaces with a continuous action of G under the discrete topology.
- Morphisms are $E[G]$-linear maps.

A general definition of categories can be found in Section 4.1 of Chapter 4. The collection $Rep_E(G)/\cong$ of isomorphism classes of objects of this category forms a set. This is obvious if G is finite. If not, writing $G = \varprojlim_j G_j$ for finite G_j, we see that $(Rep_E(G)/\cong) = \varinjlim_j (Rep_E(G_j)/\cong)$, which is a set.

We consider the Grothendieck group $R_E(G)$ of $Rep_E(G)$, which is an abelian group defined by generators and relations: $R_E(G)$ is generated by symbols $[M]$ for objects $M \in Rep_E(G)$. The only relation is $[M] = [N] + [L]$ if we have an exact sequence $0 \to N \to M \to L \to 0$ of $E[G]$-modules. In particular, $[M] = [N]$ if $M \cong N$. Thus $R_E(G)$ is generated by $[M]$ for M running over the set $Rep_E(G)/\cong$. We can easily check that $R_E(G)$ is actually a ring under the product:

$$[M] \cdot [N] = [M \otimes_E N].$$

The multiplicative identity of the ring $R_E(G)$ is the trivial G-module E.

By definition, $[0]$ is the additive identity of $R_E(G)$. If we have a filtration $0 = M_0 \subset M_1 \subset \cdots \subset M_n$ of objects in $Rep_E(G)$, we have

$[M_j] = [M_{j-1}] + [M_j/M_{j-1}]$ in $R_E(G)$ for $j = 1, \dots, n$. This shows that

$$[M_n] = \sum_{j=1}^{n} [M_j/M_{j-1}] \quad \text{in } R_E(G). \tag{2.6}$$

Let $M \in Rep_E(G)$ and $0 = M_0 \subset M_1 \subset \cdots \subset M_n = M$ be its Jordan–Hölder sequence. Then we define the semi-simplification of M by $M^{ss} = \bigoplus_{j=1}^{n} M_j/M_{j-1}$ as an object of $Rep_E(G)$. By the Brauer–Nesbitt theorem 2.8, the isomorphism class of M^{ss} is independent of the choice of the Jordan–Hölder sequence. Thus we have

$$[M] = [M^{ss}] = \sum_{j=1}^{n} [M_j/M_{j-1}]. \tag{2.7}$$

Since each component $[M_j/M_{j-1}]$ of M^{ss} is an irreducible representation, $R_E(G)$ is generated by irreducible representations.

We write $R'_E(G)$ for the space of functions generated over $\mathbb{Z}$ by the traces of irreducible objects $Rep_E(G)$. Since the trace of representations depends only on isomorphism classes and additive with respect to short exact sequences of $E[G]$-modules, we have a natural group homomorphism $\iota : R_E(G) \to R'_E(G)$ such that $\iota([M]) = \mathrm{Tr}(\rho_M)$ for the representation ρ_M on M. The homomorphism $\iota : R_E(G) \to R'_E(G)$ is surjective. If E is of characteristic 0, ι is an isomorphism by Theorem 2.6 and (2.7).

PROPOSITION 2.9 *We have a surjective group homomorphism:* $R_E(G) \to R'_E(G)$ *taking* $[M]$ *to the trace of the group action on* M. *If* E *is of characteristic* 0, *it is an isomorphism.*

For the moment, suppose that E is of characteristic 0 and that G is a finite group. Then we define a pairing

$$\langle\, ,\, \rangle_G : R'_E(G) \times R'_E(G) \to E$$

by $\langle \phi, \psi \rangle = |G|^{-1} \sum_{g \in G} \phi(g)\psi(g)$. This is a non-degenerate pairing (the orthogonality relation of characters; see (2.5) and [LRF] I.2.3). Let H be a closed subgroup of finite index in G. For an object $M \in R_E(H)$, we define the induced G-module $\mathrm{Ind}_H^G M = \mathrm{Hom}_{E[H]}(E[G], M)$. Thus $\phi \in \mathrm{Ind}_H^G M$ is an E-linear map $\phi : E[G] \to M$ such that $\phi(hx) = h\phi(x)$ for $h \in H$. We regard $\mathrm{Ind}_H^G M$ as $E[G]$-module by $g\phi(x) = \phi(xg)$. Obviously Ind_H^G induces a map $\mathrm{Ind}_H^G : Rep_E(H) \to Rep_E(G)$ and a linear map $\mathrm{Ind}_H^G : R_E(H) \to R_E(G)$, since Ind_H^G preserves short exact sequences. The following fact is well known as Frobenius reciprocity (cf. [LRF] I.7.2

and (4.27) of Chapter 4 in the text):

$$\langle \psi, \phi|_H \rangle_H = \langle \mathrm{Ind}_H^G \psi, \phi \rangle_G. \tag{Fr}$$

We now state a theorem of Artin, following Serre [LRF] II.9.2:

THEOREM 2.10 (E. ARTIN) *Let X be a family of subgroups of a finite group G. We write* Ind : $\bigoplus_{H\in X} R_E(H) \to R_E(G)$ *for* $\bigoplus_{H\in X} \mathrm{Ind}_H^G$. *Then if* $G = \bigcup_{H\in X} \bigcup_{g\in G} gHg^{-1}$, *the cokernel of* Ind : $\bigoplus_{H\in X} R_E(H) \to R_E(G)$ *is finite.*

Actually, in fact: $G = \bigcup_{H\in X} \bigcup_{g\in G} gHg^{-1}$ is equivalent to the assertion of the theorem (see [LRF] Theorem 17) if E is of characteristic 0.

Proof We first assume that E is of characteristic 0. Since $R_E(G) \cong R'_E(G)$, we prove the theorem shifting to $R'_E(G)$. We consider the restriction map $res : R'_E(G) \to \bigoplus_{H\in X} R'_E(H)$ given by $res(\phi) = \bigoplus_{H\in X} \phi|_H$. This map is obviously injective under the assumption:

$$G = \bigcup_{H\in X} \bigcup_{g\in G} gHg^{-1},$$

because elements in $R'_E(G)$ are conjugation invariant. Then by the duality (Fr), we see that

$$\mathrm{Ind} \otimes \mathrm{id} : \bigoplus_{H\in X} R'_E(H) \otimes_{\mathbb{Z}} \mathbb{Q} \to R'_E(G) \otimes_{\mathbb{Z}} \mathbb{Q}$$

is surjective, which proves the desired assertion.

Now we treat the case where E is a finite field of characteristic p. Take a finite extension K of $\mathbb{Q}_p$ so that the residue field of the p-adic integer ring O of K is E. Pick $M \in Rep_K(G)$. Then, as in the proof of Theorem 2.6, we may assume that the representation $\rho_M : G \to GL_n(K)$ attached to M actually has values in $GL_n(O)$. Thus we find an O-free lattice $L \subset M$ stable under G. We define $\overline{M} = (L \otimes_O E)^{ss}$. Then $[\overline{M}] \in R_E(G)$ is well defined and independent of the choice of L, since the isomorphism class of a semi-simple representation is determined by the trace (by Corollary 2.8). The association $[M] \mapsto [\overline{M}]$ induces a homomorphism (called the reduction map) $red : R_K(G) \to R_E(G)$. Let ρ be an irreducible representation of $E[G]$ into $M_n(E)$. Then the image $\overline{R}(\rho) = \rho(E[G])$ as in the Density theorem 2.6 is an algebra direct summand of $E[G]/\overline{I}$ for the radical $\overline{I}$ of $E[G]$. Let I be the ideal of $O[G]$ with $I \otimes_O E = \overline{I}$. Then I is topologically nilpotent (that is, $I/p^n I$ is nilpotent for all n). Take $e \in O[G]$ such that e mod I is the identity of $\overline{R}(\rho)$. Then $\varepsilon = \lim_{n\to\infty} e^{q^n}$ for $q = |E|$ is a central idempotent of $O[G]$. If $e - e' \in I$, then $\lim_{n\to\infty} e^{q^n} = \lim_{n\to\infty} e'^{q^n}$ because I is topologically nilpotent. Then we take $R(\rho) = \varepsilon O[G]$. By definition, $(R(\rho) \otimes_O E)^{ss} \cong \rho^m$ for positive multiplicity m (see (S) in 2.1.4). Thus $red([R(\rho)]) = m[\rho]$. Since the $[\rho]$'s for irreducible ρ generate $R_E(G)$, the reduction map: $red \otimes \mathrm{id} : R_K(G) \otimes_{\mathbb{Z}} \mathbb{Q} \to R_E(G) \otimes_{\mathbb{Z}} \mathbb{Q}$ is surjective. This shows the result because $\mathrm{Ind} \circ red = red \circ \mathrm{Ind}$.

Now suppose that E is a general field. Since G is a finite group, there exists a finite extension E_0 over a prime field (inside E) such that $R_E(G) \cong R_{E_0}(G)$. Then the result follows from the cases we have treated. □

It is actually known that the reduction map $red : R_K(G) \to R_E(G)$ is surjective (see [LRF] 16.1, Theorem 33).

COROLLARY 2.11 *Let E be a field and G be a finite group. Let p be the characteristic of E or 1 according as E has positive characteristic or not. Then for the set T of all cyclic subgroups of G of order prime to p,*

$$\mathrm{Ind} \otimes \mathrm{id} : \bigoplus_{H \in T} R_E(H) \otimes_{\mathbb{Z}} \mathbb{Q} \to R_E(G) \otimes_{\mathbb{Z}} \mathbb{Q}$$

is surjective.

Proof Let T_0 be the set of all cyclic subgroups of G. Since every element of G generates a cyclic subgroup, $G = \bigcup_{H \in T_0} \bigcup_{g \in G} gHg^{-1}$. Then the above theorem proves the result when E has characteristic 0.

Suppose E has characteristic $p > 0$. If M is an object of $Rep_E(H)$ for a cyclic H, then by (2.7), $[M] = [M^{ss}]$ in $R_E(G)$, and M^{ss} is isomorphic to a direct sum of characters $\chi : H \to \mathbb{F}^\times$ for finite extensions $\mathbb{F}$ of E. Since $\mathbb{F}$ has characteristic p, χ is of order prime to p, and hence it factors through $H^{(p)} = H/H_p$ for the Sylow subgroup H_p. Since $H \cong H^{(p)} \times H_p$, we may regard χ as a character of $H^{(p)}$. This shows the desired assertion in general. □

2.1.7 Representations with coefficients in Artinian rings

We now generalize our theory to representations into $M_n(A)$ for artinian rings A. What we have proved for artinian algebra R holds for artinian rings. So $A = \oplus_{\mathfrak{m}} A_{\mathfrak{m}}$ for local artinian rings $A_{\mathfrak{m}}$. So the study of representations with coefficients in an artinian ring is reduced to that for an artinian local ring. So hereafter we assume that A is an artinian local ring with residue field E. Let $\rho : G \to GL_n(A)$ be a representation. We put $\overline{\rho} = \rho \mod \mathfrak{m}_A : G \to GL_n(E)$. Let H be a closed subgroup of a profinite group G. Consider the following condition

$$\overline{\rho}|_H \textit{ is absolutely irreducible.} \qquad (\mathrm{AI}_H)$$

Of course (AI_H) implies (AI_G). We have the following generalization of Schur's lemma by Mazur.

LEMMA 2.12 (B. MAZUR) *Suppose* (AI_G), *and let R be an A-subalgebra of $M_n(A)$ generated by* $\mathrm{Im}(\rho)$. *Then $R = M_n(A)$, and in particular, if $T\rho(\sigma) = \rho(\sigma)T$ for all $\sigma \in G$ and $T \in M_n(A)$, T is a scalar matrix.*

Proof We consider the A-subalgebra R generated by $\rho(G)$ over A inside $M_n(A)$. We only need to show that $R = M_n(A)$, because then its center is scalar. We have the inclusion map $\iota : R \hookrightarrow M_n(A)$, which induces a surjection $\overline{\iota} : R/\mathfrak{m}_A R \to M_n(E)$ by Proposition 2.5. Then by Corollary 2.4, ι is surjective. This shows $R = M_n(A)$. □

Hereafter until the end of this subsection, we fix a finite extension $K/\mathbb{Q}_p$ and let $\mathcal{O} = \mathcal{O}_K$ be the p-adic integer ring of K. Local artinian algebras A are always supposed to be an $\mathcal{O}$-algebra with $A/\mathfrak{m}_A = \mathcal{O}/\mathfrak{m}_{\mathcal{O}} = \mathbb{F}$, which is a finite field with characteristic p. Since A is a finite ring, replacing G by $\mathrm{Im}(\rho)$ for a representation $\rho : G \to GL_n(A)$, we may assume that G is a finite group. Now we show that representations are characterized by their trace under (AI_G).

PROPOSITION 2.13 (CARAYOL, SERRE) *Let A be an pro-artinian local ring with finite residue field $\mathbb{F}$. Let $R = A[G]$ for a profinite group G. Let $\rho : R \to M_n(A)$ and $\rho' : R \to M_{n'}(A)$ be two continuous representations. If $\overline{\rho}$ satisfies* (AI_G) *and* $\mathrm{Tr}(\rho(\sigma)) = \mathrm{Tr}(\rho'(\sigma))$ *for all $\sigma \in G$, then $\rho \sim \rho'$.*

Proof Since we can easily reduce the pro-artinian case to the artinian case, we prove here the result only for artinian rings. Simply write $\mathfrak{m} = \mathfrak{m}_A$. Let $\ell = \ell(A)$ be the length of A. We proceed by induction on ℓ. Since $\rho \mod \mathfrak{m}$ and $\rho' \mod \mathfrak{m}$ have finite image, they factor through a finite quotient group G' of G. Applying the Brauer–Nesbitt theorem to the finite group G', if $\ell = 1$, the assertion is true, and hence $n = n'$. Suppose $\ell > 0$. Let $\mathfrak{a}$ be an ideal of A such that $\mathfrak{a} \cong \mathbb{F}$ as E-vector spaces. Then $\ell(A/\mathfrak{a}) = \ell - 1$. Since $\rho \mod \mathfrak{a}$ and $\rho' \mod \mathfrak{a}$ have values in $M_n(A/\mathfrak{a})$ and have the same trace, by the induction assumption, we find $x \in GL_n(A)$ such that $\rho \equiv x\rho' x^{-1} \mod \mathfrak{a}$. Replacing ρ' by $x\rho' x^{-1}$, we may assume that $\rho'(\sigma) = \rho(\sigma)+\delta(\sigma)$ and $\mathrm{Tr}(\delta(\sigma)) = 0$ for $\delta : R \to M_n(\mathfrak{a})$. Note that the isomorphism: $\mathfrak{a} \cong \mathbb{F}$ induces $M_n(\mathfrak{a}) \cong M_n(\mathbb{F})$. By simple computation, we get

$$\delta(\sigma\tau) = \overline{\rho}(\sigma)\delta(\tau) + \delta(\sigma)\overline{\rho}(\tau).$$

Let $\tau \in \mathrm{Ker}(\overline{\rho})$. Then

$$0 = \mathrm{Tr}(\delta(\sigma\tau)) = \mathrm{Tr}(\overline{\rho}(\sigma)\delta(\tau)).$$

By Proposition 2.5, $\mathrm{Im}(\overline{\rho}) = M_n(\mathbb{F})$. This combined with the non-degeneracy of the bilinear pairing $M_n(\mathbb{F}) \times M_n(\mathfrak{a}) \to \mathfrak{a}$ given by $(X, Y) \mapsto \mathrm{Tr}(XY)$ shows that $\delta = 0$ on $\mathrm{Ker}(\overline{\rho})$. Thus δ induces $\delta : M_n(\mathbb{F}) = \mathrm{Im}(\overline{\rho}) \to M_n(\mathfrak{a}) \cong M_n(\mathbb{F})$ and satisfies $\delta(XY) = X\delta(Y) + \delta(X)Y$ for $X, Y \in M_n(\mathbb{F})$. Then it is elementary (through a computation which is a bit complicated) to see the existence of $U \in M_n(\mathfrak{a})$ such that $\delta(\sigma) = \overline{\rho}(\sigma)U - U\overline{\rho}(\sigma)$ for all $\sigma \in R$. This shows that $\rho'(\sigma) = (1-U)\rho(\sigma)(1+U)$, which proves the assertion. □

Remark 2.14 To get an exact information of representation from the data of trace, one has to impose some condition on the representation. Here is Carayol's example exhibiting non-isomorphic representations with the same traces: Let G be the additive group of trace zero matrices in $M_n(\mathbb{F})$. Let $\rho : G \to M_n(\mathbb{F}[\varepsilon])$ be the trivial representation, that is, $\rho(\sigma) = 1_n$ for all σ, where $E[\varepsilon] = E[X]/(X^2)$ with $\varepsilon = X$, and 1_n is the $n \times n$ identity matrix. We have another $\rho'(\sigma) = 1_n + \varepsilon\sigma$. They have identical traces, identical determinants and, further, identical characteristic polynomials but are non-isomorphic.

For later use, we record the following characterization of induced representations:

Lemma 2.15 *Let $\rho : G \to GL_n(B)$ be a continuous representation for $B \in CL_{\mathcal{O}}$. Suppose that $\overline{\rho} = \rho \bmod \mathfrak{m}_B : G \to GL_n(\mathbb{F})$ is absolutely irreducible. Let $\chi : G \to B^\times$ be a continuous character of order r prime to p. Then $\rho \cong \rho \otimes \chi$ if and only if there exists a B-free local algebra B' with $\mathrm{rank}_B B' \le r$ and a representation $\varphi : H \to GL_m(B')$ for $H = \mathrm{Ker}(\chi)$ such that $(G : H)m = n$ and $\rho \cong \mathrm{Ind}_H^G \varphi$ in $GL_n(B')$.*

Proof Suppose that $\rho = \mathrm{Ind}_H^G \varphi$ for a representation $\varphi : H \to GL_m(B')$. Note that the representation space of ρ is given by

$$V(\varphi) \otimes_{B'} B'[\Delta] \cong \mathrm{Hom}_{\mathbb{Z}}(B'[\Delta], V(\varphi)).$$

Then from the isomorphism $B'[\Delta] \cong B'[\Delta] \otimes \chi$ induced by the group character: $\sigma \mapsto \chi(\sigma)\sigma$ of $\Delta = G/H$ into $B[\Delta]^{\times}$, it is obvious that the induced representation ρ satisfies $\rho \cong \rho \otimes \chi$. Conversely, we assume $\rho \cong \rho \otimes \chi$. We also write $V = V(\rho)$ for the representation space of ρ. Then by definition, we can find $C \in GL_n(B)$ such that $C(\chi(\sigma)\rho(\sigma))C^{-1} = \rho(\sigma)$ for all $\sigma \in G$. Then $C^r \rho(\sigma) C^{-r} = \rho(\sigma)$ for all σ. This combined with the absolute irreducibility of $\overline{\rho}$ shows that C^r is a scalar in $B^{\times}$ (Lemma 2.12). We take a local factor B' of $B[X]/(X^r - C^r)$, extend the scalar to B' and consider $V' = V \otimes_B B'$. Then C acts semi-simply on V'. Fixing an eigenvalue c of C on V', all other eigenvalues of C are given by ζc for an rth root of unity ζ. Note that B contains all r-th roots of unity because χ has values in B. Let $V[c\zeta]$ be the $c\zeta$-eigenspace of C, which is a direct summand of V since r is prime to p. If $v \in V'[c\zeta]$, then we see from $\rho \cong \rho \otimes \chi$ that $C\rho(\sigma)v = \chi(\sigma)^{-1}c\zeta\rho(\sigma)v$. Thus G permutes $V'[c\zeta]$ and then $V' = \oplus_{\sigma \in \Delta}\rho(\sigma)V'[c]$. From this, it is easy to construct an $B[G]$-isomorphism $V' \cong \mathrm{Hom}_{\mathbb{Z}[H]}(\mathbb{Z}[G], V'[c]) \cong B'[G] \otimes_{B'[H]} V'[c]$ sending $\sigma V'[c]$ to $\sigma \otimes V'[c]$, which proves the desired assertion. □

Exercise

(1) Show the existence of $U \in M_n(E)$ for a field E such that $\delta(X) = XU - UX$ if $\delta : M_n(F) \to M_n(F)$ satisfies $\delta(XY) = X\delta(Y) + \delta(X)Y$ for all $X, Y \in M_n(E)$.

2.2 Pseudo-representations

In this section, the coefficient ring A is always a local ring with maximal ideal $\mathfrak{m}_A$. We write $E = A/\mathfrak{m}_A$. We would like to characterize the trace of a representation of a group G.

2.2.1 *Pseudo-representations of degree 2*

We first describe in detail traces of degree 2 representations $\rho : G \to GL_2(A)$ when 2 is invertible in A and G contains c such that $c^2 = 1$ and $\det \rho(c) = -1$. Since 2 is invertible, we know that $V = V(\rho) = V_+ \oplus V_-$ for $V_{\pm} = \frac{1 \pm c}{2} V$. For $\overline{\rho} = \rho \mod \mathfrak{m}_A$, we write $\overline{V} = V(\overline{\rho})$. Then similarly as above, $\overline{V} = \overline{V}_+ \oplus \overline{V}_-$ and $\overline{V}_{\pm} = V_{\pm}/\mathfrak{m}_A V_{\pm}$. Since $\dim_E \overline{V} = 2$ and $\det \overline{\rho}(c) = -1$, $\dim_E \overline{V}_{\pm} = 1$. This shows that $\overline{V}_{\pm} = E\overline{v}_{\pm}$ for $\overline{v}_{\pm} \in \overline{V}_{\pm}$. Take $v_{\pm} \in V_{\pm}$ such that $v_{\pm} \mod \mathfrak{m}_A V_{\pm} = \overline{v}_{\pm}$, and define $\phi_{\pm} : A \to V_{\pm}$ by $\phi(a) = av_{\pm}$. Then $\phi_{\pm} \mod \mathfrak{m}_A V$ is surjective. Thus by Lemma 2.3,

$\phi_{\pm} : A \to V_{\pm}$ is surjective. Define $\phi : A^2 \to V$ by $\phi(a,b) = \phi_+(a)+\phi_-(b)$. Then ϕ is a surjective A-linear map. Identifying V with A^2, we may regard $\phi : A^2 \to A^2$ as a surjective A-linear map. Suppose that $av_+ + bv_- = 0 \Rightarrow (a \mod \mathfrak{m}_A)\overline{v}_+ + (b \mod \mathfrak{m}_A)\overline{v}_- = 0 \Rightarrow a, b \in \mathfrak{m}_A$. Thus $\mathrm{Ker}(\phi) \subset \mathfrak{m}_A(A^2)$ in A^2. Let $u, v \in V$ be a base of V over A. Pick $x, y \in A^2$ such that $\phi(x) = u$ and $\phi(y) = v$. Define $\psi : V \to A^2$ by $\psi(au+bv) = ax+by$. Then $\phi \circ \psi = \mathrm{id}_V$. Since $A^2 \ni t = \psi(\phi(t)) + t - \psi(\phi(t))$, $A^2 = \psi(V) \oplus \mathrm{Ker}(\phi)$. In particular, $\mathfrak{m}_A\psi(V) \oplus \mathfrak{m}_A \mathrm{Ker}(\phi) = \mathfrak{m}_A(A^2) \supset \mathrm{Ker}(\phi)$ and $\mathfrak{m}_A \mathrm{Ker}(\phi) = \mathrm{Ker}(\phi)$. Since $\mathrm{Ker}(\phi) = A^2/\psi(V)$, $\mathrm{Ker}(\phi)$ is an A-module of finite type. Thus by Lemma 2.3, $\mathrm{Ker}(\phi) = 0$. This shows that $\phi_{\pm} : A \cong V_{\pm}$. In other words, $\{v_-, v_+\}$ is an A-base of V. We write $\rho(r) = \begin{pmatrix} a(r) & b(r) \\ c(r) & d(r) \end{pmatrix}$ with respect to this base. Thus $\rho(c) = \begin{pmatrix} -1 & 0 \\ 0 & 1 \end{pmatrix}$. Define another function $x : G \times G \to A$ by $x(r,s) = b(r)c(s)$. Then we have

(W1) $a(rs) = a(r)a(s) + x(r,s)$, $d(rs) = d(r)d(s) + x(s,r)$ and $x(rs,tu) = a(r)a(u)x(s,t) + a(u)d(s)x(r,t) + a(r)d(t)x(s,u) + d(s)d(t)x(r,u)$;

(W2) $a(1) = d(1) = d(c) = 1$, $a(c) = -1$ and $x(r,s) = x(s,t) = 0$ if $s = 1, c$;

(W3) $x(r,s)x(t,u) = x(r,u)x(t,s)$.

These are easy to check: We have

$$\begin{pmatrix} a(r) & b(r) \\ c(r) & d(r) \end{pmatrix} \begin{pmatrix} a(s) & b(s) \\ c(s) & d(s) \end{pmatrix} = \begin{pmatrix} a(rs) & b(rs) \\ c(rs) & d(rs) \end{pmatrix}.$$

Then by computation, $a(rs) = a(r)a(s) + b(r)c(s) = a(r)a(s) + x(r,s)$. Similarly, we have $b(rs) = a(r)b(s)+b(r)d(s)$ and $c(rs) = c(r)a(s)+d(r)c(s)$. Thus

$$\begin{aligned} x(rs,tu) &= b(rs)c(tu) = (a(r)b(s) + b(r)d(s))(c(t)a(u) + d(t)c(u)) \\ &= a(r)a(u)x(s,t) + a(r)d(t)x(s,u) + a(u)d(s)x(r,t) + d(s)d(t)x(r,u). \end{aligned}$$

A triple $\{a, d, x\}$ satisfying the three conditions (W1–3) is called a *pseudo–representation* of Wiles of (G, c). For each pseudo-representation $\pi = \{a, d, x\}$, we define

$$\mathrm{Tr}(\pi)(r) = a(r) + d(r) \quad \text{and} \quad \det(\pi)(r) = a(r)d(r) - x(r,r).$$

By a direct computation using (W1–3), we see that

$$a(r) = \frac{1}{2}(\mathrm{Tr}(\pi)(r) - \mathrm{Tr}(\pi)(rc)), \quad d(r) = \frac{1}{2}(\mathrm{Tr}(\pi)(r) + \mathrm{Tr}(\pi)(rc))$$

and

$$x(r,s) = a(rs) - a(r)a(s), \quad \det(\pi)(rs) = \det(\pi)(r)\det(\pi)(s).$$

Thus the pseudo-representation π is determined by the trace of π as long as 2 is invertible in A.

PROPOSITION 2.16 (A. WILES, 1988) *Let G be a group and $R = A[G]$. Let $\pi = \{a, d, x\}$ be a pseudo-representation (of Wiles) of (G, c). Suppose either that there exists at least one pair $(r, s) \in G \times G$ such that $x(r, s) \in A^\times$ or that $x(r, s) = 0$ for all $r, s \in G$. Then there exists a representation $\rho : R \to M_2(A)$ such that $\mathrm{Tr}(\rho) = \mathrm{Tr}(\pi)$ and $\det(\rho) = \det(\pi)$ on G. If A is a topological ring, G is a topological group and all maps in π are continuous on G, then ρ is a continuous representation of G into $GL_2(A)$ under the topology on $GL_2(A)$ induced by the product topology on $M_2(A)$.*

Proof When $x(r, s) = 0$ for all $r, s \in G$, we see from (W1) that $a, d : G \to A$ satisfies $a(rs) = a(r)a(s)$ and $d(rs) = d(r)d(s)$. Thus a, d are characters of G, and we define $\rho : G \to GL_2(A)$ by $\rho(g) = \begin{pmatrix} a(g) & 0 \\ 0 & d(g) \end{pmatrix}$, which satisfies the required property. We extend ρ to $R = A[G]$ by linearity. We now suppose that $x(r, s) \in A^\times$ for $r, s \in G$. Then we define $b(g) = x(g, s)/x(r, s)$ and $c(g) = x(r, g)$ for $g \in G$. Then by (W3), $b(g)c(h) = x(r, h)x(g, s)/x(r, s) = x(g, h)$. Put $\rho(g) = \begin{pmatrix} a(g) & b(g) \\ c(g) & d(g) \end{pmatrix}$. By (W2), we see that $\rho(1)$ is the identity matrix and $\rho(c) = \begin{pmatrix} -1 & 0 \\ 0 & 1 \end{pmatrix}$. By computation,

$$\rho(g)\rho(h) = \begin{pmatrix} a(g) & b(g) \\ c(g) & d(g) \end{pmatrix} \begin{pmatrix} a(h) & b(h) \\ c(h) & d(h) \end{pmatrix} = \begin{pmatrix} a(g)a(h)+b(g)c(h) & a(g)b(h)+b(g)d(h) \\ c(g)a(h)+d(g)c(h) & d(g)d(h)+c(g)b(h) \end{pmatrix}.$$

By (W1), $a(gh) = a(g)a(h) + x(g, h) = a(g)a(h) + b(g)c(h)$ and $d(gh) = d(g)d(h) + x(h, g) = d(g)d(h) + b(h)c(g)$. Now let us look at the lower left corner:

$$c(g)a(h) + d(g)c(h) = x(r, g)a(h) + d(g)x(r, h).$$

Now apply (W1) to $(1, r, g, h)$ in place of (r, s, t, u), and we get

$$c(gh) = x(r, gh) = a(h)x(r, g) + d(g)x(r, h),$$

because $x(1, g) = x(1, h) = 0$. As for the upper right corner, we apply (W1) to $(g, h, 1, s)$ in place of (r, s, t, u). Then we get

$$b(gh)x(r, s) = x(gh, s) = a(g)x(h, s) + d(h)x(g, s) = (a(g)b(h) + d(h)b(g))x(r, s),$$

which shows that $\rho(gh) = \rho(g)\rho(h)$. We now extend ρ linearly to $R = A[G]$. This proves the first assertion. The continuity of ρ follows from the continuity of each entry, which follows from the continuity of π. $\square$

We fix an absolutely irreducible representation $\overline{\rho} : G \to GL_2(E)$ with $\det(\overline{\rho})(c) = -1$. If we have a representation $\rho : G \to GL_2(A)$

with $\rho \mod \mathfrak{m}_A \sim \overline{\rho}$, then $\det(\rho(c)) \equiv -1 \mod \mathfrak{m}_A$. Since $c^2 = 1$, if 2 is invertible in A ($\iff$ the characteristic of E is different from 2), $\det(\rho(c)) = -1$. Thus we have a well-defined pseudo-representation π_ρ of Wiles associated with ρ. Since $\overline{\rho}$ is absolutely irreducible, we find $r, s \in G$ such that $b(r) \not\equiv 0 \mod \mathfrak{m}_A$ and $c(s) \not\equiv 0 \mod \mathfrak{m}_A$. Thus π_ρ satisfies the condition of Proposition 2.16. Conversely if we have a pseudo-representation $\pi : G \to A$ such that $\pi \equiv \overline{\pi} \mod \mathfrak{m}_A$ for $\overline{\pi} = \pi_{\overline{\rho}}$, again we find $r, s \in G$ such that $x(r, s) \in A^\times$. The correspondence $\rho \mapsto \pi_\rho$ induces a bijection:

$$\{\rho : G \to GL_2(A) : \text{representation}|\rho \mod \mathfrak{m}_A \sim \overline{\rho}\} / \sim \leftrightarrow \tag{2.8}$$
$$\{\pi : G \to A : \text{pseudo-representation of Wiles}|\pi \mod \mathfrak{m}_A = \overline{\pi}\},$$

where $\overline{\pi} = \pi_{\overline{\rho}}$. The map is surjective by Proposition 2.16 and one to one by Proposition 2.13, because a pseudo-representation is determined by its trace.

2.2.2 Higher degree pseudo-representations

As we have already seen, any group representation $\rho : G \to GL_n(A)$ is induced by an A-algebra representation $\rho : A[G] \to M_n(A)$, that is, an A-algebra homomorphism. Thus without losing generality, we may restrict ourselves to A-algebra representations $\rho : R \to M_n(A)$ for an A-algebra R. We now try to characterize the trace of a representation in a general case. Let $\rho : R \to M_n(A)$ be a representation. We write $T(r) = \mathrm{Tr}(\rho(r))$. First we study obvious properties of the function T. Since ρ takes the identity to the identity matrix, we have $T(1) = n$. For any two $n \times n$ matrices X, Y, we know that $\mathrm{Tr}(XY) = \mathrm{Tr}(YX)$. More generally, for n matrices $X_1, \ldots, X_n$,

$$\mathrm{Tr}(X_1 X_2 \cdots X_n) = \mathrm{Tr}(X_n X_1 X_2 \cdots X_{n-1}),$$

and for a cyclic permutation $\sigma \in \mathfrak{S}_n$,

$$\mathrm{Tr}(X_1 X_2 \cdots X_n) = \mathrm{Tr}(X_{\sigma(1)} X_{\sigma(2)} \cdots X_{\sigma(n)}).$$

This shows that $T(rs) = T(sr)$ (symmetry) and $T(r_1 r_2 \cdots r_n) = T(r_{\sigma(1)} r_{\sigma(2)} \cdots r_{\sigma(n)})$ for a cyclic permutation $\sigma = (1, 2, \ldots, n)^k$. Then we have the following fact from the invariant theory of matrices (see a paper of Procesi [P], Theorem 4.3):

LEMMA 2.17 (C. PROCESI) *Let* $X_0, \dots, X_n \in M_n(A)$, *and let* $\sigma \in \mathfrak{S}_{n+1}$ *be a permutation on the* $n+1$*-set* $\{0, 1, \dots, n\}$. *Decompose*

$$\sigma = (i_{1,1}, \dots, i_{1,k_1})(i_{2,1}, \dots, i_{2,k_2}) \cdots (i_{m,1}, \dots, i_{m,k_m})$$

with $i_{1,1} = 0$ *as a product of disjoint cycles including all* 1*-cycles (thus* $(1,2) = (1,2)(3)$ *in* $\mathfrak{S}_3$*). Define* $Tr_\sigma(X_0, \dots, X_n) = \prod_j \mathrm{Tr}(X_{i_{j,1}} X_{i_{j,2}} \cdots X_{i_{j,k_j}})$. *Then we have*

$$\sum_{\sigma \in \mathfrak{S}_{n+1}} sgn(\sigma) Tr_\sigma(X_0, \dots, X_n) = 0$$

for every $n+1$*-tuple* $(X_0, \dots, X_n) \in M_n(A)^{n+1}$.

Proof Consider $V = A^n$. We put $W = \overbrace{V \otimes_A V \otimes \cdots \otimes_A V}^{m}$. We let $\mathfrak{S}_m$ act on W A-linearly by $\lambda_\sigma(v_1 \otimes v_2 \otimes \cdots \otimes v_n) = v_{\sigma(1)} \otimes v_{\sigma(2)} \otimes \cdots \otimes v_{\sigma(n)}$. Let $e = \sum_{\sigma \in \mathfrak{S}_m} sgn(\sigma) \lambda_\sigma \in \mathrm{End}_A(W)$. We have a natural A-linear map from $\bigwedge^m V = \overbrace{V \wedge_A V \wedge \cdots \wedge_A V}^{m}$ onto eW. Thus if $m > n$, $e = 0$. Let $V^* = \mathrm{Hom}_A(V, A)$. Then for each $v^* \otimes v \in V^* \otimes_A V$, we define an A-linear map $L : V^* \otimes_A V \to \mathrm{End}_A(V)$ by $v' \mapsto v^*(v')v$ for $v^* \in V^*$ and $v, v' \in V$. We claim that $L : V^* \otimes V \cong \mathrm{End}_A(V)$ is an isomorphism. If we choose bases (v_i) and (v_i^*) dual to each other, then we see from the definition that $L(v_i^* \otimes v_j)$ brings v_i to v_j and kills all v_k with $k \neq i$. This shows that the matrix expression of $L(v_i^* \otimes v_j)$ with respect to (v_i) is the elementary matrix $E_{i,j}$ having non-trivial entry 1 only at the (i,j)-spot. Since $E_{i,j}$ forms a basis of $M_n(A)$ and L takes the base $\{v_i^* \otimes v_j\}_{i,j}$ of $V^* \otimes_A V$ to that of $\mathrm{End}_A(V)$, L is a surjective isomorphism. Thus the equality: $\mathrm{Tr}(L(v^* \otimes v)) = (v^*, v)$ holds on the base $\{v_i^* \otimes v_j\}_{i,j}$. Since Tr is an A-linear map, we have for all $v \in V$ and $v^* \in V^*$

$$(v^*, v) = v^*(v) = \mathrm{Tr}(L(v^* \otimes v)), \tag{2.9}$$

because if $\varphi_i(v_j) = \delta_{ij}$, $L(\varphi_i \otimes v_j)$ has only one non-trivial entry 1 at the (i,j) spot. Let $W^* = \overbrace{V^* \otimes_A V^* \otimes \cdots \otimes V^*}^{m}$. Then we see that $W^* \cong \mathrm{Hom}_A(W, A)$ via the pairing $(v_1^* \otimes v_2^* \otimes \cdots \otimes v_m^*, v_1 \otimes v_2 \otimes \cdots \otimes v_m) = \prod_j v_j^*(v_j)$. We can define a perfect pairing $\langle\ ,\ \rangle : \mathrm{End}_A(W) \times (W^* \otimes W) \to A$ by

$$\langle \lambda, v_1^* \otimes v_2^* \otimes \cdots \otimes v_m^* \otimes v_1 \otimes v_2 \otimes \cdots \otimes v_m \rangle$$
$$= (v_1^* \otimes v_2^* \otimes \cdots \otimes v_m^*, \lambda(v_1 \otimes v_2 \otimes \cdots \otimes v_m)).$$

Then by definition

$$\langle \lambda_\sigma, v_1^* \otimes v_2^* \otimes \cdots \otimes v_m^* \otimes v_1 \otimes v_2 \otimes \cdots \otimes v_m \rangle = \prod_j (v_j^*, v_{\sigma(j)}).$$

Now decompose $\sigma = (i_1 i_2 \cdots i_k)(j_1 j_2 \cdots j_n) \cdots (t_1 \cdots t_e)$ as a product of disjoint cycles without omitting 1-cycles. Then we see

$$\begin{aligned}\langle \lambda_\sigma, v_1^* \otimes v_2^* \otimes \cdots \otimes v_m^* \otimes v_1 \otimes v_2 \otimes \cdots \otimes v_m \rangle &= \prod_j (v_j^*, v_{\sigma(j)}) \\ &= (v_{i_1}^*, v_{i_2})(v_{i_2}^*, v_{i_3}) \cdots (v_{i_k}^*, v_{i_1}) \cdots (v_{t_1}^*, v_{t_2})(v_{t_2}^*, v_{t_3}) \cdots (v_{t_e}^*, v_{t_1}).\end{aligned}$$

We may regard

$$W^* \otimes W = (V^* \otimes_A V) \otimes \cdots \otimes (V^* \otimes_A V) = \overbrace{\mathrm{End}(V) \otimes_A \cdots \otimes_A \mathrm{End}(V)}^{m}.$$

Then, first for $A_j = v_j^* \otimes v_j$ and $\sigma = (1, 2, \ldots, m)$, we see that

$$\begin{aligned} A_1 A_2 \cdots A_m(x) &= A_1 A_2 \cdots A_{m-1}((v_m^*, x) v_m) \\ &= A_1 A_2 \cdots A_{m-2}((v_{m-1}^*, v_m)(v_m^*, x) v_{m-1}) \\ &= \left\{ \prod_{j=1}^{m-1} (v_j^*, v_{\sigma(j)}) \right\} v_m^* \otimes v_1(x). \end{aligned}$$

Thus $\mathrm{Tr}(A_1 A_2 \cdots A_m) = Tr_\sigma(A_1, A_2, \ldots, A_m)$. Since Tr is linear, this is true for all $A_j \in \mathrm{End}_A(V)$, and (2.10) and (2.9) imply for general $\sigma \in \mathfrak{S}_m$

$$\langle \lambda_\sigma, A_1 \otimes A_2 \otimes \cdots \otimes A_m \rangle = Tr_\sigma(A_1, \ldots, A_m). \tag{2.10}$$

Applying this to $m = n+1$ and $0 = e = \sum_{\sigma \in \mathfrak{S}_m} sgn(\sigma) \lambda_\sigma \in \mathrm{End}_A(W)$, we get

$$0 = \langle e, X_0 \otimes X_1 \otimes \cdots \otimes X_n \rangle = \sum_{\sigma \in \mathfrak{S}_m} sgn(\sigma) Tr_\sigma(X_0, \ldots, X_n). \qquad \square$$

A pseudo-representation $T : R \to A$ of degree n of Taylor is a function satisfying the following property:

(T1) $T(1) = n$;
(T2) $T(rs) = T(sr)$ for all $r, s \in R$;
(T3) $\sum_{\sigma \in \mathfrak{S}_{n+1}} sgn(\sigma) T_\sigma(r_0, \ldots, r_n) = 0$ for all $r_j \in R$,

where $T_\sigma(r_0, \ldots, r_n) = \prod_j T(r_{i_{j,1}} r_{i_{j,2}} \cdots r_{i_{j,k_j}})$ under the notation of Lemma 2.17. When $n = 2$, (T3) is just:

$$\begin{aligned} T(r_0)T(r_1)T(r_2) + T(r_0 r_1 r_2) + T(r_0 r_2 r_1) - T(r_0 r_1)T(r_2) \\ - T(r_1 r_2)T(r_0) - T(r_0 r_2)T(r_1) = 0. \end{aligned}$$

If $\rho : R \to M_n(A)$ is a representation, then $T = \mathrm{Tr}(\rho)$ is a pseudo-representation of R of degree n by Lemma 2.17. For the converse, the following fact is proved by Taylor [Ty2] and Nyssen [Ny]:

THEOREM 2.18 *Suppose that A is a pro-artinian local ring. Let $T : R \to A$ be a pseudo-representation of degree n of Taylor.*

(1) (R. Taylor) *If A is an algebraically closed field of characteristic 0 and $R = A[G]$ for a group G, then there exists a semi-simple representation $\rho : R \to M_n(A)$ with $T = \mathrm{Tr}(\rho)$;*
(2) (L. Nyssen) *If there exists an absolutely irreducible representation $\overline{\rho} : R \to M_n(A/\mathfrak{m}_A)$ such that $T \mod \mathfrak{m}_A = \mathrm{Tr}(\overline{\rho})$, then there exists a unique representation $\rho : R \to A$ such that $T = \mathrm{Tr}(\rho)$;*
(3) *If T is continuous, then ρ as above is continuous.*

We fix an absolutely irreducible representation $\overline{\rho} : R \to GL_n(E)$. By the above theorem, the correspondence $\rho \mapsto \mathrm{Tr}(\rho)$ induces a bijection:

$$\{\rho : G \to GL_2(A) : \text{representation} | \rho \mod \mathfrak{m}_A \sim \overline{\rho}\} / \sim \leftrightarrow \left\{ \begin{matrix} \tau : G \to A : \text{pseudo-representation of Taylor} \\ \text{of degree } n | \tau \mod \mathfrak{m}_A = \mathrm{Tr}(\overline{\rho}) \end{matrix} \right\}. \tag{2.11}$$

2.3 Deformation of Group Representations

Let K be a finite extension of $\mathbb{Q}_p$ with p-adic integer ring $\mathcal{O}$. We write $\mathbb{F} = \mathcal{O}/\mathfrak{m}_{\mathcal{O}}$. Let G be a profinite group. We write $CL_{\mathcal{O}}$ for the category of pro-artinian local $\mathcal{O}$-algebras with residue field $\mathbb{F}$. We start from a continuous representation $\overline{\rho} : G \to GL_n(\mathbb{F})$. A continuous representation $\rho : G \to GL_n(A)$ for an object A of $CL_{\mathcal{O}}$ is called a deformation (over $\mathcal{O}$) if $\rho \mod \mathfrak{m}_A = \overline{\rho}$. We call two representations ρ and ρ' *strictly equivalent* if $\rho(g) = x\rho'(g)x^{-1}$ for every $g \in G$ with $x \in 1 + M_n(\mathfrak{m}_A)$ independent of g. We write $\rho \approx \rho'$ if they are strictly equivalent. Note that $\rho \mod \mathfrak{m}_A = \rho' \mod \mathfrak{m}_A$ if $\rho \approx \rho'$. A pair (R, ϱ) of an object R of $CL_{\mathcal{O}}$ and a deformation $\varrho : G \to GL_n(R)$ are called a *universal couple* if there exists a unique $\mathcal{O}$-algebra homomorphism $\varphi : R \to A$ for each deformation ρ of $\overline{\rho}$ such that $\varphi\varrho \approx \rho$ in $GL_n(A)$, where $\varphi\varrho(g) = (\varphi(\varrho_{ij}(g)))$ for each entry ϱ_{ij} of ϱ. For two objects A and B in $CL_{\mathcal{O}}$ and an $\mathcal{O}$-algebra homomorphism $f : A \to B$, we have automatically $f^{-1}(\mathfrak{m}_B) = \mathfrak{m}_A$, that is, f is 'local', because A and B have the same residue field $\mathbb{F}$. Thus we do not need to say that the above homomorphism inducing ρ from ϱ is local. If we further enlarge the category to all pro-artinian local $\mathcal{O}$-algebras, we then need to impose the condition that the morphisms are local. We study here the existence of a universal couple. Of course, if it exists, it is unique up to isomorphisms. Hereafter in this section, all representations and pseudo-representations are continuous under the profinite topology of G and the linear topology on $M_n(A)$ induced by the profinite topology of A.

2.3.1 Abelian deformation

Let G be a profinite abelian group. We can then split $G = G_p \times G^{(p)}$ for a p-profinite group G_p and a prime-to-p profinite group $G^{(p)}$. We call a profinite group H 'prime-to-p' if H has a system $\mathfrak{U}$ of open neighborhoods of 1 made up of normal subgroups N such that $p \nmid (H : N)$.

LEMMA 2.19 *Let $A \in CL_{\mathcal{O}}$ and $\chi : H \to A^\times$ be a character. If H is a prime-to-p profinite group, then $\chi(H)$ is a finite group.*

Proof Since any character χ factors through $H/\overline{(H,H)} = H^{ab}$, we may assume that H is an abelian prime-to-p profinite group. Write $q = p^f$ for the number of elements of $\mathbb{F}$. Then $x \to x^q$ is an automorphism of H. Write $A = \varprojlim_{\mathfrak{a}} A/\mathfrak{a}$ for artinian $\mathcal{O}$-algebras $A/\mathfrak{a}$. Then $|(A/\mathfrak{a})^\times| = p^\alpha \times (q-1)$, and hence $(A/\mathfrak{a})^\times \cong P_{\mathfrak{a}} \times \mu_{q-1}$ for an abelian p-group $P_{\mathfrak{a}}$ and $\mu_{q-1} \cong \mathbb{F}^\times$. Since $x^{q^n} = 1$ for all $x \in P_{\mathfrak{a}}$ for sufficiently large n, $((A/\mathfrak{a})^\times)^{q^n} = \mu_{q-1}$. This shows that $\omega(x) = \lim_{n\to\infty} x^{q^n}$ converges in $A^\times$ and $\omega : A^\times \to \mu_{q-1}$ gives the decomposition $A^\times = \mathrm{Ker}(\omega) \times \mu_{q-1}$. Since $\chi(H)^q = \chi(H^q) = \chi(H)$, we see $\chi(H) = \omega(\chi(H)) \subset \mu_{q-1}$, which is a finite set. □

The proof of this lemma shows that for any $A \in CL_{\mathcal{O}}$, $A^\times = \mu_{q-1}(A) \times (1 + \mathfrak{m}_A)$ and $\mu_{q-1} \cong \mathbb{F}^\times$. We start from a character $\overline{\rho} : G \to \mathbb{F}^\times$. We can lift $\overline{\rho}$ to a unique character $\tilde{\rho} : G \to \mathbb{F}^\times \cong \mu_{q-1}(A) \subset A^\times$. This character $\tilde{\rho}$ is called the Teichmüller lift of $\overline{\rho}$.

LEMMA 2.20 *Let H be a finite abelian p-group. Then for any $A \in CL_{\mathcal{O}}$, $A[H]$ is a local ring with residue field $\mathbb{F}$.*

Proof First suppose that A is an artinian ring. Then $\ell(A[H]) = \ell(A)|H|$. Thus $R = A[H]$ is artinian. Let V be an irreducible R-module. Pick $0 \neq v \in V$. Since $R \ni r \mapsto rv \in V$ is surjective because of irreducibility, V is an R-module generated by a single element. Thus V is an A-module of finite type with $\ell_A(V) \leq \ell_A(R)$. Thus by Lemma 2.3, $\mathfrak{m}_A V \neq V$.

Since $\mathfrak{m}_A V$ is R-stable, by irreducibility, $\mathfrak{m}_A V = 0$. Thus V is a finite dimensional $\mathbb{F}$-vector space. Then by Proposition 2.5, $End_R(V) = \mathbb{K}$ for a finite extension $\mathbb{K}/\mathbb{F}$. Thus $V \cong \mathbb{K}$. This shows that $V = V(\overline{\rho})$ for a character $\overline{\rho} : H \to \mathbb{K}^\times$. Since H is a p-group, every $h \in H$ satisfies $h^q = 1$ for $q = p^m$. Thus $\overline{\rho}(h)$ has to be a p-power root of unity. This shows that $\overline{\rho}(h) = 1$ because $\mathbb{F}$ is of characteristic p. Thus there is only one irreducible representation, which is trivial. Thus by (S) in Subsection 2.1.4, R is a local ring. Since $\mathbb{K}$ is generated over $\mathbb{F}$ by the values of $\overline{\rho}$, we have $\mathbb{K} = \mathbb{F}$. When $A = \varprojlim_j A_j$ for artinian rings A_j, $A[H] = \varprojlim_j A_j[H]$. Since $A_j[H]$ is local, $A_j[H]^\times = A_j[H] - \mathfrak{m}_{A_j[H]}$ by (2.3). Since the projection map takes $A_{j+1}[H] - \mathfrak{m}_{A_{j+1}[H]}$ onto $A_j[H] - \mathfrak{m}_{A_j[H]}$, $A[H]^\times = A[H] - \varprojlim_j \mathfrak{m}_{A_j[H]}$. This shows that $\mathfrak{m}_{A[H]} = \varprojlim_j \mathfrak{m}_{A_j[H]}$ is the unique maximal ideal. □

Let G be a profinite abelian group. We start from a character $\overline{\rho} : G \to \mathbb{F}^\times$ and study the universal deformation ring of $\overline{\rho}$. We write $G_p = \varprojlim_j H_j$ for finite p-groups H_j. Let $R = \mathcal{O}[[G_p]] = \varprojlim_m (\mathcal{O}/\mathfrak{m}_{\mathcal{O}}^m)[H_m] \cong \varprojlim_m \mathcal{O}[H_m]$. We define $\varrho : G \to R^\times$ by $g \mapsto \tilde{\rho}(g) \varprojlim_m [g_p]_m$, where $g = (g^{(p)}, g_p)$ with $g_p \in G_p$ and $g^{(p)} \in G^{(p)}$, and $[g_p]_m$ is the image of g_p in H_m. By the lemma, R is an object in $CL_{\mathcal{O}}$. Let $\rho : G \to A^\times$ be a deformation for an object A in $CL_{\mathcal{O}}$. We write $A = \varprojlim_j A_j$ for artinian local rings A_j. Then we write $\rho_j : G \to A_j^\times$ for the character induced by ρ. Since A_j is a finite ring, $\rho_j \tilde{\rho}^{-1}$ factors through $H_{i(j)}$ for some index $i(j)$. We may assume that $i(j) > i(j')$ if $j > j'$. Then we have a unique $\mathcal{O}$-algebra homomorphism $\iota_j : \mathcal{O}[H_{i(j)}] \to A_j$ such that $\iota_j([g_p]_j) = \rho_j(g_p)$. Taking the limit with respect to j, we get $\iota_\rho = \varprojlim_j \iota_j : R \to A$ such that $\iota_\rho \circ \varrho = \rho$. The morphism ι is uniquely determined, because R is topologically generated by $\varrho(G)$ over $\mathcal{O}$. Thus we have proved:

THEOREM 2.21 *Let G be a profinite group and $\overline{\rho} : G \to \mathbb{F}^\times$ be a character. Let G_p^{ab} be the maximal p-profinite abelian quotient of G. Then the couple $(\mathcal{O}[[G_p^{ab}]], \varrho)$ gives the universal deformation ring of $\overline{\rho}$ in the category $CL_{\mathcal{O}}$, where $\varrho(g) = \tilde{\rho}(g)[g_p]$ for $g \in G$.*

Proof Any deformation $\rho : G \to A^\times$ actually factors through the maximal abelian quotient G^{ab}. Then we apply the above argument to G^{ab} and get the result. □

Let Γ be an infinite p-profinite cyclic group. We fix a generator $\gamma \in \Gamma$. Thus $\mathbb{Z}_p \cong \Gamma$ by $s \mapsto \gamma^s$, where for $s = \lim_{n\to\infty} s_n \in \mathbb{Z}_p$ with $s_n \in \mathbb{Z}$, $\gamma^s = \lim_{n\to\infty} \gamma^{s_n}$ (Exercise 1 below). Then we consider the power series ring $\mathcal{O}[[T]]$. We look at the binomial polynomial $\binom{x}{n}$ given by $\frac{x(x-1)\cdots(x-n+1)}{n!}$ when $n > 0$, and we put $\binom{x}{0} = 1$. Then for positive integers m, $\binom{m}{n}$ is always an integer, giving a polynomial map $\binom{x}{n} : \mathbb{N} \to \mathbb{N}$. Since $\binom{x}{n}$ is a polynomial, it is continuous under the p-adic topology. Thus $\binom{x}{n}$ sends the closure $\overline{\mathbb{N}}$ of $\mathbb{N}$ in $\mathbb{Z}_p$ into $\overline{\mathbb{N}}$. Since $\mathbb{N}$ surjects onto $\mathbb{Z}/p^n\mathbb{Z}$ for all n, $\mathbb{N}$ is dense in $\mathbb{Z}_p$. Thus $s \mapsto \binom{s}{n}$ is a well-defined continuous map of $\mathbb{Z}_p$ into $\mathbb{Z}_p$. Then we consider $(1+T)^s = \sum_{n=0}^{\infty} \binom{s}{n} T^n$ and a character $\varrho : \Gamma \to \mathcal{O}[[T]]$ given by $\varrho(\gamma^s) = (1+T)^s$.

Proposition 2.22 *For the identity character $\overline{\rho} : \Gamma \to \mathbb{F}^{\times}$, the couple $(\mathcal{O}[[T]], \varrho)$ is the universal deformation ring of $\overline{\rho}$.*

Proof Let $\rho : \Gamma \to A^{\times}$ be a deformation of $\overline{\rho}$. First suppose that A is an artinian $\mathcal{O}$-algebra in $CL_{\mathcal{O}}$. Then define

$$\iota_\rho : \mathcal{O}[[T]] \to A \quad \text{by} \quad \iota(f(T)) = f(\rho(\gamma) - 1).$$

Since $\rho(\gamma) \mod \mathfrak{m}_A = 1$, $t = \rho(\gamma) - 1 \in \mathfrak{m}_A$. Since A is artinian, $t^N = 0$ for large N. Thus ι_ρ is well defined, and by definition, $\iota_\rho \circ \varrho = \rho$. For the general object $A = \varprojlim_j A_j$ of $CL_{\mathcal{O}}$ with artinian A_j, by the above argument, we find ι_j with $\iota_j \circ \varrho = \rho_j$ for the character $\rho_j : \Gamma \to A_j^{\times}$ induced by ρ. Then $\iota_\rho = \varprojlim_j \iota_j : \mathcal{O}[[T]] \to A$ satisfies the desired property. The uniqueness of ι_ρ is obvious. □

Corollary 2.23 *If $\Gamma \cong \mathbb{Z}_p$, then $\mathcal{O}[[\Gamma]] \cong \mathcal{O}[[T]]$ via $\gamma \mapsto 1 + T$. If $G \cong \mathbb{Z}_p^r$, then $\mathcal{O}[[G]] \cong \mathcal{O}[[T_1, \ldots, T_r]]$ via $\gamma_j \mapsto 1 + T_j$ for a base $\{\gamma_j\}$ of G over $\mathbb{Z}_p$.*

The algebra $\mathcal{O}[[\Gamma]]$ is called the Iwasawa algebra. If $G = \mathrm{Gal}(\mathbb{Q}^{(p,\infty)}/\mathbb{Q})$ for the maximal extension $\mathbb{Q}^{(p,\infty)}$ of $\mathbb{Q}$ unramified outside $\{p, \infty\}$, then $G_p \cong \Gamma \cong \mathbb{Z}_p$. Thus for a Galois character $\overline{\rho} : G \to \mathbb{F}^{\times}$, the universal deformation ring of $\overline{\rho}$ is isomorphic to the Iwasawa algebra $\mathcal{O}[[\Gamma]] \cong \mathcal{O}[[T]]$ if $p > 2$.

Corollary 2.24 *If a profinite group G has the maximal abelian p-profinite quotient G_p^{ab} with finitely many generators, the universal deformation ring for any character $\overline{\rho} : G \to \mathbb{F}^{\times}$ over $CL_{\mathcal{O}}$ is noetherian.*

Proof By Theorem 2.21, the universal deformation ring is isomorphic to $\mathcal{O}[[G_p^{ab}]]$. Since $G_p^{ab} \cong \Gamma^r \times \Delta$ for a finite p-group Δ, we see that $\mathcal{O}[(\Gamma^r/H) \times \Delta] = \mathcal{O}[(\Gamma^r/H))][\Delta]$ for any open subgroup $H \subset G_p^{ab}$ inside Γ^r.

Thus $\mathcal{O}[[G_p^{ab}]] = \varprojlim_H \mathcal{O}[(\Gamma^r/H))][\Delta] \cong \mathcal{O}[[\Gamma^r]][\Delta] \cong \mathcal{O}[[T_1,\dots,T_r]][\Delta]$, which is a $\mathcal{O}[[T_1,\dots,T_r]]$-free module of finite rank. Since $\mathcal{O}[[T_1,\dots,T_r]]$ is noetherian, $\mathcal{O}[[G_p^{ab}]]$ is noetherian ([CRT] Section 3). □

Let F be a number field and $F^{(p,\infty)}$ be the maximal extension unramified outside $\{p,\infty\}$. Then for $G = \mathrm{Gal}(F^{(p,\infty)}/F)$, the maximal (continuous) abelian quotient $G^{ab} = G/\overline{(G,G)}$ (for the commutator subgroup (G,G) of G) is the Galois group of the maximal abelian extension unramified outside $\{p,\infty\}$. By class field theory, we have an exact sequence:

$$U_p \longrightarrow G^{ab} \longrightarrow \mathrm{Gal}(H/F) \longrightarrow 1, \tag{2.12}$$

where $U_p = O_p^\times$ $(O_p = O \otimes_{\mathbb{Z}} \mathbb{Z}_p)$ for the integer ring O of F, and H is the Hilbert class field, that is, the maximal abelian extension unramified outside ∞. Again by class field theory, $\mathrm{Gal}(H/F)$ is isomorphic to the strict class group of F, which is a finite abelian group (see [LFE] I.1.2 for the finiteness), and U_p is isomorphic to a product of copies of $\Gamma \cong \mathbb{Z}_p$ and a finite group, because of the existence of the p-adic logarithm map $\log : U_p \to O_p$ (cf. [LFE] I.1.3 for p-adic logarithm). Thus in this case, G_p^{ab} is of finite type as a p-profinite group (Exercise 2 below).

Corollary 2.25 *For a given Galois character* $\overline{\rho} : \mathrm{Gal}(\overline{F}/F) \to \mathbb{F}^\times$ *unramified outside* $\{p,\infty\}$, *there exist a noetherian ring* $R_{\overline{\rho}} \in CL_{\mathcal{O}}$ *and a character* $\varrho : \mathrm{Gal}(\overline{F}/F) \to R_{\overline{\rho}}^\times$ *such that for any given deformation* $\rho : \mathrm{Gal}(\overline{F}/F) \to A^\times$ $(A \in CL_{\mathcal{O}})$ *unramified outside* $\{p,\infty\}$, *there exists a unique* $\mathcal{O}$*-algebra homomorphism* $\iota_\rho : R_{\overline{\rho}} \to A$ *such that* $\rho = \iota_\rho \circ \varrho$.

This ring $R_{\overline{\rho}}$ is called the universal deformation ring (unramified outside $\{p,\infty\}$), and ϱ is called the universal character unramified outside $\{p,\infty\}$.

When $F = \mathbb{Q}$, then $\chi : G^{ab} \cong \mathbb{Z}_p^\times$ by the p-adic cyclotomic character χ, and $\mathbb{Z}_p^\times \cong \Gamma \times \Delta$ for a finite group

$$\Delta = \begin{cases} \{\pm 1\} & \text{if } p = 2 \\ \mu_{p-1}(\mathbb{Z}_p) \cong \mathbb{Z}/(p-1)\mathbb{Z} & \text{if } p > 2 \end{cases}.$$

Thus $R_{\overline{\rho}}$ is isomorphic to the Iwasawa algebra if p is odd.

Exercises

(1) Show that $\Gamma \cong \mathbb{Z}_p$ if Γ is an infinite p-profinite cyclic group.
(2) Give a detailed proof of the finite generation of G_p^{ab} for the Galois group G of $F^{(p,\infty)}/F$.

2.3.2 Non-abelian deformation

We fix an absolutely irreducible representation $\overline{\rho} : G \to GL_n(\mathbb{F})$. First we consider a universal pseudo-representation. Let $\overline{\tau}$ be the pseudo-representation associated $\overline{\rho}$. It can be either that of Wiles ($n = 2$) or Taylor. A couple consisting of an object $R_{\overline{\tau}} \in CL_{\mathcal{O}}$ and a pseudo-representation $T : G \to R_{\overline{\tau}}$ is called a universal couple if the following universality condition is satisfied:

(univ) *For each pseudo-representation* $\tau : G \to A$ $(A \in CL_{\mathcal{O}})$ *with* $\tau \cong \overline{\tau} \mod \mathfrak{m}_A$, *there exists a unique* $\mathcal{O}$*-algebra homomorphism* $\iota_\tau : R_{\overline{\tau}} \to A$ *such that*

$$\tau = \iota_\tau \circ T.$$

We now show the existence of $(R_{\overline{\tau}}, T)$ for a profinite group G. Since the argument is the same for Wiles' or Taylor's pseudo-representation, we only explain the construction for Taylor's pseudo-representation. First suppose that G is a finite group. Let $\omega : \mathcal{O}^\times \to \mu_{q-1}(\mathcal{O})$ be the Teichmüller character, that is,

$$\omega(x) = \lim_{n\to\infty} x^{q^n}.$$

We also consider the following isomorphism: $\mu_{q-1}(\mathcal{O}) \ni \zeta \mapsto \zeta \mod \mathfrak{m}_{\mathcal{O}} \in \mathbb{F}^\times$. We write $\varphi : \mathbb{F}^\times \to \mu_{q-1}(\mathcal{O}) \subset \mathcal{O}^\times$ for the inverse of the above map. We look at the power series ring: $\mathcal{O}[[X_g; g \in G]]$. We put $T_g = X_g + \varphi(\overline{\tau}(g))$. We construct the ideal I so that $g \mapsto T_g \mod I$ becomes the universal pseudo-representation. Thus we consider the ideal I of $\mathcal{O}[[X_g; g \in G]]$ generated by the elements of the following type:

(1) $T_1 - n = X_1 + \varphi(\overline{\tau}(1)) - n$;
(2) $T_{gh} - T_{hg} = X_{gh} - X_{hg}$;
(3) $\sum_{\sigma\in\mathfrak{S}_{n+1}} sgn(\sigma) T_\sigma(g_0, \ldots, g_n)$ for every $(n+1)$-tuple $(g_0, \ldots, g_n)$ in G,

where for the cycle decomposition as in Lemma 2.17

$$\sigma = (i_{1,1}, i_{1,2}, \ldots, i_{1,k_1}) \cdots (i_{m,1}, i_{m,2}, \ldots, i_{m,k_m}),$$

we have put

$$T_\sigma(g_0, \ldots, g_n) = T_{g_{i_{1,1}} \cdots g_{i_{1,k_1}}} \cdots T_{g_{i_{m,1}} \cdots g_{i_{m,k_m}}}.$$

Then we put $R_{\overline{\tau}} = \mathcal{O}[[X_g; g \in G]]/I$ and define $T(g) = T_g \mod I$. By the above definition, $T(g)$ is a pseudo-representation of Taylor with

$T \mod \mathfrak{m}_{R_{\overline{\tau}}} = \overline{\tau}$. For a pseudo-representation $\tau : G \to A$ with $\tau \equiv \overline{\tau} \mod \mathfrak{m}_A$, we define

$$\iota_\tau : \mathcal{O}[[X_g; g \in G]] \to A \text{ by } f(X_g) \mapsto f(\tau(g) - \varphi(\overline{\tau}(g))).$$

Since f is a power series of X_g and $\tau(g) - \varphi(\overline{\tau}(g)) \in \mathfrak{m}_A$, the value $f(\tau(g) - \varphi(\overline{\tau}(g)))$ is well defined. Let us demonstrate this. If A is artinian, a sufficiently high power $\mathfrak{m}_A^N$ vanishes. Thus if the monomial of the X_g's is of degree higher than N, it is sent to 0 via ι_τ, and $f(\tau(g) - \varphi(\overline{\tau}(g)))$ is a finite sum of terms of degree $\leq N$. If A is pro-artinian, the morphism ι_τ is just the projective limit of the corresponding ones well defined for artinian quotients. By the axioms of pseudo-representation (T1–3) in Subsection 2.3.2.2, $\iota_\tau(I) = 0$, and hence ι_τ factors through $R_{\overline{\tau}}$. The uniqueness of ι_τ follows from the fact that $\{T_g | g \in G\}$ topologically generates $R_{\overline{\tau}}$.

Now assume that $G = \varprojlim_N G/N$ for open normal subgroups N. Since $\mathrm{Ker}(\overline{\rho})$ is an open subgroup of G, we may assume that N runs over subgroups of $\mathrm{Ker}(\overline{\rho})$. Since $\overline{\rho}$ factors through $G/\mathrm{Ker}(\overline{\rho})$, $\overline{\tau} = \mathrm{Tr}(\overline{\rho})$ factors through G/N. Therefore we can think of the universal couple $(R_{\overline{\tau}}^N, T_N)$ for $(G/N, \overline{\tau})$. If $N \subset N'$, the algebra homomorphism $\mathcal{O}[[X_{gN} | gN \in G/N]] \to \mathcal{O}[[X_{gN'} | gN' \in G/N']]$ taking X_{gN} to $X_{gN'}$ induces a surjective $\mathcal{O}$-algebra homomorphism $\pi_{N,N'} : R_{\overline{\tau}}^N \to R_{\overline{\tau}}^{N'}$ with $\pi_{N,N'} \circ T_N = T_{N'}$. We then define $T = \varprojlim_N T_N$ and $R_{\overline{\tau}} = \varprojlim_N R_{\overline{\tau}}^N$. If $\tau : G \to A$ is a pseudo-representation, by Nyssen (Theorem 2.18), we have the associated representation $\rho : G \to GL_n(A)$ such that $\tau = \mathrm{Tr}(\rho)$. If A is artinian, then $GL_n(A)$ is a finite group, and hence ρ and $\tau = \mathrm{Tr}(\rho)$ factors through G/N for a sufficiently small open normal subgroup N. Thus we have $\iota_\tau : R_{\overline{\tau}} \xrightarrow{\pi_N} R_{\overline{\tau}}^N \xrightarrow{\iota_\tau^N} A$ such that $\iota_\tau \circ T = \tau$. Since $T(g)$ generates topologically $R_{\overline{\tau}}$, ι_τ is uniquely determined.

We claim the natural map:

$$\begin{aligned} \{\rho : G \to GL_n(A) | \rho \equiv \overline{\rho} \mod \mathfrak{m}_A\} / \approx \\ \to \{\rho : G \to GL_n(A) | \rho \mod \mathfrak{m} \sim \overline{\rho}\} / \sim \end{aligned} \tag{2.13}$$

is a bijection if $\overline{\rho}$ is absolutely irreducible. The map takes the strict equivalence class of ρ to the equivalence class of ρ. If $\rho \mod \mathfrak{m} \sim \overline{\rho}$, we find $x \in GL_n(A)$ such that $x\rho x^{-1} \mod \mathfrak{m} = \overline{\rho}$, because the reduction map $x \mapsto x \mod \mathfrak{m}_A$ is a surjection of $GL_n(A)$ onto $GL_n(\mathbb{F})$. Thus (2.13) is surjective. If $\rho \equiv \rho' \mod \mathfrak{m}_A$ and $x\rho x^{-1} = \rho'$ for $x \in GL_n(A)$, then writing $\overline{x}$ for $x \mod \mathfrak{m}_A \in GL_n(\mathbb{F})$, we have $\overline{x\rho}(g) = \overline{\rho}(g)\overline{x}$. Then by Schur's lemma (Lemma 2.12), $\overline{x}$ is a scalar matrix. We pick a scalar

$y \in A$ with $y \mod \mathfrak{m}_A = \overline{x}$. Since $y \not\equiv 0 \mod \mathfrak{m}_A$, $y \in A^\times$. Then $z = y^{-1}x \in 1 + M_n(\mathfrak{m}_A)$ and $z\rho z^{-1} = \rho'$. This shows that $\rho \approx \rho'$ and hence (2.13) is injective.

The following theorem was first proved by Mazur in a paper in [GAL] in 1989.

THEOREM 2.26 (MAZUR) *Suppose that $\overline{\rho} : G \to GL_n(\mathbb{F})$ is absolutely irreducible. Then there exists the universal deformation ring $R_{\overline{\rho}}$ in $CL_{\mathcal{O}}$ and a universal deformation $\varrho : G \to GL_n(R_{\overline{\rho}})$. If we write $\overline{\tau} = \mathrm{Tr}(\overline{\rho})$, then for the universal pseudo-representation $T : G \to R_{\overline{\tau}}$ deforming $\overline{\tau}$, we have a canonical isomorphism of $\mathcal{O}$-algebras $\iota : R_{\overline{\rho}} \cong R_{\overline{\tau}}$ such that $\iota \circ \mathrm{Tr}(\varrho) = T$.*

Proof We give here a proof due to L. Nyssen. Pick a strict equivalence class of a deformation $\rho : G \to GL_n(A)$ of $\overline{\rho}$. Out of T, we can construct a representation $\varrho : G \to GL_n(R_{\overline{\tau}})$ with $\mathrm{Tr}(\varrho) = T$ (see Theorem 2.18) and $\varrho \mod \mathfrak{m}_{R_{\overline{\tau}}} = \overline{\rho}$. Then there is a unique $\mathcal{O}$-algebra homomorphism $\iota_{\mathrm{Tr}(\rho)} : R_{\overline{\tau}} \to A$ such that $\iota_{\mathrm{Tr}(\rho)} \circ T = \mathrm{Tr}(\rho)$. We see that $\mathrm{Tr}(\iota_{\mathrm{Tr}(\rho)} \circ \varrho) = \iota_{\mathrm{Tr}(\rho)} \circ T = \mathrm{Tr}(\rho)$. Thus by Carayol (Proposition 2.13), $\iota_{\mathrm{Tr}(\rho)} \circ \varrho \sim \rho$. By the bijectivity of (2.13), $\iota_{\mathrm{Tr}(\rho)} \circ \varrho \approx \rho$. This shows that $R_{\overline{\rho}} = R_{\overline{\tau}}$. □

Let $(R_{\overline{\rho}}, \varrho)$ be the universal couple for an absolutely irreducible representation $\overline{\rho} : G \to GL_n(\mathbb{F})$. We can also think of $(R_{\det(\overline{\rho})}, \nu)$, which is the universal couple for the character $\det(\overline{\rho}) : G \to GL_1(\mathbb{F}) = \mathbb{F}^\times$. As we have studied in Subsection 2.3.1, $R_{\det(\overline{\rho})} \cong \mathcal{O}[[G_p^{ab}]]$ for the maximal abelian p-profinite quotient G_p^{ab}. Note that $\det(\varrho) : G \to GL_1(R_{\overline{\rho}})$ satisfies $\det(\varrho) \mod \mathfrak{m}_{R_{\overline{\rho}}} = \det(\overline{\rho})$. Thus $\det(\varrho)$ is a deformation of $\det(\overline{\rho})$, and hence by the universality of $(\mathcal{O}[[G_p^{ab}]] \cong R_{\det(\overline{\rho})}, \nu)$, there is a unique $\mathcal{O}$-algebra homomorphism $\iota : R_{\det(\overline{\rho})} \to R_{\overline{\rho}}$ such that $\iota \circ \nu = \det(\varrho)$. In this way, $R_{\overline{\rho}}$ becomes naturally an $\mathcal{O}[[G_p^{ab}]]$-algebra via ι.

COROLLARY 2.27 *Let the notation and the assumption be as in the above theorem. Then the universal ring $R_{\overline{\rho}}$ is canonically an algebra over the Iwasawa algebra $\mathcal{O}[[G_p^{ab}]]$.*

Exercise

(1) Prove Mazur's theorem 2.26 using Wiles' pseudo-representations instead of Taylor's, when applicable.

2.3.3 Tangent spaces of local rings

We now study the case when $R_{\bar{\rho}}$ is noetherian. Here is a useful lemma:

LEMMA 2.28 *If* $t^*_{A/\mathcal{O}} = \mathfrak{m}_A/(\mathfrak{m}_A^2 + \mathfrak{m}_{\mathcal{O}})$ *is a finite dimensional vector space over* $\mathbb{F}$*, then* $A \in CL_{\mathcal{O}}$ *is noetherian. The space* $t^*_{A/\mathcal{O}}$ *is called the co-tangent space of* A *at* $\mathfrak{m}_A \in Spec(A)$ *over* $Spec(\mathcal{O})$.

Proof Define t^*_A by $\mathfrak{m}_A/\mathfrak{m}_A^2$, which is called the (absolute) tangent space of A at $\mathfrak{m}_A$. Since we have an exact sequence:

$$\mathbb{F} \cong \mathfrak{m}_{\mathcal{O}}/\mathfrak{m}_{\mathcal{O}}^2 \longrightarrow t^*_A \longrightarrow t^*_{A/\mathcal{O}} \longrightarrow 0,$$

we conclude that t^*_A is of finite dimension over $\mathbb{F}$. First suppose that $pA = 0$ and $\mathfrak{m}_A^N = 0$ for sufficiently large N. Let $\overline{x}_1, \ldots, \overline{x}_m$ be an $\mathbb{F}$-base of t^*_A. We choose $x_j \in A$ so that $x_j \mod \mathfrak{m}_A^2 = \overline{x}_j$. Then we consider the ideal $\mathfrak{a}$ generated by x_j. We have the inclusion map: $\mathfrak{a} = \sum_j Ax_j \hookrightarrow \mathfrak{m}_A$. After tensoring $A/\mathfrak{m}_A$, we have the surjectivity of the induced linear map: $\mathfrak{a}/\mathfrak{m}_A\mathfrak{a} \cong \mathfrak{a} \otimes_A A/\mathfrak{m}_A \to \mathfrak{m} \otimes_A A/\mathfrak{m}_A \cong \mathfrak{m}/\mathfrak{m}_A^2$ because $\{\overline{x}_1, \ldots, \overline{x}_m\}$ is an $\mathbb{F}$-base of t^*_A. This shows that $\mathfrak{m}_A = \mathfrak{a} = \sum_j Ax_j$. Therefore $\mathfrak{m}_A^k/\mathfrak{m}_A^{k+1}$ is generated by the monomials in x_j of degree k as an $\mathbb{F}$-vector space. In particular, $\mathfrak{m}_A^{N-1}$ is generated by the monomials in x_j of degree $N-1$. Then we define $\pi : B = \mathbb{F}[[X_1, \ldots, X_m]] \to A$ by $\pi(f(X_1, \ldots, X_m)) = f(x_1, \ldots, x_m)$. Since any monomial of degree $> N$ vanishes after applying π, π is a well-defined $\mathcal{O}$-algebra homomorphism. Let $\mathfrak{m} = \mathfrak{m}_B = (X_1, \cdots, X_m)$ be the maximal ideal of B. By the above argument, $\pi(\mathfrak{m}^{N-1}) = \mathfrak{m}_A^{N-1}$. Suppose now that $\pi(\mathfrak{m}^{N-j}) = \mathfrak{m}_A^{N-j}$, and try to prove the surjectivity of $\pi(\mathfrak{m}^{N-j-1}) = \mathfrak{m}_A^{N-j-1}$. Since $\mathfrak{m}_A^{N-j-1}/\mathfrak{m}_A^{N-j}$ is generated by monomials of degree $N-j-1$ in x_j, for each $x \in \mathfrak{m}_A^{N-j-1}$, we find $P \in \mathfrak{m}^{N-j-1}$ such that $x - \pi(P) \in \mathfrak{m}_A^{N-j} = \pi(\mathfrak{m}^{N-j})$. This proves the assertion: $\pi(\mathfrak{m}^{N-j-1}) = \mathfrak{m}_A^{N-j-1}$. Thus by induction on j, we get the surjectivity of π.

Now suppose only that $\mathfrak{m}_A^N = 0$. Then in particular, $p^N A = 0$. Thus A is an $\mathcal{O}/p^N\mathcal{O}$-module. We can still define $\pi : B = \mathcal{O}/p^N\mathcal{O}[[X_1, \ldots, X_m]] \to A$ by sending X_j to x_j. Then by the previous argument applied to B/pB and A/pA, we find that $\pi \mod p : B \otimes_{\mathcal{O}} \mathcal{O}/p\mathcal{O} \cong B/pB \to A/pA \cong A \otimes_{\mathcal{O}} \mathcal{O}/p\mathcal{O}$ is surjective. In particular, for the maximal ideal $\mathfrak{m}'$ of $\mathcal{O}/p^N\mathcal{O}$, $\pi \mod \mathfrak{m}' : B \otimes_{\mathcal{O}} \mathbb{F} \cong B/\mathfrak{m}'B \to A/\mathfrak{m}'A \cong A \otimes_{\mathcal{O}} \mathbb{F}$ is surjective. Then by Nakayama's lemma applied to the nilpotent ideal $\mathfrak{m}'$ (Corollary 2.4), π is surjective.

In general, write $A = \varprojlim_i A_i$ for artinian rings A_i. Then the projection maps induce surjections $t^*_{A_{i+1}} \to t^*_{A_i}$. Since t^*_A is of finite dimension, for sufficiently large i, $t^*_{A_{i+1}} \cong t^*_{A_i}$. Thus choosing x_j as above in A, we have its image $x_j^{(i)}$ in A_i. Use $x_j^{(i)}$ to construct $\pi_i : \mathcal{O}[[X_1,\ldots,X_m]] \to A_i$ in place of x_j. Then π_i is surjective as already shown, and $\pi = \varprojlim_i \pi_i : \mathcal{O}[[X_1,\ldots,X_m]] \to A$ remains surjective, because projective limit of surjections, if all sets involved are finite sets, remain surjective (Exercise 1). Since $\mathcal{O}[[X_1,\ldots,X_m]]$ is noetherian ([CRT] Theorem 3.3), its surjective image A is noetherian. □

Exercise

(1) Let $(G_j, \pi_{i,j})_{i\in I}$ and $(H_j, \rho_{i,j})_{i\in I}$ be two projective systems of profinite groups. Let $f_i : G_i \twoheadrightarrow H_i$ for each i be a surjective homomorphism of profinite groups satisfying $f_i \circ \pi_{j,i} = \rho_{j,i} \circ f_j$ for all $j > i$. Then show that the limit map $f = \varprojlim_i f_i : \varprojlim_i G_i \to \varprojlim_i H_i$ is surjective. Hint: First prove that f is continuous and has dense image in $\varprojlim_i H_i$. Then use the compactness of $\varprojlim_i G_i$ to show the surjectivity.

2.3.4 Cohomological interpretation of tangent spaces

Let $R = R_{\overline{\rho}}$. We let G acts on $M_n(\mathbb{F})$ by $gv = \overline{\rho}(g)v\overline{\rho}(g)^{-1}$. This G-module will be written as $ad(\overline{\rho})$.

LEMMA 2.29 *Let $R = R_{\overline{\rho}}$ for an absolutely irreducible representation $\overline{\rho} : G \to GL_n(\mathbb{F})$. Then*

$$t_{R/\mathcal{O}} = \mathrm{Hom}_{\mathbb{F}}(t^*_{R/\mathcal{O}}, \mathbb{F}) \cong H^1_{ct}(G, ad(\overline{\rho})),$$

where $H^1_{ct}(G, ad(\overline{\rho}))$ is the continuous first cohomology group of G with coefficients in the discrete G-module $V(ad(\overline{\rho}))$ (see Subsection 4.3.3 in Chapter 4 for the definition of such cohomology groups). The space $t_{R/\mathcal{O}}$ is called the tangent space of $Spec(R)_{/\mathcal{O}}$ at $\mathfrak{m}$.

Proof Let $A = \mathbb{F}[X]/(X^2)$. We write ε for the class of X in A. Then $\varepsilon^2 = 0$. We consider $\phi \in \mathrm{Hom}_{\mathcal{O}\text{-}alg}(R, A)$. Write $\phi(r) = \phi_0(a) + \phi_\varepsilon(r)\varepsilon$. Then we have from $\phi(ab) = \phi(a)\phi(b)$ that $\phi_0(ab) = \phi_0(a)\phi_0(b)$ and

$$\phi_\varepsilon(ab) = \phi_0(a)\phi_\varepsilon(b) + \phi_0(b)\phi_\varepsilon(a).$$

Thus $\mathrm{Ker}(\phi_0) = \mathfrak{m}_A = \mathbb{F}\varepsilon$ because A is local. Since ϕ is $\mathcal{O}$-linear, $\phi_0(a) = \overline{a} = a \mod \mathfrak{m}_A$, and thus ϕ kills $\mathfrak{m}_R^2$ and takes $\mathfrak{m}_R$ $\mathcal{O}$-linearly into $\mathfrak{m}_A = \mathbb{F}\varepsilon$. Moreover for $r \in \mathcal{O}$, $\overline{r} = r\phi(1) = \phi(r) = \overline{r} + \phi_\varepsilon(r)\varepsilon$, and hence ϕ_ε kills $\mathcal{O}$. Since R shares its residue field $\mathbb{F}$ with $\mathcal{O}$, any element $a \in R$ can be written as $a = r + x$ with $r \in \mathcal{O}$ and $x \in \mathfrak{m}_R$. Thus ϕ is completely determined by the restriction of ϕ_ε to $\mathfrak{m}_R$, which factors through $t^*_{R/\mathcal{O}}$. We write ℓ_ϕ for ϕ_ε regarded as an $\mathbb{F}$-linear map from $t^*_{R/\mathcal{O}}$ into $\mathbb{F}$. Then we can write $\phi(r + x) = \overline{r} + \ell_\phi(x)\varepsilon$. Thus $\phi \mapsto \ell_\phi$ induces a linear map $\ell : \mathrm{Hom}_{\mathcal{O}-alg}(R, A) \to \mathrm{Hom}_{\mathbb{F}}(t^*_{R/\mathcal{O}}, \mathbb{F})$. Note that $R/(\mathfrak{m}_R^2 + \mathfrak{m}_\mathcal{O}) = \mathbb{F} \oplus t^*_{R/\mathcal{O}}$. For any $\ell \in \mathrm{Hom}_{\mathbb{F}}(t^*_{R/\mathcal{O}}, \mathbb{F})$, we extend ℓ to $R/\mathfrak{m}_R^2$ declaring its value on $\mathbb{F}$ to be zero. Then define $\phi : R \to A$ by $\phi(r) = \overline{r} + \ell(r)\varepsilon$. Since $\varepsilon^2 = 0$, ϕ is an $\mathcal{O}$-algebra homomorphism. In particular, $\ell(\phi) = \ell$, and hence ℓ is surjective. Since algebra homomorphisms killing $\mathfrak{m}_R^2 + \mathfrak{m}_\mathcal{O}$ are determined by its values on $t^*_{R/\mathcal{O}}$, ℓ is injective.

By the universality, we have

$$\mathrm{Hom}_{\mathcal{O}-alg}(R, A) \cong \{\rho : G \to GL_n(A) | \rho \mod \mathfrak{m}_A = \overline{\rho}\}/\approx .$$

Then we can write $\rho(g) = \overline{\rho}(g) + u'_\rho(g)\varepsilon$. From the mutiplicativity, we have

$$\overline{\rho}(gh) + u'_\rho(gh)\varepsilon = \rho(gh) = \rho(g)\rho(h) = \overline{\rho}(g)\overline{\rho}(h) + (\overline{\rho}(g)u'_\rho(h) + u'_\rho(g)\overline{\rho}(h))\varepsilon.$$

Thus as a function $u' : G \to M_n(\mathbb{F})$, we have

$$u'_\rho(gh) = \overline{\rho}(g)u'_\rho(h) + u'_\rho(g)\overline{\rho}(h). \tag{2.14}$$

Define a map $u_\rho : G \to ad(\overline{\rho})$ by $u_\rho(g) = u'_\rho(g)\overline{\rho}(g)^{-1}$. Then by a simple computation, we have $gu_\rho(h) = \overline{\rho}(g)u_\rho(h)\overline{\rho}(g)^{-1}$ from the definition of $ad(\overline{\rho})$. Then from the above formula (2.14), we conclude that $u_\rho(gh) = gu_\rho(h) + u_\rho(g)$. Thus $u_\rho : G \to ad(\overline{\rho})$ is a 1-cocycle (see (4.25) in Chapter 4). Starting from a 1-cocycle u, we can reconstruct the representation reversing the above process. Then again by computation,

$$\begin{aligned} \rho \approx \rho' &\iff \overline{\rho}(g) + u'_\rho(g) = (1 + x\varepsilon)(\overline{\rho}(g) + u'_{\rho'}(g))(1 - x\varepsilon) \quad (x \in ad(\overline{\rho})) \\ &\iff u'_\rho(g) = x\overline{\rho}(g) - \overline{\rho}(g)x + u'_{\rho'}(g) \iff u_\rho(g) = (1 - g)x + u_{\rho'}(g). \end{aligned}$$

Thus the cohomology classes of u_ρ and $u_{\rho'}$ are equal if and only if $\rho \approx \rho'$. This shows:

$$\begin{aligned} &\mathrm{Hom}_{\mathbb{F}}(t^*_{R/\mathcal{O}}, \mathbb{F}) \cong \mathrm{Hom}_{\mathcal{O}-alg}(R, A) \cong \\ &\qquad \{\rho : G \to GL_n(A) | \rho \mod \mathfrak{m}_A = \overline{\rho}\}/\approx \cong H^1(G, ad(\overline{\rho})). \end{aligned}$$

In this way, we get a bijection between $\mathrm{Hom}_{\mathbb{F}}(t^*_{R/\mathcal{O}}, \mathbb{F})$ and $H^1(G, ad(\overline{\rho}))$. By tracking down (in the reverse way) our construction, one can check that the map is an $\mathbb{F}$-linear isomorphism. □

For each open subgroup H of G, we write H_p for the maximal p-profinite quotient. We consider the following condition:

(Φ) *For any open subgroup H of G, the p-Frattini quotient $\Phi(H_p)$ is a finite group,*

where $\Phi(H_p) = H_p/\overline{(H_p)^p(H_p, H_p)}$ for the commutator subgroup (H_p, H_p) of H_p.

Proposition 2.30 (Mazur) *If G satisfies (Φ), then $R_{\overline{\rho}}$ is a noetherian ring. In particular, $G = \mathrm{Gal}(F^{(p,\infty)}/F)$ satisfies (Φ), where F is a number field, and $F^{(p,\infty)}$ is the maximal extension of F unramified outside $\{p, \infty\}$.*

Proof Let $H = \mathrm{Ker}(\overline{\rho})$. Then the action of H on $ad(\overline{\rho})$ is trivial. By the inflation-restriction sequence (see Theorem 4.33 in Chapter 4), we have the following exact sequence:

$$0 \to H^1(G/H, H^0(H, ad(\overline{\rho}))) \longrightarrow H^1_{ct}(G, ad(\overline{\rho})) \longrightarrow \mathrm{Hom}_{ct}(\Phi(H_p), M_n(\mathbb{F})),$$

where the subscript '*ct*' indicates the continuous cohomology (see Subsection 4.3.3 in Chapter 4). From this, it is clear that $\dim_{\mathbb{F}} H^1(G, ad(\overline{\rho})) < \infty$. The fact that $\mathrm{Gal}(F^{(p,\infty)}/F)$ satisfies (Φ) follows from class field theory (see (2.12)). □

3

Representations of Galois Groups and Modular Forms

The purpose of this chapter is to identify the $GL(2)$-Hecke algebras with universal deformation rings with certain additional structures. This fact was first conjectured by B. Mazur and now is a theorem of Wiles in many cases (see Subsection 3.2.7 for a description of the present knowledge to date: October 1999), which is one of the key points of his proof of Fermat's last theorem. In this chapter, we will prove the theorem in a typical case (which covers the case when the weight is bigger than or equal to 2), assuming the knowledge of the modular two-dimensional Galois representations, control theorems of Hecke algebras and the Poitou–Tate duality theorem on Galois cohomology. We will come back later to the duality theorems used here and give a full exposition of them in Chapter 4. As for modular Galois representations and control theorems, we content ourselves only by describing the precise result necessary for the proof and giving some indication of further reading (see Theorem 3.15, Corollary 3.19 and Theorem 3.26). These two results left untouched here will be covered in my forthcoming book [GMF].

3.1 Modular Forms on Adele Groups of $GL(2)$

We first recall a general theory of elliptic modular forms in the language of adeles.

3.1.1 Elliptic modular forms

Let $\mathfrak{H} = \{z \in \mathbb{C} \mid \mathrm{Im}(z) > 0\}$ be the upper half complex plane. If $\mathrm{Im}(z) \neq 0$, $(1, z)$ is a base of $\mathbb{C}$ over $\mathbb{R}$; so, $j(\gamma, z) = cz+d \neq 0$ for $\gamma = \left(\begin{smallmatrix} a & b \\ c & d \end{smallmatrix}\right) \in GL_2(\mathbb{R})$. We can therefore consider $\gamma(z) = \frac{az+b}{cz+d} \in \mathbb{C}$ for $z \in \mathbb{C} - \mathbb{R}$. We have

$$\gamma\begin{pmatrix} z & w \\ 1 & 1 \end{pmatrix} = \begin{pmatrix} az+b & aw+b \\ cz+d & cw+d \end{pmatrix} = \begin{pmatrix} \gamma(z) & \gamma(w) \\ 1 & 1 \end{pmatrix}\begin{pmatrix} cz+d & 0 \\ 0 & cw+d \end{pmatrix}.$$

Taking the determinant of the above identity, we get

$$(\det\gamma)(z-w) = (\gamma(z)-\gamma(w))(cz+d)(cw+d).$$

Replacing w by the complex conjugate $\overline{z}$ of z, we have

$$\det\gamma\,\mathrm{Im}(z) = \mathrm{Im}(\gamma(z))|cz+d|^2.$$

This, in particular, implies that $\gamma(\mathfrak{H}) = \mathfrak{H}$ if $\det\gamma > 0$. Again by the above formulas, we get $\gamma(\delta(z)) = (\gamma\delta)(z)$ and the 1-cocycle relation $j(\gamma\delta,z) = j(\gamma,\delta(z))j(\delta,z)$. In particular, $G(\mathbb{R})_+ = \{\gamma \in GL_2(\mathbb{R}) | \det\gamma > 0\}$ acts on $\mathfrak{H}$. Now define, for a function $f : \mathfrak{H} \to \mathbb{C}$ and an integer k, another function $f|_k\gamma : \mathfrak{H} \to \mathbb{C}$ by

$$f|_k\gamma(z) = \det\gamma^{k/2} f(\gamma(z))j(\gamma,z)^{-k}.$$

By the cocycle relation, $f|_k\gamma\delta = (f|_k\gamma)|_k\delta$. From this, we see that

$$f|_k \left(\begin{smallmatrix} a & 0 \\ 0 & a \end{smallmatrix}\right) = \left(\frac{|a|^k}{a^k}\right) f(z). \tag{3.1}$$

We call a function $f : \mathfrak{H} \to \mathbb{C}$ *fast decreasing* as $\mathrm{Im}(z) \to \infty$ if

$$|f(z)| \le C\,\mathrm{Im}(z)^M \exp(-M'\,\mathrm{Im}(z)) \quad \text{as} \quad \mathrm{Im}(z) \to \infty$$

for three positive constants C, M, M'.

We note that $SL_2(\mathbb{Z}) = \{\gamma \in M_2(\mathbb{Z}) | \det\gamma = 1\}$ is given by the intersection $G(\mathbb{R})_+ \cap GL_2(\mathbb{Z})$. We consider the following (standard) congruence subgroups:

$$\Gamma_0(N) = \left\{\left(\begin{smallmatrix} a & b \\ c & d \end{smallmatrix}\right) \in SL_2(\mathbb{Z}) | c \equiv 0 \mod N\right\}, \tag{3.2}$$
$$\Gamma_1(N) = \left\{\left(\begin{smallmatrix} a & b \\ c & d \end{smallmatrix}\right) \in \Gamma_0(N) | a \equiv d \equiv 1 \mod N\right\}, \tag{3.3}$$
$$\Gamma(N) = \left\{\left(\begin{smallmatrix} a & b \\ c & d \end{smallmatrix}\right) \in \Gamma_1(N) | b \equiv 0 \mod N\right\}, \tag{3.4}$$

of level N for a positive integer N. We consider the space of holomorphic cusp forms $S_k(\Gamma_0(N),\chi)$ for a character $\chi : (\mathbb{Z}/N\mathbb{Z})^\times \to \mathbb{C}^\times$. As we know from [IAT] Chapter 2 (or [LFE] Sections 5.2–3), this space of cusp forms is finite dimensional. Thus $f \in S_k(\Gamma_0(N),\chi)$ is a function $f : \mathfrak{H} \to \mathbb{C}$ satisfying the following conditions:

$$f|_k\gamma = \chi(d)f(z) \quad \text{for all } \gamma = \left(\begin{smallmatrix} a & b \\ c & d \end{smallmatrix}\right) \in \Gamma_0(N); \tag{E1}$$

f is a holomorphic function defined on $\mathfrak{H}$; (E2)

$f|_k\gamma$ is fast decreasing as $\mathrm{Im}(z) \to \infty$ for all $\gamma = \left(\begin{smallmatrix} a & b \\ c & d \end{smallmatrix}\right) \in SL_2(\mathbb{Q})$. (E3)

More generally, for any subgroup Γ of finite index in $\Gamma_1(N)$, we write $S_k(\Gamma)$ for the space of functions $f : \mathfrak{H} \to \mathbb{C}$ satisfying (E1,2,3) for all elements in Γ. Note here that $\chi(a)$ in (E1) is trivial because $\Gamma \subset \Gamma_1(N)$. If $f \in S_k(\Gamma_0(N), \chi)$, we see that

$$f(z) = f(-1_2(z)) = \chi(-1)f(z)(-1)^k$$

because $-1_2 = -\left(\begin{smallmatrix}1 & 0\\ 0 & 1\end{smallmatrix}\right) \in \Gamma_0(N)$. Thus if $f \neq 0$, we have

$$\chi(-1) = (-1)^k. \tag{P}$$

As seen in [IAT] Chapter 2, $S_k(\Gamma)$ is of finite dimension over $\mathbb{C}$, $S_k(\Gamma) = 0$ if $k < 0$, and $S_k(\Gamma) = \mathbb{C}$ if $k = 0$.

Since $\frac{\partial^2}{\partial z \partial \bar{z}} = \frac{y^2}{4}\Delta$ for $y^2\Delta = \frac{\partial^2}{\partial x^2} + \frac{\partial^2}{\partial y^2}$ $(z = x + \sqrt{-1}y)$, the holomorphy of f implies that $\Delta f = 0$. Conversely, it is known that $\Delta\phi = 0$ implies that $\phi = f + f'$ for holomorphic f and anti-holomorphic f' under (E3). One could think of a real analytic cusp form f of weight $k = 0$ replacing the holomorphy condition (E2) by the following condition:

$$\Delta f = \lambda f \tag{E$'$2}$$

for a given real number λ. The modular forms with positive eigenvalues λ are equally important as holomorphic ones, although we only look into holomorphic cusp forms in this book.

We would like to relate cusp forms $f : \mathfrak{H} \to \mathbb{C}$ with certain functions on the adele group $G(\mathbb{A}) = GL_2(\mathbb{A})$. Here the adele ring $\mathbb{A}$ is a subring of the product $(\prod_p \mathbb{Q}_p) \times \mathbb{R}$ over all places of $\mathbb{Q}$ made up of elements (x_p) such that $x_p \in \mathbb{Z}_p$ except for finitely many primes p (see, for example, [LFE] Chapter 8 for the basic properties of the adele ring). We need to do some preparation, which will be given in the following subsections.

3.1.2 Structure theorems on $GL(\mathbb{A})$

In this section, we would like to prove the density of $SL_n(\mathbb{Q})$ in $SL_n(\mathbb{A}^{(\infty)})$; that is, the *strong approximation theorem.* A similar statement holds for $SL(n)$ over number fields. However, the validity of strong approximation is subtle, for example, we know that $GL_n(\mathbb{Q})$ is not dense in $GL_n(\mathbb{A}^{(\infty)})$ (see Exercise 3 below).

Let $\widehat{G} = \varprojlim_{N \lhd G} G/N$ be a profinite group. The topology of $\widehat{G}$ is given as follows: The system of open neighborhoods of $x \in \widehat{G}$ is given by the closure of $x\overline{N}$ of xN for the normal subgroups N appearing in the projective limit (thus $\widehat{G}/\overline{N}$ is a finite group). Thus we can embed $\widehat{G}$ into $\prod_{N \lhd G} G/N$ by $x \mapsto (xN)_N$. It is easy to see that this map is a topological embedding onto a closed subset if we equip $\prod_{N \lhd G} G/N$ with the product topology of the discrete topology on G/N. Since $\prod_{N \lhd G} G/N$ is compact, $\widehat{G}$ is compact. By definition, a subset G is dense in $\widehat{G}$ if and only if G surjects onto G/N for all N appearing in the projective limit (see Proposition 2.1 in Chapter 2 for details).

We put $\widehat{\mathbb{Z}} = \varprojlim_{N>1} \mathbb{Z}/N\mathbb{Z}$ with respect to the projection $\mathbb{Z}/N'\mathbb{Z} \twoheadrightarrow \mathbb{Z}/N\mathbb{Z}$ given by $x \mod N' \mapsto x \mod N$ if $N|N'$. Then $\widehat{\mathbb{Z}}$ is a profinite ring, and the Chinese remainder theorem tells us: $\widehat{\mathbb{Z}} = \prod_p \mathbb{Z}_p$ for the p-adic integer ring $\mathbb{Z}_p = \varprojlim_{n>1} \mathbb{Z}/p^n\mathbb{Z}$, where p runs over all primes. In particular, the integer ring $\mathbb{Z}$ is dense in $\widehat{\mathbb{Z}}$.

For a given commutative ring A, we write $M_n(A)$ for the (non-commutative) ring of $n \times n$-matrices with coefficients in A. Then $M_n(\widehat{\mathbb{Z}}) = \varprojlim_N M_n(\mathbb{Z}/N\mathbb{Z})$, which is a compact profinite ring. The group $GL_n(A)$ is a subset of $M_n(A)$ made up of invertible matrices. The group structure is given by matrix multiplication. In particular, $GL_n(\widehat{\mathbb{Z}}) = \varprojlim_N GL_n(\mathbb{Z}/N\mathbb{Z})$ is again a profinite compact group. Since we have $\det \left(\begin{smallmatrix} x & 0 \\ 0 & 1 \end{smallmatrix}\right) = x$, we have the surjectivity of the determinant map in the following sequence:

$$1 \longrightarrow SL_n(A) \longrightarrow GL_n(A) \xrightarrow{\det} A^\times \longrightarrow 1. \tag{3.5}$$

The subgroup $SL_n(A) \lhd GL_n(A)$ is defined by the exactness of the above sequence.

Since $\det(GL_n(\mathbb{Z})) = \mathbb{Z}^\times = \{\pm 1\}$, the exactness of (3.5) tells us that $GL_n(\mathbb{Z})$ is not dense in $GL_n(\widehat{\mathbb{Z}})$. However, we have

Lemma 3.1 *$SL_n(\mathbb{Z})$ is dense in $SL_n(\widehat{\mathbb{Z}})$.*

Proof We first give a proof when $n = 2$. The general case will be treated similarly by induction on n. Since $\widehat{\mathbb{Z}} = \varprojlim_N \mathbb{Z}/N\mathbb{Z}$ and $M_2(\widehat{\mathbb{Z}}) = \varprojlim_N M_2(\mathbb{Z}/N\mathbb{Z})$ as profinite rings, the system of neighborhoods of the zero matrix $0_2 \in M_2(\widehat{\mathbb{Z}})$ is given by $\{NM_2(\widehat{\mathbb{Z}}) | N \in \mathbb{N}\}$. Thus the system of neighborhoods of 1_2 in $SL_2(\widehat{\mathbb{Z}})$ is given by $U_N = (1_2 + NM_2(\widehat{\mathbb{Z}})) \cap SL_2(\widehat{\mathbb{Z}})$. Here we note that $1_2 + NM_2(\widehat{\mathbb{Z}})$ is not a subset of $SL_2(\widehat{\mathbb{Z}})$ (see Exercise 1). Note that U_N is the kernel of the natural group homomorphism $\pi_N : SL_2(\widehat{\mathbb{Z}}) \to SL_2(\mathbb{Z}/N\mathbb{Z})$ given by $\left(\begin{smallmatrix} a & b \\ c & d \end{smallmatrix}\right) \mapsto \left(\begin{smallmatrix} a & b \\ c & d \end{smallmatrix}\right) \mod N\widehat{\mathbb{Z}}$. If π_N restricted to $SL_2(\mathbb{Z})$ is a surjection for all N, then for any given $x \in SL_2(\widehat{\mathbb{Z}})$, we can choose $\gamma_N \in SL_2(\mathbb{Z})$ such that $\pi_N(\gamma_N) = \pi_N(x)$. Then $x^{-1}\gamma_N \in U_N$. In other words, $\gamma_N \in xU_N$. By choosing N large, we can make U_N as small as we like; so, γ_N converges to x as N grows. Thus we need to show the surjectivity of $\pi_N|_{SL_2(\mathbb{Z})}$. Let

$$V = \{v \in (\mathbb{Z}/N\mathbb{Z})^2 | \text{the order of } v \text{ is equal to } N\}.$$

Since $SL_2(\mathbb{Z}/N\mathbb{Z}) \subset GL_2(\mathbb{Z}/N\mathbb{Z}) = \mathrm{Aut}(V)$, $SL_2(\mathbb{Z}/N\mathbb{Z})$ acts on V by matrix multiplication. Let $v_0 = \left(\begin{smallmatrix}1\\0\end{smallmatrix}\right)$. Then for any given $g \in SL_2(\mathbb{Z}/N\mathbb{Z})$, $v = gv_0 \in V$. If $v = \left(\begin{smallmatrix}p\\q\end{smallmatrix}\right) \mod N$ and d is a common divisor of (p, q, N), then $\frac{N}{d}v = 0$. Thus $d = 1$. From this, it is easy to see that v can be represented by $\left(\begin{smallmatrix}p\\q\end{smallmatrix}\right) \mod N$ for relatively prime integers p and q. Then by the Euclidean algorithm, we find two integers x, y such that $py - qx = 1$. Thus we see $\alpha = \left(\begin{smallmatrix}p & x\\q & y\end{smallmatrix}\right) \in SL_2(\mathbb{Z})$ and $\alpha v_0 = v = gv_0$. Thus $\alpha^{-1}g$ stabilizes v_0 and hence it is of the form $\left(\begin{smallmatrix}1 & u\\0 & t\end{smallmatrix}\right)$. Since $\det(\alpha^{-1}g) = 1$, $t = 1$. Then for any integer m with $m \mod N = u$, putting $\delta = \left(\begin{smallmatrix}1 & m\\0 & 1\end{smallmatrix}\right) \in SL_2(\mathbb{Z})$, we see that $\delta^{-1}\alpha^{-1}g \equiv 1_2 \mod N$, which shows that $\pi_N(\alpha\delta) = g$ as desired.

We now prove the lemma for general n by induction on n. Put

$$V = \{v = {}^t(a_1, a_2, \ldots, a_n) \in (\mathbb{Z}/N\mathbb{Z})^n | ord(v) = N\},$$

where $ord(v)$ is the order of v in the additive group $(\mathbb{Z}/N\mathbb{Z})^n$. Let $e = {}^t(1, 0, \ldots, 0)$. If $\overline{\gamma}e = e$ for $\overline{\gamma} \in SL_n(\mathbb{Z}/N\mathbb{Z})$, then $\overline{\gamma}$ has the following form:

$$\overline{\gamma} = \begin{pmatrix}1 & \overline{u}\\0 & \overline{\gamma'}\end{pmatrix} \quad \text{for } \overline{u} \in (\mathbb{Z}/N\mathbb{Z})^{n-1} \text{ and } \overline{\gamma'} \in SL_{n-1}(\mathbb{Z}/N\mathbb{Z}).$$

By induction assumption, we can lift $\overline{\gamma'}$ to $\gamma' \in SL_{n-1}(\mathbb{Z})$ so that $\gamma' \equiv \overline{\gamma'} \mod N$. We can also lift $\overline{u}$ to $u \in \mathbb{Z}^{n-1}$ by the density of $\mathbb{Z}$ in $\widehat{\mathbb{Z}}$. Then we form $\gamma = \left(\begin{smallmatrix}1 & u\\0 & \gamma'\end{smallmatrix}\right)$. Thus we only need to prove

$$V = SL_n(\mathbb{Z}/N\mathbb{Z})e = SL_n(\mathbb{Z})e.$$

Pick $\overline{v} = {}^t(\overline{a}_1, \ldots, \overline{a}_n)$. We lift $\overline{a}_j$ to $a_j \in \mathbb{Z}$ so that $v = {}^t(a_1, \ldots, a_n) \equiv \overline{v} \mod N$ and entries of v do not have a non-trivial common divisor. If $aw \in \mathbb{Z}v$ with $w = {}^t(b_1, \ldots, b_n) \in L = \mathbb{Z}^n$, then $ab_i = a_i$ and a is a common divisor of entries of v, which shows that $w \in \mathbb{Z}v$. Thus $L/\mathbb{Z}v$ is free of rank n. Choose a base $\{\overline{v}_2, \ldots, \overline{v}_n\}$ of $L/\mathbb{Z}v$ and lift it to $v_j \in L$. Then $\{v, v_2, \ldots, v_n\}$ is a base of L over $\mathbb{Z}$. Thus $\gamma = (v, v_2, \ldots, v_n) \in GL_n(\mathbb{Z})$, and hence $\det\gamma = \pm 1$. Replacing γ by $(v, v_2, \ldots, -v_n)$ if necessary, we may assume that $\gamma \in SL_n(\mathbb{Z})$, and hence $\overline{v} = \gamma e$; so, we are done. □

A proof similar to the above works well for split semi-simple groups like

$$S(A) = \{\alpha \in SL_n(A) | \alpha J {}^t\alpha = J\}$$

for

$$J = \begin{pmatrix} 0 & -1_m \\ 1_m & 0 \end{pmatrix}, \begin{pmatrix} 0 & 1_m \\ 1_m & 0 \end{pmatrix} \ (n = 2m) \text{ and } \begin{pmatrix} 0 & 0 & 1_m \\ 0 & 1 & 0 \\ 1_m & 0 & 0 \end{pmatrix} \ (n = 2m+1 \geq 3).$$

In these cases, $S(\mathbb{Z})$ is dense in $S(\widehat{\mathbb{Z}})$.

THEOREM 3.2 (STRONG APPROXIMATION THEOREM) *The diagonally embedded subgroup $SL_n(\mathbb{Q})$ is dense in $SL_n(\mathbb{A}^{(\infty)})$, where $\mathbb{A}^{(\infty)} = \{x \in \mathbb{A} | x_\infty = 0\}$ is the finite part of $\mathbb{A}$.*

Proof Let $L = \widehat{\mathbb{Z}}^n \subset (\mathbb{A}^{(\infty)})^n = V$ be the standard $\widehat{\mathbb{Z}}$-lattice. We consider the elements in L as column vectors. Then via matrix multiplication, $GL_n(\mathbb{A}^{(\infty)})$ acts on V. Let $g \in GL_n(\mathbb{A}^{(\infty)})$. Then gL is a $\widehat{\mathbb{Z}}$-free module of rank n. Choose a base $\{x_1, \ldots, x_n\}$ of gL. Since $\mathbb{Q}$ is dense in $\mathbb{A}$, we may assume that $x_i \in \mathbb{Q}^n$. Let $X = (x_1, \ldots, x_n)$ be the $n \times n$-matrix with column vectors x_i. Then $XL = gL$. Thus $g^{-1}X \in GL_n(\widehat{\mathbb{Z}}) = \mathrm{Aut}_{\widehat{\mathbb{Z}}}(L)$. Suppose that $\det(g) = 1$. Then

$$\det(X) = \det(g^{-1}X) \in \widehat{\mathbb{Z}}^\times \cap \mathbb{Q} = \{\pm 1\}.$$

Thus changing x_1 with $-x_1$ if necessary, we may assume that $\det(X) = 1$. Then $X^{-1}g \in SL_n(\widehat{\mathbb{Z}})$. Since $SL_n(\mathbb{Z})$ is dense in $SL_n(\widehat{\mathbb{Z}})$, for any small neighborhood U of 1 in $SL_n(\mathbb{A}^{(\infty)})$, we can find $\gamma \in SL_n(\mathbb{Z})$ such that $g^{-1}X \in U\gamma$. In other words, $SL_n(\mathbb{Q}) \ni X\gamma^{-1} \in gU$, and hence $SL_n(\mathbb{Q})$ is dense in $SL_n(\mathbb{A}^{(\infty)})$. □

COROLLARY 3.3 *Let S be an open compact subgroup of $GL_n(\mathbb{A}^{(\infty)})$. Then $\det S \subset \widehat{\mathbb{Z}}^\times$, and if they are equal,*

$$\begin{aligned} GL_n(\mathbb{A}) &= S \cdot GL_n(\mathbb{R})_+ \cdot GL_n(\mathbb{Q}) = GL_n(\mathbb{Q}) \cdot S \cdot GL_n(\mathbb{R})_+, \\ GL_n(\mathbb{A}^{(\infty)}) &= GL_n(\mathbb{Q}) \cdot S = S \cdot GL_n(\mathbb{Q}), \\ GL_n(\mathbb{A})_+ &= GL_n(\mathbb{Q})_+ \cdot S \cdot GL_n(\mathbb{R})_+ = S \cdot GL_n(\mathbb{R})_+ \cdot GL_n(\mathbb{Q})_+, \end{aligned}$$

where

$$\begin{aligned} GL_n(\mathbb{Q})_+ &= \{x \in GL_n(\mathbb{Q}) | \det(x) > 0\}, \\ GL_n(\mathbb{R})_+ &= \{x \in GL_n(\mathbb{R}) | \det(x) > 0\}, \text{ and} \\ GL_n(\mathbb{A})_+ &= GL_n(\mathbb{A}^{(\infty)}) \cdot GL_n(\mathbb{R})_+ \subset GL_n(\mathbb{A}). \end{aligned}$$

Proof If C is a compact subgroup of $\mathbb{Q}_p^\times$ not contained in $\mathbb{Z}_p^\times$, then we have $x \in C$ with $|x|_p \neq 1$. Interchanging x and x^{-1} if necessary, we may assume that $|x|_p > 1$. Then $\lim_{n\to\infty} |x^n|_p = \infty$, which contradicts the compactness of C. Thus $\mathbb{Z}_p^\times$ is the unique maximal compact subgroup of $\mathbb{Q}_p^\times$, and $\widehat{\mathbb{Z}}^\times = \prod_p \mathbb{Z}_p^\times$ is the unique maximal compact subgroup of $(\mathbb{A}^{(\infty)})^\times$. Since $\det : GL_n(\mathbb{A}^{(\infty)}) \to (\mathbb{A}^{(\infty)})^\times$ is a continuous map, $\det S$ is a compact subgroup of $\widehat{\mathbb{Z}}^\times$.

We now prove $GL_n(\mathbb{A}^{(\infty)}) = GL_n(\mathbb{Q}) \cdot S$ if $\det(S) = \widehat{\mathbb{Z}}^\times$. We use the notation introduced in the proof of the approximation theorem. Take $g \in GL_n(\mathbb{A}^{(\infty)})$. Then $gL = XL$ for $X \in GL_n(\mathbb{Q})$. Then $g^{-1}X \in GL_n(\widehat{\mathbb{Z}})$. In particular, $\det(g^{-1}X) \in \widehat{\mathbb{Z}}^\times = \det S$. Thus we can find $s \in S$ such that $\det(g^{-1}X) = \det s$. Then $s^{-1}g^{-1}X \in SL_n(\widehat{\mathbb{Z}})$. Since $U = S \cap SL_n(\widehat{\mathbb{Z}})$ is a neighborhood of 1_n, by the density of $SL_n(\mathbb{Z})$ in $SL_n(\widehat{\mathbb{Z}})$, we can find $\gamma \in SL_n(\mathbb{Z})$ such that $\gamma \in Us^{-1}g^{-1}X$, which shows that $g \in GL_n(\mathbb{Q})S$. Thus $GL_n(\mathbb{A}^{(\infty)}) = GL_n(\mathbb{Q}) \cdot S$. Since the automorphism $x \mapsto x^{-1}$ of the set $GL_n(\mathbb{A}^{(\infty)})$ reverses the order of multiplication, $GL_n(\mathbb{A}^{(\infty)}) = S \cdot GL_n(\mathbb{Q})$.

In the above argument, by changing $-x_1$ for x_1 if necessary, we may assume that $\det X > 0$. Thus, $GL_n(\mathbb{A}^{(\infty)}) = S \cdot GL_n(\mathbb{Q})_+ = GL_n(\mathbb{Q})_+ \cdot S$. Since $GL_n(\mathbb{A})_+ \cap GL_n(\mathbb{Q}) = GL_n(\mathbb{Q})_+$, for each $g \in GL_n(\mathbb{A})_+$, choosing $\gamma \in GL_n(\mathbb{Q})_+$ such that $g^{(\infty)} = \gamma^{(\infty)}s$ with $s \in S$, we see that

$$g = g^{(\infty)}g_\infty = \gamma\gamma_\infty^{-1}g_\infty s \in GL_n(\mathbb{Q})_+ \cdot GL_n(\mathbb{R})_+ \cdot S.$$

This shows that

$$GL_n(\mathbb{A})_+ = GL_n(\mathbb{Q})_+ \cdot GL_n(\mathbb{R})_+ \cdot S = GL_n(\mathbb{R})_+ \cdot S \cdot GL_n(\mathbb{Q})_+.$$

Similarly, it can be shown from $GL_n(\mathbb{A}^{(\infty)}) = GL_n(\mathbb{Q})_+ \cdot S = S \cdot GL_n(\mathbb{Q})_+$ that

$$GL_n(\mathbb{A}) = S \cdot GL_n(\mathbb{R})_+ \cdot GL_n(\mathbb{Q}) = GL_n(\mathbb{Q}) \cdot S \cdot GL_n(\mathbb{R})_+. \qquad \square$$

Exercises

(1) Show that $1_2 + NM_2(\widehat{\mathbb{Z}}) \not\subset SL_2(\widehat{\mathbb{Z}})$.
(2) Show that $SL_2(\mathbb{Q})$ is not dense in $SL_2(\mathbb{A})$.
(3) Show that $GL_2(\mathbb{Q})$ is not dense in $GL_2(\mathbb{A}^{(\infty)})$.
(4) Show that $\widehat{\mathbb{Z}}^\times \cap \mathbb{Q} = \{\pm 1\}$.

3.1.3 Maximal compact subgroups

For simplicity, we write $G(A) = GL_2(A)$ for each commutative algebra A with identity, where $GL_2(A)$ is the group of 2×2 matrices with an unit determinant. Thus if A is not a field, even if $\det(\gamma) \neq 0$, it is possible to have $\gamma = \left(\begin{smallmatrix} a & b \\ c & d \end{smallmatrix}\right) \notin G(A)$ (see Exercise 1 below). We also write $Z(A)$ for the center of $G(A)$. Then

$$Z(A) = \left\{ \left(\begin{smallmatrix} a & 0 \\ 0 & a \end{smallmatrix}\right) | a \in A^\times \right\} \cong A^\times. \tag{3.6}$$

We put $G(\mathbb{R})_+ = \{g \in G(\mathbb{R}) | \det g > 0\}$. Then $G(\mathbb{R})_+$ is the connected component of $G(\mathbb{R})$ containing the identity element $1_2 = \left(\begin{smallmatrix} 1 & 0 \\ 0 & 1 \end{smallmatrix}\right)$ (see Exercise 7). For $z \in \mathfrak{H}$ and $g = \left(\begin{smallmatrix} a & b \\ c & d \end{smallmatrix}\right) \in G(\mathbb{R})_+$, we define $g(z) = \frac{az+b}{cz+d}$. This is well defined because $cz + d \neq 0$ as long as $\mathrm{Im}(z) \neq 0$. Then $gh(z) = g(h(z))$ and $1_2(z) = z$. Since we have

$$\left(\begin{smallmatrix} a & b \\ c & d \end{smallmatrix}\right) \left(\begin{smallmatrix} z & \bar{z} \\ 1 & 1 \end{smallmatrix}\right) = \begin{pmatrix} g(z) & \overline{g(z)} \\ 1 & 1 \end{pmatrix} \begin{pmatrix} cz+d & 0 \\ 0 & c\bar{z}+d \end{pmatrix}, \tag{3.7}$$

taking determinant, we have $\mathrm{Im}(g(z)) = \det(g)\,\mathrm{Im}(z)|cz + d|^{-2}$, and thus $g(z) \in \mathfrak{H}$ as long as $z \in \mathfrak{H}$. In this way, $G(\mathbb{R})_+$ acts on $\mathfrak{H}$. Again by (3.7), if we write $j(g, z) = cz + d$, we have the cocycle relation:

$$j(gh, z) = j(g, h(z))j(h, z). \tag{3.8}$$

First consider the map: $\pi : GL_2(\mathbb{R}) \to \mathfrak{H}$ given by $\left(\begin{smallmatrix} a & b \\ c & d \end{smallmatrix}\right) \mapsto \frac{ai+b}{ci+d}$ for $i = \sqrt{-1}$. The map is surjective, because $\left(\begin{smallmatrix} y & x \\ 0 & 1 \end{smallmatrix}\right)(i) = x + iy$. The fiber of π at i can be computed as

$$\pi^{-1}(i) = Z(\mathbb{R})SO_2(\mathbb{R}), \tag{3.9}$$

where

$$SO_2(A) = \left\{ g \in G(\mathbb{R}) | g^t g = 1 \text{ and } \det g = 1. \right\}.$$

By (3.9), the stabilizer of $i \in \mathfrak{H}$ is given by $C_\infty = Z(\mathbb{R})SO_2(\mathbb{R})$. Thus for $g, h \in Z(\mathbb{R})$, we have from the cocycle relation (3.8) that

$$j(gh, i) = j(g, i)j(h, i).$$

This shows that $g \mapsto j(g, i)$ is a group homomorphism of C_∞ into $\mathbb{C}^\times$. We define for each integer k,

$$J_k(g, z) = \det(g)^{-1} j(g, z)^k. \tag{3.10}$$

Then $g \mapsto J_k(g, i)$ is again a character of C_∞.

Exercises

(1) Find an example of $\gamma = \left(\begin{smallmatrix} a & b \\ c & d \end{smallmatrix}\right)$ with $\det \gamma \neq 0$ but $\gamma \notin GL_2(A)$ for $A = \mathbb{Q}[X]/(X^2)$.

(2) Show (3.6) and (3.9).

(3) Show that $SO_2(\mathbb{R}) \cong S^1$ as Lie groups, where S^1 is a one-dimensional unit circle. Here two Lie groups X and Y are isomorphic if we have a real analytic bijection $f : X \to Y$ which preserves the group structure.
(4) Show that $SO_2(\mathbb{R})$ is a maximal compact subgroup of $G(\mathbb{R})_+$. That is, if there is a compact subgroup $C \subset G(\mathbb{R})_+$ such that $C \supset SO_2(\mathbb{R})$, then $C = SO_2(\mathbb{R})$.
(5) Show that for any given maximal compact subgroup C of $G(\mathbb{R})_+$, there exists $g \in G(\mathbb{R})_+$ such that $gCg^{-1} = SO_2(\mathbb{R})$.
(6) Show that $g \mapsto j(g,i)$ is an isomorphism of Lie groups from C_∞ onto $\mathbb{C}^\times$.
(7) Show that $G(\mathbb{R})_+$ is connected. Hint: Use $\pi : G(\mathbb{R})_+ \to \mathfrak{H}$.

3.1.4 Open compact subgroups of $GL_2(\mathbb{A})$ and Dirichlet characters

We consider the following subgroups of $GL_2(\widehat{\mathbb{Z}})$:

$$S_0(N) = \left\{ \left(\begin{smallmatrix} a & b \\ c & d \end{smallmatrix}\right) \in GL_2(\widehat{\mathbb{Z}}) | c \equiv 0 \mod N\widehat{\mathbb{Z}} \right\}, \tag{3.11}$$

$$S_1(N) = \left\{ \left(\begin{smallmatrix} a & b \\ c & d \end{smallmatrix}\right) \in S_0(N) | d \equiv 1 \mod N\widehat{\mathbb{Z}} \right\}, \tag{3.12}$$

$$S(N) = \left\{ \left(\begin{smallmatrix} a & b \\ c & d \end{smallmatrix}\right) \in S_1(N) | b \equiv 0 \mod N\widehat{\mathbb{Z}} \text{ and } a \equiv d \equiv 1 \mod N\widehat{\mathbb{Z}} \right\}, \tag{3.13}$$

Then inside $G(\mathbb{A}^{(\infty)})$,

$$\begin{aligned} S_0(N) \bigcap G(\mathbb{Q})_+ &= \Gamma_0(N), \\ S_1(N) \bigcap G(\mathbb{Q})_+ &= \Gamma_1(N), \\ S(N) \bigcap G(\mathbb{Q})_+ &= \Gamma(N). \end{aligned} \tag{3.14}$$

From this, it is easy to conclude that inside $G(\mathbb{A})$,

$$\begin{aligned} S_0(N) \cdot G(\mathbb{R})_+ \bigcap G(\mathbb{Q}) &= \Gamma_0(N), \\ S_1(N) \cdot G(\mathbb{R})_+ \bigcap G(\mathbb{Q}) &= \Gamma_1(N), \\ S(N) \cdot G(\mathbb{R})_+ \bigcap G(\mathbb{Q}) &= \Gamma(N). \end{aligned} \tag{3.15}$$

Note that $\det S_0(N) = \det S_1(N) = \widehat{\mathbb{Z}}^\times$, but $\det S(N) \subsetneq \widehat{\mathbb{Z}}^\times$ if $N > 2$. Actually

$$\det S(N) = \left\{ x \in \widehat{\mathbb{Z}}^\times | x \equiv 1 \mod N\widehat{\mathbb{Z}} \right\}.$$

We write the above subgroup of $\widehat{\mathbb{Z}}^\times$ as $U(N)$. Then

$$\widehat{\mathbb{Z}}^\times/U(N) \cong (\mathbb{Z}/N\mathbb{Z})^\times. \tag{3.16}$$

LEMMA 3.4 *We have a canonical group isomorphism:*

$$\iota : \mathbb{A}^\times/\mathbb{Q}^\times U(N)\mathbb{R}_+^\times \cong (\mathbb{Z}/N\mathbb{Z})^\times,$$

where $\mathbb{R}_+^\times = \{x \in \mathbb{R} | x > 0\}$. *The isomorphism* ι *sends* $p_p \in \mathbb{Q}_p^\times \subset \mathbb{A}^\times$ *to* p mod N *if a prime* p *is not a factor of* N.

Proof Pick $x \in \mathbb{A}^\times$. For each component $x_p \in \mathbb{Q}_p^\times$ for a prime p, we define $v_p(x) \in \mathbb{Z}$ so that $x_p p^{-v_p(x)} \in \mathbb{Z}_p^\times$. Thus, in particular, $v_p(p^n) = n$ for $n \in \mathbb{Z}$. Then we define $rat(x) \in \mathbb{Q}^\times$ by $rat(x) = \frac{x_\infty}{|x_\infty|}\prod_p p^{v_p(x)}$. Note that $v_p(x) \geq 0$ except for finitely many primes p because $x \in \mathbb{A}$. Since $x^{-1} \in \mathbb{A}$ also, $-v_p(x) = v_p(x^{-1}) \geq 0$ except for finitely many primes p. Thus $v_p(x) = 0$ except for finitely many primes p. We have shown that

$$\mathbb{A}^\times = \left\{ \begin{array}{c} x = (x_p) \in (\prod_p \mathbb{Q}_p^\times) \times \mathbb{R}^\times | v_p(x) = 0 \\ \text{except for finitely many primes } p \end{array} \right\}. \tag{3.17}$$

Then by definition $x = rat(x)u$ for $u \in U(1)\mathbb{R}_+^\times$, where $U(1) = \widehat{\mathbb{Z}}^\times$. Thus $\mathbb{A}^\times = \mathbb{Q}^\times \cdot U(1) \cdot \mathbb{R}_+^\times$. Pick $x \in (U(1) \cdot \mathbb{R}_+^\times) \bigcap \mathbb{Q}^\times$. Then $x > 0$ because $x_\infty > 0$, and $x \in \mathbb{Z}^\times = \{\pm 1\}$ because $x \in U(1) \bigcap \mathbb{Q}$. Thus $x = 1$. This shows

$$(U(1) \cdot \mathbb{R}_+^\times) \bigcap \mathbb{Q}^\times = \{1\},$$

which tells us that

$$\mathbb{A}^\times = \mathbb{Q}^\times \cdot (U(1)\mathbb{R}_+^\times) \cong \mathbb{Q}^\times \times (U(1) \cdot \mathbb{R}_+^\times).$$

Then

$$\mathbb{A}^\times/(\mathbb{Q}^\times \cdot U(N) \cdot \mathbb{R}_+^\times) \cong \frac{\mathbb{Q}^\times \times U(1) \cdot \mathbb{R}_+^\times}{\mathbb{Q}^\times \times U(N) \cdot \mathbb{R}_+^\times} \cong U(1)/U(N) \cong (\mathbb{Z}/N\mathbb{Z})^\times. \quad \square$$

By this lemma, we may regard a Dirichlet character $\chi : (\mathbb{Z}/N\mathbb{Z})^\times \to \mathbb{C}^\times$ as a finite order Hecke character $\chi^* = \chi \circ \iota : \mathbb{A}^\times/\mathbb{Q}^\times \to \mathbb{C}^\times$. Note that $\chi^*(p_p) = \chi(p)$ if p is not a factor of N. We can restrict χ^* to $U(1) \subset \mathbb{A}^\times$. Since $\chi^*(p) = 1$ for a prime $p \nmid N$ and $\chi^*(p_\infty) = 1$ ($\chi^*(\mathbb{R}_+^\times) = 1$; see Exercise 4), $1 = \chi^*(p^{(p\infty)})\chi^*(p_p)$ for $p^{(p\infty)} = pp_p^{-1}p_\infty^{-1}$. Since $p^{(p\infty)} \in U(1)$ and $p^{(p\infty)}$ mod $U(N) = p$ mod N (identifying $U(1)/U(N)$ with $(\mathbb{Z}/N\mathbb{Z})^\times$), we find that

$$\chi^*(p^{(p\infty)} \bmod U(N)) = \chi^{-1}(p \bmod N) \text{ if } p \nmid N. \tag{3.18}$$

By definition, $\left(\begin{smallmatrix} a & b \\ c & d \end{smallmatrix}\right) \mapsto d \mod N\widehat{\mathbb{Z}}$ induces an isomorphism of groups

$$S_0(N)/S_1(N) \cong (\mathbb{Z}/N\mathbb{Z})^\times.$$

For a given χ, we then define

$$\chi_N\left(\begin{smallmatrix} a & b \\ c & d \end{smallmatrix}\right) = \chi^*(d_N) = \chi^{-1}(d \mod N), \tag{3.19}$$

where d_N is the component of d in $\prod_{p|N} \mathbb{Q}_p$. In the last term of the above equation, we have χ^{-1} instead of χ because of (3.18). By the above argument, $d_N \in \prod_{p|N} \mathbb{Z}_p^\times$, although d itself may not be contained in $\mathbb{A}^\times$. In any case. $u \mapsto \chi_N(u) \in \mathbb{C}^\times$ gives a character of $S_0(N)$.

We start from a continuous character $\psi : \mathbb{A}^\times/\mathbb{Q}^\times \to \mathbb{C}^\times$. If ψ is of finite order, then ψ has values in a finite group μ_n of nth roots of unity. Thus ψ induces a continuous character from $\mathbb{A}^\times$ to a discrete finite group μ_n, because the topology of μ_n induced from $\mathbb{C}$ is the discrete topology. Then Ker(ψ) contains

(1) The connected component of $\mathbb{A}^\times$ containing 1, which is $\mathbb{R}_+^\times \subset \mathbb{A}^\times$ (see Exercise 6);
(2) An open neighborhood of 1 in $(\mathbb{A}^{(\infty)})^\times \subset \mathbb{A}^\times$, which is $U(N)$ for a suitable positive integer N.

Thus ψ factors through $\mathbb{A}^\times/\mathbb{Q}^\times U(N)\mathbb{R}_+^\times \stackrel{\iota}{\cong} (\mathbb{Z}/N\mathbb{Z})^\times$. We write $\psi_* : (\mathbb{Z}/N\mathbb{Z})^\times \to \mathbb{C}^\times$ for the induced character. Then we have $\psi|_{U(1)} = \psi_*^{-1}$. Even if ψ is not of finite order, the restriction of ψ to $U(1)$ is of finite order because $U(1)$ is a profinite group (see Proposition 2.2 in Chapter 2). Thus we can again define the Dirichlet character $\psi_* : (\mathbb{Z}/N\mathbb{Z})^\times \to \mathbb{C}^\times$ associated to ψ by $\psi_* = (\psi|_{U(1)})^{-1}$.

For a Hecke character $\psi : \mathbb{A}^\times/\mathbb{Q}^\times \to \mathbb{C}^\times$, the conductor C is defined to be the smallest positive integer so that $U(C) \subset \mathrm{Ker}(\psi)$. Then $C|N$ for an integer $N > 0$ if and only if $U(N) \subset \mathrm{Ker}(\psi)$ (see Exercise 7).

Here we give an example of infinite order Hecke characters. Define $|x|_\mathbb{A} = |x_\infty| \prod_p p^{-v_p(x)}$ for $x \in \mathbb{A}^\times$. Then by (3.17), $v_p(x) = 0$ except for finitely many primes p. Thus $|x|_\mathbb{A} \in \mathbb{C}^\times$ is well defined. By definition, $x \mapsto |x|_\mathbb{A}$ is an infinite order continuous character of $\mathbb{A}^\times$, but $|U(1)|_\mathbb{A} = \{1\}$; thus ψ_* is the identity character for $\psi = |\ |_\mathbb{A}$. If $x \in \mathbb{Q}^\times$, $|x_\infty| = \prod_p p^{v_p(x)} = |x^{(\infty)}|_\mathbb{A}^{-1}$, and hence $|\mathbb{Q}^\times|_\mathbb{A} = \{1\}$; thus, $|\ |_\mathbb{A} : \mathbb{A}^\times/\mathbb{Q}^\times \to \mathbb{C}^\times$, giving an infinite order Hecke character. By definition, we have

$$|p_p|_\mathbb{A} = \frac{1}{p} \quad \text{and} \quad |p_\infty|_\mathbb{A} = p. \tag{3.20}$$

Exercises

(1) Show (3.14), (3.15) and (3.16).

(2) Following the proof of Lemma 3.4, show that $\iota(p_p) = p \mod N \in (\mathbb{Z}/N\mathbb{Z})^\times$ if a prime p is not a factor of N.

(3) Compute $\chi^*(p_p)$ for a prime $p|N$.

(4) Find an example χ of Dirichlet characters such that $\chi^*(\mathbb{R}^\times) \neq \{1\}$. In that case, what is $\chi^*(\mathbb{R}^\times)$?

(5) Find $s = \left(\begin{smallmatrix} a & b \\ c & d \end{smallmatrix}\right) \in S_0(N)$ such that $d \notin \mathbb{A}^\times$.

(6) Show that the connected component of $\mathbb{A}^\times$ containing 1 is $\mathbb{R}_+^\times \subset \mathbb{A}^\times$.

(7) Let C be the conductor of a Hecke character ψ. Show that $C|N$ for an integer $N > 0$ if and only if $U(N) \subset \mathrm{Ker}(\psi)$.

3.1.5 Adelic and classical modular forms

Let $\chi : (\mathbb{Z}/N\mathbb{Z})^\times \to \mathbb{C}^\times$ be a Dirichlet character and k be an integer. We write $\psi : \mathbb{A}^\times/\mathbb{Q}^\times \to \mathbb{C}^\times$ for the (possibly infinite order) Hecke character given by $\psi(x) = |x|_{\mathbb{A}}^{2-k}\chi^*(x)$; thus $\psi_* = \chi$.

Consider a function $f : G(\mathbb{A}) \to \mathbb{C}$ satisfying the following condition:

$$f(\gamma zxuc) = \psi(z)\chi_N(u)f(x)J_k(c,i)^{-1}$$
$$\text{for all } z \in Z(\mathbb{A}),\ u \in S_0(N),\ c \in C_\infty \text{ and } \gamma \in G(\mathbb{Q}). \quad \text{(A1)}$$

By (A1), f is actually a function on $G(\mathbb{Q})\backslash G(\mathbb{A})$, generalizing the notion of Hecke characters in the case of $GL(1)$. Note that $\psi(z_\infty) = \chi^*(z_\infty)|z_\infty|^{2-k}$ (see (3.20)) and that $J_k(z_\infty, i)^{-1} = z_\infty^{2-k}$ (see (3.10)) for $z_\infty \in Z(\mathbb{R}) = Z(\mathbb{A}) \cap C_\infty$. Thus, to have a non-zero function f satisfying (A1), we need to assume $\psi(z_\infty) = J_k(z_\infty, i)^{-1}$ for all $z_\infty \in Z(\mathbb{R})$, which in turn implies (see Exercise 1 below) that

$$\chi(-1) = (-1)^k. \quad \text{(P)}$$

We now relate functions f satisfying (A1) to modular forms on $\mathfrak{H}$. For each $z = x + iy \in \mathfrak{H}$, choose $u_z \in G(\mathbb{R})_+$ such that $u_z(i) = z$ for $i = \sqrt{-1}$. By the surjectivity of $\pi : G(\mathbb{R})_+ \to \mathfrak{H}$, this is always possible. There is a canonical choice of u_z: $\left(\begin{smallmatrix} y & x \\ 0 & 1 \end{smallmatrix}\right)$. We then define $f_*(z) = f(u_z)J_k(u_z, i)$.

We claim here that $f_*(z)$ is well defined, independent of the choice of $u = u_z$. To show this, we choose another $u' \in G(\mathbb{R})_+$ such that $u'(i) = z$. Then $u(i) = z = u'(i)$ and hence, $u^{-1}u'(i) = i$ and $u' = uc$ with $c \in C_\infty$.

Then we see that

$$\begin{aligned} f(u')J_k(u',i) &= f(uc)J_k(uc,i) = f(u)J_k(c,i)^{-1}J_k(u,c(i))J_k(c,i) \\ &= f(u)J_k(u,i) = f_*(z) \end{aligned}$$

because of $c(i) = i$. This proves the claim.

Proposition 3.5 *Suppose* (A1) *for* $f : G(\mathbb{A}) \to \mathbb{C}$. *Then for* $\gamma = \left(\begin{smallmatrix} a & b \\ c & d \end{smallmatrix}\right) \in \Gamma_0(N)$, *we have* $f_*(\gamma(z)) = \chi(d)f_*(z)j(\gamma,z)^k$.

Proof Take $\gamma \in (S_0(N)G(\mathbb{R})_+ \cap G(\mathbb{Q})) = \Gamma_0(N)$ in $G(\mathbb{A})$. Then $f(\gamma u) = f(u)$ for $u = u_z$. We first compute $f(\gamma u)$. We see that

$$f(\gamma u) = f(\gamma^{(\infty)}\gamma_\infty u) = f(\gamma_\infty u \gamma^{(\infty)}) = \chi_N(\gamma^{(\infty)})f(\gamma_\infty u),$$

since $\gamma^{(\infty)} \in S_0(N)$. Since $(\gamma u)_\infty(i) = \gamma(z)$, we see that

$$\begin{aligned} f_*(\gamma(z)) &= f((\gamma u)_\infty)J_k((\gamma u)_\infty, i) \\ &= \chi_N^{-1}(\gamma^{(\infty)})f((\gamma u)_\infty)J_k(\gamma, u(i))J_k(u,i) = \chi_N^{-1}(\gamma^{(\infty)})f_*(z)j(\gamma,z)^k, \end{aligned}$$

because $u(i) = z$ and $\det\gamma = 1$. Thus we need to show that $\chi_N^{-1}(\gamma^{(\infty)}) = \chi(d)$. By definition, $\psi = |\ |_{\mathbb{A}}^{2-k}\chi^*$, and by (3.19), we have $\chi_N(\gamma^{(\infty)}) = \chi^{-1}(d \bmod N)$. This gives the desired equality. □

We can generalize the above argument a little to $\gamma \in G(\mathbb{Q})_+$. We compute

$$\begin{aligned} f_*(\gamma(z)) &= f(\gamma_\infty u)J_k(\gamma u, i) = f(\gamma'^{-1}\gamma_\infty u)J_k(\gamma u, i) \\ &= f((\gamma^{(\infty)})^{-1}u)J_k(u,i)J_k(\gamma,z) = f(u(\gamma^{(\infty)})^{-1})J_k(u,i)J_k(\gamma,z), \end{aligned}$$

because $(\gamma^{(\infty)})^{-1}u = u(\gamma^{(\infty)})^{-1}$ ($u \in G(\mathbb{R})$ and $(\gamma^{(\infty)})^{-1} \in G(\mathbb{A}^{(\infty)})$). Thus defining $g(x) = f(x(\gamma^{(\infty)})^{-1})$, we see that

$$f_*(\gamma(z)) = g_*(z)J_k(\gamma,z). \tag{3.21}$$

Now we suppose that $f : G(\mathbb{A}) \to \mathbb{C}$ to satisfy the following two conditions in addition to (A1):

$$f_* : \mathfrak{H} \to \mathbb{C} \text{ is a holomorphic function on } \mathfrak{H}; \tag{A2}$$

$$\lim_{\mathrm{Im}(z)\to\infty} f_*\left(\frac{az+b}{cz+d}\right) = 0 \quad \text{for all} \quad \left(\begin{smallmatrix} a & b \\ c & d \end{smallmatrix}\right) \in SL_2(\mathbb{Q}). \tag{A3}$$

Let $\mathscr{S}_k(N,\chi)$ be the space of functions on $G(\mathbb{A})$ satisfying (A1,2,3). Then we have

Theorem 3.6 *Suppose* $\psi = |\ |_{\mathbb{A}}^{2-k}\chi^*$ *for an integer k. Then* $f \mapsto f_*$ *induces an isomorphism of vector spaces:* $\mathscr{S}_k(N,\chi) \cong S_k(\Gamma_0(N),\chi)$.

Proof Starting from $f \in S_k(\Gamma_0(N),\chi)$, we define $f^* : G(\mathbb{A}) \to \mathbb{C}$ in the following way. For $x \in G(\mathbb{A})$, we write $x = \gamma s$ for $s \in S_0(N) \cdot G(\mathbb{R})_+$ and $\gamma \in G(\mathbb{Q})$. This is possible by Corollary 3.3. Then we define

$$f^*(x) = \chi_N(s) f(s_\infty(i)) J_k(s_\infty, i)^{-1}.$$

We claim that this is independent of the choice of γ and s. If we have $x = \gamma s = \gamma' s'$ for $s' \in S_0(N) \cdot G(\mathbb{R})_+$ and $\gamma' \in G(\mathbb{Q})$. Then

$$S_0(N) \cdot G(\mathbb{R})_+ \ni s' s^{-1} = (\gamma')^{-1}\gamma \in G(\mathbb{Q}),$$

which shows that $\delta = (\gamma')^{-1}\gamma \in \Gamma_0(N)$. As in the proof of Proposition 3.5, we see that $\chi_N(\delta) = \chi^{-1}(d)$ if $\delta = \left(\begin{smallmatrix} a & b \\ c & d \end{smallmatrix}\right)$. Then we have

$$\begin{aligned} \chi_N(s') f(s'_\infty(i)) J_k(s'_\infty, i)^{-1} &= \chi_N(\delta s) f(\delta s_\infty(i)) J_k(\delta s_\infty, i)^{-1} \\ &= \chi_N(\delta)\chi_N(s) f(\delta s_\infty(i)) J_k(\delta, s_\infty(i))^{-1} J_k(s_\infty, i) \\ &= \chi_N(s)\chi^{-1}(d) f(\delta(z)) J_k(\delta, z)^{-1} J_k(s_\infty, i) = f^*(x), \end{aligned}$$

where we have written $z = s_\infty(i)$. This proves the claim.

By definition, f^* satisfies (A1)for $\gamma \in G(\mathbb{Q})$, $u \in S_0(N)$ and $c \in C_\infty$, because $\psi = \chi^*$ on $\widehat{\mathbb{Z}}^\times$. Since $\mathbb{A}^\times = \mathbb{Q}^\times U(1)\mathbb{R}_+^\times$, any Hecke character $\mathbb{A}^\times/\mathbb{Q}^\times \to \mathbb{C}^\times$ is determined by its restriction to $U(1)\mathbb{R}_+^\times$. Since $Z(\mathbb{R}) \subset C_\infty$ and $Z(\widehat{\mathbb{Z}}) \subset S_0(N)$, the above information is enough to conclude that $f^*(zx) = \psi(z) f^*(x)$ for $z \in Z(\mathbb{A})$. Thus $f^* \in \mathscr{S}_k(N,\chi)$.

By definition, $(f^*)_* = f$ and for $g \in \mathscr{S}_k(N,\chi)$, $(g_*)^* = g$. This proves the desired assertion. □

Let S be an open subgroup of $GL_2(\widehat{\mathbb{Z}})$. Then it is easy to see that the intersection $\Gamma = (S \cdot G(\mathbb{R})_+) \cap G(\mathbb{Q})$ in $G(\mathbb{A})$ is a subgroup of finite index of $SL_2(\mathbb{Z})$. We define $\mathscr{S}(S)$ to be the space of functions $f : G(\mathbb{A}) \to \mathbb{C}$ satisfying the following condition in addition to (A2–3):

$$f(\gamma u x c) = f(x) J_k(c,i)^{-1} \text{for all } z \in Z(\mathbb{A}),\ u \in S,\ c \in C_\infty \text{ and } \gamma \in G(\mathbb{Q}). \tag{A$'$1}$$

We can prove by a similar argument, the following result:

Corollary 3.7 *Let S be an open subgroup of* $GL_2(\widehat{\mathbb{Z}})$. *Then the intersection* $\Gamma = (S \cdot G(\mathbb{R})_+) \cap G(\mathbb{Q})$ *in* $G(\mathbb{A})$ *is a subgroup of finite index in* $SL_2(\mathbb{Z})$. *If* $\det S = \widehat{\mathbb{Z}}^\times$, *then* $\mathscr{S}(S) \cong S_k(\Gamma)$ *via* $f \mapsto f_*$.

Exercises

(1) Show that $\psi(z_\infty) = J_k(z_\infty, i)^{-1}$ implies (P): $\chi(-1) = (-1)^k$.
(2) Prove Corollary 3.7.

3.1.6 Hecke algebras

We would like to define Hecke operators acting on the space of cusp forms: $\mathscr{S}_k(N,\chi)$. We fix $\psi = |\ |_{\mathbb{A}}^{2-k}\chi^*$ for a Dirichlet character χ of $(\mathbb{Z}/N\mathbb{Z})^\times$. The operator is associated with a double coset $SaS = \{sas'|s, s' \in S\}$ for $S = S_0(N)$ and $a \in G(\mathbb{A}^{(\infty)})$.

We study the structure of SaS. The double coset SaS is compact since S is compact, and the right coset space SaS/S is compact, but tS is open since S is open. Thus SaS/S is a discrete and compact set. This shows that SaS/S is a finite set. We write $\deg(SaS) = |SaS/S|$. We can therefore decompose it as a disjoint union of right cosets: $SaS = \bigsqcup_{b\in R} bS$, where $R = R_a$ is a finite subset of $G(\mathbb{A}^{(\infty)})$.

First we define an action of individual cosets bS for $b \in R_a$. We know that $f(xu) = \chi_N(u)f(x)$ for $f \in \mathscr{S}_k(N,\chi)$ for the character $\chi_N : S \to \mathbb{C}^\times$ defined in 3.1.4. If we can extend the homomorphism χ_N to a subgroup $\Delta \subset G(\mathbb{A}^{(\infty)})$ containing S, then we can define an action on f of a coset bS for $b \in \Delta$ by $[bS]f(x) = \chi_N(b)^{-1}f(xb)$. This is well defined because

$$[(bs)S]f(x) = \chi_N(bs)^{-1}f(xbs) = \chi_N(b)^{-1}\chi_N(s)^{-1}\chi_N(s)f(xb) = [bS]f(x)$$

for any $s \in S$. However, $[bS]f$ may not be in $\mathscr{S}_k(N,\chi)$. Since

$$[bS]f(xs) = \chi_N(b)^{-1}f(xsb),$$

without knowing that s commutes with b or at least that b normalizes S, we cannot show that $[bS]f(xs) = \chi_N(s)[bS]f(x)$. In any case, this shows that if $b = z \in Z(\mathbb{A}^{(\infty)})$, then $[zS]f \in \mathscr{S}_k(N,\chi)$ is well defined. If χ_N is the identity character (equivalently if χ is the identity character), we can take Δ to be the total group $G(\mathbb{A}^{(\infty)})$. We will see later how to find a good Δ for non-trivial characters χ.

To compensate for the non-commutativity of b and $s \in S$, we create an average: Supposing $SaS \subset \Delta$, define

$$[SaS]f = \sum_{b\in R_a}[bS]f = \sum_{bS\in SaS/S}[bS]f. \tag{Av}$$

Then $[SaS]f(xs) = \sum_b \chi_N(b)^{-1}f(xsb)$. However $sb \in b'S$ for $b' \in R_a$. The association: $b \mapsto b'$ induces a permutation of R_a. Thus writing $sb = b's_b$ with $s_b \in S$, we see that

$$\begin{aligned}[SaS]f(xs) &= \sum_b \chi_N(b)^{-1} f(xsb) = \sum_b \chi_N(b)^{-1} f(xb's_b) \\ &= \sum_{b'} \chi_N(s^{-1}b's_b)^{-1}\chi_N(s_b) f(xb') = \chi_N(s)([SaS]f)(x).\end{aligned}$$

Then the condition (A1) can easily be verified for $[SaS]f$, because $b \in G(\mathbb{A}^{(\infty)})$ commutes with any $t \in C_\infty$.

We need to check (A2–3). We compute $([aS]f)_*$ for $aS \subset \Delta$ choosing $u_z \in G(\mathbb{R})_+$ with $u_z(i) = z$: Since $G(\mathbb{A}) = G(\mathbb{Q}) \cdot S \cdot G(\mathbb{R})_+$, we can write $a = \gamma s u$ with $\gamma \in G(\mathbb{Q})$, $s \in S$ and $u \in G(\mathbb{R})_+$. Then automatically, we see that $\gamma \in G(\mathbb{Q})_+$, because $1 = a_\infty = \gamma_\infty u_\infty$. We now compute

$$\begin{aligned}([aS]f)_*(z) &= \chi_N(a)^{-1} f(au_z) J_k(u_z, i) = \chi_N(a)^{-1} f(\gamma u u_z s) J_k(u_z, i) \quad (3.22)\\ &= \chi_N(a)^{-1}\chi_N(s) f(\gamma_\infty^{-1} u_z) J_k(u_z, i) z \\ &= \chi_N(a)^{-1}\chi_N(s) f(\gamma_\infty^{-1} u_z) J_k(\gamma^{-1} u_z, i) J_k(\gamma^{-1}, z)^{-1} \\ &= \chi_N(a)^{-1}\chi_N(s) \det(\gamma)^{(k/2)-1} f_* | \gamma^{-1}(z),\end{aligned}$$

where $g|\alpha(z) = \det(\alpha)^{k/2} g(\alpha(z)) j(\alpha, z)^{-k}$ for $g : \mathfrak{H} \to \mathbb{C}$ and $\alpha \in G(\mathbb{R})_+$. This shows (A2–3) for $([aS]f)_*$ and hence for $[SaS]f$.

We define

$$\begin{aligned}\Delta &= \Delta_0(N) \quad (3.23)\\ &= \left\{ \left(\begin{smallmatrix} a & b \\ c & d \end{smallmatrix}\right) \in M_2(\widehat{\mathbb{Z}}) \cap G(\mathbb{A}^{(\infty)}) | d_p \in \mathbb{Z}_p^\times \text{ for all } p|N \text{ and } c \in N\widehat{\mathbb{Z}} \right\}.\end{aligned}$$

Obviously the set Δ is stable under matrix multiplication and contains S, and hence $S\Delta S = \Delta$, but it is not a group. It is a semi-group, that is, Δ is stable under multiplication but the inverse of some elements of Δ may get out of Δ. This does not affect the above argument establishing the action of $[SaS]$ for $a \in \Delta$. We then define $\chi_N : \Delta \to \mathbb{C}$ by

$$\chi_N \left(\begin{smallmatrix} a & b \\ c & d \end{smallmatrix}\right) = \chi^{-1}(d \mod N\widehat{\mathbb{Z}}),$$

which plainly extends the homomorphism $\chi_N : S_0(N) \to \mathbb{C}^\times$.

We define $R(\Delta) = \mathbb{Z}[S\backslash\Delta/S]$ to be the set of formal linear combinations (with coefficients in $\mathbb{Z}$) of double cosets (SaS) in Δ. We would like to make $R(\Delta)$ into a ring. For that, we look at the set of formal linear combinations $\mathbb{Z}[\Delta/S]$ of right cosets tS in Δ. Since $SaStS$ is compact and S is open, $SaSbS/S$ is discrete compact and hence finite. Thus we define

$$(SaS)(tS) = \sum_{b \in R_a} btS = \sum_{xS \in SaSbS/S} m_x(xS),$$

where the multiplicity m_x is given by $\#\{b \in R_a | btS = xS\}$. We extend this action to the entire $R(\Delta)$ and $\mathbb{Z}[\Delta/S]$ by linearity. Then we see by definition that

$$(SaS)(ScS) = (SaS)\sum_{t\in R_c} tS = \sum_{b\in R_a}\sum_{t\in R_c} btS = \sum_{y\in S\backslash SaScS/S} m_y(SyS),$$

where $m_y = \#\{(b,t) \in R_a \times R_c | btS = yS\}$.

To make the above formula valid, we need to show that m_y is independent of $y \in SyS$. This can be shown as follows:

$$\begin{aligned} m_y &= \#\{(b,t) \in R_a \times R_c | btS = yS\} = \#\{b \in R_a | b^{-1}y \in ScS\} \\ &= \#\{b \in R_a | y^{-1}b \in Sc^{-1}S\} = \#\{b \in R_a | b \in ySc^{-1}S\} \\ &= \#\{b | bS \subset SaS \cap ySc^{-1}S\}/S = \#(SaS \cap ySc^{-1}S)/S. \end{aligned}$$

If $SyS = Sy'S$, then $y' = sys''$ and

$$\#(SaS \cap y'Sc^{-1}S)/S = \#(SaS \cap sySc^{-1}S)/S$$

which is equal to $\#(SaS \cap ySc^{-1}S)/S$ through multiplying s^{-1} from the left. This shows the independence of m_y on $y \in SyS$. Since $SbtS = SyS$ if and only if there exists a unique $w \in R_y$ such that $btS = wS$. This shows that

$$\begin{aligned} \#\{(b,t) \in R_a \times R_c | SbtS = SyS\} &= \sum_{w\in R_y} \#\{(b,t) \in R_a \times R_c | btS = wS\} \\ &= \deg(SyS) \cdot m_y, \end{aligned}$$

where $\deg(SyS) = \#(SyS/S)$. Then we have

$$(SaS)(SbS) = \sum_{ScS\in S\backslash SaSbS/S} m_{ScS}(ScS), \tag{3.24}$$

where

$$\deg(ScS)m_{ScS} = |\{(tS, sS) \in (SaS/S) \times (SbS/S) | StsS = ScS\}|$$

with $m_{ScS} \in \mathbb{Z}$.

The above arguments show that

$$(ScS)((SaS)(tS)) = ((ScS)(SaS))(tS)$$

and the associativity of the multiplication in $R(\Delta)$. The identity of $R(\Delta)$ is given by $S = S1S$.

Extend the association $(SaS) \mapsto [SaS] \in \mathrm{End}_{\mathbb{C}}(\mathcal{S}_k(N,\chi))$ to a linear map $R(\Delta) \to \mathrm{End}_{\mathbb{C}}(\mathcal{S}_k(N,\chi))$ by linearity. By definition, we see that

$$[SaS][SbS]f = [SaS]\left(\sum_{cS \in S\backslash SbS/S} [cS]f\right)$$
$$= \left(\sum_{cS \in S\backslash SbS/S} [(SaS)(cS)]f\right) = [(SaS)(SbS)]f.$$

This shows that the linear map: $R(\Delta) \to \mathrm{End}_{\mathbb{C}}(\mathcal{S}_k(N,\chi))$ is actually a ring homomorphism. The image of this map is written as $h_k(N,\chi;\mathbb{Z})$ and is called the Hecke algebra for $\mathcal{S}_k(N,\chi)$ with coefficients in $\mathbb{Z}$. In other words, $h_k(N,\chi;\mathbb{Z})$ is a subalgebra of $\mathrm{End}_{\mathbb{C}}(\mathcal{S}_k(N,\chi))$ generated over $\mathbb{Z}$ by $[SaS]$ for all double cosets $SaS \subset \Delta$. We can also define for any subalgebra $A \subset \mathbb{C}$, the Hecke algebra $h_k(N,\chi;A)$ with coefficients in A as a subalgebra of $\mathrm{End}_{\mathbb{C}}(\mathcal{S}_k(N,\chi))$ generated over A by operators $[SaS]$ for all $a \in \Delta$.

For a general open subgroup $S \subset GL_2(\widehat{\mathbb{Z}})$, we can think of the Hecke operator $[SaS]$ acting on $\mathcal{S}_k(S)$ in the same manner as above. In this case, we do not need to restrict ourselves to double cosets in Δ.

Although we defined $h_k(N,\chi;\mathbb{Z})$ as a subalgebra of $\mathrm{End}_{\mathbb{C}}(\mathcal{S}_k(N,\chi)) \cong M_d(\mathbb{C})$ for $d = \dim_{\mathbb{C}}(\mathcal{S}_k(N,\chi))$, we do not know at this moment if the algebra is finitely generated or not. By definition, we have a surjective A-algebra homomorphism

$$h_k(N,\chi;\mathbb{Z}) \otimes_{\mathbb{Z}} A \twoheadrightarrow h_k(N,\chi;A) \tag{3.25}$$

given by $[SaS] \otimes b \mapsto b[SaS]$ for $b \in A$ and $a \in \Delta$. We ask whether this is an isomorphism. To answer these questions, we start by identifying a canonical set of generators of $R(\Delta)$. For a positive integer n, we first put

$$\mathbb{T}(n) = \left\{x \in \Delta | (\det x)\widehat{\mathbb{Z}} = n\widehat{\mathbb{Z}}\right\}.$$

Then obviously, $S\mathbb{T}(n)S = \mathbb{T}(n)$. Thus we can think of the set of double cosets $S\backslash\mathbb{T}(n)/S$, which is a finite set (see Exercise 2). We then put

$$T(n) = \sum_{SaS \in S\backslash\mathbb{T}(n)/S} [SaS] \in h_k(N,\chi;A).$$

THEOREM 3.8 *The algebra $h_k(N,\chi;A)$ is commutative with identity $T(1)$. As long as A contains the values of χ, the algebra $h_k(N,\chi;A)$ is generated by $T(n)$ for $n = 1,2,\ldots$ over A as an A-module.*

Here we prove a weaker assertion that $h_k(N,\chi;A)$ is commutative and is generated by $T(p)$ for primes p as a ring, which follows from the following lemma, because $[SpS] = \psi(p) = p^{k-2}\chi(p) \in A$ in the Hecke algebra. We will later prove the full assertion in 3.1.9 after analyzing how Fourier coefficients behave under Hecke operators.

LEMMA 3.9 *The ring $R(\Delta)$ is commutative and is generated (as a ring) by $T(p)$ for all primes p together with $S(p) = SpS$ for primes $p \nmid N$.*

This is proved in [IAT] Chapter 3 in a classical way, but we shall give a sketch of an adelic proof.

Proof Write $S = S_0(N)$. Let $\Delta_p = \{x_p \in GL_2(\mathbb{Q}_p)|x \in \Delta\} = \Delta \cap GL_2(\mathbb{Q}_p)$ in $G(\mathbb{A}^{(\infty)})$ and $S_p = S \cap GL_2(\mathbb{Q}_p)$ in $G(\mathbb{A}^{(\infty)})$. Then it is easy to see that $S = \prod_p S_p$, and $\Delta = \prod'_p \Delta_p = (\prod_p \Delta_p) \cap G(\mathbb{A}^{(\infty)})$. This shows that

$$\bigotimes_p R(\Delta_p) \cong R(\Delta),$$

where $R(\Delta_p) = \mathbb{Z}[S_p\backslash\Delta_p/S_p]$, and the identification is given by $\otimes_p(S_pa_pS_p) \mapsto \prod_p(Sa_pS)$ for $a \in \Delta$. The product $\prod_p(Sa_pS)$ is well defined because $S_pa_pS_p$ is equal to the multiplicative identity S_p of $R(\Delta_p)$ except for finitely many primes p. Thus we only need to show that $R(\Delta_p)$ is generated by $S(p) = S_ppS_p$ and $T(p) = \{x_p|\det x_p\mathbb{Z}_p = p\mathbb{Z}_p\}$ if $p \nmid N$ and by $T(p)$ if $p|N$. We write $T(p^n) = \{x_p|(\det x_p)\mathbb{Z}_p = p^n\mathbb{Z}_p\} \subset GL_2(\mathbb{Q}_p)$.

First suppose that $p \nmid N$. Then $S_p = GL_2(\mathbb{Z}_p)$. Let $GL_2(\mathbb{Q}_p)$ act on the column vector space $V = \mathbb{Q}_p^2$. Then $S_p = \{x \in GL_2(\mathbb{Q}_p)|xL = L\}$ for $L = \mathbb{Z}_p^2 \subset V$ and $\Delta_p = \{x \in GL_2(\mathbb{Q}_p)|xL \subset L\}$. If $\Delta_p = \bigsqcup_b bS_p$, then $bS_pL = bL$. If $bL = cL$, then $c^{-1}bL = L$ and hence $c^{-1}b \in S_p \Leftrightarrow cS_p = bS_p$. This shows that the correspondence $bS \mapsto bL$ is an injection from right cosets of S_p in Δ_p into the set of additive subgroups $L' \subset L$ of finite index. If L' is such a subgroup, we see that $\mathbb{Q}_pL' = V$ because $p^rL \subset L'$ for some $r > 0$. Thus L' has a base (x,y), that is, $L' = \mathbb{Z}_px + \mathbb{Z}_py$, because $\mathbb{Z}_p$ is a DVR (in particular, PID). Writing $g = (x,y) \in GL_2(\mathbb{Q}_p)$, we have $gL = L'$. Thus we have

$$\Delta_p/S_p \cong \{L' \colon \text{additive subgroups of } L|[L:L'] < \infty\}.$$

Let $\alpha = \left(\begin{smallmatrix} p & 0 \\ 0 & 1 \end{smallmatrix}\right)$. Since L is generated by two elements, the minimal number of generators of L/L' is at most two. Suppose that L/L' is cyclic of order p^r. Then we find $x \in L$ such that $x \bmod L'$ generates L/L'. Then we can choose $y \in L'$ such that $L = \mathbb{Z}_px + \mathbb{Z}_py$. This shows that

$L' = \mathbb{Z}_p p^r x + \mathbb{Z}_p y$ and $L' = \beta\alpha^r L$ for $\beta = (x, y) \in S_p$. Since any $b \in S_p$ induces a group automorphism of L, L/bL' is again cyclic of order p^r. This shows that $b\beta\alpha^r = \beta'\alpha^r$ for $\beta' \in S_p$, that is,

$$a \in S_p\alpha^r S_p \iff L/aL \text{ is cyclic of order } p^r.$$

Now suppose that L/L' is non-cyclic and thus is generated by x mod L' and y mod L'. We know (by Nakayama's lemma: Chapter 2 Lemma 2.3 in the text or [CRT] Theorem 2.2) that $L = \mathbb{Z}_p x + \mathbb{Z}_p y$. By interchanging x and y if necessary, we may assume that the order of x mod L' is p^{r+s} and that of y mod L' is p^s for positive integers r, s. Then $L/p^{-s}L'$ is cyclic of order p^r, and $L' = p^s\alpha^r\beta$ for some $\beta \in S_p$. Thus we have found that

$$a \in S_p p^s \alpha^r S_p \iff L/aL \text{ is isomorphic to } (\mathbb{Z}/p^{r+s}\mathbb{Z}) \oplus (\mathbb{Z}/p^s\mathbb{Z}).$$

Combining what we have proved, we get

$$T(p^n) = \bigsqcup_{r \geq s; r+s=n} S_p \begin{pmatrix} p^r & 0 \\ 0 & p^s \end{pmatrix} S_p = S_p\alpha^n S_p \sqcup \bigsqcup_{r' \geq s'; r'+s'=n-2} pS_p \begin{pmatrix} p^{r'} & 0 \\ 0 & p^{s'} \end{pmatrix} S_p;$$

$$\Delta_p = M_2(\mathbb{Z}_p) \cap GL_2(\mathbb{Q}_p) = \bigsqcup_{r \geq s \geq 0} S_p \begin{pmatrix} p^r & 0 \\ 0 & p^s \end{pmatrix} S_p = \bigsqcup_{r \geq 0, s \geq 0} S_p p^s \alpha^r S_p. \quad (3.26)$$

From this we conclude:

$$(S_p\alpha S_p) = T(p) \text{ and } (S_p\alpha^r S_p) = T(p^r) - S(p)T(p^{r-2}) \quad (r \geq 2). \quad (3.27)$$

As a set, we have

$$S_p\alpha^n S_p \subset \overbrace{S_p\alpha S_p\alpha S_p \cdots S_p\alpha S_p}^{n}.$$

We claim that in the product $(S_p\alpha S_p)^n$ the multiplicity of $S_p\alpha^n S_p$ is equal to 1. To show this, it is enough to show that the multiplicity m of $S_p\alpha^{r+s}S_p$ in the product $(S_p\alpha^r S_p)(S_p\alpha^s S_p)$ is equal to 1. By the definition of the multiplication, we get

$$m = \#\{(t, b) \in R_{\alpha^r} \times R_{\alpha^s} | tbS_p = \alpha^{r+s}S_p\}.$$

We have a bijection

$$R_{\alpha^r} \leftrightarrow \{X \subset L/p^r L = (\mathbb{Z}/p^r\mathbb{Z})^2 | X \text{ is cyclic of order } p^r\}$$

via $b \mapsto \mathrm{Ker}(b : L/p^r L \to L/p^r L)$. Then it is easy to see that

$$\mathrm{Ker}(tb) = 0 \oplus (\mathbb{Z}/p^{r+s}\mathbb{Z}) \iff$$
$$\mathrm{Ker}(t) = 0 \oplus (\mathbb{Z}/p^r\mathbb{Z}) \text{ and } \mathrm{Ker}(b) = 0 \oplus (\mathbb{Z}/p^s\mathbb{Z}),$$

and hence $m = 1$. This shows that

$$T(p)^n = S_p\alpha^n S_p + \sum_{r\geq s>0, r+s=n} c_{r,s} S_p p^s \alpha^{r-s} S_p$$
$$= S_p\alpha^n S_p + S(p) \sum_{r\geq s>0, r+s=n} c_{r,s} S_p p^{s-1} \alpha^{r-s} S_p$$

with $c_{r,s} \in \mathbb{Z}$. Then by induction on n, $S_p\alpha^n S_p$ is a polynomial of $T(p)$ and $S(p)$ because $T(p)$ and $S(p)$ commute with each other. Hence $R(\Delta_p)$ is commutative and generated by $T(p)$ and $S(p)$.

Now suppose that $p|N$. If $\gamma = \left(\begin{smallmatrix} a & b \\ p^e c & d \end{smallmatrix}\right) \in T(p^r)$ with $d \in \mathbb{Z}_p^\times$ and $e > 0$, then $ad - p^e bc = \det\gamma \equiv 0 \mod p^r$. We compute

$$\begin{pmatrix} p^{-r} & p^{-r}d^{-1}b \\ 0 & 1 \end{pmatrix} \begin{pmatrix} a & b \\ p^e c & d \end{pmatrix} = \begin{pmatrix} p^{-r}a & 0 \\ p^e c & d \end{pmatrix},$$

which has p-adic unit determinant $p^{-r}ad \in \mathbb{Z}_p^\times$. This shows that

$$\gamma \in \begin{pmatrix} p^r & d^{-1}b \\ 0 & 1 \end{pmatrix} S_p.$$

From this, we conclude that

$$T(p^r) = \bigsqcup_{u \bmod p^r\mathbb{Z}_p} \begin{pmatrix} p^r & u \\ 0 & 1 \end{pmatrix} S_p, \tag{3.28}$$
$$T(p^r) = T(p)^r \quad \text{and} \quad T(p^r) = S_p\alpha^r S_p.$$

This proves the desired assertion in this case. □

We would like to compute $\deg T(p^n) = |\mathbb{T}(p^n)/S|$. Since we already know the number when $p|N$ (3.28), we suppose that $p \nmid N$. Then by the argument in the proof, we have

$$\deg S_p\alpha^n S_p = \#\left\{X \subset (\mathbb{Z}/p^n\mathbb{Z})^2 | X \text{ is cyclic of order } p^n\right\}$$
$$= \frac{\#\{x \in (\mathbb{Z}/p^n\mathbb{Z})^2 | ord(x) = p^n\}}{\#(\mathbb{Z}/p^n\mathbb{Z})^\times} = \frac{p^{2n} - p^{2n-2}}{p^n - p^{n-1}} = p^n\left(1 + \frac{1}{p}\right).$$

Then by induction on n using (3.27), we conclude that

$$\deg S_p p^s \alpha^r S_p = p^r\left(1 + \frac{1}{p}\right) \quad \text{and} \quad \deg T(p^n) = 1 + p + \cdots + p^n \tag{3.29}$$

if $p \nmid N$. We see easily that $aS \neq bS$ for two distinct elements a, b in the following set

$$R_r = \left\{\begin{pmatrix} p^m & u \\ 0 & p^n \end{pmatrix} | m \geq n,\ m + n = r,\ u = 0, 1, 2, \ldots, p^m - 1\right\}.$$

Since $|R_r| = \deg T(p^r)$, we conclude that

$$T(p^r) = \bigsqcup_{a \in R_r} aS. \tag{3.30}$$

From the argument proving the theorem, we know that $T(m)T(n) = T(mn)$ if m and n are co-prime. More generally we have (cf. [MFM] Chapter 4)

$$T(m)T(n) = \sum_{d|m,d|n} dS(d)T\left(\frac{mn}{d^2}\right), \tag{3.31}$$

where we agree to put $S(d) = 0$ if d is not prime to N. We will see this relation later in a slightly different way.

By (3.30), (3.28) and (3.22), we can state the effect of $T(n)$ on the classical modular form f_* associated with $f \in \mathcal{S}_k(N,\chi)$:

$$(T(p)f)_* = p^{(k/2)-1}\left\{\sum_{u=0}^{p-1} f_*|_k \left(\begin{smallmatrix} 1 & u \\ 0 & p \end{smallmatrix}\right) + f_*|_k \left(\begin{smallmatrix} p & 0 \\ 0 & 1 \end{smallmatrix}\right)\right\} \quad \text{and} \tag{3.32}$$

$$(S(p)f)_* = p^{k-2}\chi(p)f_* \quad \text{if } p \nmid N;$$

$$(T(p)f)_* = p^{(k/2)-1}\left\{\sum_{u=0}^{p-1} f_*|_k \left(\begin{smallmatrix} 1 & u \\ 0 & p \end{smallmatrix}\right)\right\} \quad \text{if } p|N. \tag{3.33}$$

Exercises

(1) Define $\deg : R(\Delta) \to \mathbb{Z}$ by

$$\deg\left(\sum_{SaS} m_{SaS}(SaS)\right) = \sum_{SaS} m_{SaS} \deg(SaS)$$

with $\deg(SaS) = |SaS/S|$. Show that $\deg : R(\Delta) \to \mathbb{Z}$ is a ring homomorphism.

(2) Show that $\mathbb{T}(n)$ is a compact set, and using this, show further that $S\backslash\mathbb{T}(n)/S$ is a finite set.

3.1.7 Fourier expansion

Each cusp form $f \in S_k(\Gamma_1(N))$ has a Fourier expansion of the form

$$f(z) = \sum_{n=1}^{\infty} a(n;f)q^n$$

for $q = \exp(2\pi iz)$ with $z \in \mathfrak{H}$. This can be seen as follows: Let $\alpha = \left(\begin{smallmatrix} 1 & 1 \\ 0 & 1 \end{smallmatrix}\right) \in \Gamma_1(N)$. Then $\alpha(z) = z + 1$ for $z \in \mathfrak{H}$. Thus $f(z+1) = f(z)$. From calculus, we know that $\mathbb{C}/\mathbb{Z} \cong \mathbb{C}^\times$ via $z \mapsto \exp(2\pi iz) = q$. Thus we may regard $f(z)$ as a holomorphic function of q. Then its Laurent expansion around 0 gives:

$$f(q) = \sum_{n\in\mathbb{Z}} a(n; f)q^n.$$

By (A3), we conclude that $\lim_{q\to 0} f(q) = 0$ and hence $a(n; f) = 0$ if $n \le 0$.

We would like to relate this expansion to an expansion of adelic modular form $f^* : G(\mathbb{A}) \to \mathbb{C}$. Namely, we compute $f^*\left(\begin{smallmatrix} y & x \\ 0 & 1\end{smallmatrix}\right)$. Note that

$$\widehat{\mathbb{Z}} \cap (\mathbb{A}^{(\infty)})^\times = \bigsqcup_{n=1}^{\infty} n\widehat{\mathbb{Z}}^\times.$$

Thus we may extend the coefficients $n \mapsto a(n; f)$ to $\mathbb{A}^\times$ so that $a(y; f) = a(n; f)$ if $y \in n\widehat{\mathbb{Z}}^\times$ for some $n > 0$, otherwise we put $a(y; f) = 0$.

We need to introduce one more fact. For each $x \in \mathbb{A}$, we expand x_p into a p-adic expansion $x_p = \sum_n c_n p^n$ with $0 \le c_n \le p-1$. We then define $[x_p] = \sum_{n<0} c_n p^n$, which is a finite sum and hence is a rational number with a p-power denominator. Finally we put $[x] = -\sum_p [x_p] + x_\infty \in \mathbb{R}$, which is again a finite sum because $x_p \in \mathbb{Z}_p$ for almost all p. We then define $\mathbf{e} : \mathbb{A} \to \mathbb{C}^\times$ by $\mathbf{e}(x) = \exp(2\pi i[x])$. If $x \in \mathbb{Q}$, then $[x_p] - x \in \mathbb{Z}_p$ and $\sum_{\ell\ne p}[x_\ell] \in \mathbb{Z}_p$. This shows that $[x] = -x + \sum_p [x_p] \in \widehat{\mathbb{Z}} \cap \mathbb{Q} = \mathbb{Z}$. Thus we have $\mathbf{e}(x) = \exp(2\pi i[x]) = 1$, and $\mathbf{e} : \mathbb{A}/\mathbb{Q} \to \mathbb{C}^\times$ is a continuous character.

THEOREM 3.10 *We have for $f \in S_k(\Gamma_0(N), \chi)$,*

$$f^*\left(\begin{smallmatrix} y & x \\ 0 & 1\end{smallmatrix}\right) = |y|_{\mathbb{A}} \sum_{0<n\in\mathbb{Q}} a(ny; f)\exp(-2\pi n y_\infty)\mathbf{e}(nx)$$

for all $x \in \mathbb{A}$ and $y \in \mathbb{A}^\times$ with $y_\infty > 0$.

Proof Let $a = \begin{pmatrix} y^{(\infty)} & x^{(\infty)} \\ 0 & 1\end{pmatrix} \in G(\mathbb{A}^{(\infty)})$. Then $\left(\begin{smallmatrix} y & x \\ 0 & 1\end{smallmatrix}\right) = a\left(\begin{smallmatrix} y_\infty & x_\infty \\ 0 & 1\end{smallmatrix}\right)$. Thus

$$f^*\left(\begin{smallmatrix} y & x \\ 0 & 1\end{smallmatrix}\right) = f^*\left(a\left(\begin{smallmatrix} y_\infty & x_\infty \\ 0 & 1\end{smallmatrix}\right)\right) = f^*\left(\left(\begin{smallmatrix} y_\infty & x_\infty \\ 0 & 1\end{smallmatrix}\right)a\right) = [aS]f^*\left(\begin{smallmatrix} y_\infty & x_\infty \\ 0 & 1\end{smallmatrix}\right).$$

We claim that $a = \gamma s u$ with $\gamma = \left(\begin{smallmatrix} \eta & \xi \\ 0 & 1\end{smallmatrix}\right) \in G(\mathbb{Q})_+$, $s = \left(\begin{smallmatrix} w & t \\ 0 & 1\end{smallmatrix}\right) \in S = S_0(N)$ and $u \in G(\mathbb{R})_+$. Then $u = \gamma^{-1}$ in $G(\mathbb{R})_+$.

The claim can be shown as follows: Since $\mathbb{A}^\times = \mathbb{Q}^\times\widehat{\mathbb{Z}}^\times\mathbb{R}_+^\times$, $y^{(\infty)} = \eta w w_\infty$ with $\eta \in \mathbb{Q}^\times$, $w \in \widehat{\mathbb{Z}}^\times$ and $w_\infty \in \mathbb{R}_+^\times$. Then $\begin{pmatrix} \eta^{-1} & 0 \\ 0 & 1\end{pmatrix} a = \begin{pmatrix} w w_\infty & \eta^{-1}x^{(\infty)} \\ 0 & 1\end{pmatrix}$ with $v = \eta^{-1}x^{(\infty)} \in \mathbb{A}$. Since $\mathbb{A} = \mathbb{Q} + \widehat{\mathbb{Z}} + \mathbb{R}$ (Exercise 1), $v = \xi' + t + t_\infty$ with $\xi' \in \mathbb{Q}$, $t \in \widehat{\mathbb{Z}}$ and $t_\infty \in \mathbb{R}$. Then $s = \left(\begin{smallmatrix} w & t \\ 0 & 1\end{smallmatrix}\right) \in S = S_0(N)$ and $\gamma = \left(\begin{smallmatrix} \eta & 0 \\ 0 & 1\end{smallmatrix}\right)\left(\begin{smallmatrix} 1 & \xi' \\ 0 & 1\end{smallmatrix}\right)$ do the job.

Writing $z = x_\infty + iy_\infty \in \mathfrak{H}$, we see from (3.22) that

$$\begin{aligned} f^*\left(\begin{smallmatrix} y & x \\ 0 & 1 \end{smallmatrix}\right) &= [aS]f^*\left(\begin{smallmatrix} y_\infty & x_\infty \\ 0 & 1 \end{smallmatrix}\right) = J_k(\left(\begin{smallmatrix} y & x \\ 0 & 1 \end{smallmatrix}\right), z)^{-1}\det(\gamma)^{(k/2)-1} f|\gamma^{-1}(z) \quad (3.34)\\ &= \eta^{-1} y_\infty f(\eta^{-1}z - \eta^{-1}\xi)\\ &= \eta^{-1} y_\infty \sum_{m=1}^{\infty} a(m;f)\exp(2\pi i\eta^{-1}mz)\exp(-2\pi i\eta^{-1}m\xi). \end{aligned}$$

Here we used the fact that $J_k(\left(\begin{smallmatrix} y & x \\ 0 & 1 \end{smallmatrix}\right), z) = y_\infty^{-1}$, and the convergence of the last summation follows from the convergence of the Fourier expansion of f, because $|\exp(-2\pi i\eta^{-1}m\xi)| = 1$.

Make a variable change: $0 < n = \eta^{-1}m \in \mathbb{Q}^\times$ in the above summation (3.34). Then we have

$$\exp(2\pi i\eta^{-1}mz)\exp(-2\pi i\eta^{-1}\xi) = \exp(-2\pi n y_\infty)\mathbf{e}(nx_\infty)\exp(-2\pi i\eta^{-1}m\xi).$$

From $a = \gamma s u$, writing $s = \left(\begin{smallmatrix} w & t \\ 0 & 1 \end{smallmatrix}\right)$ with $w \in \widehat{\mathbb{Z}}^\times$ and $t \in \widehat{\mathbb{Z}}$, we see that $\eta^{(\infty)}w = y^{(\infty)}$, and thus $a(m;f) = a(ny;f)$ and $|y|_\mathbb{A} = |y_\infty\eta^{(\infty)}w|_\mathbb{A} = y_\infty\eta^{-1}$. Here we use the fact that

$$1 = |\eta|_\mathbb{A} = |\eta^{(\infty)}|_\mathbb{A}|\eta_\infty| = |\eta^{(\infty)}|_\mathbb{A}\eta.$$

We claim that

$$\exp(-2\pi i\eta^{-1}m\xi) = \mathbf{e}(nx^{(\infty)}).$$

Then by definition, the desired assertion follows from the claim.

We now prove the claim: Again by $a = \gamma s u$, we have $x^{(\infty)} = \eta^{(\infty)}t + \xi^{(\infty)}$ and hence

$$\mathbf{e}(nx^{(\infty)}) = \mathbf{e}(n(\eta^{(\infty)}t + \xi^{(\infty)})) = \mathbf{e}(\eta^{-1}m(\eta^{(\infty)}t + \xi^{(\infty)})) = \mathbf{e}(mt)\mathbf{e}(\eta^{-1}m\xi^{(\infty)}).$$

Since $\mathbf{e}(\eta^{-1}m\xi) = 1$, $\mathbf{e}(\eta^{-1}m\xi^{(\infty)}) = \mathbf{e}(-\eta^{-1}m\xi_\infty) = \exp(-2\pi i\eta^{-1}m\xi)$. Thus we need to show that $\mathbf{e}(mt) = 1$, which follows from the fact that $mt \in \widehat{\mathbb{Z}}$. □

Corollary 3.11 (Hecke, 1938) *Let $f \in S_k(\Gamma_0(N), \chi)$. Then we have, for two positive integers m and n,*

$$a(m;T(n)f) = \sum_{d|m,d|n} d^{k-1}\chi(d)a(\frac{mn}{d^2};f).$$

Here we have used the convention that $\chi(d) = 0$ if d has a non-trivial common factor with N.

Proof This just follows from (3.31) or an easy computation using the explicit decomposition of $T(p^r)$ (for a prime p) given in (3.30) and (3.28). For general n, we remark (by our definition of $T(n)$) that $T(n) = \prod_{p|n} T(p^{e(p)})$ if $n = \prod_{p|n} p^{e(p)}$ is the prime decomposition of n (see [IAT] Chapter 3).

Here we describe the computation in the simplest case of $T(p)$. We need to show

$$a(m; T(p)f) = \begin{cases} a(mp; f) + p^{k-1}\chi(p)a(\frac{m}{p}; f) & \text{if } p \nmid N; \\ a(mp; f) & \text{if } p|N, \end{cases}$$

where we used the convention that $a(\xi; f) = 0$ if the idele $\xi \notin \widehat{\mathbb{Z}}$. Write $S = S_0(N)$.

We first take care of the case when $p \nmid N$. We have from (3.30) that

$$T(p) = \left(\begin{smallmatrix} 1 & 0 \\ 0 & p_p \end{smallmatrix}\right) S_p \sqcup \bigsqcup_{u=0}^{p-1} \left(\begin{smallmatrix} p_p & u \\ 0 & 1 \end{smallmatrix}\right) S_p.$$

Then

$$\begin{aligned} T(p)f^* \left(\begin{smallmatrix} y & x \\ 0 & 1 \end{smallmatrix}\right) &= f^*\left(p\left(\begin{smallmatrix} y & x \\ 0 & 1 \end{smallmatrix}\right)\left(\begin{smallmatrix} p_p^{-1} & 0 \\ 0 & 1 \end{smallmatrix}\right)\right) + \sum_u f^*\left(\begin{smallmatrix} p_p y & yu+x \\ 0 & 1 \end{smallmatrix}\right) \\ &= \psi(p_p) f^*\left(\begin{smallmatrix} \frac{y}{p_p} & x \\ 0 & 1 \end{smallmatrix}\right) + \sum_u f^*\left(\begin{smallmatrix} p_p y & yu+x \\ 0 & 1 \end{smallmatrix}\right). \end{aligned}$$

This shows that

$$a(y; T(p)f^*) = |p_p|_{\mathbb{A}}^{-1}\psi(p_p)a\left(\frac{y}{p}; f\right) + |p_p|_{\mathbb{A}}\left(\sum_u \mathbf{e}(yu)\right)a(py; f).$$

Since $a(y; T(p)f^*) \neq 0$ only if $y \in \widehat{\mathbb{Z}}$, $\mathbf{e}(yu) = 1$ for all u. Since $\psi(p_p) = |p_p|_{\mathbb{A}}^{2-k}\chi(p)$, this shows that

$$a(y; T(p)f^*) = p^{k-1}\chi(p)a\left(\frac{y}{p}; f\right) + a(py; f),$$

which is equivalent to the desired assertion.

When $p|N$, we only have the second term (see (3.28)):

$$a(y; T(p)f^*) = |p_p|_{\mathbb{A}}\left(\sum_u \mathbf{e}(yu)\right)a(py; f) = a(py; f),$$

which finishes the proof. □

Exercise

(1) Show that $\mathbb{A} = \mathbb{Q} + \widehat{\mathbb{Z}} + \mathbb{R}$. Hint: Use $[x] = x_\infty - \sum_p [x_p]$ in the definition of $\mathbf{e} : \mathbb{A}/\mathbb{Q} \to \mathbb{C}^\times$.

3.1.8 Rationality of modular forms

We state several results on the rationality of the space of modular forms. Some of them are profound, and others are elementary.

Let $\mathbb{Z}[\chi]$ be the subring of $\mathbb{C}$ generated over $\mathbb{Z}$ by the values of the Dirichlet character χ. We define for a $\mathbb{Z}[\chi]$-subalgebra A of $\mathbb{C}$,

$$S_k(\Gamma_0(N), \chi; A) = \{f \in S_k(\Gamma_0(N), \chi) | a(n; f) \in A \text{ for all } n \in \mathbb{N}.\}.$$

More generally, for a subgroup Γ with $\Gamma_1(N) \subset \Gamma \subset \Gamma_0(N)$, we put

$$S_k(\Gamma; A) = \{f \in S_k(\Gamma) | a(n; f) \in A \text{ for all } n \in \mathbb{N}\} \tag{3.35}$$

$$S_k(\Gamma, \chi; A) = \{f \in S_k(\Gamma, \chi) | a(n; f) \in A \text{ for all } n \in \mathbb{N}\}. \tag{3.36}$$

Here $S_k(\Gamma, \chi)$ is a subspace of $S_k(\Gamma_1(N))$ on which $\Gamma/\Gamma_1(N) \subset (\mathbb{Z}/N\mathbb{Z})^\times$ acts by the character χ. We have a natural A-linear map:

$$S_k(\Gamma; \mathbb{Z}) \otimes_{\mathbb{Z}} A \to S_k(\Gamma; A)$$

given by $f \otimes a \mapsto af$. We first note the integrality theorem:

THEOREM 3.12 (SHIMURA, DELIGNE–RAPOPORT, KATZ) *Take a congruence subgroup Γ with $\Gamma_1(N) \subset \Gamma \subset \Gamma_0(N)$ for an integer $N \geq 1$. Then, for each subalgebra A of $\mathbb{C}$, we have*

$$S_k(\Gamma; \mathbb{Z}) \otimes_{\mathbb{Z}} A \cong S_k(\Gamma; A)$$

by the above natural map. Similarly

$$S_k(\Gamma, \chi; \mathbb{Z}[\chi]) \otimes_{\mathbb{Z}[\chi]} A \cong S_k(\Gamma, \chi; A)$$

if A contains $\mathbb{Z}[\chi]$.

For an elementary proof with some assumption on the level and the weight, see [LFE] Chapter 5. For the proof without any assumptions, see [H86a] Section 1 and [H88a] Section 4. Another proof based on the solution of moduli problems of elliptic curves can be found in [GMF] Chapter III.

By the above integrality theorem, we may define $S_k(\Gamma; A)$ by $S_k(\Gamma, \mathbb{Z}) \otimes_{\mathbb{Z}} A$ for any algebra A (not necessarily in $\mathbb{C}$) and $S_k(\Gamma, \chi; A)$ by $S_k(\Gamma, \chi, \mathbb{Z}[\chi]) \otimes_{\mathbb{Z}[\chi]} A$ for any $\mathbb{Z}[\chi]$-algebra A.

We explain briefly, using the solution to the moduli problem of elliptic curves, how the above integrality result is obtained: We write $\mathscr{V}$ for the localizaion of $\mathbb{Z}$ at a prime $p \geq 5$ prime to N; thus, we have $\mathscr{V} = \mathbb{Q} \cap \mathbb{Z}_p$ inside $\mathbb{Q}_p$. We assume that A is a $\mathscr{V}$-algebra, and we would like to prove the slightly weaker assertion:

$$S_k(\Gamma_1(N), \mathscr{V}) \otimes_{\mathscr{V}} A \cong S_k(\Gamma_1(N); A) \text{ if } k \geq 2 \text{ and } p \nmid N. \qquad (3.37)$$

Proof For any given $\mathscr{V}$-algebra A, consider a pair (E, ω) of an elliptic curve $E_{/A}$ and a differential $\omega \in H^0(E, \Omega_{E/A})$, such that $0 \neq \omega_x \in \Omega_{E/A} \otimes k(x)$ for any geometric point $x \in E$ (such a differential will be called *nowhere vanishing*). Write $[0]$ for the divisor on E given by the 0-section $0 : Spec(A) \hookrightarrow E$. Then we can find global sections $x \in H^0(E, \mathscr{L}(-2[0]))$ and $y \in H^0(E, \mathscr{L}(-3[0]))$ so that $(1, x)$ is a base of $H^0(E, \mathscr{L}(-2[0]))$ and $(1, x, y)$ is a base of $H^0(E, \mathscr{L}(-3[0]))$. We can normalize (x, y) uniquely by translation (since 6 is invertible in A) so that $y^2 = 4x^3 - g_2 x - g_3$ and $\omega = \frac{dx}{y}$ for a unique choice of $g_2(E), g_3(E) \in A$. Since E is smooth, $\Delta(E) = (g_2(E))^3 - 27(g_3(E))^2$ is in $A^\times$. This shows that the functor from the category of $\mathscr{V}$-algebras to sets given by

$$\mathscr{P}(A) = \left\{(E, \omega)_{/A}\right\} / \cong$$

is represented by an affine scheme $Spec(\mathscr{A})$ for $\mathscr{A} = \mathscr{V}[g_2, g_3, \frac{1}{\Delta}]$, where g_2 and g_3 are variables and $\Delta = (g_2)^3 - 27(g_3)^2$. In other words, we have the universal elliptic curve $(\mathbb{E}, \boldsymbol{\omega} = \frac{dx}{y})_{/\mathscr{A}}$ defined by the equation $y^2 = 4x^3 - g_2 - g_3 \in \mathscr{A}[x, y]$, and for each $(E, \omega) \in \mathscr{P}(A)$, there is a unique $\mathscr{V}$-algebra homomorphism $\phi : \mathscr{A} \to A$ such that $\phi(g_k) = g_k(E)$, or more precisely, $(\mathbb{E}, \boldsymbol{\omega}) \times_{Spec(\mathscr{A}), \phi} A \cong (E, \omega)$. See Subsection 4.1.3 in Chapter 4 for more details of representability of a functor, and see [AME] II.2.2 (page 67) for details of computation about g_2 and g_3.

The action $\omega \mapsto \lambda\omega$ of $\lambda \in \mathbf{G}_m(A) = A^\times$ gives rise to an action of $\mathbf{G}_m$ on $\mathscr{A}$ such that $g_k(E, \lambda\omega) = \lambda^{-2k} g_k(E, \omega)$. Thus $\mathscr{A}$ is a graded algebra $\mathscr{A} = \oplus_j \mathscr{A}_k$ with $\mathscr{A}_{2j+1} = 0$, where $\mathbf{G}_m$ acts on $\mathscr{A}_k$ by the character $\lambda \mapsto \lambda^{-k}$ on $\mathscr{A}_k$. We define

$$\overline{\mathscr{A}} = \mathscr{A} \cap \mathscr{V}[g_2, g_3] = \bigoplus_{k \geq 0} \overline{\mathscr{A}}_k.$$

Then $Proj(\overline{\mathscr{A}})[\frac{1}{\Delta}]$ affine j-line $\mathbb{A}^1(j)$ which is the coarse moduli scheme classifying elliptic curves (without differential ω).

We now consider the following functor of level N:

$$\mathscr{P}_N(A) = \{(E, \omega, \phi : \mathbb{Z}/N\mathbb{Z} \hookrightarrow E)\} / \cong,$$

where ϕ is a closed immersion of the constant group scheme $\mathbb{Z}/N\mathbb{Z}$. The finite group $(\mathbb{Z}/N\mathbb{Z})^\times$ acts naturally on $\mathscr{P}_N$ via inner multiplication of ϕ by an element of $(\mathbb{Z}/N\mathbb{Z})^\times$. This is an étale covering of $\mathscr{P}$ and hence, it is representable by an affine ring $\mathscr{A}^1_N$ étale finite over $\mathscr{A}$. Similarly, we can think of

$$\mathscr{P}^0_N(A) = \{(E, \omega, C \subset E[N])\} / \cong,$$

where C is a cyclic subgroup scheme of order N in the kernel $E[N] = \mathrm{Ker}(x \mapsto Nx)$. Then these functors are representable by an affine graded algebra $\mathscr{A}^0_N$ finite étale over $\mathscr{A}$. By taking the integral closure of $\overline{\mathscr{A}}$ in these algebras, we have $\overline{\mathscr{A}}^1_N$ and $\overline{\mathscr{A}}^0_N$. We define

$$X_1(N) = Proj(\overline{\mathscr{A}}^1_N),\ X_0(N) = Proj(\overline{\mathscr{A}}^0_N) \tag{3.38}$$

and

$$Y_1(N) = Spec(\overline{\mathscr{A}}^1_N),\ Y_0(N) = Spec(\overline{\mathscr{A}}^0_N),$$

all over $\mathscr{V}$. Note that $X_?(N)(\mathbb{C}) \cong \Gamma_?(N)\backslash(\mathfrak{H}\cup\{cusps\})$. To avoid technicality, we *pretend* that these schemes are (smooth) fine moduli schemes (over $\mathscr{V}$) of elliptic curves with the corresponding level structure. When $X_0(N)$ is not smooth, we replace $\Gamma_?(N)$ by the intersection $\Gamma'_?(N) = \Gamma_?(N) \cap \Gamma(3)$ for $\Gamma'_?(N) = \{\gamma \in \Gamma_0(N) | \gamma \equiv 1 \mod 3\}$. Then the modular curve associated with $\Gamma'_?(N)$ ($? = 0, 1$) is smooth over $\mathscr{V}$, and we proceed in the same manner as in the case where $X_0(N)$ is smooth over $\mathscr{V}$ (at the end, we need to take $\Gamma_0(N)/\Gamma(3)$-invariants to prove the assertion in general).

Since j^{-1} has a q expansion in $\mathbb{Z}[[q]]$ whose leading term is q, by construction, q gives rise to the local parameter at ∞ of $\mathbf{P}^1(j) = Proj(\overline{\mathscr{A}}) = X_0(1)$. By the theory of Tate curves ([T2]), q gives the local parameter at ∞ of $X_1(N)_{/\mathscr{V}}$ and $X_0(N)_{/\mathscr{V}}$. Thus all these curves are finite étale over $\mathbf{P}^1(j) - \{\infty\}$. If N is a prime $\ell \neq p$, $X_0(\ell)$ has two cusps represented by 0 and ∞. The local parameter of $X_?(\ell)$ at 0-cusp is $q^{\frac{1}{\ell}}$ over $\mathscr{V}$ (so, the 0-cusp ramifies over the infinity cusp of $\mathbf{P}^1(j)$ by ramification index ℓ). The local parameter at the infinity cusp of $X_?(\ell)$ is just q. These conclusions follow from the fact that the Tate curve is well defined over $\mathscr{V}$ and the analysis of the level structure of the Tate curve sitting at the 0- and ∞-cusps (see [AME] VIII 8.8–8.11). Thus $X_1(\ell)_{/\mathscr{V}}$ is an étale finite Galois covering of $X_0(\ell)_{/\mathscr{V}}$ with Galois group $G_\ell = (\mathbb{Z}/\ell\mathbb{Z})^\times/\{\pm 1\}$. Now by induction on the number of primes $\neq p$ dividing N, we can prove the étaleness of $X_1(N)$ over $X_0(N)$ as long as N is square-free:

$$X_0(N) = Proj((\overline{\mathscr{A}}^1_N)^{(\mathbb{Z}/N\mathbb{Z})^\times}) \cong Proj(\overline{\mathscr{A}}^0_N) \cong X_1(N)/G_N; \qquad (3.39)$$

$$\text{$X_1(N)/X_0(N)$ is a finite étale Galois covering with Galois group G_N if N is square-free.} \qquad (3.40)$$

Since $Y_?(N)$ is a $\mathbf{G}_m$-torsor over $X_?(N)$, we also get

$$Y_1(N)/(\mathbb{Z}/N\mathbb{Z})^\times = Spec((\overline{\mathscr{A}}^1_N)^{(\mathbb{Z}/N\mathbb{Z})^\times}) \cong Spec(\overline{\mathscr{A}}^0_N) = Y_0(N). \qquad (3.41)$$

Let D_j be the reduced divisor on $X_j(N)$ supported on $X_j(N) - X_j(N)[\frac{1}{\Delta}]$. We write $\omega^{\otimes k}$ for the invertible sheaf on $X_j(N)$ associated to the graded component $\overline{\mathscr{A}}^j_{N,k}$ and put

$$\omega^{\otimes k}_{cusp} = \omega^{\otimes k}(-D_j) \cong \omega^{\otimes k-2} \otimes \Omega_{X_j(N)/\mathscr{V}}.$$

Here the last isomorphism comes from the Kodaira–Spencer isomorphism. By definition, $H^0(X_j(N)_{/A}, \omega^{\otimes k}_{cusp/A})$ is the space of cusp forms of weight k on $\Gamma_j(N)$ defined over A. We then have

$$S_k(\Gamma_1(N); A) = H^0(X_1(N)_{/A}, \omega^{\otimes k}_{cusp/A}) \subset \overline{\mathscr{A}}^1_{N,k} \otimes_{\mathscr{V}} A. \qquad (3.42)$$

Then the integrality assertion (3.37) follows from the ampleness of $\omega^{\otimes k}_{cusp}$ when $p \geq 5$ and $k \geq 2$:

$$\begin{aligned} H^0(X_1(N)_{/B}, \omega^{\otimes k}_{cusp/B}) &\cong H^0(X_1(N)_{/A}, \omega^{\otimes k}_{cusp/A}) \otimes_A B \\ S_k(\Gamma_1(N); B) &= S_k(\Gamma_1(N); A) \otimes_A B \end{aligned} \qquad (3.43)$$

for all A-algebra B. □

By Corollary 3.11, the space $S_k(\Gamma_0(N), \chi; A)$ is stable under $T(n)$ for all n. Thus $S_k(\Gamma_0(N), \chi; A)$ is a module over $h_k(N, \chi; A)$. When $|(\mathbb{Z}/N\mathbb{Z})^\times|$ is invertible in A, we have (see Theorem 2.6 in Chapter 2)

$$S_k(\Gamma_1(N); A) = \oplus_\chi S_k(\Gamma_0(N), \chi; A). \qquad (3.44)$$

In other words, $\gamma \in \Gamma_0(N)/\Gamma_1(N) \cong (\mathbb{Z}/N\mathbb{Z})^\times$ acts on $S_k(\Gamma_1(N))$ by $f \mapsto f|\gamma$. Then $S_k(\Gamma_0(N), \chi)$ is the χ-eigenspace of the action. We write $[a]$ for the action of $a \in (\mathbb{Z}/N\mathbb{Z})^\times$ on $S_k(\Gamma_1(N))$ and $\langle p \rangle f = p^{k-1}[p]f$ for a positive integer p prime to N. Then by Corollary 3.11, we have

$$\begin{aligned} a(m; T(n)f) &= \sum_{d|m, d|n, d\mathbb{Z}+N\mathbb{Z}=\mathbb{Z}} a(\frac{mn}{d^2}; \langle d \rangle f); \\ \langle p \rangle &= pS(p) = T(p)^2 - T(p^2) \quad \text{for a prime } p \nmid N. \end{aligned} \qquad (3.45)$$

This is also a consequence of (3.31). Here is a somewhat deeper theorem:

THEOREM 3.13 (DELIGNE–RAPOPORT, KATZ) *Suppose that, for a positive integer N, $\Gamma_1(N) \subset \Gamma \subset \Gamma_0(N)$. Then for each subring $A \subset \mathbb{C}$, $S_k(\Gamma; A)$ is stable under the action of $(\mathbb{Z}/N\mathbb{Z})^\times$. Hence by (3.45), $S_k(\Gamma; A)$ is stable under the action of $T(n)$ for all n.*

A proof can be found in [GMF] Chapter III (or [H86a] Section 1). This is based on the existence of the Tate curve and due to Deligne–Rapoport [DR] and N. M. Katz [K2], since the action of $(\mathbb{Z}/N\mathbb{Z})^\times$ coincides with the action of $\mathrm{Gal}(X_1(N)/X_0(N))$ on $\underline{\omega}^{\otimes k}$. We define $h_k(N; A)$ for the A-subalgebra of $\mathrm{End}_A(S_k(\Gamma_1(N)))$ generated by $T(n)$ for all n. Thus by Theorem 3.13, we know that $S_k(\Gamma_1(N); A)$ is a faithful $h_k(N; A)$-module.

For any algebra A not necessarily in $\mathbb{C}$, we define $S_k(\Gamma; A)$ by $S_k(\Gamma; \mathbb{Z}) \otimes_{\mathbb{Z}} A$ for Γ as in Theorem 3.13. Since we can embed $S_k(\Gamma; \mathbb{Z})$ into $\mathbb{Z}[[q]]$ by $f \mapsto f(q) = \sum_n a(n; f)q^n$, we have a natural map: $S_k(\Gamma; A) \to A[[q]]$ given by $f \otimes a \mapsto af(q)$.

THEOREM 3.14 (q-EXPANSION PRINCIPLE) *Suppose that $\Gamma_1(N) \subset \Gamma \subset \Gamma_0(N)$ for a positive integer N. Then the natural map $S_k(\Gamma; A) \to A[[q]]$ is injective. If A is an integral domain with quotient field K, then*

$$S_k(\Gamma; A) = S_k(\Gamma; K) \cap A[[q]].$$

Proof By definition, if $A \subset \mathbb{C}$, then $K \subset \mathbb{C}$, and

$$S_k(\Gamma; A) = S_k(\Gamma; \mathbb{C}) \cap A[[q]] = S_k(\Gamma; K) \cap A[[q]]$$

in $\mathbb{C}[[q]]$, which shows the result. Let $f \in \mathbb{Z}[[q]]$ and $S = S_k(\Gamma; \mathbb{Z}) \subset \mathbb{Z}[[q]]$. If $nf = g \in S$, then $f = n^{-1}g \in S_k(\Gamma; \mathbb{Q})$. Thus $f \in (S_k(\Gamma; \mathbb{Q}) \cap \mathbb{Z}[[q]]) = S$. This shows that $\mathbb{Z}[[q]]/S$ is torsion-free, and hence $\mathbb{Z}$-flat (see [CRT] Theorem 7.7). Thus tensoring an arbitrary ring A to the exact sequence:

$$0 \to S \to \mathbb{Z}[[q]] \to \mathbb{Z}[[q]]/S \to 0,$$

we get another exact sequence (see [CRT] Appendix B):

$$0 \to S \otimes_{\mathbb{Z}} A \to \mathbb{Z}[[q]] \otimes_{\mathbb{Z}} A \to (\mathbb{Z}[[q]]/S) \otimes_{\mathbb{Z}} A \to 0.$$

We see easily from $\mathbb{Z} \otimes_{\mathbb{Z}} A \cong A$ that $\mathbb{Z}[[q]] \otimes_{\mathbb{Z}} A$ naturally injects into $A[[q]]$, which may not be surjective if A is infinite (Exercise 1). This shows the first assertion. We prove the last assertion only when $A[[q]] = \mathbb{Z}[[q]] \otimes_{\mathbb{Z}} A$, which holds if A is finite and hence if A is profinite and also if A is a $\mathbb{Z}$-module of finite type (see Exercise 2). Since $\mathbb{Z}[[q]]/S$ is $\mathbb{Z}$-flat, $(\mathbb{Z}[[q]]/S) \otimes_{\mathbb{Z}} A$ is A-flat. In particular, it is A-torsion-free. If

$f \in (A[[q]] \cap S_k(\Gamma;K))$, then $af \in S_k(\Gamma;A)$ for $0 \neq a \in A$, because $S_k(\Gamma;K) = S_k(\Gamma;A) \otimes_A K$. Then $f \mod S \otimes_{\mathbb{Z}} A$ is a torsion-element in $(\mathbb{Z}[[q]]/S) \otimes_{\mathbb{Z}} A$. Since $(\mathbb{Z}[[q]]/S) \otimes_{\mathbb{Z}} A$ is A-torsion-free, we have $f \in S_k(\Gamma;A) = S \otimes_{\mathbb{Z}} A$. This shows the desired assertion. □

Split N so that $N = N'p^r$ with $p \nmid N'$. The Hecke operator $T(p)$ acts on $S_k(\Phi;A)$, where Φ is either $\Gamma_1(N)$ or $\Gamma_0(N') \cap \Gamma_1(p^r)$ (so, $\Gamma_0(N) \supset \Phi \supset \Gamma_1(N)$), simply because $a(n;f|T(p)) = a(np;f)$ if $r \geq 1$ (of course, when $r = 0$, we need the full force of Theorem 3.13 for the stability of the space under $T(p)$). Let A be a discrete valuation ring finite flat over $\mathbb{Z}_p$ with residue field $\mathbb{F}$. The algebra $(A/p^nA)[T(p)] \subset \mathrm{End}_A(S_k(\Phi;A/p^nA))$ is a finite ring. Thus it has only finitely many maximal ideals. We write the set of maximal ideals as Ω_n. We then have

$$(A/p^nA)[T(p)] = \bigoplus_{\mathfrak{m}\in\Omega_n} (A/p^nA)[T(p)]_{\mathfrak{m}}$$

for the localization $A/p^nA[T(p)]_{\mathfrak{m}}$ at $\mathfrak{m}$. Since p is in all maximal ideals, $\mathfrak{m} \mapsto \mathfrak{m} \bmod p^n$ induces a bijection $\Omega_{n+1} \cong \Omega_n$. Thus taking the projective limit with respect to n (and identifying all Ω_n with the set of maximal ideals Ω of $A[T(p)] = \varprojlim_n A/p^nA[T(p)]$), we have $A[T(p)] = \bigoplus_{\mathfrak{m}\in\Omega} A[T(p)]_{\mathfrak{m}}$. We define $A[T(p)]^{ord}$ by the direct sum of all $A[T(p)]_{\mathfrak{m}}$ with $T(p) \notin \mathfrak{m}$. Thus $A[T(p)]^{ord}$ is the maximal algebra direct summand of $A[T(p)]$ in which $T(p)$ is invertible. Let e be the idempotent of $A[T(p)]$. Since $T(p)$ is topologically nilpotent on the complementary direct summand in $A[T(p)]$, we have

$$e = \lim_{n\to\infty} T(p)^{n!}.$$

For any A-module B, $A[T(p)]$ acts naturally on $S_k(\Phi;B) = S_k(\Phi;A) \otimes_A B$. Then we define $S_k^{ord}(\Phi;B) = e(S_k(\Phi;A))$ and call it the *p-ordinary* part of $S_k(\Phi;A)$.

As already seen, $S_k(\Phi;B)$ is a module over a finite group $(\mathbb{Z}/N\mathbb{Z})^\times$. Since $T(p)$ commutes with the action, $S_k^{ord}(\Phi;B)$ is naturally a module over $(\mathbb{Z}/N\mathbb{Z})^\times$. We recall that $H^0(H, S_k^{ord}(\Gamma_1(N);A))$ for a subgroup $H \subset (\mathbb{Z}/N\mathbb{Z})^\times$ is the submodule of $S_k^{ord}(\Gamma_1(N);A)$ invariant (element by element) under the action of H. We have the following important result on this action:

THEOREM 3.15 (CONTROL THEOREM) *Let $k \geq 2$ and A be a valuation ring* *finite flat over* $\mathbb{Z}_p$. *Split* N *so that* $N = N'p^r$ *with* $p \nmid N'$, *and suppose that* $p^r > 3$ *and that* N' *is square-free. Then we have*

$$H^0((\mathbb{Z}/N'\mathbb{Z})^\times, S_k^{ord}(\Gamma_1(N); A/p^nA)) = S_k^{ord}(\Gamma_0(N') \cap \Gamma_1(p^r); A/p^nA).$$

More generally, for a character $\chi : (\mathbb{Z}/N'\mathbb{Z})^\times \to A^\times$, *we twist the action of* $(\mathbb{Z}/N'\mathbb{Z})^\times$ *by* χ^{-1}, *that is, the new action is given by* $f \mapsto \chi(a)^{-1}[a]f$. *Then we have*

$$H^0((\mathbb{Z}/N'\mathbb{Z})^\times, S_k^{ord}(\Gamma_1(N); A/p^nA)) = S_k^{ord}(\Gamma_0(N') \cap \Gamma_1(p^r), \chi; A/p^nA),$$

where

$$S_k^{ord}(\Gamma_0(N') \cap \Gamma_1(p^r), \chi; \mathbb{Z}) = \{f \in S_k^{ord}(\Gamma_1(N); \mathbb{Z}) | [a]f = \chi(a)\ \forall a \in (\mathbb{Z}/N'\mathbb{Z})^\times\}$$

and

$$S_k^{ord}(\Gamma_0(N') \cap \Gamma_1(p^r), \chi; A/p^nA) = S_k^{ord}(\Gamma_0(N') \cap \Gamma_1(p^r), \chi; \mathbb{Z}) \otimes_{\mathbb{Z}} A/p^nA.$$

For the proof of this, we need the modular interpretation of the analytically defined space of cusp forms, given in (1.42), and more details can be found in [GMF] Chapter III. Then this follows from the fact that (i) the ordinary loci of the affine modular curves $X_1(N)$ (with $X_1(N)(\mathbb{C}) = \Gamma_1(N)\backslash(\mathfrak{H} \cup \{cusps\})$ and X associated with $\Gamma_0(N') \cap \Gamma_1(p^r)$) are smooth over $\mathbb{Z}/p^n\mathbb{Z}$ (if $p^r > 3$); (ii) $X_1(N)$ is étale over X, and (iii) the line bundle $\underline{\omega}^k(\chi)$ on the étale site over X whose global sections over X are $M_k(\Gamma_0(N') \cap \Gamma_1(p^r), \chi; A/p^nA)$ is ample if $k \geq 2$ and $p^r > 3$; (iv) the q-expansion principle for cusp forms. The control theorem was first proved in a slightly different context in [K2] 1.6 and [K1] and later formulated as above in [H86a] Section 1. An exposition of the proof can be found in [GMF] Chapters II and III.

The theorem only takes care of the p-ordinary part. However, under the assumption that $p \nmid N$, we have a similar assertion including the non-ordinary part:

PROPOSITION 3.16 *Let* $k \geq 2$ *and* A *be a valuation ring finite flat over* $\mathbb{Z}_p$. *Suppose that* $p \nmid N$ *and that* N *is square-free. Split* N *so that* $N = N'q$ *with* $q \nmid N'$ *for a prime* $q > 3$, *and suppose that* $N' > 3$. *Then we have for a character* $\chi : (\mathbb{Z}/q\mathbb{Z})^\times \to A^\times$, *we twist the action of* $(\mathbb{Z}/q\mathbb{Z})^\times$ *by* χ^{-1}, *that is, the new action is given by $f \mapsto \chi(a)^{-1}[a]f$. Then*

$$H^0((\mathbb{Z}/q\mathbb{Z})^\times, S_k(\Gamma_1(N); A/p^nA)) = S_k(\Gamma_0(q) \cap \Gamma_1(N'), \chi; A/p^nA),$$

where

$$S_k(\Gamma_0(q) \cap \Gamma_1(N'), \chi; \mathbb{Z}_p) = \{f \in S_k(\Gamma_1(N); \mathbb{Z}_p) | [a]f = \chi(a)\ \forall a \in (\mathbb{Z}/q\mathbb{Z})^\times\}$$

and

$$S_k(\Gamma_0(q) \cap \Gamma_1(N'), \chi; A/p^nA) = S_k(\Gamma_0(q) \cap \Gamma_1(N'), \chi; \mathbb{Z}_p) \otimes_{\mathbb{Z}_p} A/p^nA.$$

The proof is similar to the one for the control theorem ([GMF] Chapters II and III), but we consider the moduli problem of pairs $(E, \phi_N : \mathbb{Z}/N\mathbb{Z} \hookrightarrow E)$ over A. Since $p \nmid N$, we do not need to impose the p-ordinarity condition.

Proof We shall give a brief sketch of the proof of the above proposition, in terms of the notation and definitions introduced in the proof of (3.37). Thus we first prove a similar assertion for $\Gamma_0(N)$ and $\Gamma_1(N)$ a little more generally for $\mathscr{V}$-algebras A ($\mathscr{V} = \mathbb{Z}_p \cap \mathbb{Q}$ in place of $\mathbb{Z}_p$-algebras A). After doing this, we explain how to deal with the specific case in the proposition. Since $Y_1(N)_{/A} = Spec(\overline{\mathscr{A}}^1_N \otimes_{\mathscr{V}} A)$ is an étale Galois covering of $Y_0(N)_{/A} = Spec(\overline{\mathscr{A}}^0_N \otimes_{\mathscr{V}} A)$ with Galois group $(\mathbb{Z}/N\mathbb{Z})^\times$ (see (3.40)), we have

$$\begin{aligned} H^0((\mathbb{Z}/N\mathbb{Z})^\times, H^0(X_1(N), \omega^{\otimes k})) &= H^0((\mathbb{Z}/N\mathbb{Z})^\times, \overline{\mathscr{A}}^1_{N,k}) \\ &\cong \overline{\mathscr{A}}^0_{N,k} = H^0(X_0(N), \omega^{\otimes k}). \end{aligned}$$

Note that $\omega^{\otimes k}_{cusp/X_1(N)} = \pi^*\omega^{\otimes k}_{cusp/X_0(N)}$ for the projection $\pi : X_1(N) \to X_0(N)$. This fact follows from the unramifiedness of the cusps of $X_0(N)$ in $X_1(N)$ (or the Kodaira–Spencer isomorphism: $\omega^{\otimes k}_{cusp} \cong \omega^{\otimes k-2} \otimes \Omega_{X_j(N)/\mathscr{V}}$). Then, we also have

$$H^0((\mathbb{Z}/N\mathbb{Z})^\times, H^0(X_1(N), \omega^{\otimes k}_{cusp})) = H^0(X_0(N), \omega^{\otimes k}_{cusp}).$$

Then for each character $\varepsilon : (\mathbb{Z}/N\mathbb{Z})^\times \to \mathscr{O}^\times$, we can twist an invertible sheaf $\mathscr{L}$ on $X_0(N)$ in the following way. For each open set $O \subset X_0(N)$, we consider $H^0(\pi^{-1}(O), \pi^*\mathscr{L})$ on which $(\mathbb{Z}/N\mathbb{Z})^\times$ acts naturally. Then we twist the Galois action by ε and define $H^0(O, \mathscr{L} \otimes \varepsilon) = H^0(\pi^{-1}(O), \pi^*\mathscr{L})[\varepsilon]$, where '$[\varepsilon]$' indicates the ε-eigenspace. Then $\mathscr{L} \otimes \varepsilon$ is an invertible sheaf well defined over $X_0(N)$. Then we have

$$H^0(X_1(N)_{/A}, \pi^*\mathscr{L})[\varepsilon] = H^0(X_0(N)_{/A}, \mathscr{L} \otimes \varepsilon),$$

which is an A-flat module for any $\mathcal{O}$-algebra A. Applying this to $\mathcal{L} = \omega_{cusp}^{\otimes k}$, for each character $\varepsilon : (\mathbb{Z}/N\mathbb{Z}) \to \mathcal{O}^\times$ and an integer $n > 0$, the space of level N cusp forms $S_k(\Gamma_1(N); A/p^nA)[\varepsilon]$ is A/p^nA-free of finite type. Moreover, we have

$$S_k(\Gamma_1(N); A/p^nA)[\varepsilon] \cong S_k(\Gamma_1(N); A)[\varepsilon] \otimes_A A/p^nA.$$

for every $\mathcal{V}$-algebra A, as desired. To get the assertion for each prime factor q, we look into the intermediate surfaces: $Y_1(N) \twoheadrightarrow Y \twoheadrightarrow Y_0(N)$ corresponding to $\Gamma_0(q) \cap \Gamma_1(N')$. Then $Y_1(N)/Y$ is a finite étale covering with Galois group $(\mathbb{Z}/q\mathbb{Z})^\times$, and hence the same argument as above works well for $Y_1(N)/Y$ in place of $Y_1(N)/Y_0(N)$. This proves the assertion. □

Exercises

(1) Give an example of a ring A such that $A \otimes_{\mathbb{Z}} \mathbb{Z}[[q]] \to A[[q]]$ given by $a \otimes f(q) \mapsto af(q)$ is not surjective.

(2) Show the above map: $A \otimes_{\mathbb{Z}} \mathbb{Z}[[q]] \to A[[q]]$ is surjective if either A is a finite ring or A is a $\mathbb{Z}$-module of finite type.

3.1.9 p-adic Hecke algebras

For any A-module M with an action of $R(\Delta)$, like $S_k(\Gamma_1(N); A)$, we define $h(M) \subset \mathrm{End}_A(M)$ to be the A-subalgebra generated by $T(n)$ for all $n = 1, 2, \ldots$. Then we put

$$h_k(N; A) = h(S_k(\Gamma_1(N); A)) \quad \text{and} \quad h_k^{ord}(N; A) = h(S_k^{ord}(\Gamma_1(N); A)).$$

Similarly, we define for any $\mathbb{Z}[\chi]$-algebra A,

$$h_k(N, \chi; A) = h(S(\Gamma_0(N), \chi; A)) \quad \text{and} \quad h_k^{ord}(N, \chi; A) = h(S_k^{ord}(\Gamma_0(N), \chi; A)).$$

Here for the definition of h_k^{ord}, we assume that A is an algebra over $\mathbb{Z}_p$ (to have the space S_k^{ord} well defined). The action of $T(n)$ on $S_k(\Gamma_1(N); A)$ and $S_k(\Gamma_0(N), \chi; A)$ is described by the formula (3.45) of q-expansion coefficients.

Let $\mathcal{S}$ be either $S_k(\Gamma, \chi; A)$ or $S_k(\Gamma; A)$ for Γ satisfying $\Gamma_1(N) \subset \Gamma \subset \Gamma_0(N)$ for a positive integer N. Then we write $\mathcal{H}$ for the subalgebra of $\mathrm{End}_A(\mathcal{S})$ generated by $T(n)$ for all positive integer n. Thus $\mathcal{H} = h(S_k(\Gamma; A))$ or $h(S_k(\Gamma, \chi; A))$ according as $\mathcal{S} = S_k(\Gamma; A)$ or $S_k(\Gamma, \chi; A)$. We define the following pairing:

$$\langle\ ,\ \rangle : \mathcal{H} \times \mathcal{S} \to A$$

by $\langle h, f\rangle = a(1; h(f))$. We see from (3.45) that

$$\langle T(n), f\rangle = a(n; f). \tag{3.46}$$

THEOREM 3.17 (DUALITY) *The pairing $\langle\ ,\ \rangle$ is perfect. In other words, it induces the following isomorphisms:*

$$\mathrm{Hom}_A(\mathcal{S}, A) \cong \mathcal{H} \quad \text{and} \quad \mathrm{Hom}_A(\mathcal{H}, A) \cong \mathcal{S}.$$

Moreover, we have $h(S_k(\Gamma, \chi; A)) = h(S_k(\Gamma, \chi, \mathbb{Z}[\chi])) \otimes_{\mathbb{Z}[\chi]} A$; in particular,

$$h_k(N; A) = h_k(N; \mathbb{Z}) \otimes_{\mathbb{Z}} A \quad \text{and} \quad h_k(N, \chi; A) = h_k(N, \chi; \mathbb{Z}[\chi]) \otimes_{\mathbb{Z}[\chi]} A.$$

Proof We first prove this when A is a field. Since in this case, $\mathcal{S}$ and $\mathcal{H}$ are of finite dimension, we only need to show the non-degeneracy of the pairing. If $\langle h, f\rangle = 0$ for all h, then by (3.46), $a(n; f) = 0$ for all n. This implies $f = 0$ by the q-expansion principle: Theorem 3.14. Conversely if $\langle h, f\rangle = 0$ for all $f \in \mathcal{S}$, we see that

$$a(n, hf) = \langle T(n), hf\rangle = a(1, T(n)hf) = \langle T(n)h, f\rangle = 0.$$

Thus $hf = 0$ for all f implies that h vanishes as an operator acting on $\mathcal{S}$.

Now assume that A is a PID with field K of fractions. Since $\mathcal{H}$ and $\mathcal{S}$ is torsion-free, they are A-free of finite rank. Since A is a PID, writing $M^* = \mathrm{Hom}_A(M, A)$ for an A-module M, $(M^*)^* \cong M$ canonically. Thus we need to show only one identity, say, $\mathrm{Hom}_A(\mathcal{H}, A) \cong \mathcal{S}$. The natural map $\phi : \mathcal{S} \to \mathrm{Hom}_A(\mathcal{H}, A)$ is given by $\phi(f)(h) = \langle h, f\rangle$. Thus if $\phi(f) = 0$, then $\phi(f)(T(n)) = \langle T(n), f\rangle = a(n; f) = 0$, and hence $f = 0$. Thus ϕ is injective. We need to show ϕ is surjective. We write $\mathcal{S}(K) = \mathcal{S} \otimes_A K$ and $\mathcal{H}(K) = \mathcal{H} \otimes_A K$. As we have seen, $\langle\ ,\ \rangle : h(K) \times \mathcal{S}(K) \to K$ is perfect. Thus for each $\varphi : \mathcal{H} \to A \in \mathrm{Hom}_A(\mathcal{H}, A)$, by extending $\varphi : \mathcal{H}(K) \to K$ by linearity, we find $f \in \mathcal{S}(K)$ such that $\varphi(h) = \langle h, f\rangle$ for all $h \in \mathcal{H}(K)$. In particular, $a(n; f) = \langle T(n), f\rangle = \varphi(T(n)) \in A$ for all n. Thus $f \in \mathcal{S}(K) \cap A[[q]] = \mathcal{S}$ by the q-expansion principle. Thus $\varphi = \phi(f)$, and ϕ is surjective.

Let A be a general ring. Suppose that $\mathcal{S} = S_k(\Gamma; A)$. Then writing $\mathcal{S}(\mathbb{Z}) = S_k(\Gamma_1(N), \mathbb{Z})$ and $\mathcal{H}(\mathbb{Z}) = h_k(N; \mathbb{Z})$, we have $\mathcal{S}(A) = \mathcal{S}(\mathbb{Z}) \otimes_{\mathbb{Z}} A$ and claim to have $\mathcal{H} = \mathcal{H}(\mathbb{Z}) \otimes_{\mathbb{Z}} A$. Since $\mathbb{Z}$ is a PID, we have $\mathrm{Hom}_{\mathbb{Z}}(\mathcal{S}(\mathbb{Z})A, \mathbb{Z}) \cong \mathcal{H}(\mathbb{Z})$.

Note that the natural map: $h(S_k(\Gamma; \mathbb{Z})) \otimes_{\mathbb{Z}} A \to h(S_k(\Gamma; A))$ given by $h \otimes a \mapsto ah$ is surjective, because the two sides of the map are both generated by $T(n)$. Since $\mathcal{H}(\mathbb{Z})$ is the dual of $\mathcal{S}(\mathbb{Z})$, for the inclusion $i : \mathcal{H}(\mathbb{Z}) \hookrightarrow \mathrm{End}_{\mathbb{Z}}(\mathcal{S}(\mathbb{Z}))$, $\mathrm{Coker}(i)$ is $\mathbb{Z}$-free. Thus $\mathcal{H}(\mathbb{Z}) \otimes_{\mathbb{Z}} A$ injects into $\mathrm{End}_{\mathbb{Z}}(\mathscr{S}(\mathbb{Z})) \otimes_{\mathbb{Z}} A \hookrightarrow \mathrm{End}_{\mathbb{Z}}(\mathscr{S}(A))$. Therefore the injectivity follows from the fact: $S_k(\Gamma; A) = S_k(\Gamma; \mathbb{Z}) \otimes_{\mathbb{Z}} A$. Similarly, we can prove that $h(S_k(\Gamma, \chi; A)) = h(S_k(\Gamma, \chi; \mathbb{Z}[\chi])) \otimes_{\mathbb{Z}[\chi]} A$.

Then

$$\begin{aligned}\mathrm{Hom}_A(\mathscr{S}, A) &\cong \mathrm{Hom}_A(\mathscr{S}(\mathbb{Z}) \otimes_{\mathbb{Z}} A, \mathbb{Z} \otimes_{\mathbb{Z}} A) \\ &\cong \mathrm{Hom}_{\mathbb{Z}}(\mathscr{S}(\mathbb{Z}), \mathbb{Z}) \otimes_{\mathbb{Z}} A \cong \mathscr{H}(\mathbb{Z}) \otimes_{\mathbb{Z}} A \cong \mathscr{H}.\end{aligned}$$

Similarly we conclude that $\mathrm{Hom}_A(\mathscr{H}, A) \cong \mathscr{S}$.

Although $\mathbb{Z}[\chi]$ may not be a PID, it is a Dedekind domain (see [CRT] Chapter 4). In this case, $\langle\ ,\ \rangle$ is perfect over $\mathbb{Z}[\chi]$ if and only if it is perfect over localizations $A_{\mathfrak{p}}$ of $\mathbb{Z}[\chi]$ at all maximal ideals $\mathfrak{p}$ of $\mathbb{Z}[\chi]$. Since $A_{\mathfrak{p}}$ is a valuation ring (hence a PID), the assertion holds for $\mathbb{Z}[\chi]$. Then, for any $\mathbb{Z}[\chi]$-algebra A, the assertion holds in the same way as above. □

We now prove Theorem 3.8. Let $M = M(A) = \sum_n AT(n) \subset h_k(N, \chi; A)$. Then by the above proof of the duality theorem, we see that M is the A-dual of $\mathscr{S}$. However, $h_k(N, \chi; A)$ is also an A-dual of $\mathscr{S}$. Thus if A is a field, this is enough to conclude that $M = h_k(N, \chi; A)$. If A is a valuation ring with residue field $\mathbb{F}$, we have

$$M(A) \otimes_A \mathbb{F} = M(\mathbb{F}) = h_k(N, \chi; \mathbb{F}) = h_k(N, \chi; A) \otimes_A \mathbb{F}.$$

Then, we have $M(A) = h_k(N, \chi; A)$ by Nakayama's lemma (Lemma 2.3 in Chapter 2). For an Dedekind domain A, if the assertion holds for all localizations, it holds for A. Since localization of A is either a field or a discrete valuation ring, the assertion holds for Dedekind domains A. Since $\mathbb{Z}[\chi]$ is a Dedekind domain, it holds for $\mathbb{Z}[\chi]$ and for general A, because $h_k(N, \chi; A) = h_k(N, \chi; \mathbb{Z}[\chi]) \otimes_{\mathbb{Z}[\chi]} A$. □

THEOREM 3.18 *Let K be a finite extension of $\mathbb{Q}_p$ and $\mathcal{O}$ be the p-adic integer ring of K. Then the pairing $\langle\ ,\ \rangle$ for $A = \mathcal{O}$ induces a Pontryagin duality*

$$\mathrm{Hom}_{\mathcal{O}}(\mathscr{S} \otimes_{\mathcal{O}} K/\mathcal{O}, K/\mathcal{O}) \cong \mathscr{H}.$$

For profinite or discrete $\mathbb{Z}_p$-modules M, their Pontryagin dual $\widehat{M}$ is defined by $\widehat{M} = \mathrm{Hom}_{\mathbb{Z}_p}(M, \mathbb{Q}_p/\mathbb{Z}_p)$. If M is an $\mathcal{O}$-module, $\widehat{M}$ is isomorphic to $\mathrm{Hom}_{\mathcal{O}}(M, K/\mathcal{O})$. It is known that the duality is perfect, that is, $\widehat{\widehat{M}} \cong M$ canonically for all M as above (cf. [LFE] Section 8.3).

Proof If M is an $\mathcal{O}$-free module of finite rank, it is known that

$$M^* = \mathrm{Hom}_{\mathcal{O}}(M, \mathcal{O}) \cong \mathrm{Hom}_{\mathcal{O}}(M \otimes_{\mathcal{O}} K/\mathcal{O}, K/\mathcal{O}).$$

Just applying this general result to $M = \mathscr{S}$, we get the theorem. □

Note that $\mathcal{O}/p^n\mathcal{O}$ is free of finite rank r over $\mathbb{Z}/p^n\mathbb{Z}$. This shows that

$$\begin{aligned} S_k(\Gamma_1(N), p^{-n}\mathcal{O}/\mathcal{O}) &\overset{f\mapsto p^n f}{\cong} S_k(\Gamma_1(N), \mathcal{O}/p^n\mathcal{O}) \\ &\cong S_k(\Gamma_1(N); \mathbb{Z}/p^n\mathbb{Z}) \otimes \mathcal{O}/p^n\mathcal{O} \cong S_k(\Gamma_1(N), \mathbb{Z}/p^n\mathbb{Z})^r \end{aligned}$$

as $(\mathbb{Z}/N\mathbb{Z})^\times$-modules. Thus from the control theorem (Theorem 3.15), we conclude that

$$H^0((\mathbb{Z}/q\mathbb{Z})^\times, S_k^{ord}(\Gamma_1(N); p^{-n}\mathcal{O}/\mathcal{O})) = S_k^{ord}(\Gamma_1(N') \cap \Gamma_0(q); p^{-n}\mathcal{O}/\mathcal{O}),$$

if $N = N'q$ with $q \nmid N'$ and $p \nmid q$ (under either $N' > 3$ or $p > 3$). Since $S_k(\Gamma_1(N); K/\mathcal{O}) = \bigcup_n S_k(\Gamma_1(N); p^{-n}\mathcal{O}/\mathcal{O})$, we see

$$H^0((\mathbb{Z}/q\mathbb{Z})^\times, S_k^{ord}(\Gamma_1(N); K/\mathcal{O})) = S_k^{ord}(\Gamma_1(N') \cap \Gamma_0(q); K/\mathcal{O}).$$

Under Pontryagin duality, invariants correspond to co-invariants, that is, for a locally compact module M with a continuous action of a group G, $\widehat{M^G} \cong (\widehat{M})_G$, where $\widehat{M}$ is the Pontryagin dual of M and $M_G = H_0(G, M) = M/\sum_{g\in G}(g-1)M$ (the maximal quotient on which G acts trivially). We can twist the action by changing the action from: $x \mapsto [a]x$ to $x \mapsto \chi(a)^{-1}[a]x$ for a character $\chi : (\mathbb{Z}/q\mathbb{Z})^\times \to \mathcal{O}^\times$. This shows:

Corollary 3.19 *Split N so that $N = N'q$ with $q \nmid N'$ and $p \nmid q$, and suppose either $N' > 3$ or $q > 3$. Let $\chi : (\mathbb{Z}/q\mathbb{Z})^\times \to \mathcal{O}^\times$ be a character. Then, for $k \geq 2$, the algebra*

$$h_k^{ord}(N; \mathcal{O}) / \sum_{a \in (\mathbb{Z}/q\mathbb{Z})^\times} ([a] - \chi(a)) h_k^{ord}(N; \mathcal{O})$$

is isomorphic to the Hecke algebra $h_k^{ord}(\Gamma_1(N') \cap \Gamma_0(q), \chi; \mathcal{O})$ of

$$S_k^{ord}(\Gamma_1(N') \cap \Gamma_0(q), \chi; \mathcal{O}).$$

We can remove the p-ordinarity assumption from the above corollary if N is square-free, $N' > 3$ and $p \nmid N$ (using Proposition 3.16 instead of Theorem 3.15).

Now suppose that q is a prime with $q \equiv 1 \mod p$. Then we split $(\mathbb{Z}/q\mathbb{Z})^\times = \Delta_q \times \Delta'_q$ for the p-Sylow subgroup Δ_q. For the moment, we assume that $\mathcal{O}$ is sufficiently large to contain all characters of $(\mathbb{Z}/q\mathbb{Z})^\times$ with values in $\overline{\mathbb{Q}}_p$. Then the group algebra $\mathcal{O}[\Delta'_q]$ is just the product of $|\Delta'_q|$-copies of $\mathcal{O}$ (see the proof of Theorem 2.6 in Chapter 2). For each character $\chi' : \Delta'_q \to \mathcal{O}^\times$, we have an idempotent $1_{\chi'} = |\Delta'_q|^{-1} \sum_{\delta' \in \Delta'_q} \chi'^{-1}(\delta')\delta'$. This is well defined in $\mathcal{O}[\Delta'_q]$, because $|\Delta'_q|$ is prime to p and hence has an inverse in $\mathcal{O}$. Then by definition, for any $\mathcal{O}$-module X with an action of Δ'_q, Δ'_q acts by the character χ' on $X[\chi'] = 1_{\chi'}X$. If we have two characters $\chi' \neq \psi'$, this shows that $1_{\chi'} = 0$ on $1_{\psi'}X$. On the other hand, on $X[\chi']$, the multiplication by $1_{\chi'}$ is the identity map. Note that $\mathcal{O}[\Delta'_q]$ is an Δ'_q-module by left multiplication. Applying the above facts to $X = \mathcal{O}[\Delta'_q]$, we find that $1_{\chi'}^2 = 1_{\chi'}$ and $1_{\chi'}1_{\psi'} = 0$ if $\psi' \neq \chi'$. Then counting the rank, we conclude that

$$\mathcal{O}[\Delta'_q] \cong \prod_{\chi'} 1_{\chi'}\mathcal{O}[\Delta'_q] = \prod_{\chi'} \mathcal{O}1_{\chi'}, \tag{3.47}$$

where χ' runs over all characters $\chi' : \Delta'_q \to \mathcal{O}^\times$.

Let $S = S_k(\Gamma_1(N);\mathcal{O})$ on which $\Delta_q \times \Delta'_q$ acts. Let

$$S[\chi'] = \{f \in S | [a]f = \chi'(a)f \text{ for all } a \in \Delta'_q\}.$$

Note that the action of Δ_q is not specified and hence Δ_q still acts on $S[\chi']$ non-trivially. On the other hand, Δ'_q acts on $S[\chi']$ by the character χ'. Thus $S[\chi'] = 1_{\chi'}S$ and $S = \bigoplus_{\chi'} S[\chi']$. We put $h_k(N;\mathcal{O})[\chi']$ for the $\mathcal{O}$-subalgebra of $\mathrm{End}_{\mathcal{O}}(S[\chi'])$ generated by $T(n)$ for all n. Then $h_k(N;\mathcal{O}) = \bigoplus_{\chi'} h_k(N;\mathcal{O})[\chi']$ as a ring direct sum, and $h_k(N;\mathcal{O})[\chi'] = 1_{\chi'}h_k(N;\mathcal{O})$ by the duality theorem. The p-group Δ_q still acts on $h_k(N;\mathcal{O})[\chi']$, which is different from $h_k(\Gamma_1(N') \cap \Gamma_0(q), \chi';\mathcal{O})$, because the action of Δ_q may not be trivial.

Contrary to $\mathcal{O}[\Delta'_q]$, the group algebra $\mathcal{O}[\Delta_q]$ is a local ring (cf. Lemma 2.20 in Chapter 2). Since $(\mathbb{Z}/q\mathbb{Z})^\times$ is cyclic (by the existence of a primitive root modulo q), Δ_q is cyclic of order p^ℓ for $\ell \geq 1$. Let δ be a generator of Δ_q. Then $\mathcal{O}[\Delta_q] \cong \mathcal{O}[S]/((1+S)^{p^\ell} - 1)$ via $\delta \mapsto 1 + S$. Then we have

$$\mathcal{O}[\Delta_q] \otimes_{\mathbb{Z}} \mathbb{F}_p \cong \mathbb{F}[\Delta_q] \cong \mathbb{F}[S]/((1+S)^{p^\ell} - 1) \cong \mathbb{F}[S]/(S^{p^\ell}),$$

which is obviously a local ring. Thus $\mathcal{O}[\Delta_q]$ is a local ring.

Corollary 3.20 *Suppose $p > 2$ and $N' > 3$. For a character $\chi' : \Delta'_q \to \mathcal{O}^\times$, the algebra $h_k^{ord}(N;\mathcal{O})[\chi']$ as a Δ_q-module is free of finite rank over $\mathcal{O}[\Delta_q]$. The rank is given by $r = \mathrm{rank}_{\mathcal{O}}\, h_k^{ord}(\Gamma_1(N') \cap \Gamma_0(q), \chi';\mathcal{O})$. In particular, any local ring h of $h_k^{ord}(N;\mathcal{O})$ is $\mathcal{O}[\Delta_q]$-free of finite rank.*

A similar result for modular cohomology groups, like $H^1(X_1(N), \mathbb{Z}_p)$, is also true (under some additional assumptions) without assuming the p-ordinarity condition. This result is often attributed to de Shalit (cf. [TW] Section 3). The attribution is correct for non-ordinary cases, but in the p-ordinary case, the above fact for p-ordinary cohomology groups was proved much earlier in [H89a] Lemma 3.10 in a more general Hilbert modular setting. We shall give a version of the proof given for [H89a] Lemma 3.10.

Proof If the result is proved for sufficiently large $\mathcal{O}$, we can descend the freeness to smaller rings because $h_k^{ord}(N,\chi;\mathcal{O}) \otimes_{\mathcal{O}} A = h_k^{ord}(N,\chi;A)$. Thus we may assume that $\mathcal{O}$ contains the values of all characters of $(\mathbb{Z}/q\mathbb{Z})^\times$. Write H for $h_k^{ord}(N;\mathcal{O})[\chi']$. Let $R = \mathcal{O}[\Delta_q]$, and write $\mathfrak{m}$ for the unique maximal ideal of R. Then $\mathfrak{m}$ contains the ideal $\mathfrak{a}_\chi$ generated by $\delta - \chi(\delta)$ for a character $\chi : \Delta_q \to \mathcal{O}^\times$. By the control theorem (and Corollary 3.19), we know that $H/\mathfrak{a}_\chi H \cong h_k^{ord}(\Gamma_1(N') \cap \Gamma_0(q), \chi\chi'; \mathcal{O})$, which is free of finite rank r over $\mathcal{O}$. By Nakayama's lemma (Lemma 2.3 in Chapter 2), the rank r is given by $\dim_{\mathbb{F}} H/\mathfrak{m}H$ which is independent of χ. Thus we pick a base $\overline{x}_1, \ldots, \overline{x}_r$ of $H/\mathfrak{m}H$ and lift it to $x_1, \ldots, x_r$ in H. Then we have an R-linear map: $\pi : R^r \to H$ given by $(a_1, \ldots, a_r) \mapsto \sum_j a_i x_i$. By Nakayama's lemma, π is surjective (because after tensoring $R/\mathfrak{m} = \mathbb{F}$ it is subjective by our choice of x_i). After tensoring $R/\mathfrak{a}_\chi$ ($\cong \mathcal{O}$), we get a surjection:

$$\mathcal{O}^r \cong (R/\mathfrak{a}_\chi)^r \to H/\mathfrak{a}_\chi H = h_k^{ord}(\Gamma_1(N') \cap \Gamma_0(q), \chi\chi'; \mathcal{O}).$$

Since the right-hand side is $\mathcal{O}$-free of rank r as already remarked, π mod $\mathfrak{a}_\chi$ is an isomorphism, showing $\mathrm{Ker}(\pi) \subset \mathfrak{a}_\chi^r$. Since $\bigcap_\chi \mathfrak{a}_\chi = \{0\}$, we find that π is a surjective isomorphism. □

3.2 Modular Galois Representations

In this section, we briefly recall the theory of Galois representations attached to elliptic Hecke eigenforms and prove the universality of the couple of the Hecke algebra and its Galois representation in an appropriate sense.

3.2.1 Hecke eigenforms

Let A be a commutative ring (with identity) and f be a cusp form in $S_k(\Gamma_0(N), \chi; A)$ which is an eigenvector of all Hecke operators $T(n)$. Then we have a map $\lambda : h_k(N, \chi; A) \to A$ giving the eigenvalues: $h(f) = \lambda(h)f$. So λ is an A-algebra homomorphism. A cusp form $f \in S_k(\Gamma_0(N), \chi; A)$ is called a Hecke eigenform if $a(1, f) = 1$ and $h(f) = \lambda(h)f$ for all $h \in h_k(N, \chi; A)$ with an eigenvalue λ. The integer N is called the level of f.

Writing $\mathrm{Hom}_{A\text{-}alg}(B,C)$ for the set of all A-algebra homomorphisms from an A-algebra B into another A-algebra C, we see that

$$\mathrm{Hom}_{A\text{-}alg}(h_k(N,\chi;A),A) \subset \mathrm{Hom}_A(h_k(N,\chi;A),A),$$

where the right-hand side is the A-module of all A-linear maps of $h_k(N,\chi;A)$ into A. As already seen in the duality theorem,

$$\mathrm{Hom}_A(h_k(N,\chi;A),A) \cong S_k(\Gamma_0(N),\chi;A)$$

via $\phi \mapsto \sum_{n=1}^{\infty} \phi(T(n))q^n$ because $\phi(T(n)) = (\phi, T(n)) = a(1, T(n)g) = a(n,g)$ if g corresponds to ϕ. In particular, $a(n,f) = a(1,T(n)f) = \lambda(T(n))a(1,f)$. From this, we see that $f = a(1,f)\sum_{n=1}^{\infty}\lambda(T(n))q^n$ for $\lambda \in \mathrm{Hom}_{A\text{-}alg}(h_k(N,\chi;A),A)$. Thus if f is an eigenvector of all Hecke operators, it is a constant multiple of a Hecke eigenform $\sum_{n=1}^{\infty}\lambda(T(n))q^n$, and we obtain the following multiplicity one result.

Proposition 3.21 *If $f \in S_k(\Gamma_0(N),\chi;A)$ is a Hecke eigenform with eigenvalues given by an A-algebra homomorphism $\lambda : h_k(N,\chi;A) \to A$, then*

$$f = \sum_{n=1}^{\infty} \lambda(T(n))q^n,$$

because we have assumed an extra condition: $a(1,f) = 1$ for f to be a Hecke eigenform. In particular, the λ-eigenspace in $S_k(\Gamma_0(N),\chi;A)$ is A-free of rank 1 generated by a unique Hecke eigenform.

There is another type of multiplicity one result when $A = \mathbb{C}$ for which we refer to Miyake's book [MFM]. To describe this, we say that two cusp forms f and g are *equivalent* and write $f \sim g$ if f and g are eigenvectors of $T(\ell)$ for almost all primes ℓ (that is, except for finitely many) and have the same eigenvalues again for almost all ℓ. In each equivalence class, there is a Hecke eigenform f° in $S_k(\Gamma_1(N))$ with minimal level N. This N is called the conductor of each element in the class and f° is called a *primitive* form in the class.

Theorem 3.22 (Strong multiplicity one theorem) *Suppose that f is an element of $S_k(\Gamma_0(N),\chi;\mathbb{C})$ and is an eigenvector of the operator $T(\ell)$ for almost all primes ℓ. Then we have*

(1) *The primitive form f° in the class of f is uniquely determined by f;*
(2) *If $f \in S_k(\Gamma_0(N),\chi)$, the level N is a multiple of the conductor C of f;*

(3) *The cusp form f is a linear combination of elements in*

$$\left\{ f^\circ(tz) | 0 \le t | \frac{N}{C} \right\};$$

(4) *The forms $\{f^\circ(tz) | t \in \mathbb{N}\}$ are linearly independent over $\mathbb{C}$;*

(5) *If N is equal to the conductor of χ, then $N = C$ and f is a constant multiple of a primitive form f°.*

This theorem tells us that basically a Hecke eigenform is determined by the eigenvalues of $T(p)$ for almost all primes p. Thus this theorem is strong, but it is valid only over $\mathbb{C}$ (equivalently, over A of characteristic 0). On the other hand, the prior proposition is valid over any A but requires knowledge of all eigenvalues of $T(n)$. A modular form in the class of f whose level N is not equal to the conductor C is called an *old* form. An old form is a linear combination of $f^\circ(tz)$ with $t > 0$. On the other hand, a linear combination of primitive forms of conductor N is sometimes called a new form of level N.

Now we want to show that the subalgebra $h'_k(N, \chi; \mathbb{C})$ of $h_k(N, \chi; \mathbb{C})$ generated by $T(n)$ for n prime to N is semi-simple, that is, isomorphic to a product of $\mathbb{C}$ as $\mathbb{C}$-algebras. The idea is to introduce a positive definite inner product: $(\, , \,)$ on $S_k(\Gamma_0(N), \chi)$ such that $(T'(n)f, g) = (f, T'(n)g)$ for $T'(n) = \sqrt{\chi(n)}^{-1} T(n)$ for n prime to N. Since $T'(n)$ is self adjoint, it is semi-simple (that is, diagonalizable).

If an operator $T : V \to V$ on a K-vector space V is diagonalizable, then for its eigenvalue λ, we write $V[\lambda]$ for the λ-eigenspace of T. By the diagonalizability of T, $V = \oplus_\lambda V[\lambda]$. If we have another diagonalizable operator $S : V \to V$ which commutes with T, $TSv = STv = \lambda Sv$ for $v \in V[\lambda]$. Thus S induces $S : V[\lambda] \to V[\lambda]$. Applying the above argument to $S : V[\lambda] \to V[\lambda]$, $V[\lambda]$ is further decomposed into a direct sum of eigenspaces of S. In this way, if we have a set $\Sigma = \{T\}$ of mutually commutative (diagonalizable) linear operators, we can simultaneously diagonalize all operators in Σ. Then the subalgebra of $\mathrm{End}(V)$ generated by Σ over the field K is isomorphic to the product of copies of K, just associating T to the tuple of eigenvalues in K. Applying this argument to $\Sigma = \{T(n) | n \text{ is prime to } N\}$, we get the semi-simplicity of $h'_k(N, \chi; \mathbb{C})$, once we find an appropriate inner product as above.

Here is how to define an inner product: By an easy computation, we find for $\gamma \in G(\mathbb{R})_+$, $d(\gamma(z)) = \det(\gamma) j(\gamma, z)^{-2} dz$ and $d(\gamma(\bar{z})) = \det(\gamma) j(\gamma, \bar{z})^{-2} d\bar{z}$. Thus

$$d\gamma(z) \wedge d\gamma(\bar{z}) = \det(\gamma)^2 |j(\gamma, z)|^{-4} dz \wedge d\bar{z}.$$

Since $y = \mathrm{Im}(z)$ satisfies $y \circ \gamma = \det(\gamma)|j(\gamma,z)|^{-2}y$, $y^{-2}dz \wedge d\bar{z} = -2iy^{-2}dx \wedge dy$ is $G(\mathbb{R})_+$-invariant. Let $\Gamma \subset SL_2(\mathbb{Z})$ be a subgroup of finite index, and choose an open set Φ in $\mathfrak{H}$ such that the projection $\pi : \mathfrak{H} \twoheadrightarrow \Gamma\backslash\mathfrak{H}$ brings Φ into a dense open subset of the quotient $\Gamma\backslash\mathfrak{H}$ (Φ is called a fundamental domain of Γ). For any function $f : \Gamma\backslash\mathfrak{H} \to \mathbb{C}$, we consider the integral $\int_\Phi f(x+iy)y^{-2}dxdy$. Since $y^{-2}dx \wedge dy$ is $G(\mathbb{R})_+$-invariant, the measure $y^{-2}dxdy$ is again $G(\mathbb{R})_+$-invariant. Thus the integral $\int_\Phi f(x+iy)y^{-2}dxdy$ is well defined, independent of the choice of Φ. Thus we write $\int_{\Gamma\backslash\mathfrak{H}} f(x+iy)y^{-2}dxdy = \int_\Phi f(x+iy)y^{-2}dxdy$; in other words, $y^{-2}dxdy$ induces a positive measure on $\Gamma\backslash\mathfrak{H}$. Since $f(\gamma(z)) = \chi(a)f(z)j(\gamma,z)^k$ for $f \in S_k(\Gamma_0(N),\chi)$ (where a is the upper left shoulder entry of γ), the complex conjugate $\bar{f}$ satisfies $\bar{f}(\gamma(z)) = \bar{\chi}(a)\bar{f}(z)\overline{j(\gamma,z)}^k$. Thus for $f, g \in S_k(\Gamma_0(N),\chi)$, $f\bar{g}(z)\,\mathrm{Im}(z)^k$ is $\Gamma_0(N)$-invariant. Then we define the Petersson inner product $(\ ,\ )$ by

$$(f,g) = (f,g)_N = \int_{\Gamma_0(N)\backslash\mathfrak{H}} f(z)\overline{g(z)}y^{k-2}dxdy. \tag{3.48}$$

This is obviously positive definite if it converges. Its convergence follows from the fact that f decreases exponentially towards cusps of $\Gamma_0(N)$. The fact that $(T'(n)f,g) = (f,T'(n)g)$ is a plain computation using the formula (3.22) (see [IAT] Chapter 3, [MFM] or [LFE] Section 5.3).

PROPOSITION 3.23 *Suppose that A is an integral domain of characteristic 0. The A-subalgebra $h'_k(N,\chi;A)$ of $h_k(N,\chi;A)$ generated by $T(n)$ for all n prime to N is reduced, that is, its nilradical is reduced to $\{0\}$. Suppose further that the conductor of χ coincides with N. Then $h_k(N,\chi;A)$ itself is reduced if A is an integral domain of characteristic 0, and $h_k(N,\chi;A) = h'_k(N,\chi;A)$ if A is a $\mathbb{Q}$-algebra.*

Proof We already know that $h'_k(N,\chi;\mathbb{C})$ is reduced. Since $h'_k(N,\chi;\mathbb{Z}[\chi]) \subset h'_k(N,\chi;\mathbb{C})$, $h'_k(N,\chi;\mathbb{Z}[\chi])$ is reduced. If A is an integral domain of characteristic 0, it is $\mathbb{Z}$-flat; so, $h'_k(N,\chi;A)$ is a subalgebra of $h'_k(N,\chi;K)$ for the field of fractions K of A. Since K is a separable extension of $\mathbb{Q}$, $h'_k(N,\chi;K) = h'_k(N,\chi;\mathbb{Q}) \otimes_\mathbb{Q} K$ is reduced (see [CRT] Section 26). Then the subalgebra $h'_k(N,\chi;A)$ is reduced.

Since all operators in $h'_k(N,\chi;\mathbb{C})$ are simultaneously diagonalizable, we have a base of $S_k(\Gamma_0(N),\chi)$ made up of common eigenvectors of all operators of $h'_k(N,\chi;\mathbb{C})$. Then by the strong multiplicity one theorem, if N coincides with the conductor of χ, all elements of this base are primitive; so, they are Hecke eigenforms, that is, eigenvectors of all $T(n)$

not just for $T(n)$ with n prime to N. This shows that $h_k(N,\chi;\mathbb{C})$ is semi-simple. Then by the same argument as above, $h_k(N,\chi;A)$ is reduced. Since a primitive form is determined by the eigenvalues of $T(n)$ for n prime to N by the strong multiplicity one theorem, the dimension over $\mathbb{C}$ of $h'_k(N,\chi;\mathbb{C})$ is equal to the number of elements in the base made of primitive forms. Thus

$$\dim_{\mathbb{C}} h'_k(N,\chi;\mathbb{C}) = \dim_{\mathbb{C}} S_k(\Gamma_0(N),\chi).$$

By the duality theorem,

$$\dim_{\mathbb{C}} h_k(N,\chi;\mathbb{C}) = \dim_{\mathbb{C}} S_k(\Gamma_0(N),\chi).$$

Thus $h'_k(N,\chi;\mathbb{C}) = h_k(N,\chi;\mathbb{C})$. Since

$$h'_k(N,\chi;\mathbb{Q}) \otimes_{\mathbb{Q}} \mathbb{C} = h'_k(N,\chi;\mathbb{C}) = h_k(N,\chi;\mathbb{C}) = h_k(N,\chi;\mathbb{Q}) \otimes_{\mathbb{Q}} \mathbb{C},$$

we know the two Hecke algebras with coefficients in $\mathbb{Q}$ have the same dimension over $\mathbb{Q}$, and hence $h'_k(N,\chi;\mathbb{Q}) = h_k(N,\chi;\mathbb{Q})$. This shows that

$$h'_k(N,\chi;A) = h'_k(N,\chi;\mathbb{Q}) \otimes_{\mathbb{Q}} A = h_k(N,\chi;\mathbb{Q}) \otimes_{\mathbb{Q}} A = h_k(N,\chi;A)$$

as long as A is a $\mathbb{Q}$-algebra. □

Let $\lambda' : h'_k(N,\chi;\mathbb{C}) \to \mathbb{C}$ be a $\mathbb{C}$-algebra homomorphism. Since $h' = h'_k(N,\chi;\mathbb{C})$ is a subalgebra of $h = h_k(N,\chi;\mathbb{C})$, we can extend λ' to a $\mathbb{C}$-algebra homomorphism $\lambda : h \to \mathbb{C}$. The point is that $\mathrm{Ker}(\lambda')$ is a maximal ideal of h'. Take any maximal ideal $\mathfrak{m}$ of h containing $\mathrm{Ker}(\lambda')$ (that exists, see [CRT] Theorem 1.1). Then $h/\mathfrak{m}$ is a finite extension of $h'/\mathrm{Ker}(\lambda') = \mathbb{C}$; therefore, $h/\mathfrak{m} = \mathbb{C}$, and $\lambda(t) = t \bmod \mathfrak{m}$ is the extension of λ'. Then the Hecke eigenform associated with λ belongs to the λ'-eigenspace. Thus writing $Spec(h')(\mathbb{C}) = \mathrm{Hom}_{\mathbb{C}-alg}(h',\mathbb{C})$, we have the eigenspace decomposition:

$$S_k(\Gamma_0(N),\chi) = \bigoplus_{\lambda' \in Spec(h')(\mathbb{C})} S[\lambda']$$

for λ'-eigenspaces $S[\lambda']$. Then we have, by Proposition 3.23 combined with the strong multiplicity one theorem, the following fact:

COROLLARY 3.24 *Let the notation be as above. Let $f_{\lambda'}$ be the primitive form associated with λ' and $C(\lambda')$ be the conductor of $f_{\lambda'}$. Then $C(\lambda')$ is a divisor of N, and $\{f_{\lambda'}(tz)\}$ for all non-negative divisors t of $\frac{N}{C(\lambda')}$ gives a base of $S[\lambda']$. In particular, $h_k(N,\chi;\mathbb{C}) \neq h'_k(N,\chi;\mathbb{C})$ if there exists a primitive form $f_{\lambda'}$ whose conductor is a proper divisor of N.*

Here we add a remark on the relation between local components of two Hecke algebras $h_k(N;\mathcal{O})$ and $h_k(\Gamma_1(N)\cap\Gamma_0(q);\mathcal{O})$ when q is a prime outside N. Here $\mathcal{O}$ is the p-adic integer ring of a finite extension of $\mathbb{Q}p$. By definition, $T(q)$ acting on $S_k(\Gamma_1(N)\cap\Gamma_0(q);\mathcal{O})$ and $T(q)$ acting on $S_k(\Gamma_1(N);\mathcal{O})$ are different; so, to avoid confusion, we sometimes write $U(q)$ for the operator with level q. Then if f is of level N, we have $f, f(qz) \in S_k(\Gamma_1(N)\cap\Gamma_0(q))$. Let h be a local ring of $h_k(N;\mathcal{O})$ and write $S \subset S_k(\Gamma_1(N);\mathcal{O})$ for the corresponding direct summand (under the duality theorem). Thus h acts on S. Let S_q be the space generated by $f \in S$ and $[q]f = f(qz) = \sum_n a(n,f)q^{nq}$. We consider the Hecke algebra h_q of S_q, that is, the $\mathcal{O}$-subalgebra of $\mathrm{End}_{\mathcal{O}}(S_q)$ generated by $T(n)$ for n prime to q and $U(q)$. By definition, we have a surjective $\mathcal{O}$-algebra homomorphism $\pi : h_k(\Gamma_1(N)\cap\Gamma_0(q);\mathcal{O}) \to h_q$ sending $T(n)$ to $T(n)$ and $U(q)$ to $U(q)$.

LEMMA 3.25 *Let $\mathfrak{m}$ be the maximal ideal of h. Write $t_q(n)$ and $u(q)$ (resp. $t(n)$) for the projection of the Hecke operator $T(n)$ and $U(q)$ in h_q (resp. h). Then $h_q \cong h[X]/(X^2 - t(q)X + \langle q\rangle)$. The isomorphism sends $u(q)$ to X and $t_q(n)$ to $t(n)$ for n outside q.*

Proof By the strong multiplicity one theorem (3)–(4), the map $[q] : S \to [q]S$ is an isomorphism and $S \cap [q]S = \{0\}$. Thus $S_q = S \oplus [q]S$. We have $a(n, U(q)f) = a(nq, f)$. Thus $U(q) \circ [q]$ is the identity. By the duality theorem, we see that $h_q \otimes_{\mathcal{O}} K$ is dual to $S_q \otimes_{\mathcal{O}} K$. Let $h' = \mathcal{O}[t(n)|q \nmid n] \subset h$. Since $t_q(n)[q] = [q]t(n)$, we have an embedding of the subalgebra of h' into h_q sending $t(n)$ to $t_q(n)$. Thus $h_q = h'[u(q)]$. We first prove that $u(q)$ satisfies the equation $X^2 - t(q)X + \langle q\rangle$. By (3.45), we see that that $a(n, T(q)f) = a(nq, f) + a(n/q, \langle q\rangle f)$. Thus $U(q) = T(q) - \langle q\rangle[q]$ on S and on $[q]S$, it is $[q]^{-1}$. Thus the matrix form of $U(q)$ on $S \oplus [q]S$ (regarding S and $[q]S$ as h-modules) is

$$\begin{pmatrix} t(q) & -\langle q\rangle[q] \\ [q]^{-1} & 0 \end{pmatrix}.$$

Thus $U(q)$ satisfies $X^2 - t(q)X + \langle q\rangle$. This, in particular, shows that $h \subset h_q$ because $Tr_h(u(q)) = t(q) \in h_q$, and thus we have a surjective h-algebra homomorphism $h[X]/(X^2 - t(q)X + \langle q\rangle)$ onto h_q. Since $h_q \otimes_{\mathcal{O}} K$ is dual to $S_q \otimes_{\mathcal{O}} K$ for $K = Frac(\mathcal{O})$, we see that after tensoring K, the two algebras have the same dimension over K. Since the two algebras are $\mathcal{O}$-free, we conclude the desired assertion. □

If $q = p$ and if $t(p) \in h^\times$, then $X^2 - t(p)X + p^{k-1}\langle p\rangle$ has two distinct roots, one unit and the other 0 modulo $\mathfrak{m}$ (if $k \geq 2$). Thus $h_p \cong h \oplus h$, and the isomorphism sends $u(q)$ to the unit root for the first factor and non-unit for the second factor h. When we refer to a p-ordinary factor h of h_p, it means h in h_p with $u(p) \in h^\times$, which is uniquely determined if it exists. Similarly, if $q \neq p$ but $X^2 - t(q)X + \langle q\rangle$ has two distinct roots modulo $\mathfrak{m}$, we again have $h_q \cong h \oplus h$.

Exercises

(1) Give a detailed proof of Corollary 3.24 using the strong multiplicity one theorem.

(2) Let $\lambda : h_k'(N,\chi;\mathbb{C}) \to \mathbb{C}$ be an algebra homomorphism. Find a formula for the dimension of the λ-eigenspace $S[\lambda]$ in $S_k(\Gamma_0(N),\chi)$ in terms of the conductor C of an element of $S[\lambda]$. (By definition, the conductor of any element of $S[\lambda]$ is equal to C.)

3.2.2 Galois representation of Hecke eigenforms

Let $\mathcal{O}$ be a discrete valuation ring finite flat over $\mathbb{Z}_p$. In other words, $\mathcal{O}$ is the p-adic integer ring of a finite extension K of $\mathbb{Q}_p$. Let $\mathfrak{m}$ be the maximal ideal of $\mathcal{O}$. Then $\mathcal{O} = \varprojlim_n \mathcal{O}/\mathfrak{m}^n$ and hence is a profinite ring.

Let $\mathbb{Q}^{(N,\infty)}$ be the maximal extension of $\mathbb{Q}$ in which all primes $p \nmid N$ are unramified (∞ may ramify, that is, $\mathbb{Q}^{(N,\infty)}$ can be imaginary). Write I_p for the inertia group at p inside $\mathrm{Gal}(\overline{\mathbb{Q}}/\mathbb{Q})$. Then letting U be the closure of the subgroup generated by all conjugates of I_p for all $p \nmid N$, $\mathbb{Q}^{(N,\infty)}$ is the fixed field of U inside $\overline{\mathbb{Q}}$. By Hermite's theorem, $\mathbb{Q}^{(\infty)} = \mathbb{Q}$, but if $N > 1$, $\mathbb{Q}^{(N,\infty)}/\mathbb{Q}$ is an infinite extension, because it contains the field of all p-power roots of unity for $p|N$.

Let $\mathfrak{G} = \mathfrak{G}_N = \mathrm{Gal}(\mathbb{Q}^{(N,\infty)}/\mathbb{Q})$. This is a profinite group:

$$\mathfrak{G} = \varprojlim_{K \subset \mathbb{Q}^{(N,\infty)}} \mathrm{Gal}(K/\mathbb{Q}),$$

where K runs over all finite Galois extensions inside $\mathbb{Q}^{(N,\infty)}$ over $\mathbb{Q}$; the order is given by $K > K' \iff K \supset K'$ and the projection $\mathrm{Gal}(K/\mathbb{Q}) \twoheadrightarrow \mathrm{Gal}(K'/\mathbb{Q})$ is given by $\sigma \mapsto \sigma|_{K'}$ if $K \supset K'$. Thus $\mathfrak{G}$ is a compact topological group. For each prime p, we pick a prime ideal $\mathfrak{P}$ of the integer ring $O^{(N,\infty)} = \bigcup_{K \subset \mathbb{Q}^{(N,\infty)}} O_K$ of $\mathbb{Q}^{(N,\infty)}$ such that $\mathfrak{P} \cap \mathbb{Z} = (p)$. Then the decomposition group $D_{\mathfrak{P}}$ of $\mathfrak{P}$ is just the set of all $\sigma \in \mathfrak{G}$ such that $\mathfrak{P}^\sigma = \mathfrak{P}$. Writing $\mathbb{F}^{(N,\infty)} = O^{(N,\infty)}/\mathfrak{P}$, we have an isomorphism

$D_{\mathfrak{P}} \cong \mathrm{Gal}(\mathbb{F}^{(N,\infty)}/\mathbb{F}_p)$ if $p \nmid N$ by $\sigma \mapsto (x \mod \mathfrak{P} \mapsto x^\sigma \mod \mathfrak{P})$. Since $\mathrm{Gal}(\mathbb{F}^{(N,\infty)}/\mathbb{F}_p)$ is generated by $Frob_{\mathfrak{P}} : x \mapsto x^p$, we can regard $Frob_{\mathfrak{P}}$ as an element of $\mathfrak{G}$. Since $Frob_{\mathfrak{P}^\sigma} = \sigma^{-1} Frob_{\mathfrak{P}} \sigma$, the Frobenius elements for p make a conjugacy class $Frob_p$ in $\mathfrak{G}$. We put

$$\Sigma = \{Frob_{\mathfrak{P}} | \mathfrak{P} \nmid N\} = \bigcup_{p \nmid N} Frob_p.$$

By the Chebotarev density theorem ([CFN] Theorem 6.4), the restriction map $\mathfrak{G} \to \mathrm{Gal}(K/\mathbb{Q})$ given by $\sigma \mapsto \sigma|_K$ induces a surjection of Σ onto $\mathrm{Gal}(K/\mathbb{Q})$ for all finite Galois extensions $K \subset \mathbb{Q}^{(N,\infty)}$, and hence, Σ is dense in $\mathfrak{G}$.

Let A be a profinite $\mathcal{O}$-algebra sharing the same residue field $\mathbb{F}$ with $\mathcal{O}$. Let $\rho : \mathfrak{G} \to GL_n(A)$ be a continuous representation. By the density of Σ, the values $\rho(Frob_{\mathfrak{P}})$ over Σ determine the representation ρ (not just its isomorphism class). If $\overline{\rho} = \rho \mod \mathfrak{m}_A$ for the maximal ideal $\mathfrak{m}_A$ of A is absolutely irreducible, by Carayol's theorem (Chapter 2, Proposition 2.13), ρ is uniquely determined by $\mathrm{Tr}(\rho)$ up to isomorphisms. Thus the isomorphism class over A of ρ is determined by $\mathrm{Tr}(\rho)$ over Σ if $\overline{\rho}$ is absolutely irreducible.

If A is an integral domain, regarding $\rho : \mathfrak{G} \to GL_n(A)$ as having values in $GL_2(Frac(A))$ for the field of fractions $Frac(A)$ of A, if it is absolutely irreducible (or more generally, if semi-simple), by the Brauer–Nesbitt theorem (Corollary 2.8 in Chapter 2), the isomorphism class of ρ over $Frac(A)$ is again determined by $\mathrm{Tr}(\rho)$ over Σ.

We write h'_k for

$$h'_k(N;\mathbb{Z}_p) = \mathbb{Z}_p[T(n)|n \text{ is prime to } N] \subset h_k(N;\mathbb{Z}_p).$$

We know that

$$S_k(\Gamma_1(N);K) = \bigoplus_{\chi} S_k(\Gamma_0(N),\chi;K)$$

for any field K of characteristic 0, where χ runs over all characters of $(\mathbb{Z}/N\mathbb{Z})^\times$. By duality,

$$h_k(N;K) = \bigoplus_{\chi} h_k(N,\chi;K).$$

This remains true for a general ring K if its characteristic is prime to the order of $(\mathbb{Z}/N\mathbb{Z})^\times$. Thus if A is an integral domain of characteristic 0 with $K = Frac(A)$, $h_k(N;A) \subset h_k(N;A) \otimes_A K = h_k(N;K)$, and h'_k is reduced. Then we consider the algebraic closure $\overline{\mathbb{Q}}_p$ of $\mathbb{Q}_p$, and write $Spec(h'_k)(\overline{\mathbb{Q}}_p) = \mathrm{Hom}_{\mathbb{Z}_p\text{-}alg}(h'_k, \overline{\mathbb{Q}}_p)$. For each $\lambda' \in Spec(h'_k)(\overline{\mathbb{Q}}_p)$, let $\mathbb{Q}_p(\lambda')$

be the subfield of $\overline{\mathbb{Q}}_p$ generated by $\lambda'(T(n))$ for all n prime to N, which is a finite extension of $\mathbb{Q}_p$ (Exercise 3 below). Let $\mathbb{Z}_p(\lambda')$ be the p-adic integer ring of $\mathbb{Q}_p(\lambda')$. The homomorphism λ' is associated with a primitive form f of conductor C. Then we define an $\mathcal{O}$-algebra homomorphism $\lambda : h_k(C,\chi;\mathcal{O}) \to \overline{\mathbb{Q}}_p$ by $h(f) = \lambda(h)f$. We define $\mathbb{Z}_p(\lambda)$ and $\mathbb{Q}_p(\lambda)$ in a similar way to $\mathbb{Z}_p(\lambda')$ and $\mathbb{Q}_p(\lambda')$ using all $T(n)$ instead of $T(n)$ with n prime to N. Similarly, we write $\mathbb{Q}(\lambda)$ for the subfield of $\overline{\mathbb{Q}}$ generated by $\lambda(T(n))$ for all n. Since $h_k(N;\mathbb{Z})$ is free of finite rank over $\mathbb{Z}$, $\lambda(h_k(N;\mathbb{Z})) \subset \overline{\mathbb{Q}}_p$ is free of finite rank over $\mathbb{Z}$. Thus $\lambda(T(n))$ is an algebraic integer inside the finite extension $\mathbb{Q}(\lambda)/\mathbb{Q}$. Thus we may think (if we want) that $\lambda(T(n)) \in \overline{\mathbb{Q}} \subset \mathbb{C}$.

THEOREM 3.26 *Let $\lambda' \in Spec(h'_k(N;\mathbb{Z}_p))(\overline{\mathbb{Q}}_p)$ with $\lambda'(\langle \ell \rangle) = \ell^{k-1}\chi(\ell)$ for a Dirichlet character χ modulo N. Then we have:*

(1) (Shimura, Deligne, Serre) *There exists a continuous absolutely irreducible Galois representation $\rho_{\lambda'} : \mathfrak{G}_{pN} \to GL_2(\mathbb{Q}_p(\lambda'))$ such that $\operatorname{Tr}\rho(Frob_\ell) = \lambda'(T(\ell))$ and $\det\rho(Frob_\ell) = \chi(\ell)\ell^{k-1}$ for all primes ℓ outside pN. Moreover, ρ has values in $GL_2(\mathbb{Z}_p(\lambda'))$.*

(2) (Deligne, Mazur–Wiles) *Suppose $k \geq 2$ and that $\lambda(T(p))$ is a unit in $\mathbb{Z}_p(\lambda)$. Then the restriction of ρ to $D_{\mathfrak{P}}$ for $\mathfrak{P}|p$ is isomorphic to an upper triangular representation*

$$\sigma \mapsto \begin{pmatrix} \varepsilon(\sigma) & * \\ 0 & \delta(\sigma) \end{pmatrix},$$

where δ is unramified and $\delta(Frob_{\mathfrak{P}})$ is the unique p-adic unit root of $X^2 - \lambda(T(p))X + \chi(p)p^{k-1} = 0$. Here we have used the convention that $\chi(p) = 0$ if $p|N$. This, in particular, implies that $\mathbb{Z}_p(\lambda')$ contains $\lambda(T(p))$.

(3) (Langlands) *Let q be a prime different from p, and let C (resp. N) be the conductor of χ (resp. λ'). Write $N = q^e N'$ (resp. $C = q^{e'} C'$) so that $q \nmid N'$ (resp. $q \nmid C'$).*

(a) *If $e = e' > 0$, ρ restricted to the inertia group $I_{\mathfrak{Q}}$ for $\mathfrak{Q}|q$ is equivalent to a diagonal representation:*

$$\sigma \mapsto \begin{pmatrix} \chi(\sigma) & 0 \\ 0 & 1 \end{pmatrix},$$

where we regard the Dirichlet character χ as a Galois character $\chi : \mathfrak{G} \to \mathbb{Z}_p(\lambda)^\times$ by class field theory. Moreover, ρ restricted to the decomposition group at q is still diagonal, and writing δ_q for the unique unramified character appearing in $\rho|_{D_q}$, we have a unique extension of λ' to $h'_k[U(q)]$ such that $\delta_q(Frob_q) = \lambda(U(q))$.

(b) *If $e = 1$ and $e' = 0$, ρ restricted to the decomposition group $D_{\mathfrak{Q}}$ for $\mathfrak{Q}|q$ is ramified and is equivalent to a diagonal representation:*

$$\sigma \mapsto \begin{pmatrix} \eta(\sigma)\nu(\sigma) & * \\ 0 & \eta(\sigma) \end{pmatrix},$$

where $\nu : D_{\mathfrak{Q}} \to \mathbb{Z}_p^{\times}$ is the p-adic cyclotomic character and η is an unramified character taking $Frob_{\mathfrak{Q}}$ to $\lambda(T(q))$ (and hence $\lambda(T(q)(-\mathbb{Z}_p^{\times}(\lambda'))$.

(4) (Deligne–Serre) *If $k = 1$, then there exists a complex continuous representation $\rho_0 : \mathfrak{G}_N \to GL_2(\mathbb{Q}(\lambda')) \subset GL_2(\mathbb{C})$ with finite image, which is isomorphic to $\rho_{\lambda'}$ over $\mathbb{Q}_p(\lambda')$.*

Here is some history of this difficult result: The existence of $\rho : \mathrm{Gal}(\overline{\mathbb{Q}}/\mathbb{Q}) \to GL_2(\mathbb{Q}_p(\lambda))$ in (1) was first shown by Shimura in the 50's when $k = 2$ (see [IAT] Chapter 7). The unramifiedness outside pN (that is, ρ factoring through $\mathfrak{G}_{pN}$) was shown by Igusa in 1959 (see [DR]). Then the existence of ρ for $k > 2$ was shown directly from weight k Hecke eigenform by Deligne in 1969 (see [D]). A little earlier than Deligne, Shimura found a way to reduce the case of $k > 2$ to $k = 2$ (see Ohta [O1–3]), and a description of this idea (combined with the technique of pseudo-representations) can be found in [LFE] Section 7.5. The case of weight 1 was treated by Deligne and Serre in 1974 [DS]. The analysis of ramification is due to Deligne (1974; see [E]) and Mazur–Wiles (1986; [MW1]) for p and to Langlands for $q \neq p$ in N (1973; [L]). Although we described in the above theorem the result when the local component π_q of the automorphic representation π of λ is either principal or special at q, the analysis of ramification can be done also for supercuspidal π_q (see Carayol [C1]). The above facts are now generalized to holomorphic modular forms on $GL_2(F_{\mathbb{A}})$ for totally real fields F and cohomological cusp forms on $GL_2(F_{\mathbb{A}})$ for imaginary quadratic fields F. The work for totally real fields was started again by Shimura (see Ohta [O1–3]) in the 60's and finished by Blasius–Rogawski [BR] and R. Taylor [Ty1], independently, by different methods in the late 80's. The imaginary quadratic case was settled by R. Taylor [Ty3] in 1994. An exposition covering the construction of the Galois representations in the elliptic modular case can be found in [GMF] Chapter IV. In any case, we will take for granted this theorem, whose proof is a little outside the scope of this book.

Let $\chi_{\mathfrak{G}} : \mathfrak{G} \to \overline{\mathbb{Q}}^{\times}$ be the Galois character associated with χ^*. Write $[x, \mathbb{Q}] \in \mathrm{Gal}(\mathbb{Q}_{ab}/\mathbb{Q})$ for the Artin symbol of $x \in \mathbb{A}^{\times}$, where $\mathbb{Q}_{ab}$ is the maximal abelian extension. Then $\chi_{\mathfrak{G}}([x,\mathbb{Q}]) = \chi^*(x)$ and in particular, $\chi_{\mathfrak{G}}(Frob_\ell) = \chi^*(\ell_\ell) = \chi(\ell)$ for primes $\ell \nmid N$. Thus, by the assertion (1), $\det \rho_{\lambda'}$ coincides with $\nu^{k-1}\chi_{\mathfrak{G}}$ for the p-adic cyclotomic character $\nu : \mathfrak{G}_{pN} \to \mathbb{Z}_p^\times$, because $\nu(Frob_\ell) = \ell$. Here, recall that ν is defined so that $\zeta^\sigma = \zeta^{\nu(\sigma)}$ for all p-power roots of unity $\zeta \in \mathbb{Q}^{(pN,\infty)}$. In particular, writing c for a complex conjugation, $\nu(c) = -1$, because $\zeta^c = \zeta^{-1}$. By the parity condition (P), $\chi(-1) = (-1)^k$. By class field theory, the Artin symbol $x \mapsto [x,\mathbb{Q}] \in \mathrm{Gal}(\mathbb{Q}_{ab}/\mathbb{Q})$ brings $(-1)_\infty$ to c. Note that $\chi_{\mathfrak{G}}(c) = \chi^*((-1)_\infty) = \chi^*((-1)^{(\infty)}) = \chi(-1) = (-1)^k$. Thus we have

$$\det \rho_{\lambda'}(c) = (-1)^k(-1)^{k-1} = -1. \tag{3.49}$$

A sketch of a proof of the irreducibility of $\rho_{\lambda'}$ can be given as follows: If $\rho_{\lambda'}$ is reducible, its semi-simplification is isomorphic to a sum of two characters $\eta, \xi : \mathfrak{G} \to \mathbb{Q}_p(\lambda')$. Since $\eta(Frob_\ell)$ and $\xi(Frob_\ell)$ are roots of $X^2 - \lambda'(T(\ell))X + \chi(\ell)\ell^{k-1}$, they are algebraic integers. From this, one can conclude the existence of two arithmetic Hecke characters $\xi, \eta : \mathbb{A}^\times/\mathbb{Q}^\times \to \mathbb{C}^\times$ such that $\xi(\ell_\ell) = \xi(Frob_\ell)$ and $\eta(\ell_\ell) = \eta(Frob_\ell)$ (see Chapter 1 Section 1, the theorem of Weil and [GMF] Corollary 4.2.3). Thus for any other Hecke character ψ,

$$L(s,\lambda\otimes\psi) = \sum_{n=1}^{\infty} \lambda(T(n))\psi((n))n^{-s} = L(s,\xi\psi)L(s,\eta\psi).$$

It is known that $L(s,\lambda\otimes\psi)$ has an analytic continuation to the whole complex plane as an entire function of s for all ψ. Making ψ either ξ^{-1} or η^{-1}, the above L-function would contain the Riemann zeta function as a factor, which has singularity at $s = 1$. This is a contradiction.

There is one more remark. The Frobenius $Frob_{\mathfrak{P}}$ taking $x \mod \mathfrak{P}$ to $x^p \mod \mathfrak{P}$ is called the Arithmetic Frobenius. This induces an endomorphism of any algebraic variety (or scheme) V defined over $\mathbb{F}_p$. Thus we can think of its pull back map on the algebraic functions on V. If we consider the algebraic closure $\overline{\mathbb{F}}_p$ of $\mathbb{F}_p$ as the set of algebraic functions on $Spec(\overline{\mathbb{F}}_p)$, the pull back map is just $\phi_{\mathfrak{P}} = Frob_{\mathfrak{P}}^{-1}$ which is called the geometric Frobenius. We may regard $\phi_{\mathfrak{P}} \in \mathfrak{G}$ if $p \nmid N$. With each $\lambda' \in Spec(h'_k(N,\chi;\mathbb{Z}_p))(\overline{\mathbb{Q}}_p)$, we have associated a Galois representation $\rho_{\lambda'}$. Representations have a natural duality: the 'contragredient' $\widetilde{\rho}$ of a given representation ρ of a group G in $GL_n(A)$. Thus we can associate with λ' $\widetilde{\rho}_{\lambda'}$ instead of $\rho_{\lambda'}$. This can be formulated nicely using geometric Frobenii instead of arithmetic ones.

To explain this, we first recall the definition of the 'contragredient'. Let $g \in G$ act on $V(\rho) = A^n$ by the matrix multiplication of $\rho(g)$, where

$\rho : G \to GL_n(A)$ is a representation. Then $V^*(\rho) = \mathrm{Hom}_A(V(\rho), A)$ is again an A-free module of rank n. We can let $g \in G$ act on $V^*(\rho)$ by $v^*g(v) = v^*(\rho(g)v)$ for $v^* : V(\rho) \to A$. Obviously this is a right action. To make it into a left action, we define $\widetilde{\rho}(g)v^* = v^*g^{-1}$. In this way, we get an isomorphism class of $\widetilde{\rho}$ (over A). To make a physical construction of an element in the isomorphism class, we can take the standard inner product on $V(\rho) = A^n$, that is, $({}^t(x_1, \ldots, x_n), {}^t(y_1, \ldots, y_n)) = \sum_i x_i y_i$ and identify $V(\rho)$ with $V^*(\rho)$. Then the above action of $\widetilde{\rho}(g)$ is given by the matrix multiplication of ${}^t\rho(g)^{-1}$. Thus $g \mapsto {}^t\rho(g)^{-1}$ gives one physical contragredient representation. By definition, $\mathrm{Tr}(\widetilde{\rho}(g^{-1})) = Tr(\rho(g))$ and $\det(\widetilde{\rho}(g^{-1})) = \det(\rho(g))$.

Since $\phi_{\mathfrak{P}} = Frob_{\mathfrak{P}}^{-1}$, $\widetilde{\rho}_{\lambda'}$ satisfies the assertion (1) of Theorem 3.26 for ϕ_ℓ in place of $Frob_\ell$: $\mathrm{Tr}(\widetilde{\rho}_{\lambda'}(\phi_\ell)) = \lambda'(T(\ell))$ and $\det(\widetilde{\rho}_{\lambda'}(\phi_\ell)) = \ell^{k-1}\chi(\ell)$. The assertion (2) looks the same for $\widetilde{\rho}_{\lambda'}$ but in this case, the character at the upper left corner entry is unramified.

If one uses the jacobian of modular curves to construct modular Galois representations attached to weight 2 forms (as is done in Shimura's book [IAT]), we obtain $\rho_{\lambda'}$ directly, because the association of the jacobian variety to a curve is a covariant functor. If one uses the étale cohomology group of modular curves to construct representations, we get actually $\widetilde{\rho}_{\lambda'}$, because the cohomology functor is contravariant. Thus depending on the situation, the two different associations are used in the literature.

Exercises

(1) For the algebraic closure $\overline{\mathbb{F}}_p$ of $\mathbb{F}_p$, is it true that $\mathbb{F}^{(N,\infty)} = \overline{\mathbb{F}}_p$? What do you think?

(2) Let $G = \{\gamma \in GL_2(\mathbb{Z}_p) | \gamma \mod p \text{ is diagonal}\}$. Find two representations $\rho, \rho' : G \to GL_2(\mathbb{Z}_p)$ such that $\rho \cong \rho'$ as representations into $GL_2(\mathbb{Q}_p)$, but are non-isomorphic as representations into $GL_2(\mathbb{Z}_p)$.

(3) Show that $\mathbb{Q}_p(\lambda) = \mathbb{Q}_p(\lambda')$ if $\lambda(T(p))$ is a unit in $\mathbb{Z}_p(\lambda)$ and $N = 1$ or p.

3.2.3 Galois representation with values in the Hecke algebra

We now try to construct a Galois representation ρ having values in $GL_2(h'_k)$ for $h'_k = h'_k(N, \chi; \mathbb{Z}_p[\chi])$ such that $\rho \mod \mathrm{Ker}(\lambda') \cong \rho_{\lambda'}$ for all $\lambda' \in Spec(h'_k)(\overline{\mathbb{Q}}_p))$. This is a generalization of Theorem 3.26, because after knowing this fact, we will have a Galois representation $\rho_{\mathfrak{a}} : \mathfrak{G} \to GL_2(h'_k/\mathfrak{a})$ satisfying (1) in Theorem 3.26 for all ideals $\mathfrak{a}$ of h'_k.

The ring h'_k is a direct sum of (finitely many) local rings. This can be seen as follows: Let $\mathfrak{m}$ be a maximal ideal of h'_k. Then $\mathfrak{m} \cap \mathbb{Z}_p$ is a maximal ideal of $\mathbb{Z}_p$, which is $p\mathbb{Z}_p$. Thus $\mathfrak{m}$ is a pull back of a maximal ideal of the finite ring h'_k/ph'_k. Thus there are finitely many such maximal ideals. Let Ω be this finite set of all maximal ideals of h'_k. Then regarding $\mathfrak{m} \in \Omega$ as a maximal ideal of $h'_k/p^m h'_k$, we have a descending sequence: $\mathfrak{m} \supset \mathfrak{m}^2 \supset \mathfrak{m}^3 \supset \cdots$. Nakayama's lemma (Lemma 2.3 in Chapter 2) tells us that $I^\alpha = I^{\alpha+1}$ implies $I^\alpha = 0$ for $I = \bigcap_{\mathfrak{m}\in\Omega} \mathfrak{m}$. Thus for large enough α, $I^\alpha = 0$, and $\mathfrak{m}^\alpha + \mathfrak{n}^\alpha = h'_k/p^m h'_k$ for any two distinct $\mathfrak{m}, \mathfrak{n} \in \Omega$ and $\bigcap_{\mathfrak{m}} \mathfrak{m}^\alpha = 0$. Then by the Chinese remainder theorem, $h'_k/p^m h'_k = \bigoplus_{\mathfrak{m}} h'_k/\mathfrak{m}^\alpha$. The α grows as m grows. Since h'_k is free of finite rank over $\mathbb{Z}_p$, it is p-adically complete, and hence

$$h'_k = \varprojlim_m h'_k/p^m h'_k = \bigoplus_{\mathfrak{m}} \varprojlim_\alpha h'_k/\mathfrak{m}^\alpha = \bigoplus_{\mathfrak{m}} h_{\mathfrak{m}},$$

where $h_{\mathfrak{m}} = \varprojlim_\alpha h'_k/\mathfrak{m}^\alpha$ which is a local ring and $\mathfrak{m}h_{\mathfrak{m}}$ is the unique maximal ideal of $h_{\mathfrak{m}}$. Since $GL_2(h'_k) = \prod_{\mathfrak{m}} GL_2(h_{\mathfrak{m}})$, we may concentrate on a local ring $h_{\mathfrak{m}}$.

Hereafter, we write h for one local ring of h'_k and try to construct a Galois representation $\rho_h : \mathfrak{G} \to GL_2(h)$ satisfying $\mathrm{Tr}(\rho(Frob_\ell)) = T_h(\ell)$ and $\det(\rho(Frob_\ell)) = \ell^{k-1}\chi(\ell)$ for all primes $\ell \nmid pN$. The method we use is the method of pseudo-representations of Wiles.

We recall the technique from Chapter 2 Section 2.2: Let A be a local ring with maximal ideal $\mathfrak{m}_A$. We write $E = A/\mathfrak{m}_A$. Let G be a group. We want to characterize the trace of a degree 2 representation $\rho : G \to GL_2(A)$, when 2 is invertible in A and G contains an element c such that $c^2 = 1$ and $\det \rho(c) = -1$. Since 2 is invertible, we know that $V = V(\rho) = V_+ \oplus V_-$ for $V_\pm = \frac{1\pm c}{2} V$. We can show either by the Hensel lemma or the Krull–Schmidt theorem (for A-modules) that $V_\pm \cong A$ (see Chapter 2 Section 2.2 for details). Taking a base $\{v_\pm\}$ of $V_\pm$, we can write $\rho(r) = \begin{pmatrix} a(r) & b(r) \\ c(r) & d(r) \end{pmatrix}$ with respect to this base. Thus $\rho(c) = \begin{pmatrix} -1 & 0 \\ 0 & 1 \end{pmatrix}$. Define another function $x : G \times G \to A$ by $x(r,s) = b(r)c(s)$. Then we have

(W1) $a(rs) = a(r)a(s) + x(r,s)$, $d(rs) = d(r)d(s) + x(s,r)$ and $x(rs,tu) = a(r)a(u)x(s,t) + a(u)d(s)x(r,t) + a(r)d(t)x(s,u) + d(s)d(t)x(r,u)$;

(W2) $a(1) = d(1) = d(c) = 1$, $a(c) = -1$ and $x(r,s) = x(s,t) = 0$ if $s = 1, c$;

(W3) $x(r,s)x(t,u) = x(r,u)x(t,s)$.

These are easy to check (see Subsection 2.2.1 in Chapter 2). Forgetting about the representation ρ, a triple $\{a,d,x\}$ satisfying the three conditions

(W1–3) is called a pseudo representation of Wiles of (G,c). For each pseudo-representation $\pi = \{a,d,x\}$, we define

$$\mathrm{Tr}(\pi)(r) = a(r) + d(r) \quad \text{and} \quad \det(\pi)(r) = a(r)d(r) - x(r,r).$$

By a direct computation using (W1–3), we see that

$$a(r) = \frac{1}{2}(\mathrm{Tr}(\pi)(r) - \mathrm{Tr}(\pi)(rc)), \quad d(r) = \frac{1}{2}(\mathrm{Tr}(\pi)(r) + \mathrm{Tr}(\pi)(rc))$$

and

$$x(r,s) = a(rs) - a(r)a(s), \quad \det(\pi)(rs) = \det(\pi)(r)\det(\pi)(s).$$

Thus the pseudo-representation π is determined by the trace of π as long as 2 is invertible in A. We recall here Proposition 2.16 in Chapter 2:

PROPOSITION 3.27 (A. WILES) *Let G be a group and $R = A[G]$. Let $\pi = \{a,d,x\}$ be a pseudo-representation (of Wiles) of (G,c). Suppose either that there exists at least one pair $(r,s) \in G \times G$ such that $x(r,s) \in A^\times$ or that $x(r,s) = 0$ for all $r,s \in G$. Then there exists a representation $\rho : R \to M_2(A)$ such that $\mathrm{Tr}(\rho) = \mathrm{Tr}(\pi)$ and $\det(\rho) = \det(\pi)$ on G. If A is a topological ring, G is a topological group and all maps in π are continuous on G, then ρ is a continuous representation of G into $GL_2(A)$ under the topology on $GL_2(A)$ induced by the product topology on $M_2(A)$.*

Hereafter we assume that $p > 2$ so that in all $\mathbb{Z}_p$-algebras A, 2 is invertible. Let $\mathfrak{a}$ and $\mathfrak{b}$ be two ideals of A.

LEMMA 3.28 *Let G be a topological group. Let $\pi_\mathfrak{a} : G \to A/\mathfrak{a}$ and $\pi_\mathfrak{b} : G \to A/\mathfrak{b}$ be continuous pseudo-representations of Wiles, respectively. Suppose that these two pseudo-representations are compatible; that is, there exist functions* Tr *and* det *on a dense subset $\Sigma \subset G$ with values in $A/\mathfrak{a} \cap \mathfrak{b}$ such that for all $\sigma \in \Sigma$,*

$$\mathrm{Tr}(\pi_\mathfrak{a}(\sigma)) \equiv \mathrm{Tr}(\sigma) \bmod \mathfrak{a} \quad \textit{and} \quad \mathrm{Tr}(\pi_\mathfrak{b}(\sigma)) \equiv \mathrm{Tr}(\sigma) \bmod \mathfrak{b} \quad (3.50)$$
$$\det(\pi_\mathfrak{a}(\sigma)) \equiv \det(\sigma) \bmod \mathfrak{a} \quad \textit{and} \quad \det(\pi_\mathfrak{b}(\sigma)) \equiv \det(\sigma) \bmod \mathfrak{b}.$$

Then there exists a continuous pseudo-representation $\pi_{\mathfrak{a}\cap\mathfrak{b}} : G \to A/(\mathfrak{a}\cap\mathfrak{b})$ such that

$$\mathrm{Tr}(\pi_{\mathfrak{a}\cap\mathfrak{b}}(\sigma)) = \mathrm{Tr}(\sigma) \quad \textit{and} \quad \mathrm{Tr}(\pi_{\mathfrak{a}\cap\mathfrak{b}}(\sigma)) = \mathrm{Tr}(\sigma).$$

Proof We look at the following exact sequence:

$$0 \to A/(\mathfrak{a}\cap\mathfrak{b}) \xrightarrow{\beta} A/\mathfrak{a} \oplus A/\mathfrak{b} \xrightarrow{\alpha} A/(\mathfrak{a}+\mathfrak{b}) \to 0,$$

where $\beta(a) = (a \mod \mathfrak{a}) \oplus (a \mod \mathfrak{b})$ and $\alpha(a \oplus b) = a - b \mod (\mathfrak{a} + \mathfrak{b})$. Define a pseudo-representation $\pi_{\mathfrak{a}} \oplus \pi_{\mathfrak{b}} : G \to A/\mathfrak{a} \oplus A/\mathfrak{b}$ by the direct sum of the two pseudo-representations. Then $\alpha(\pi_{\mathfrak{a}} \oplus \pi_{\mathfrak{b}}) = 0$ by our assumption (3.50). Thus the pseudo-representation $\pi_{\mathfrak{a}\cap\mathfrak{b}} = \beta^{-1} \circ (\pi_{\mathfrak{a}} \oplus \pi_{\mathfrak{b}})$ has values in $A/(\mathfrak{a} \cap \mathfrak{b})$ and satisfies the desired requirement. □

Let $\lambda' : h'_k \to \overline{\mathbb{Q}}_p$ be a $\mathbb{Z}_p[\chi]$-algebra homomorphism and $\rho = \rho_{\lambda'} : \mathfrak{G}_{pN} \to GL_2(K)$ for $K = \mathbb{Q}_p(\lambda')$. Since $\mathfrak{G}$ is compact, $\rho(\mathfrak{G})$ lands in a maximal compact subgroup $C \subset GL_2(K)$. Since $gCg^{-1} = GL_2(\mathcal{O})$ for $\mathcal{O} = \mathbb{Z}_p(\lambda')$ (see Exercise 1), by changing ρ to $g\rho g^{-1}$, we may assume that $\rho : \mathfrak{G} \to GL_2(\mathcal{O})$. Let $\mathfrak{m}_{\mathcal{O}}$ be the maximal ideal of $\mathcal{O}$, and write $\mathbb{F}$ for $\mathcal{O}/\mathfrak{m}_{\mathcal{O}}$, which is a finite field of characteristic p. Let $\overline{\rho} = (\rho \mod \mathfrak{m}_{\mathcal{O}}) : \mathfrak{G} \to GL_2(\mathbb{F})$. Then $\overline{\rho}$ satisfies $\mathrm{Tr}(\overline{\rho}(Frob_\ell)) = \lambda'(T(\ell)) \mod \mathfrak{m}_{\mathcal{O}}$ and $\det(\overline{\rho}(Frob_\ell)) = \ell^{k-1}\chi(\ell) \mod \mathfrak{m}_{\mathcal{O}}$ for all primes $\ell \nmid pN$. Let $\mathfrak{m}$ be the unique maximal ideal of h'_k containing $\mathrm{Ker}(\lambda')$. Then $\lambda'(\mathfrak{m}) = \mathfrak{m}_{\mathcal{O}} \cap \mathrm{Im}(\lambda')$, and we may identify $\mathrm{Im}(\lambda')$ with $h'_k/\mathrm{Ker}(\lambda')$. Under this identification, $\lambda'(T(\ell)) \mod \mathfrak{m}_{\mathcal{O}} = T(\ell) \mod \mathfrak{m}$. Thus $\overline{\rho} : \mathfrak{G} \to GL_2(\mathbb{F})$ satisfies $\mathrm{Tr}(\overline{\rho}(Frob_\ell)) = T(\ell) \mod \mathfrak{m}$ and $\det(\overline{\rho}(Frob_\ell)) = \ell^{k-1}\chi(\ell) \mod \mathfrak{m}$ for all primes $\ell \nmid pN$. If $\overline{\rho}$ is semi-simple, this condition determines the isomorphism class of $\overline{\rho}$ uniquely. If not, $\overline{\rho}$ is isomorphic to an upper triangular representation:

$$\overline{\rho} \cong \begin{pmatrix} \varepsilon & * \\ 0 & \delta \end{pmatrix}.$$

In this case, we replace $\overline{\rho}$ by $\left(\begin{smallmatrix} \varepsilon & 0 \\ 0 & \delta \end{smallmatrix}\right)$, which is called the semi-simplification of the original $\overline{\rho}$ and has the same trace and determinant as the original. The semi-simple representation $\overline{\rho} = \overline{\rho}_{\mathfrak{m}}$ satisfying $\mathrm{Tr}(\overline{\rho}(Frob_\ell)) = T(\ell) \mod \mathfrak{m}$ and $\det(\overline{\rho}(Frob_\ell)) = \ell^{k-1}\chi(\ell) \mod \mathfrak{m}$ for all primes $\ell \nmid pN$ is called the *residual* representation attached to $\mathfrak{m}$ (or $h_{\mathfrak{m}}$).

THEOREM 3.29 *Assume that $p > 2$. Let h be a local ring of $h'_k(N, \chi; \mathbb{Z}_p[\chi])$ whose residual representation $\overline{\rho}$ is absolutely irreducible. Then there exists a unique Galois representation $\rho = \rho_h : \mathfrak{G}_{pN} \to GL_2(h)$ (up to isomorphism) such that* $\mathrm{Tr}(\rho(Frob_\ell)) = T_h(\ell)$ *and* $\det(\rho(Frob_\ell)) = \ell^{k-1}\chi(\ell)$ *for all primes $\ell \nmid pN$, where $T_h(\ell)$ is the image of $T(\ell)$ in h.*

Proof We apply Lemma 3.28 in the following way. Let $\lambda \in Spec(h)(\overline{\mathbb{Q}}_p) = \mathrm{Hom}_{\mathbb{Z}_p[\chi]-alg}(h, \overline{\mathbb{Q}}_p)$. Since h is a direct summand of h'_k, we can consider $\lambda \in Spec(h'_k) = \sqcup_{\mathfrak{m}\in\Omega} Spec(h_{\mathfrak{m}})(\overline{\mathbb{Q}}_p)$. Thus we have a Galois representation $\rho_\lambda : \mathfrak{G} \to GL_2(\overline{\mathbb{Q}}_p)$. Since h is free of finite rank over $\mathbb{Z}_p$, $Spec(h)(\overline{\mathbb{Q}}_p)$

is a finite set (Exercise 3). Write the elements as λ_j $j = 1, 2, \ldots, m$. Let $P_j = \mathrm{Ker}(\lambda_j)$, which is a prime ideal. We may assume that $P_1, \ldots, P_m$ are all distinct. Since the nilradical of h is an intersection of all prime ideals ([CRT] Section 1 page 3), $\bigcap_j P_j = \{0\}$ because the nilradical of h is trivial (Proposition 3.23). Since λ_j induces an isomorphism of h/P_j with $\mathrm{Im}(\lambda_j)$, the pseudo-representation π_j associated with ρ_{λ_j} has values in $h/P_j = \mathrm{Im}(\lambda_j)$ on $\Sigma = \{Frob_\ell | \ell \nmid pN\}$. Since Σ is dense in $\mathfrak{G}$ and h/P_j is compact, π_j has values in h/P_j. We write $\mathfrak{a}_j = P_1 \cap P_2 \cap \cdots \cap P_j$.

We construct by induction on j, pseudo-representation $\Pi_j : \mathfrak{G} \to h/\mathfrak{a}_j$ such that $Tr(\Pi_j(Frob_\ell)) = T_h(\ell) \mod \mathfrak{a}_j$ and $\det(\Pi_j(Frob_\ell)) = \ell^{k-1}\chi(\ell) \mod \mathfrak{a}_j$ for all primes $\ell \nmid pN$. We put $\Pi_1 = \pi_1$. Suppose that we have constructed Π_{j-1}. Then we define functions $\mathrm{Tr} : \Sigma \to h/\mathfrak{a}_j$ and $\det : \Sigma \to h/\mathfrak{a}_j$ by $\mathrm{Tr}(Frob_\ell) = T_h(\ell) \mod \mathfrak{a}_j$ and $\det(Frob_\ell) = \ell^{k-1}\chi(\ell) \mod \mathfrak{a}_j$. Then apply Lemma 3.28 to the following situation: $\mathfrak{a} = \mathfrak{a}_{j-1}$, $\mathfrak{b} = P_j$, $\pi_\mathfrak{a} = \Pi_{j-1}$ and $\pi_\mathfrak{b} = \pi_j$. Then we have Π_j. By definition, $\Pi = \Pi_m : \mathfrak{G} \to h$ satisfies $\mathrm{Tr}(\Pi(Frob_\ell)) = T_h(\ell)$ and $\det(\Pi(Frob_\ell)) = \ell^{k-1}\chi(\ell)$. The representation associated with Π mod $\mathfrak{m}$ is $\overline{\rho}$ which is absolutely irreducible. Thus we can find $r, s \in \mathfrak{G}$ such that $x(r,s) \in h^\times$ (writing $\Pi = (a, d, x)$), and thus we can lift the pseudo-representation Π to a representation $\rho : \mathfrak{G} \to GL_2(h)$ keeping the formula of traces and determinants. Then by the result of Carayol (Chapter 2 Proposition 2.13), the isomorphism class of ρ is unique. □

Exercises

(1) Let C be a maximal compact subgroup of $GL_n(\mathbb{Q}_p)$. Show that $C = gGL_n(\mathbb{Z}_p)g^{-1}$ for some $g \in GL_2(\mathbb{Q}_p)$. Hint: Let $L = \mathbb{Z}_p^n \subset \mathbb{Q}_p^n = V$ (column vector space). Show first that $L' = \bigcap_{c \in C} cL \subset V$ is a $\mathbb{Z}_p$-free module of rank n using compactness of C and openness of L. Then show that, for a base $x_1, \ldots, x_n$ of L' over $\mathbb{Z}_p$, $C = gGL_n(\mathbb{Z}_p)g^{-1}$ for $g = (x_1, \ldots, x_n)$.

(2) Let $\mathfrak{m}, \mathfrak{n}$ be two distinct maximal ideals of h'_k. If $T(p) \not\equiv 0 \mod \mathfrak{m}$ and $T(p) \not\equiv 0 \mod \mathfrak{n}$, show that $\overline{\rho}_\mathfrak{m}$ is not isomorphic to $\overline{\rho}_\mathfrak{n}$.

(3) Let h is a $\mathbb{Z}_p$-algebra free of finite rank over $\mathbb{Z}_p$. Show that $Spec(h)(\overline{\mathbb{Q}}_p)$ is a finite set.

(4) Find a profinite group G and a representation $\pi : G \to GL_2(\mathbb{Z}_p)$ such that π as a representation into $GL_2(\mathbb{Q}_p)$ is absolutely irreducible but $\overline{\pi} = \pi \mod p$ is reducible as a representation into $GL_2(\mathbb{F}_p)$.

3.2.4 Universal deformation rings

Starting with an absolutely irreducible residual representation $\overline{\rho} : \mathfrak{G}_p \to GL_2(\mathbb{F})$ associated with a Hecke eigenform $f \in S_k(\Gamma_1(p))$ ordinary at p, we want to show (under suitable assumptions) that all representations $\rho : \mathfrak{G}_p \to GL_2(A)$ for a p-profinite local ring A with residue field $\mathbb{F}$ are actually isomorphic to $\rho_h \mod \mathfrak{a}$ for some ideal $\mathfrak{a} \subset h$ if $\rho \mod \mathfrak{m}_A = \overline{\rho}$, ρ is ordinary and $\det(\rho) = \nu^{k-1}\chi_{\mathfrak{G}}$. This is called the universality of the pair (h, ρ_h), and this fact was first conjectured by Mazur and proven by Wiles in many cases.

We fix the notation: Let $\chi : (\mathbb{Z}/N\mathbb{Z})^\times \to \mathcal{O}^\times$ be a character and $\nu : \mathfrak{G}_{pN} \to \mathcal{O}^\times$ be the p-adic cyclotomic character. For any $\mathcal{O}$-algebra A, we have the structure homomorphism $\iota : \mathcal{O} \to A$ giving the $\mathcal{O}$-algebra structure on A. Then we regard $\chi_{\mathfrak{G}}$ and ν as characters having values in $A^\times$ by composing ι. We consider the following property of Galois representations $\rho : \mathfrak{G}_{pN} \to GL_2(A)$ for local profinite $\mathcal{O}$-algebras A:

(det) $\det(\rho) = \nu^{k-1}\chi_{\mathfrak{G}}$;
(ord) For the decomposition group $D = D_p$ at p, $\rho|_{D_p}$ is isomorphic (over A) to an upper triangular representation $\left(\begin{smallmatrix} \varepsilon & * \\ 0 & \delta \end{smallmatrix}\right)$ with unramified δ.

The decomposition group $D_{\mathfrak{P}}$ depends on the choice of a prime ideal $\mathfrak{P}|p$ of the integer ring $O^{(N,\infty)}$ of $\mathbb{Q}^{(N,\infty)}$. However, its conjugacy class is independent of p. The statement (ord) only depends on the conjugacy class, that is, if (ord) is satisfied for one choice of $D_{\mathfrak{P}}$, then it is satisfied by all other choices. In this sense, the decomposition group is written as D_p in (ord). Hereafter we write $\phi = \nu^{k-1}\chi_{\mathfrak{G}}$.

Let N be a positive integer prime to p and $k \geq 2$. Let $h'_k(Np;\mathcal{O})$ be the subalgebra of $h_k(Np;\mathcal{O})$ generated by $T(n)$, n prime to Np. If a prime ℓ is prime to Np, the action $T_{Np}(\ell)$ on $S_k(\Gamma_1(Np);A)$ induces $T_N(\ell)$ on $S_k(\Gamma_1(N);A)$, because the right coset representatives $T_{Np}(\ell)/S_0(pN)$ and $T_N(\ell)/S_0(N)$ are the same. Thus by restricting Hecke operators of $h'_k(Np;A)$ to $S_k(\Gamma_1(N);A)$, we get an algebra homomorphism $\pi : h'_k(Np;A) \twoheadrightarrow h'_k(N;A)$, which sends $T_{Np}(\ell)$ to $T_N(\ell)$ for all $\ell \nmid pN$. We fix a local ring h of $h'_k(Np;\mathcal{O})[U(p)]$. Here we write $U(p)$ for $T_{Np}(p)$ defined on $S_k(\Gamma_1(Np);A)$, which does not match with $T_N(p)$ on $S_k(\Gamma_1(N))$. Here is a technical result which follows from Lemma 3.25 (cf. [H85] Lemma 3.3):

LEMMA

(red) If the image of $U(p)$ is a unit in h, then h is reduced;
(ism) If $k \geq 3$ and $U(p)$ is a unit in h, π extends to h and sends it isomorphically to a local ring h' of $h'_k(N;\mathcal{O})$. This isomorphism sends

$U(p)$ to a unique unit root of $X^2 - T_{h'}(p)X + pS_{h'}(p) = 0$, where $T_{h'}(p)$ and $S_{h'}(p)$ are the images of $T(p)$ and $S(p)$ in h', and $pS(p) = \langle p \rangle$.

Proof Let us prove these two statements: We have the action of $(\mathbb{Z}/p\mathbb{Z})^\times = \Gamma_1(N) \cap \Gamma_0(p)/\Gamma_1(Np)$ on $S_k(\Gamma_1(Np))$ via the character $(\mathbb{Z}/p\mathbb{Z})^\times \ni a \mapsto [a] \in h_k(Np; \mathcal{O})$ (see Corollary 3.19). Since the order $p-1$ of $(\mathbb{Z}/p\mathbb{Z})^\times$ is prime to p, for each character $\xi : (\mathbb{Z}/p\mathbb{Z})^\times \to \mathbb{Z}_p^\times$, the idempotent $\frac{1}{p-1}\sum_a \xi(a)^{-1}[a]$ splits the algebra $h'_k(Np; \mathcal{O})[U(p)]$ into the direct sum of algebras on which $(\mathbb{Z}/p\mathbb{Z})^\times$ acts by different characters. Thus h is inside the direct summand of $h_k(Np; \mathcal{O})[U(p)]$ on which $(\mathbb{Z}/p\mathbb{Z})^\times$ acts trivially. Let h' be the image in $h'_k(N; \mathcal{O})$ of $h \cap h'_k(Np; \mathcal{O})$. By Proposition 3.23, h' is reduced. Thus the only possibility of getting nilpotents is to have old forms in the h-eigenspaces: $h \cdot S_k(\Gamma_1(Np); \mathcal{O})$, and the nilpotent elements possibly reside in $h'_p = h'[X]/(P(X))$ for the Hecke polynomial $P(X) = X^2 - T(p)X + \chi(p)p^{k-1}$ (see Lemma 3.25). If $U(p)$ is a unit in h, the Hecke polynomial $X^2 - T(p)X + \chi(p)p^{k-1} = 0$ has two distinct roots, one is unit and another is non-unit. Thus h'_p has to be reduced. This shows (red). As for (ism), if $k \geq 3$, all primitive forms, whose conductor is divisible by p and whose Neben character is trivial, have non-p-adic-unit eigenvalue for $U(p)$ (see [MFM] or [H85] Lemma 3.3). This shows (ism). □

We suppose, by taking sufficiently large $\mathcal{O}$, $\mathcal{O}$ and h share the same residue field $\mathbb{F}$. Let $\overline{\rho}$ be the residual representation of h. We suppose that $U(p)$ is a unit in h, and hence h is reduced. We consider the following condition for a subfield F of $\mathbb{Q}^{(Np,\infty)}$:

(ai$_F$) $\overline{\rho}$ restricted to $\mathfrak{H}_F = \mathrm{Gal}(\mathbb{Q}^{(Np,\infty)}/F)$ is absolutely irreducible;

(rg$_p$) Suppose (ord) for $\overline{\rho}$ and write $\overline{\rho}|_{D_p} \cong \begin{pmatrix} \overline{\varepsilon} & * \\ 0 & \overline{\delta} \end{pmatrix}$. Then $\overline{\varepsilon}$ is ramified; in particular, $\overline{\varepsilon} \neq \overline{\delta}$ on I_p.

It is obvious that (ai$_F$) implies (ai$_\mathbb{Q}$).

Let $CL_\mathcal{O}$ be the category of p-profinite local $\mathcal{O}$-algebras A with $A/\mathfrak{m}_A = \mathbb{F}$. Hereafter, we always assume that $\mathcal{O}$-algebra is an object of $CL_\mathcal{O}$. A Galois representation $\rho : \mathfrak{G}_{pN} \to GL_2(A)$ is called a deformation of $\overline{\rho}$ if ρ mod $\mathfrak{m}_\mathbb{A} = \overline{\rho}$. Two deformations $\rho, \rho' : \mathfrak{G}_{pN} \to GL_2(A)$ are called strictly equivalent if $x\rho(\sigma)x^{-1} = \rho'(\sigma)$ for all $\sigma \in \mathfrak{G}_{pN}$ for an element $x \in GL_2(A)$ such that $x \equiv 1_2 \mod \mathfrak{m}_A$ ($\Leftrightarrow x \in 1_2 + M_2(\mathfrak{m}_A)$). We write $\rho \approx \rho'$ if they are strictly equivalent.

A couple $(R \in CL_\mathcal{O}, \varrho : \mathfrak{G}_{pN} \to GL_2(R))$ is called a universal couple (over $\mathfrak{G}_{pN}$) if for any deformation $\rho : \mathfrak{G}_{pN} \to GL_2(A)$ of $\overline{\rho}$ ($A \in CL_{\mathcal{O}}$), there exists a unique $\mathcal{O}$-algebra homomorphism $\varphi = \varphi_\rho : R \to A$ such that $\varphi \circ \varrho \approx \rho$ in $GL_2(A)$. If the uniqueness of φ does not hold, we just call (R, ϱ) a versal couple.

A couple $(R^{ord,\phi} \in CL_{\mathcal{O}}, \varrho^{ord,\phi} : \mathfrak{G}_{pN} \to GL_2(R^{ord,\phi}))$ is called a p-ordinary universal couple (over $\mathfrak{G}_{pN}$) with determinant ϕ if $\varrho^{ord,\phi}$ satisfies (ord) and (det) and for any deformation $\rho : \mathfrak{G}_{pN} \to GL_2(A)$ of $\overline{\rho}$ ($A \in CL_{\mathcal{O}}$) satisfying (ord) and (det), there exists a unique $\mathcal{O}$-algebra homomorphism $\varphi = \varphi_\rho : R^{ord,\phi} \to A$ such that $\varphi \circ \varrho^{ord,\phi} \approx \rho$ in $GL_2(A)$. If the uniqueness of φ does not hold, we just call $(R^{ord,\phi}, \varrho^{ord,\phi})$ a versal p-ordinary couple with determinant ϕ.

Similarly a couple $(R^{ord} \in CL_{\mathcal{O}}, \varrho^{ord} : \mathfrak{G}_{pN} \to GL_2(R^{ord}))$ (resp. (R^ϕ, ϱ^ϕ)) is called a p-ordinary universal couple (over $\mathfrak{G}_{pN}$) (resp. a universal couple with determinant ϕ) if ϱ^{ord} satisfies (ord) (resp. $\det(\varrho^\phi) = \phi$) and for any deformation $\rho : \mathfrak{G}_{pN} \to GL_2(A)$ of $\overline{\rho}$ ($A \in CL_{\mathcal{O}}$) satisfying (ord) (resp. $\det(\rho) = \phi$), there exists a unique $\mathcal{O}$-algebra homomorphism $\varphi = \varphi_\rho : R^{ord} \to A$ (resp. $\varphi = \varphi_\rho : R^\phi \to A$) such that $\varphi \circ \varrho^{ord} \approx \rho$ (resp. $\varphi \circ \varrho^\phi \approx \rho$) in $GL_2(A)$.

By the universality, if a universal couple exists, it is unique up to isomorphisms in $CL_{\mathcal{O}}$. Even if we only have a versal couple, it is (non-canonically) unique if it is minimal and noetherian. Here minimality of (R, ϱ) means that if (R', ϱ') is another versal couple, the morphism $\alpha : R' \to R$ such that $\alpha \circ \varrho' \approx \varrho$ is always surjective. Non-canonicality means that there could be an automorphism α of R such that $\varrho \approx \alpha \circ \varrho$. This follows from the fact that if we have a surjective algebra endomorphism of a noetherian ring R, it has to be an automorphism (Exercise 1).

Proposition 3.30 (Mazur) *The minimal versal couples (R, ϱ), (R^{ord}, ϱ^{ord}) and $(R^{ord,\phi}, \varrho^{ord,\phi})$ always exist in $CL_{\mathcal{O}}$, and they are all noetherian. Under* ($\mathrm{ai}_{\mathbb{Q}}$)*, universal couples (R, ϱ) and (R^ϕ, ϱ^ϕ) exist. Under* (rg_p) *and* ($\mathrm{ai}_{\mathbb{Q}}$)*, universal couples $(R^{ord}, \varrho^{ord} : \mathfrak{G}_{pN} \to GL_2(R))$ and $(R^{ord,\phi}, \varrho^{ord,\phi})$ exist (as long as $\overline{\rho}$ satisfies* (ord) *and* (det)*). All these universal rings are noetherian if they exist.*

This fact is proved in Mazur's paper in [GAL]. The existence of the universal couple $(R, \varrho : \mathfrak{G}_{pN} \to GL_2(R))$ (and its noetherian property) is proved also in Theorem 2.26 in Chapter 2 by a different method. Here we prove the existence of the universal couples (R^ϕ, ϱ^ϕ), (R^{ord}, ϱ^{ord}) and $(R^{ord,\phi}, \varrho^{ord,\phi})$ assuming the existence of a universal couple (R, ϱ).

Proof An ideal $\mathfrak{a} \subset R$ is called ordinary if ϱ mod $\mathfrak{a}$ satisfies (ord). Let $\mathfrak{a}^{ord}$ be the intersection of all ordinary ideals, and put $R^{ord} = R/\mathfrak{a}^{ord}$ and $\varrho^{ord} = \varrho$ mod $\mathfrak{a}^{ord}$. If $\rho : \mathfrak{G}_{pN} \to GL_2(A)$ satisfies (ord), we have a unique morphism $\varphi_\rho : R \to A$ such that (ϱ mod Ker(φ_ρ)) $\approx \varphi_\rho \circ \varrho \approx \rho$. So Ker($\varphi_\rho$) is ordinary, and hence Ker(φ_ρ) $\supset \mathfrak{a}^{ord}$. So φ_ρ factors through R^{ord}. All we need to show is the ordinarity of ϱ mod $\mathfrak{a}^{ord}$. Since $\mathfrak{a}^{ord}$ is an intersection of ordinary ideals, we need to show that if $\mathfrak{a}$ and $\mathfrak{b}$ are ordinary, then $\mathfrak{a} \cap \mathfrak{b}$ is ordinary.

To show this, we prepare some notation. Let V be an A-module with an action of $\mathfrak{G}$. Let $I = I_{\mathfrak{P}}$ be an inertia group at p, and put $V_I = V/\sum_{\sigma \in I}(\sigma - 1)V$. Then by (rg$_p$), ρ is ordinary if and only if $V(\rho)_I$ is A-free of rank 1. The point here is that, writing $\pi : V(\rho) \twoheadrightarrow V(\rho)_I$ for the natural projection, then Ker(π) is an A-direct summand of $V(\rho)$ and hence $V(\rho) \cong \text{Ker}(\pi) \oplus V(\rho)_I$ as A-modules (but not necessarily as $\mathfrak{G}$-modules). Since $V(\rho) \cong A^2$, the Krull–Schmidt theorem tells us that Ker(π) is free of rank 1. Then taking an A-base (x, y) of $V(\rho)$ so that $x \in \text{Ker}(\pi)$, we write the matrix representation ρ with respect to this base. We have desired upper triangular form with $V(\rho)_I/\mathfrak{m}_A V(\rho)_I = V(\overline{\delta})$.

Now suppose that $\rho = \varrho$ mod $\mathfrak{a}$ and $\rho' = \varrho$ mod $\mathfrak{b}$ are both ordinary. Let $\rho'' = \varrho$ mod $\mathfrak{a} \cap \mathfrak{b}$, and write $V = V(\rho)$, $V' = V(\rho')$ and $V'' = V(\rho'')$. By definition, $V''/\mathfrak{a}V'' = V$ and $V''/\mathfrak{b}V'' = V'$. This shows by definition: $V''_I/\mathfrak{a}V''_I = V_I$ and $V''_I/\mathfrak{b}V''_I = V'_I$. Then by Nakayama's lemma, V''_I is generated by one element, thus a surjective image of $A = R/\mathfrak{a} \cap \mathfrak{b}$. Since in A, $\mathfrak{a} \cap \mathfrak{b} = 0$, we can embed A into $A/\mathfrak{a} \oplus A/\mathfrak{b}$ by the Chinese remainder theorem. Since $V_I \cong A/\mathfrak{a}$ and $V'_I \cong A/\mathfrak{b}$, the kernel of the diagonal map $V''_I \to V_I \oplus V'_I \cong A/\mathfrak{a} \oplus A/\mathfrak{b}$ has to be zero. Thus $V''_I \cong A$, which was desired.

As for R^ϕ and $R^{ord,\phi}$, we see easily that

$$R^\phi = R/\sum_{\sigma \in \mathfrak{G}} R(\det \varrho(\sigma) - \phi(\sigma))$$

$$R^{ord,\phi} = R^{ord}/\sum_{\sigma \in \mathfrak{G}} R^{ord}(\det \varrho^{ord}(\sigma) - \phi(\sigma)),$$

which completes the proof. □

Let h be a local ring of $h_k(p, \chi; \mathcal{O})$. We suppose that $U_h(p) \in h^\times$. Let h' be the subalgebra of h generated by $T(n)$ for n prime to p. Then we have a Galois representation $\rho_h : \mathfrak{G}_p \to GL_2(h')$ (if $\overline{\rho}$ is absolutely irreducible). By the assumption $U_h(p) \in h^\times$, Theorem 3.26 (2) tells us that $U_h(p) \in h'$.

Thus $h = h'$. In the following several paragraphs, we try to prove the next important result of Wiles and Taylor ([W2] Theorems 3.1 and 3.3).

THEOREM 3.31 (A. WILES, R. TAYLOR, 1995) *Let p be an odd prime. Let $\kappa/\mathbb{Q}$ be the unique quadratic extension in $\mathbb{Q}^{(p,\infty)}$, that is $\mathbb{Q}(\sqrt{(-1)^{(p-1)/2}p})$. Let h be a local ring of $h_k(p,\chi;\mathcal{O})$ for $k \geq 2$ such that h and $\mathcal{O}$ share the residue field $\mathbb{F}$. Assume $U_h(p) \in h^\times$, (ai$_\kappa$) and (rg$_p$) for $\overline{\rho}$. Then $(h, \rho_h) \cong (R^{ord,\phi}, \varrho^{ord,\phi})$ for $\phi = \chi\nu^{k-1}$. Here the isomorphism is the inverse of the natural morphism $\alpha : R^{ord,\phi} \to h$ with $\alpha\varrho^{ord,\phi} \approx \rho_h$.*

Actually in [W2] Wiles gave a detailed proof of the above result for $k = 2$ assuming the complete intersection property of the Hecke algebra which was proved in [TW] Theorem 1. There are several ways to prove it for $k > 2$. One is to use the theory of p-ordinary Hecke algebra (cf. [HM] Section 4) to reduce the problem to the knowledge of weight 2. Another is to prove directly the result, using an axiomatized generalization of the method of Taylor–Wiles due to Diamond and Fujiwara, which we will describe in the following sections.

One can ease the condition (rg_p) in the above theorem to $\overline{\epsilon} \neq \overline{\delta}$ on the decomposition group at P; however this is actually equivalent to the original (rg_p) in our p-power level case.

The result in the case of weight 2 of Wiles actually covers more cases where $T_h(p)$ can be non-unit, but in that case, they suppose that the Galois representation $\overline{\rho}$ is *flat*, that is, $\overline{\rho}$ is associated with a finite flat group scheme over $\mathbb{Z}_p$ (in other words, crystalline). Also they allow general level N not necessarily p.

Exercises

(1) If R is a noetherian ring and $\alpha : R \to R$ is a surjective algebra endomorphism, show that α is an automorphism.

(2) If $\rho : \mathfrak{G}_p \to GL_2(A)$ $(A \in CL_{\mathcal{O}})$ satisfies (ord) for one $D_{\mathfrak{P}}$, show that it satisfies (ord) for all $D_{\mathfrak{P}}$.

3.2.5 Local deformation ring

Let $q \neq p$ be a prime. We study here the deformation problem for $D = D_q = \mathrm{Gal}(\overline{\mathbb{Q}}_q/K)$ for a finite extension K of $\mathbb{Q}_q$. Let $\overline{\rho} : D \to GL_2(\mathbb{F})$ be an unramified representation. Then $\overline{\rho}$ factors through the Galois group of the maximal unramified extension K^{ur}/K. Let O be the q-adic integer ring of K and $\mathfrak{q}$ be the maximal ideal of O. We write $k = O/\mathfrak{q}$, which is a finite field of characteristic q. Then $\mathrm{Gal}(K^{ur}/K) \cong \mathrm{Gal}(\overline{k}/k)$ for the algebraic closure $\overline{k}$ of k. Since $\mathrm{Gal}(\overline{k}/k)$ is generated by the Frobenius $\phi : x \mapsto x^Q$ for $Q = q^f = |k|$, K^{ur}/K is a $\widehat{\mathbb{Z}}$-extension with Galois group $\langle\phi\rangle = \phi^{\widehat{\mathbb{Z}}}$. Then $\overline{\rho}$ factors through this group $\phi^{\widehat{\mathbb{Z}}}$, and therefore, it is determined by $\overline{\rho}(\phi)$. We suppose

(rg_q) *$\overline{\rho}(\phi)$ is semi-simple having two distinct eigenvalues; in other words, up to conjugation, $\overline{\rho}(\phi) = \begin{pmatrix} \overline{\alpha} & 0 \\ 0 & \overline{\beta} \end{pmatrix}$ with $\overline{\alpha} \neq \overline{\beta}$.*

By the local class field theory, the inertia group of the maximal abelian extension K^{ab}/K is isomorphic to $O^\times$. Thus the p-Sylow part of the inertia group is isomorphic to the p-Sylow subgroup $\Delta_\mathfrak{q}$ of $k^\times$. We would like to prove the following theorem of Faltings (see [TW] Appendix) in this section:

THEOREM 3.32 *We fix a character $\xi : D \to \mathcal{O}^\times$ such that $\xi \mod \mathfrak{m}_\mathcal{O} = \det\overline{\rho}$. Suppose that $\overline{\rho}$ is unramified and satisfies (rg_q). Let $\mathcal{O}[\Delta_\mathfrak{q}]$ be the group algebra of $\Delta_\mathfrak{q}$ and $\delta_\mathfrak{q} : \Delta_\mathfrak{q} \to \mathcal{O}[\Delta_\mathfrak{q}]$ be the tautological character sending σ to the group element $[\sigma]$ in $\mathcal{O}[\Delta_\mathfrak{q}]$. Then we have, if $Q \equiv 1 \mod p$,*

(1) *For any deformation $\rho : D \to GL_2(A)$ of $\overline{\rho}$ ($A \in CL_\mathcal{O}$) with $\det\rho = \xi$, there exists a unique $\mathcal{O}$-algebra homomorphism $\varphi_\rho : \mathcal{O}[\Delta_\mathfrak{q}] \to A$ such that $\rho \sim \varphi_\rho \circ \begin{pmatrix} \xi\delta_\mathfrak{q} & 0 \\ 0 & \delta_\mathfrak{q}^{-1} \end{pmatrix}$ on the inertia group $I = I_\mathfrak{q}$;*

(2) *For any deformation $\rho : D \to GL_2(A)$ of $\overline{\rho}$ ($A \in CL_\mathcal{O}$) with $\rho \sim \begin{pmatrix} \xi & * \\ 0 & * \end{pmatrix}$, there exists a unique $\mathcal{O}$-algebra homomorphism $\varphi_\rho : \mathcal{O}[\Delta_\mathfrak{q}] \to A$ such that $\rho \sim \varphi_\rho \circ \begin{pmatrix} \xi & 0 \\ 0 & \delta_\mathfrak{q} \end{pmatrix}$ on the inertia group I.*

Before proving the theorem, we shall do some preparation. Let L/K be a finite Galois extension. We introduce a filtration on the inertia group $I(L/K)$ of $\mathrm{Gal}(L/K)$. Let $\mathfrak{Q}$ be the maximal ideal of the q-adic integer ring of L. By definition,

$$I(L/K) = I_0(L/K) = \{\sigma \in \mathrm{Gal}(L/K) | x^\sigma \equiv x \mod \mathfrak{Q}\ \forall x \in O_L\}.$$

Thus for each integer $j > 0$, we define

$$I_j(L/K) = \{\sigma \in \mathrm{Gal}(L/K) | x^\sigma \equiv x \mod \mathfrak{Q}^{j+1}\ \forall x \in O_L\}.$$

Then by definition, the subgroups $I_j(L/K)$ of $D(L/K) = \mathrm{Gal}(L/K)$ are normal. Let $U_0 = O_L^\times$ and $U_j = 1 + \mathfrak{Q}^j \subset U_0$ for $j > 0$. Then we claim that the map:

$$I_j(L/K)/I_{j+1}(L/K) \to U_j/U_{j+1}$$

given by $\sigma \mapsto \varpi^{\sigma-1}$ for a generator ϖ of $\mathfrak{Q}$ is an injective homomorphism. The injectivity immediately follows from the existence of a $\mathfrak{Q}$-adic expansion of any $x \in O_L^\times$: $\sum_{j\geq 0} a_j\varpi^j$ with $a_j^\sigma = a_j$ for all $\sigma \in I(L/K)$, which implies

$$\varpi^{\sigma-1} \in 1 + \mathfrak{Q}^{j+1} \iff \sigma \in I_{j+1}(L/K).$$

Note that $(\sigma-1)(\tau-1) = (\sigma\tau-1)-(\sigma-1)-(\tau-1)$. Then $\varpi^{(\sigma-1)(\tau-1)} \in U_{j+1}$ because again by $\mathfrak{Q}$-adic expansion, $(\sigma - 1)$ for $\sigma \in I_j(L/K)$ takes U_j into U_{j+1} for all j. Thus it is a homomorphism.

The above claim shows that $(I_j : I_{j+1})$ is a q-power for $j > 0$ and for $r = \dim_{O_K/\mathfrak{q}} O_L/\mathfrak{Q}$,

$$(I_0 : I_1)|(O_L/\mathfrak{Q})^\times| = Q^r - 1$$

which is prime to q. Since $I_\infty(L/K) = \{1\}$, $I_1(L/K)$ is the q-Sylow subgroup of $I(L/K)$. In particular, $I(L/K)/I_1(L/K) \hookrightarrow k_L^\times$ for $k_L = O_L/\mathfrak{Q}$, which is an abelian group. By the theory of Lubin–Tate formal groups (see [CFN]), there is an explicit way of constructing such L fully ramified and abelian over K with a given Galois group isomorphic to $k_K^\times$, and if $K = \mathbb{Q}_q$, $\mathbb{Q}_q(\mu_q)$ does the job. Thus $D(L/K)$ is a soluble group, and $I_1(L/K)$ is nilpotent (because $(I_j, I_j) \subset I_{j+1}$; Exercise 1).

By class field theory, there always exists a totally ramified extension of K with a Galois group isomorphic to $k^\times$. Let $I_q \subset I$ be the maximal q-profinite subgroup. Since $I = \mathrm{Gal}(\overline{\mathbb{Q}}_q/K^{ur})$,

$$I^{(q)} = I/I_q = \varprojlim_{L\subset K^{ur}} I(L/K)/I_1(L/K) \cong \varprojlim_{L\subset K^{ur}} k_L^\times \cong \widehat{\mathbb{Z}}^{(q)},$$

where $\widehat{\mathbb{Z}}^{(q)} = \prod_{\ell\neq q} \mathbb{Z}_\ell$. In particular, if we write $I_{(p)}$ for the maximal p-profinite quotient of $I^{(q)}$ (and hence of I), it is isomorphic to $\mathbb{Z}_p$. Thus we have a unique q-totally ramified $\mathbb{Z}_p$-extension $K^{(p)}/K^{ur}$. Let $K_n^{(p)} \subset K^{(p)}$ be the subfield such that $\mathrm{Gal}(K_n^{(p)}/K^{ur}) \cong \mathbb{Z}/p^n\mathbb{Z}$. Since K^{ur} contains all p-power roots of unity, by Kummer theory ([CFN] 1.5), $K_n^{(p)} = K^{ur}(\sqrt[p^n]{\varpi})$ for a prime element ϖ in K^{ur}. Thus if $\phi \in \mathrm{Gal}(K^{(p)}/K)$ induces the Frobenius and $\sigma \in \mathrm{Gal}(K^{(p)}/K^{ur})$, we see from this that

$$\phi\sigma\phi^{-1} = \sigma^Q = \sigma^{\nu(\phi)},$$

for the p-adic cyclotomic character $\nu : D \to \mathbb{Z}_p^\times$ (Exercise 2). This shows that

$$I_{(p)} \cong \mathbb{Z}_p(1) \quad \text{as } \mathrm{Gal}(K^{ur}/K)\text{-module}, \tag{3.51}$$

where $\mathbb{Z}_p(m)$ is a Galois module isomorphic to $\mathbb{Z}_p$ as (plain) modules on which the Galois group acts via ν^m. In particular, $D/I^{(p)} \cong \phi^{\widehat{\mathbb{Z}}} \ltimes \mathbb{Z}_p(1)$, where $I^{(p)}$ is the maximal prime-to-p-profinite subgroup of I.

Proof of Theorem 3.32. Since $\overline{\rho}$ is unramified, $\rho(I) \subset 1_2 + M_2(\mathfrak{m}_A)$. Since $1_2 + M_2(\mathfrak{m}_A)$ is p-profinite, ρ factors through the abelian quotient $I_{(p)} = I/I^{(p)} \cong \mathbb{Z}_p$. Thus the restriction of ρ to $I_{(p)}$ is determined by the value of the generator γ of $I_{(p)}$. By (3.51), the matrix $X = \rho(\gamma)$ has to satisfy $YXY^{-1} = X^Q$ for $Y = \rho(\phi)$. By (rg_q), by choosing a base of $V(\rho)$, we may assume that $Y = \left(\begin{smallmatrix}\alpha & 0\\ 0 & \beta\end{smallmatrix}\right)$ (Exercise 3).

We now prove by induction on n that $X \equiv \left(\begin{smallmatrix}a & 0\\ 0 & b\end{smallmatrix}\right) \mod \mathfrak{m}_A^n$ for $a, b \in A$. Thus $a \equiv b \equiv 1 \mod \mathfrak{m}_A$. When $n = 1$, $X \equiv 1_2 \mod \mathfrak{m}_A$; so, there is nothing to prove. Suppose that the assertion is true for n. Then we write

$$X = \left(\begin{smallmatrix}a & 0\\ 0 & b\end{smallmatrix}\right)(1_2 + Z)$$

for $Z \in M_2(\mathfrak{m}_A^n)$. Write $T = \left(\begin{smallmatrix}a & 0\\ 0 & b\end{smallmatrix}\right)$. Then $T^{-1}(1_2 + Z)T = 1_2 + T^{-1}ZT$. Write $z_\pm(Z)$ for the right shoulder element of Z for $+$ and the left bottom corner element for $-$. Then $z_\pm(T^{-1}ZT) = a^{\mp 1} z_\pm(Z) b^{\pm 1} \equiv z_\pm(Z) \mod \mathfrak{m}_A^{n+1}$ because $a \equiv b \equiv 1 \mod \mathfrak{m}_A$ and $z_\pm(Z) \equiv 0 \mod \mathfrak{m}_A^n$. Thus T commutes with Z modulo $\mathfrak{m}_A^{n+1}$. Then we see that

$$\begin{aligned} 1_2 + \left(\begin{smallmatrix}\alpha & 0\\ 0 & \beta\end{smallmatrix}\right) Z \left(\begin{smallmatrix}\alpha & 0\\ 0 & \beta\end{smallmatrix}\right)^{-1} &= Y \left(\begin{smallmatrix}a & 0\\ 0 & b\end{smallmatrix}\right)^{-1} XY^{-1} = \left(\begin{smallmatrix}a & 0\\ 0 & b\end{smallmatrix}\right)^{-1} X^Q \\ &\equiv \begin{pmatrix} a^{Q-1} & 0 \\ 0 & b^{Q-1}\end{pmatrix}(1+Z)^Q \equiv \begin{pmatrix} a^{Q-1} & 0 \\ 0 & b^{Q-1}\end{pmatrix}(1+QZ) \\ &\equiv \begin{pmatrix} a^{Q-1} & 0 \\ 0 & b^{Q-1}\end{pmatrix} + Z \mod \mathfrak{m}_A^{n+1}. \end{aligned} \tag{3.52}$$

The last equality follows from the fact that $Q - 1 \in \mathfrak{m}_A$ (because of the assumption $Q \equiv 1 \mod p$ and $p \in \mathfrak{m}_A$). This shows that Z has to be diagonal.

By the above argument, we now know that $\rho \sim \left(\begin{smallmatrix}\eta & 0\\ 0 & \varepsilon\end{smallmatrix}\right)$ for two characters $\eta, \varepsilon : D \to A^\times$. Since the values of these characters at ϕ are already given by α, β, the characters are determined by restriction to $I_{(p)}$. Since the representation factors through $\mathrm{Gal}(K^{ab}/K)$ for the maximal abelian extension K^{ab}/K, the characters restricted to $I_{(p)}$ factor through Δ_q. If $\det(\rho) = \xi$, then writing $\eta = \xi\overline{\delta}$, we have $\varepsilon = \overline{\delta}^{-1}$. If $\eta = \xi$, then we write

ε as $\overline{\delta}$. Then $\delta(\sigma) \mapsto \overline{\delta}(\sigma)$ induces the unique algebra homomorphism $\varphi_\rho : \mathcal{O}[\Delta_{\mathrm{q}}] \to A$ with the required property. $\square$

REMARK 3.33 The assertion of Theorem 3.32 can be generalized to $\rho : D \to G(A)$ for a split reductive group G defined over $\mathcal{O}$. For example, if $G = GL(n)$, we have the following assertion: Suppose that $\overline{\rho} : D \to GL_n(\mathbb{F})$ is unramified and that $\rho(\phi)$ has n distinct eigenvalues. Then we consider the group algebra $\mathcal{O}[\Delta_{\mathrm{q}}^{n-1}]$. We have n-distinct tautological characters $\delta_j : \Delta_{\mathrm{q}} \to \mathcal{O}[\Delta_{\mathrm{q}}^{n-1}]^\times$ which brings σ to the group element of the j-th component of Δ_{q}. We write $[a_1, \ldots, a_n]$ for the diagonal matrix with the diagonal entry a_j at the (j, j)-place. Then we have, if $Q \equiv 1 \mod p$,

(1) For any deformation $\rho : D \to GL_n(A)$ of $\overline{\rho}$ ($A \in CL_{\mathcal{O}}$) with $\det \rho = \xi$, there exists a unique $\mathcal{O}$-algebra homomorphism $\varphi_\rho : \mathcal{O}[\Delta_{\mathrm{q}}^{n-1}] \to A$ such that $\rho \sim \varphi_\rho \circ [\xi\delta_1, \delta_2, \ldots, \delta_n]$ on the inertia group $I = I_{\mathrm{q}}$, where $\delta_n = \prod_{j=1}^{n-1} \delta_j^{-1}$;
(2) For any deformation $\rho : D \to GL_2(A)$ of $\overline{\rho}$ ($A \in CL_{\mathcal{O}}$) with $\rho \sim \varphi_\rho \circ [\xi, *, \ldots, *]$, there exists a unique $\mathcal{O}$-algebra homomorphism $\varphi_\rho : \mathcal{O}[\Delta_{\mathrm{q}}] \to A$ such that $\rho \sim \varphi_\rho \circ [\xi, \delta_1, \ldots, \delta_{n-1}]$ on the inertia group I.

Exercises

(1) Show that $I(L/K)$ is nilpotent.
(2) Show that $\phi\sigma\phi^{-1} = \sigma^Q = \sigma^{\nu(\phi)}$ for $\sigma \in \mathrm{Gal}(K^{(p)}/K^{ur})$ and $\phi \in \mathrm{Gal}(K^{(p)}/K)$ which lifts the Frobenius in $\mathrm{Gal}(K^{ur}/K)$. (Hint: Apply Kummer theory to show that if $K^{ur} \subset K_i \subset K^{(p)}$ and $[K_i : K^{ur}] = p^i$, then $K_i = K^{ur}[\sqrt[p^i]{\varpi}]$ for a prime element ϖ of K.)
(3) If an $n \times n$ matrix $Y \mod \mathfrak{m}_A$ has n-distinct eigenvalues, show that $Y \in M_n(A)$ can be diagonalized in $M_n(A)$, where A is a local ring in $CL_{\mathcal{O}}$. Hint: Use Hensel's lemma in the algebra $A[Y] \subset M_n(A)$.

3.2.6 Taylor–Wiles systems

We keep the assumption and the notation of Theorem 3.31. Since the representation $\rho_h : \mathfrak{G}_p \to GL_2(h)$ satisfies (ord) and (det) by Theorem 3.26 (1)–(2), we have a unique algebra homomorphism $\alpha : R^{ord,\phi} \to h$ such that $\alpha\varrho^{ord,\phi} \approx \rho_h$. Since h is generated by traces of the representation: $\mathrm{Tr}(\rho_h(Frob_\ell)) = T_h(\ell)$ for ℓ prime to N (by the Chebotarev density theorem) and $\alpha(\mathrm{Tr}(\varrho^{ord,\phi}(Frob_\ell))) = \mathrm{Tr}(\rho_h(Frob_\ell)) = T_h(\ell)$, α is surjective. Thus we need to prove the injectivity of α.

The idea of Taylor–Wiles is to introduce deformation rings which allow ramification at a finite set of primes $Q = \{q_1, \ldots, q_r\}$. We assume that $\overline{\rho}$ is unramified at Q and satisfies (rg$_q$) for all $q \in Q$ and $q \equiv 1 \mod p$ for all $q \in Q$. We write Δ_q for the p-Sylow subgroup of $(\mathbb{Z}/q\mathbb{Z})^\times$ and we put $\Delta_Q = \prod_{q \in Q} \Delta_q$. Writing $\mathfrak{G}_Q = \mathfrak{G}_{pq_1q_2\cdots q_r}$, we look at the universal couple (R_Q, ϱ_Q) with the following property: For each deformation $\rho : \mathfrak{G}_Q \to GL_2(A)$ $(A \in CL_{\mathcal{O}})$ satisfying (det), (ord) and the following condition:

$$\det \rho|_{D_q} = \phi|_{D_q}, \qquad (\det_q)$$

there exists a unique $\mathcal{O}$-algebra homomorphism $\varphi_\rho : R_Q \to A$ such that $\varphi_\rho \circ \varrho_Q \approx \rho$. The existence of (R_Q, ϱ_Q) can be proved in the same manner as Mazur's proposition: Proposition 3.30.

If $Q' \supset Q$, we have the restriction map $res : \mathfrak{G}_{Q'} \twoheadrightarrow \mathfrak{G}_Q$. If ρ is a deformation of $\overline{\rho}$ on $\mathfrak{G}_Q$, then $\rho \circ res$ is a deformation on $\mathfrak{G}_{Q'}$. Thus we have a natural $\mathcal{O}$-algebra homomorphism $\pi_{Q',Q} : R_{Q'} \to R_Q$ such that $\varrho_Q \circ res \approx \pi_{Q',Q} \circ \varrho_{Q'}$. Since $\overline{\rho}$ is absolutely irreducible, by the theory of pseudo-representation (see Proposition 2.16 in Chapter 2), the subring R' of R_Q generated (topologically) by the trace of ϱ_Q carries a representation $\varrho' : \mathfrak{G}_Q \to GL_2(R')$ such that $Tr(\varrho') = Tr(\varrho_Q)$. Then by Carayol's theorem (Chapter 2 Proposition 2.13), $\varrho_Q \cong \varrho'$ over R_Q. Therefore $R_Q = R'$, and R_Q is generated by $\mathrm{Tr}(\varrho_Q)$ (Exercise 1). Hence $\pi_{Q',Q}$ is surjective.

By Falting's theorem (Theorem 3.32), there is a unique algebra homomorphism $\varphi_q : \mathcal{O}[\Delta_q] \to R_Q$ such that $\varrho_Q|_{D_q} \sim \varphi_q \circ \begin{pmatrix} \delta_q^{-1} & 0 \\ 0 & \delta_q \end{pmatrix}$ on I_q. Thus R_Q is naturally an $\mathcal{O}[\Delta_Q]$-algebra. Let $\varrho' = \varrho_{Q'} \mod \mathfrak{a}$ for $\mathfrak{a} = \sum_{\sigma \in \Delta_{Q'-Q}} R_{Q'}(\sigma - 1)$ and $R' = R_{Q'}/\sum_{\sigma \in \Delta_{Q'-Q}} R_{Q'}(\sigma - 1)$. Then the restriction of ϱ' to $I_{q'}$ for $q' \in Q' - Q$ is isomorphic to a diagonal representation with diagonal entries $\varphi_{q'} \circ \delta_{q'}^{\pm 1} \mod \mathfrak{a}$, which is trivial. Thus ϱ' is unramified at $q' \in Q' - Q$. For any given deformation $\rho : \mathfrak{G}_Q \to GL_2(A)$, by definition of the universal couple, we have a unique $\mathcal{O}$-algebra homomorphism $\iota_\rho : R_{Q'} \to A$ such that $\iota_\rho \circ \varrho_{Q'} \approx \rho$. Since ρ is unramified at $q \in Q' - Q$, $\iota_\rho \circ \varphi_{q'} \circ \delta_{q'}$ is trivial, and hence $\iota_\rho(\mathfrak{a}) = 0$. In other words, ι_ρ factors through R'. This shows that (R', ϱ') is a universal couple for $(\mathfrak{G}_Q, \overline{\rho})$. By the uniqueness of the universal ring, we get

$$R_{Q'}/\sum_{\sigma \in \Delta_{Q'-Q}} R_{Q'}(\sigma - 1) \cong R_Q \quad \text{via } \pi_{Q',Q}. \qquad (3.53)$$

Since Δ_Q is a p-group, $\mathcal{O}[\Delta_Q]$ is a local ring with residue field $\mathbb{F}$, that is, it is an object in $CL_{\mathcal{O}}$ (see Lemma 2.20 in Chapter 2).

Actually we need to introduce a similar system on the Hecke side. Write $h_k(Q;\mathcal{O})$ for the A-subalgebra of $h_k(pq_1\cdots q_r;\mathcal{O})$ generated by $T(\ell)$ for all primes ℓ outside Q (thus $U(p)$ is included), and write $\Gamma_j(Q)$ for $\Gamma_j(pq_1\cdots q_r)$ $(j=0,1)$. We will see later that $h_k(Q;\mathcal{O})$ is reduced. Since $\Gamma_1(Q)\supset\Gamma_1(Q')$ for $Q\subset Q'$, we have an inclusion $S_k(\Gamma_1(Q);\mathcal{O})\subset S_k(\Gamma_1(Q');\mathcal{O})$. On the other hand, locally $T(\ell)=S_\ell\left(\begin{smallmatrix}\ell&0\\0&1\end{smallmatrix}\right)S_\ell$ for the open subgroups $S=S_1(p)$, $S_1(Q)=S_1(pq_1\cdots q_r)$ and $S_1(Q')$ for all primes ℓ outside Q'. Thus $T(\ell)$ acting on $S_k(\Gamma_1(Q');\mathcal{O})$ restricted to $S_k(\Gamma_1(Q);\mathcal{O})$ is equal to $T(\ell)$ on the latter (smaller) space, and we have the following commutative diagram:

$$\begin{array}{ccc} S_k(\Gamma_1(Q);\mathcal{O}) & \xrightarrow{\subset} & S_k(\Gamma_1(Q');\mathcal{O}) \\ {\scriptstyle T(\ell)}\downarrow & & \downarrow{\scriptstyle T(\ell)} \\ S_k(\Gamma_1(Q);\mathcal{O}) & \xrightarrow{\subset} & S_k(\Gamma_1(Q');\mathcal{O}) \end{array}$$

for primes $\ell\notin Q'$. Then the restriction of operators of $h_k(Q';\mathcal{O})$ to the subspace $S_k(\Gamma_1(Q);\mathcal{O})$ induces an algebra homomorphism $\pi^h_{Q',Q}$ of $h_k(Q';\mathcal{O})$ into $h_k(Q;\mathcal{O})$. Let h_Q be a local ring of $h_k(Q;\mathcal{O})$ which is sent to h by $\pi^h_{Q,\emptyset}$. By Theorem 3.26 (3)(a) and (b), the inclusion map

$$h_Q\hookrightarrow h_k(\Gamma_1(q_1\cdots q_r)\cap\Gamma_0(p),\chi;\mathcal{O})\subset h_k(N_Q;\mathcal{O})\ \text{ for } N_Q=pq_1\cdots q_r$$

sends h_Q isomorphically to a unique local ring of $h_k(\Gamma_1(q_1\cdots q_r)\cap\Gamma_0(p),\chi;\mathcal{O})$ and also $h_k(N_Q;\mathcal{O})$. Then by Corollary 3.19 applied inductively to the q's, $h_0=h_Q/\sum_{\sigma\in\Delta_Q}(\sigma-1)h_Q$ is a local ring of $h_k(N_Q,\chi;\mathcal{O})$. The Galois representation $\overline{\rho}$ satisfies (rg_q) at $q\in Q$, thus the ratio α/β of two eigenvalues of $\overline{\rho}(Frob_q)$ is not equal to 1. By Theorem 3.26 (3)(a) and (b), there are three possibilities:

(1) $\overline{\rho}$ ramifies at q (the case of Theorem 3.26 (3)(a));
(2) $\alpha/\beta=1$ (the case of Theorem 3.26 (3)(b)) because $\nu(Frob_q)=q\equiv 1 \bmod p$ for the p-adic cyclotomic character ν;
(3) $h_Q/\sum_{\sigma\in\Delta_q}(\sigma-1)h_Q$ is a local ring of $h_k(N_{Q-\{q\}},\chi;\mathcal{O})$.

Since $\overline{\rho}$ is unramified at q, Case (1) cannot occur. As already explained, by (rg_q), Case (2) does not occur. Thus we are in Case (3). Applying this argument to each $q\in Q$, we know that h_0 is associated with a residual Galois representation of a local ring of $h_k(p,\chi;\mathcal{O})$. Then by Lemma 3.25, h_0 is isomorphic to $h=h_\emptyset$. Thus we get from Corollary 3.20, the following facts:

(fr) *h_Q is $\mathcal{O}[\Delta_Q]$-free of rank $d=\mathrm{rank}_{\mathcal{O}}\,h$;*
(ct) $h_Q/\sum_{\sigma\in\Delta_Q}(\sigma-1)h_Q\cong h$.

Thus $\pi^h_{Q,\emptyset}$ induces a surjective $\mathcal{O}$-algebra homomorphism $h_Q \to h$ with kernel given by $\sum_{\sigma\in\Delta_Q}(\sigma - 1)h_Q$. Let $\rho : \mathfrak{G}_Q \to GL_2(h_Q)$ be the Galois representation constructed similarly to Theorem 3.29. Then $\det\rho \equiv \phi \mod \mathfrak{m}_{h_Q}$, and $\xi = (\det\rho^{-1})\phi$ has values in p-profinite group. Since $p > 2$, we have a unique square root $\sqrt{\xi}$. We then define $\rho_Q = \sqrt{\xi}\otimes\rho$, which is a deformation of $\overline{\rho}$ satisfying ($\det_q$) for all $q \in Q$.

Since $GL_2(h_Q)$ is generated by the trace of its Galois representation $\rho_Q : \mathfrak{G}_Q \to GL_2(h_Q)$, it is generated by the trace of a dense subset of $\mathfrak{G}_Q$. Since $Frob_\ell$ for $\ell \notin Q$ is still dense in $\mathfrak{G}_p$, this shows again the surjectivity of $\pi^h_{Q,\emptyset} : h_Q \to h$. We also have a surjection $\alpha_Q : R_Q \to h_Q$ such that $\alpha_Q\varrho_Q \approx \rho_Q$. Thus h_Q is a R_Q-algebra; in particular, h_Q is an R_Q-module and is free of finite rank over $\mathcal{O}[\Delta_Q]$, whose rank d is independent of Q.

The algebra h_Q has a natural action of $(\mathbb{Z}/p\prod_{q\in Q} q\mathbb{Z})^\times$ induced by the action of $\Gamma_0(Q)/\Gamma_1(Q)$. By Theorem 3.26 (3), the induced action of Δ_Q coincides with the $\mathcal{O}[\Delta_Q]$-algebra structure induced by the surjection $R_Q \twoheadrightarrow h_Q$. We can decompose $(\mathbb{Z}/p\prod_{q\in Q} q\mathbb{Z})^\times \cong \Gamma_0(Q)/\Gamma_1(Q)$ into a product $\Delta_Q \times \Delta'_Q$. Then the order $|\Delta'_Q|$ is prime to p. In particular, the action of $\Gamma_0(Q)/\Gamma_1(Q)$ on h_Q induces a character $\xi : \Delta'_Q \to h_Q^\times$. Then $\mathrm{Im}(\xi) \to \mathrm{Im}(\xi \mod \mathfrak{m}_{h_Q})$ is a bijection, because the order of Δ'_Q is prime to p. In other words, the character coincides with $\chi|_{\Delta'_Q}$. Thus, as already remarked, we can view h_Q as a local ring of $h_k(\Gamma_0(p)\cap\Gamma_1(\prod_{q\in Q} q), \chi; \mathcal{O})$.

The local and global analysis of the rings (R_Q, ϱ_Q) and R_Q-modules $\{h_Q\}_Q$ yields the injectivity of α_Q. Fujiwara [Fu] axiomatized what we need to prove to show the isomorphism: $R^{ord,\phi} \cong h$. To describe the formalism, we first define Taylor–Wiles systems as follows.

DEFINITION 3.34 Let $\mathcal{Q}$ be a set of finite sets Q of primes congruent to 1 modulo p. A Taylor–Wiles system is a triple $\{R = R^{ord,\phi}, R_Q, M_Q\}_{Q\in\mathcal{Q}}$ (indexed by $Q \in \mathcal{Q}$) made of R_Q-module M_Q such that (1) M_Q is free of finite rank d over $\mathcal{O}[\Delta_q]$ for d independent of Q, and (2) $R \cong R_Q/\sum_{\sigma\in\Delta_Q} R_Q(\sigma - 1)$.

In our application, M_Q is given by h_Q.

Since $R = R_\emptyset$, we have a natural surjection: $\pi_{Q,\emptyset} : R_Q \twoheadrightarrow R$ inducing $R \cong R_Q/(\delta_q - 1)_{q\in Q}$ for a generator δ_q of Δ_q; so in our case, the above condition (2) is satisfied. Here is a fundamental theorem (see [TW], [W2], [Dd] and [Fu]):

THEOREM 3.35 (TAYLOR–WILES, DIAMOND, FUJIWARA, 1995–6) *Suppose that* $p > 2$. *Let* $\{R = R^{ord,\phi}, R_Q, M_Q\}_{Q\in\mathcal{Q}}$ *be a Taylor–Wiles system for*

$\mathcal{Q} = \{Q_m | m \in \mathbb{N}\}$. We write simply R_m for R_{Q_m}, Δ_m for Δ_{Q_m} and M_m for M_{Q_m}. Suppose the following four conditions:

(tw1) If $q \in Q_m$, then $q \equiv 1 \mod p^m$;

(tw2) The cardinality $|Q_m|$ is independent of m, which we write r;

(tw3) R_m is generated by at most r elements as an $\mathcal{O}$-algebra;

(tw4) The image T of R in $\mathrm{End}_{\mathcal{O}}(M_m/\sum_{\sigma\in\Delta_m}(\sigma-1)M_m)$ is independent of m as an R-algebra, that is, $Ker(R \to End_{\mathcal{O}}(M_m/\sum_\sigma(\sigma-1)M_m))$ is independent of m.

Then R is a complete intersection over $\mathcal{O}$, is $\mathcal{O}$-free of finite rank and is isomorphic to T. If, further, the following condition is satisfied:

(tw5) $M_m/\sum_{\sigma\in\Delta_m}(\sigma-1)M_m$ is isomorphic to an R-module M for all m,

then M is a free R-module.

We say that an algebra A is a *complete intersection over $\mathcal{O}$* if A is free of finite rank over $\mathcal{O}$ and $A \cong \mathcal{O}[[X_1,\dots,X_r]]/(f_1,\dots,f_r)$ for the r generators f_j (here the number r has to match the number of variables r). Actually the freeness of A over $\mathcal{O}$ is redundant (which follows from the complete intersection property if A is in $CL_{\mathcal{O}}$ and is an $\mathcal{O}$-module of finite type; see Lenstra [Lt] Lemma 1).

In our application, $M_Q = h_Q$, thus obviously $T = h$. Therefore to prove Theorem 3.31, we need to check the four conditions (tw1–4). In our application, the last assertion is trivial, but also follows from (tw5), which is basically a consequence of Corollary 3.19 and a result of Langlands (Theorem 3.26 (3)).

Proof We prove the theorem following Fujiwara [Fu]. Let δ_q be a generator of the cyclic subgroup Δ_q. If $q \equiv 1 \mod p^m$, for $n \le m$, $\Delta_q/\langle\delta_q^{p^n}\rangle$ is cyclic of order p^n. Thus

$$\mathcal{O}[\Delta_q]/(\delta_q^{p^n}-1) \cong \mathcal{O}[\Delta_q/\langle\delta_q^{p^n}\rangle] \cong \mathcal{O}[[S]]/(1+S)^{p^n}-1)$$

by $\delta_q \mapsto 1+S$. Write $Q_m = \{q_1,\dots,q_r\}$. Let I_n be the ideal of $\mathcal{O}[\Delta_m]$ generated by $\{p^n, \delta_{q_1}^{p^n}-1,\dots,\delta_{q_r}^{p^n}-1\}$. Then we have an isomorphism

$$\mathcal{O}[\Delta_m]/I_n \cong \mathcal{O}[[S_1,\dots,S_r]]/(p^n,(1+S_1)^{p^n}-1,\dots,(1+S_r)^{p^n}-1) =: A_n$$

via $\delta_{q_j} \mapsto 1+S_j$. Note that $|A_n| = p^{ntp^{nr}}$ for $t = \mathrm{rank}_{\mathbb{Z}_p}\mathcal{O}$.

Write $R_{n,m}$ for the image of R_m/I_nR_m in $\mathrm{End}_{\mathcal{O}[\Delta_m]/I_n}(M_m/I_nM_m)$ and $\widetilde{R}_{n,m}$ for $R_{n,m}/(\delta_q-1)_{q\in Q_m}$. The idea is to find an increasing sequence $\{m(n)|n=1,2,\dots\}$ such that $(R_{n,m(n)},\pi_{n,n'})$ with appropriate surjections $\pi_{n,n'} : R_{n,m(n)} \twoheadrightarrow R_{n',m(n')}$ is a projective system of $\mathcal{O}[[T_1,\dots,T_r]]$-algebras such that $R_\infty = \varprojlim_n R_{n,m(n)} \cong \mathcal{O}[[T_1,\dots,T_r]]$ and the kernel of the natural projection $R_\infty \to T$ is again generated by r elements, proving the complete-intersection property.

Before starting to make such a projective system, we remark that we have the following two algebra homomorphisms α and β:

$$A_n \xrightarrow{\alpha} R_{n,m} \xrightarrow{\beta} \mathrm{End}_{A_n}\left(\frac{M_m}{I_nM_m}\right) \cong M_d(A_n)\text{: } d\times d \text{ matrices with entries in } A_n.$$

Since M_m is $\mathcal{O}[\Delta_m]$-free of rank d, $\frac{M_m}{I_nM_m}$ is A_n-free of rank d, and hence the composite $\beta\circ\alpha$ is injective, proving the injectivity of α (β is injective by definition of $R_{n,m}$). Thus the cardinality of $R_{n,m}$ is bounded by $N_n = |A_n|^{d^2}$, which is a finite number.

Since R_m is generated by r elements over $\mathcal{O}$ and $R_{n,m}$ is covered surjectively by R_m, $R_{n,m}$ is generated by r elements. Choose r generators $f_1,\dots,f_r$ in the maximal ideal of $R_{n,m}$. We now consider triples made of $((R_{n,m},\alpha,\beta),\widetilde{R}_{n,m},(f_1,\dots,f_r))$. We say that two triples

$$((R_{n,m},\alpha,\beta),\widetilde{R}_{n,m},(f_1,\dots,f_r)) \quad\text{and}\quad ((R_{n,m'},\alpha',\beta'),\widetilde{R}_{n,m'},(f'_1,\dots,f'_r))$$

are isomorphic if there is an isomorphism of A_n-algebras: $\iota : R_{n,m} \to R_{n,m'}$ inducing an isomorphism $\widetilde{R}_{n,m} \cong \widetilde{R}_{n,m'}$ such that $\iota(f_j) = f'_j$ for all j and the following two diagrams are commutative:

$$\begin{array}{ccccc} A_n & \xrightarrow{\alpha} & R_{n,m} & \xrightarrow{\beta} & M_d(A_n) \\ \| \downarrow & & \iota\downarrow & & \|\downarrow \\ A_n & \xrightarrow{\alpha'} & R_{n,m'} & \xrightarrow{\beta'} & M_d(A_n) \end{array}$$

and

$$\begin{array}{ccc} R_{n,m} & \longrightarrow & \widetilde{R}_{n,m} \\ \iota\downarrow & & \wr\|\downarrow \\ R_{n,m'} & \longrightarrow & \widetilde{R}_{n,m'}. \end{array}$$

Since $|R_{n,m}| \le N_n$ (which is independent of m) and we have infinitely many choices of m, even if we move around m, there are finitely many isomorphism classes of such triples. Thus starting from $n = 1$, we will have an infinite sequence $\mathbb{N}_1 \subset \mathbb{N}$ such that

$$((R_{1,m},\alpha,\beta),\widetilde{R}_{1,m},(f_1,\dots,f_r))$$

for all $m \in \mathbb{N}_1$ are all isomorphic to each other for some choice of f_j. Suppose we have constructed a sequence $\mathbb{N}_n \subset \mathbb{N}_{n-1} \subset \cdots \subset \mathbb{N}_1 \subset \mathbb{N}$

made of infinite sets $\mathbb{N}_j$ such that $((R_{n,m},\alpha,\beta),\widetilde{R}_{n,m},(f_1,\dots,f_r))$ for all $m \in \mathbb{N}_n$ are isomorphic to each other. When we move around

$$((R_{n+1,m},\alpha,\beta),\widetilde{R}_{n+1,m},(f_1,\dots,f_r))$$

for all $m \geq n+1$ in $\mathbb{N}_n$, there are only finitely many isomorphism classes; so, we can choose an infinite sequence $\mathbb{N}_{n+1} \subset \mathbb{N}_n$ such that

$$((R_{n+1,m},\alpha,\beta),\widetilde{R}_{n+1,m},(f_1,\dots,f_r))$$

for $m \in \mathbb{N}_{n+1}$ are all isomorphic to each other.

Let $m(n)$ be the minimal element in $\mathbb{N}_n$, and write the triple for this choice as $((R_{n,m(n)},\alpha_n,\beta_n),\widetilde{R}_{n,m(n)},(f_1^{(n)},\dots,f_r^{(n)}))$. By definition, we have a surjection

$$\left(\frac{R_{n+1,m(n+1)}}{I_nR_{n+1,m(n+1)}},\frac{\widetilde{R}_{n+1,m(n+1)}}{I_n\widetilde{R}_{n+1,m(n+1)}},(f_1^{(n+1)},\dots,f_1^{(n+1)}) \bmod I_n\right)$$
$$\twoheadrightarrow \left(R_{n,m(n+1)},\widetilde{R}_{n,m(n+1)},(\overline{f}_1^{(n+1)},\dots,\overline{f}_1^{(n+1)})\right)$$

for $(R_{n,m(n+1)},\widetilde{R}_{n,m(n+1)},(\overline{f}_1^{(n+1)},\dots,\overline{f}_1^{(n+1)}))$ isomorphic to

$$((R_{n,m(n)},\alpha_n,\beta_n),\widetilde{R}_{n,m(n)},(f_1^{(n)},\dots,f_r^{(n)})).$$

Thus we have a projective system of triples:

$$\left\{((R_{n,m(n)},\alpha_n,\beta_n),\widetilde{R}_{n,m(n)},(f_1^{(n)},\dots,f_r^{(n)}))|n \in \mathbb{N}\right\}.$$

Take the projective limit:

$$((R_\infty,\alpha_\infty,\beta_\infty),\widetilde{R}_\infty,(f_1^{(\infty)},\dots,f_r^{(\infty)}))$$
$$= \varprojlim_n((R_{n,m(n)},\alpha,\beta),\widetilde{R}_{n,m(n)},(f_1^{(n)},\dots,f_r^{(n)})).$$

Since $(f_1^{(\infty)},\dots,f_r^{(\infty)})$ generates R_∞, we have a surjection:

$$\mathcal{O}[[T_1,\dots,T_r]] \twoheadrightarrow R_\infty$$

taking T_j to $f_j^{(\infty)}$ for $j = 1,2,\dots,r$. Since α_n brings A_n injectively into $R_{n,m(n)}$, $\alpha_\infty : \mathcal{O}[[S_1,\dots,S_r]] = \varprojlim_n A_n \to R_\infty$ is injective. Similarly by β_∞, we have $R_\infty \subset M_d(\mathcal{O}[[S_1,\dots,S_r]])$. Thus R_∞ is a torsion-free $\mathcal{O}[[S_1,\dots,S_r]]$-module of finite type. Thus by the validity of the going-up and down theorem ([CRT] Theorem 9.4), the Krull dimension of R_∞ is $r+1$. Recall that the Krull dimension s of a ring A is the length of a maximal chain of prime ideals $P_0 \subset P_1 \subset \cdots \subset P_s$ in A (see [CRT] Section 5). For each prime ideal $P \subset A$, the height of P is the maximal length of the sequence of prime ideals inside P.

If the surjection $\pi : \mathcal{O}[[T_1,\dots,T_r]] \twoheadrightarrow R_\infty$ has a non-trivial kernel, then $\mathrm{Ker}(\pi)$ contains an element $f \neq 0$. Thus a minimal prime ideal P containing f is of height 1, and hence

$$\dim \frac{\mathcal{O}[[T_1,\dots,T_r]]}{\mathrm{Ker}(\pi)} \le \dim \mathcal{O}[[T_1,\dots,T_r]]/(f) = \dim \mathcal{O}[[T_1,\dots,T_r]]/P = r,$$

which is a contradiction. Thus $\mathrm{Ker}(\pi) = 0$, and hence $R_\infty \cong \mathcal{O}[[T_1,\dots,T_r]]$.

By definition, we have the following exact sequence:

$$R_{n,m}^{Q_m} \xrightarrow{\varphi} R_{n,m} \longrightarrow \widetilde{R}_{n,m} \longrightarrow 0,$$

where φ sends $(a_q)_{q\in Q_m}$ to $\sum_{q\in Q_m} a_q(\delta_q - 1)$. Note that $|Q_m| = r$ is independent of m. Taking the projective limit of this sequence, we get another exact sequence:

$$R_\infty^r \longrightarrow R_\infty \longrightarrow \widetilde{R}_\infty = \varprojlim_n \widetilde{R}_{n,m(n)} \longrightarrow 0.$$

Thus $\widetilde{R}_\infty \cong R_\infty/\mathfrak{a}$ for an ideal $\mathfrak{a}$ generated by the r elements $S_1,\dots,S_r$. Since R_∞ is regular, it has to be free of finite rank over $\mathcal{O}[[S_1,\dots,S_r]]$ ([CRT] Theorem 23.1; see also Remark 3.36 below). This shows that $\widetilde{R}_\infty$ is free of finite rank over $\mathcal{O}$ and is a complete intersection.

By the compatibility of the projective system with maps $\beta : R_{n,m} \hookrightarrow M_d(A_n) = \mathrm{End}_{A_n}(M_m/I_nM_m)$, we have a projective system of $R_{n,m(n)}$-modules:

$$\{M_{m(n)}/I_nM_{m(n)} \cong A_n^d | n \in \mathbb{N}\}.$$

Thus $L = \varprojlim_n \frac{M_{m(n)}}{I_nM_{m(n)}}$. Since $M_{m(n)}/I_nM_{m(n)}$ is $\mathcal{O}[\Delta_m]/I_n$-free of rank d, L is $\mathcal{O}[[S_1,\dots,S_r]]$-free of rank d. In this situation, it is known that L is free of finite rank over $R_\infty = \mathcal{O}[[T_1,\dots,T_r]]$ by abstract ring theory (Auslander Buchsbaum theorem, see Remark 3.36 after the proof). Under (tw5), we have

$$M \cong \varprojlim_n M/p^nM \cong \varprojlim_n M_{m(n)}/(I_n + \mathfrak{a})M_{m(n)} \cong L/\mathfrak{a}L.$$

Since L is a free R_∞-module, $M \cong L/\mathfrak{a}L$ is a free $\widetilde{R}_\infty$-module of finite rank, because $\widetilde{R}_\infty = R_\infty/\mathfrak{a}R_\infty$.

Let $J_n = (p^n, (1+S_1)^{p^n} - 1, \dots, (1+S_r)^{p^n} - 1)$ inside $\mathcal{O}[[S_1,\dots,S_r]]$. Then $M_{m(n)}/I_nM_{m(n)}$ is isomorphic to L/J_nL and hence is free of finite rank over $R_\infty/J_nR_\infty = R_{n,m(n)}$, using freeness of L over $\mathcal{O}[[S_1,\dots,S_r]]$. From this, we also have $R_{n,m(n)} = R_{m(n)}/I_n$.

Now we want to show that $T \cong \widetilde{R}_\infty$. We claim that the projection $R_\infty \twoheadrightarrow \widetilde{R}_\infty$ factors through T. By definition, T is the image of R_∞ in

$$\mathrm{End}_{\mathcal{O}}(M_{m(n)}/(S_1,\dots,S_r)M_{m(n)}).$$

Since $M_{m(n)}/I_nM_{m(n)}$ is $R_{n,m(n)}$-free as remarked above, $\widetilde{R}_{n,m(n)}$ is the subalgebra of

$$\mathrm{End}_{\mathcal{O}}(M_{m(n)}/(I_n + \mathfrak{a})M_{m(n)})$$

and therefore, it is a T-algebra. After taking the limit with respect to n, the algebra homomorphism $R_\infty \to \widetilde{R}_\infty$ factors through T. Since naturally $\widetilde{R}_\infty = R_\infty/\mathfrak{a}$ covers T (because $S_j = 0$ in T), we have $T \cong \widetilde{R}_\infty$. By the same argument, we also see that

$$R_{n,m(n)}/\mathfrak{a}R_{n,m(n)} \cong T/p^nT.$$

We now prove that $R \cong T$. Since $R_\infty = \varprojlim_n R_{n,m(n)} \cong \mathcal{O}[[T_1, \ldots, T_r]]$, we have a surjective $\mathcal{O}[[S_1, \ldots, S_r]]$-algebra homomorphism $\mathcal{O}[[T_1, \ldots, T_r]] \twoheadrightarrow R_{m(n)}$ taking T_i to the generator $f_i^{(n)}$. Thus, writing $\mathfrak{m}_\infty$ for $\mathfrak{m}_{R_\infty}$, for each given $N > 0$, we find $n(N) > 0$ such that $\mathrm{Ker}(R_\infty \to R_{n(N),m(n(N))}) \subset \mathfrak{m}_\infty^N$. Then tensoring $R_\infty/\mathfrak{m}_\infty^N$ to the following commutative diagram with exact rows:

$$\begin{array}{ccccccc}
R^r_{m(n(N))} & \xrightarrow{a} & R_{m(n(N))} & \longrightarrow & R & \longrightarrow & 0 \\
\downarrow & & \downarrow & & \downarrow & & \\
R^r_{n(N),m(n(N))} & \xrightarrow{b} & R_{n(N),m(n(N))} & \longrightarrow & T/p^{n(N)}T & \longrightarrow & 0,
\end{array}$$

we get another commutative diagram with exact rows:

$$\begin{array}{ccccccc}
(R_{m(n(N))}/\mathfrak{m}_\infty^N)^r & \xrightarrow{a} & R_{m(n(N))}/\mathfrak{m}_\infty^N & \longrightarrow & R/\mathfrak{m}_R^N & \longrightarrow & 0 \\
{\scriptstyle c}\downarrow & & {\scriptstyle d}\downarrow & & \downarrow & & \\
(R_{n(N),m(n(N))}/\mathfrak{m}_\infty^N)^r & \xrightarrow{b} & R_{n(N),m(n(N))}/\mathfrak{m}_\infty^N & \longrightarrow & T/\mathfrak{m}_T^N & \longrightarrow & 0.
\end{array}$$

Here a and b take tuples $(a_1, \ldots, a_r)$ to $\sum_i a_iS_i$. Since by our choice, c and d are surjective isomorphisms, and hence we have $R/\mathfrak{m}_R^N \cong T/\mathfrak{m}_T^N$ for all N, which shows that $R \cong T$. This finishes the proof. □

REMARK 3.36 The proof of the freeness of L over $B = \mathcal{O}[[T_1, \ldots, T_r]]$ goes as follows. Write $A = \mathcal{O}[[S_1, \ldots, S_r]]$. The number of a maximal sequence of elements $x_1, \ldots, x_s \in \mathfrak{m}_A$ for an A-module M such that the multiplication by x_{j+1} on $M/(x_1M + \cdots + x_jM)$ is injective (for $j = 0, 1, \ldots$) is called the depth of M and is written as $\mathrm{depth}_A M$. Similarly the minimal length of the resolution $0 \to P_j \to \cdots \to P_0 \to M \to 0$ by A-free module P_j is called the homological dimension of M and is written as $\mathrm{hdim}_A M$. Since $(S_1, \ldots, S_r, p)$ is a maximal regular sequence for the A-module L as well as the B-module L, we have $\mathrm{depth}_A L = \mathrm{depth}_B L =$

$\mathrm{depth}_A A = \mathrm{depth}_B B$ for $B = \mathcal{O}[[T_1, \ldots, T_r]]$ with the notation of the above proof of the theorem. By a theorem of Auslander–Buchsbaum ([CRT] Theorem 19.1),

$$\mathrm{depth}_A L + \mathrm{hdim}_A L = \mathrm{depth}_A A = \mathrm{depth}_B B = \mathrm{depth}_B L + \mathrm{hdim}_B L.$$

Since L is A-free, $\mathrm{hdim}_A L = 0$ and hence, $\mathrm{hdim}_B L = 0$ and hence L is B-free.

The original proof of Taylor–Wiles of the identity: $R^{ord,\phi} \cong h$ does not use explicitly the Taylor–Wiles system introduced in this section. The introduction of the system by K. Fujiwara substantially simplified the proof. Here I basically followed Fujiwara's treatment with modification.

The difficult point in application is to find convenient R_Q-modules M_Q, which have to be $\mathcal{O}[\Delta_Q]$-free. Fujiwara applied this technique to the Hilbert modular forms on $GL_2(F_{\mathbb{A}})$. In his case, he took M_Q to be the cohomology groups with coefficients in $\mathcal{O}$ over either a Shimura curve over F associated with a (partially definite) quarternion algebra over F or a finite modular variety associated with a definite quarternion algebra over F. The study of Hecke algebras acting on such cohomology groups (on definite quarternions) was initiated in the late 80's in [H88a], in which the relation of the quarternionic Hecke algebras and the Hecke algebras of $GL_2(F_{\mathbb{A}})$ is clarified.

In the following subsections, we will show that the Hecke algebra itself can be used as M_Q in the elliptic modular case. By the duality theorem, the weight 2 Hecke algebra is the dual of $S_2(\Gamma_1(N); \mathbb{Q}_p/\mathbb{Z}_p)$, which is equal to

$$H^0(X_1(N)_{/\mathbb{Z}_p}, \Omega) \otimes_{\mathbb{Z}_p} (\mathbb{Q}_p/\mathbb{Z}_p)$$

for the sheaf of 1-differentials on the modular curve $X_1(N)_{/\mathbb{Z}}$ of level N. Note that $X_1(N)(\mathbb{C}) = \Gamma_1(N)\backslash(\mathfrak{H} \cup \{cusps\})$ over $\mathbb{C}$. In general, probably one might be able to use the Pontryagin dual of $H^0(X, \Omega^q_X) \otimes \mathbb{Q}_p/\mathbb{Z}_p$ as M_Q for general Shimura varieties carrying a canonical family of abelian varieties, in order to generalize the result of Taylor–Wiles to reductive algebraic groups [H99c].

Exercise

(1) Show that R_Q is generated by the values of $\mathrm{Tr}(\varrho_Q)$ over $\mathcal{O}$.

3.2.7 Taylor–Wiles system of Hecke algebras

Here we briefly sketch how to prove that the Taylor–Wiles system $(R = R^{ord,\phi}, R_Q, M_Q = h_Q)_Q$ satisfies the conditions (tw1–5) of Theorem 3.35. The details will be given in the following subsection assuming the Poitou–Tate duality theorem, which will be proved in Chapter 4.

We look at $h_k(p;\mathcal{O}) = \bigoplus_\chi h_k(p,\chi;\mathcal{O})$, where χ runs over all characters of $(\mathbb{Z}/p\mathbb{Z})^\times$. We pick a local ring $h \subset h_k(p,\chi;\mathcal{O})$. As already remarked just before the Taylor–Wiles theorem, h is generated by $T(n)$ for n prime to p if $h \subset h^{ord}$. Suppose that $k \geq 2$.

First choose a finite set Q of primes $q \equiv 1 \mod p$. We also suppose (rg_q) for all $q \in Q$. Then we look at the p-divisible module

$$S_k(\Gamma_1(Q);K/\mathcal{O}) = S_k(\Gamma_1(Q);\mathcal{O}) \otimes_{\mathcal{O}} K/\mathcal{O}$$

for the field of fractions K of $\mathcal{O}$. By the duality theorem (Theorem 3.18), this module is the Pontryagin dual of $h_k(Q;\mathcal{O}) = h_k(pq(Q);\mathcal{O})$ for $q(Q) = \prod_{q\in Q} q$. Since we have a surjective algebra homomorphism $\pi_Q : h'_k(Q;\mathcal{O}) \to h'_k(p;\mathcal{O})$ taking $T(n)$ to $T(n)$ for n prime to $pq(Q)$, we claim to have a unique local ring h'_Q of $h'_k(Q;\mathcal{O})$ projecting down to h. Here we have used notations introduced in the earlier subsections; in particular, the algebra $h'_k(Q;\mathcal{O})$ is the $\mathcal{O}$-subalgebra of $h_k(Q;\mathcal{O})$ generated by the $T(n)$'s for n prime to $pq(Q)$.

We prove this claim as follows: Since $\overline{\rho}(Frob_q) = \begin{pmatrix} \overline{\alpha}_q & 0 \\ 0 & \overline{\beta}_q \end{pmatrix}$ with $\overline{\alpha}_q \neq \overline{\beta}_q$ by (rg_q), we claim again (the second claim) that there is a unique reduced local ring $h(Q)$ of $h_k(\Gamma_0(Q),\chi;\mathcal{O})$ such that $U(q) \equiv \overline{\alpha}_q \mod \mathfrak{m}_{h_0(Q)}$. When $Q = \{q\}$, this follows from the following two facts:

(1) $h(Q)$ surjects down to a local ring of $h_q = h[X]/(L_q(X))$ for $L_q(X) = X^2 - T(q)X + q^{k-1}\chi(q)$, where $h_q = h \oplus h$ (and hence reduced) so that X is the image of $U(q)$ and $X = \alpha_q \oplus \beta_q$ for $\alpha_q \equiv \overline{\alpha}_q \mod \mathfrak{m}_h$ and $\beta_q \equiv \overline{\beta}_q \mod \mathfrak{m}_h$;

(2) $h(\{q\}) \subset h_q$, because $q \equiv 1 \mod p$. Here by (rg_q), the Galois representation $\rho_{h(\{q\})}$ attached to the local ring $h(\{q\})$ cannot specialize to representations satisfying Theorem 3.26 (3)(b), which is associated with a primitive form in $S_k(\Gamma_0(qp),\chi)$.

Write $Q = \{q_1 = q,\ldots,q_m\}$. The above statements are valid, replacing h by $h(\{q_1,\ldots,q_{m-1}\})$ and $h(\{q\})$ by $h(Q)$, by the same argument. From this, by induction on $|Q|$, the second claim follows. Since $h'_k(Q;\mathcal{O})$ is deprived of the $U(q)$'s, the local ring h'_Q of $h'_k(Q;\mathcal{O})$ is unique and actually isomorphic to $h(Q)$ for any choice of $(\overline{\alpha}_q)_{q\in Q}$, because this ring is generated by $T(n)$ for n prime to $q(Q)p$ (see Theorem 3.32).

We make a choice of $(\overline{\alpha}_q)_{q\in Q}$, and hence a choice of $h(Q)$. We write this $h(Q)$ as h_Q, which is a local ring of $h_k(Q;\mathcal{O})$. As already remarked, h_Q is generated by $T(n)$ for n prime to $q(Q)$ and p. Then $h_Q \cong h'_Q$ and in this case, $U(q)$ gives the one $\overline{\alpha}_q$ of the eigenvalues of $\rho(Frob_q)$. This type of analysis of h_Q and h'_Q was done meticulously by Wiles, occupying a large part of the long paper of Wiles ([W2] Section 2) to include everything known together in his terminology, although the result was known earlier more or less by specialists. Let 1_Q be the idempotent of h_Q in $h_k(Q;\mathcal{O})$, and write $S_Q = 1_Q S_k(\Gamma_1(Q);K/\mathcal{O})$. Thus S_Q is a p-divisible module (Pontryagin) dual to h_Q.

Then by Corollary 3.19, $S_Q^{\Delta_Q} = H^0(\Delta_Q, S_Q) = S_Q \cap S_k(\Gamma_0(Q),\chi;K/\mathcal{O})$. However by Theorem 3.26 (3)(b), if S_Q is not contained in $S_k(\Gamma_0(p),\chi;K/\mathcal{O})$, the Galois representation does not satisfy (rg$_q$) for at least one $q \in Q$. Thus we know that $S_Q^{\Delta_Q} = S_\emptyset$. By the Pontryagin duality, this tells us $h_Q/\sum_{\sigma\in\Delta_Q}(\sigma-1)h_Q = h$, which implies (tw4–5). The $\mathcal{O}[\Delta_Q]$-freeness of h_Q is proved in Lemma 3.25, and as explained before Theorem 3.35, the rank is independent of Q. In [TW] (and also in Fujiwara's work), this control theorem and freeness over $\mathcal{O}[\Delta_Q]$ is proved using modular cohomology groups $H^1(X,\mathcal{O})$ with constant coefficients. This requires us to use Lemma 3 of [DT] at one prime. This causes some trouble in Fujiwara's work (and therefore, he needs to assume some extra conditions). In our work, we use coherent cohomology $S_k(\Gamma,\mathcal{O}) = H^0(X,\underline{\omega}^{k-2}\otimes\Omega_{X/\mathbb{Z}_p})$; so, we have not invoked Lemma 3 of [DT]. If one can follow this path even in the Hilbert modular case, the linear disjointness from $\mathbb{Q}(\mu_p)$ might not be necessary, but it would be a study in the future.

For a given number s, the existence of infinitely many Q with $|Q| = s$ is shown by using the Chebotarev density, theorem using the assumption (ai$_\kappa$) for $\kappa = \mathbb{Q}(\sqrt{(-1)^{(p-1)/2}p})$, which is done in Lemmas 1.10 and 1.12 and (3.9) of [W2] (see the following subsection). Basically, from (ai$_\kappa$), we can find a semi-simple matrix X with two distinct eigenvalues α and β in $\overline{\rho}(\mathrm{Gal}(\overline{\mathbb{Q}}/\mathbb{Q}(\mu_{p^n})))$. Then by the Chebotarev density, we can find infinitely many primes $q \neq p$ such that $\overline{\rho}(Frob_{\mathfrak{Q}}) = X$ for a prime $\mathfrak{Q}|q$. Since $Frob_{\mathfrak{Q}}$ is trivial on $\mathbb{Q}(\mu_{p^n})$, by class field theory, $q \equiv 1 \mod p^n$. Thus these q's do the job.

The most difficult part is to bound the number of generators of R_Q. This is done as follows: Note here that the dimension $\dim_{\mathbb{F}} t^*_{R_Q}$

is the minimum number of generators of R_m over $\mathcal{O}$ (see Chapter 2 Lemma 2.28). Here $t^*_{R_Q} = \mathfrak{m}_{R_Q}/\mathfrak{m}_{\mathcal{O}} + \mathfrak{m}^2_{R_Q}$ is the co-tangent space of R_Q (see Subsection 5.2.3 in Chapter 5 for more general definitions of tangent and co-tangent spaces of schemes). Starting with large enough Q with $|Q|$ exceeding a certain $r \geq \dim_{\mathbb{F}} t^*_{R_Q}$ (see the definition of r just below (3.58) in the following subsection), one throws away suitable $q \in Q$ to attain the minimal $|Q| = r$. For that, Wiles uses the Galois cohomological interpretation of the tangent space t_{R_Q} (see Lemma 2.29 in Chapter 2 and Lemma 3.38 in the following subsection) in terms of Greenberg type Selmer groups $\mathrm{Sel}(Ad(\overline{\rho}))$ (see Section 5.1 in Chapter 5 for a general definition of the Selmer groups) and studies its dimension using the Poitou–Tate duality theorem to find the minimal Q with $|Q| = r$. This key part again essentially uses the assumption (ai_κ) and is due to Wiles. In this way, he (and Taylor) could prove a more general version of Theorem 3.31 (removing the ordinarity assumption: $U_h(p) \in h^\times$ and introducing an auxiliary level N with $p \nmid N$) in the following cases:

(FL) $\overline{\rho}$ is flat at p (and require flatness also for deformations, the Selmer group is not a Greenberg type in this case but is that of Bloch–Kato);

(ST) $\det \overline{\rho} \equiv \nu \mod \mathfrak{m}$ and $\overline{\rho}|_{I_p}$ is not semi-simple (purely a multiplicative reduction case: this case is called a 'strict' case in [W2]);

(SE) $\det \overline{\rho} \not\equiv \nu \mod \mathfrak{m}$ (this case is called the minimal Selmer (non-flat) case).

After proving (FL) above, Wiles further estimates the difference of the Bloch–Kato Selmer group in (FL) and that of Greenberg, and finally gets Theorem 3.31 in the mixed case of flat and non-flat deformation (that is, the Selmer case with $\det \overline{\rho} \equiv \nu \mod \mathfrak{m}$ but not 'strict'). This type of argument occupies Section 3 of [W2], another large chunk of it. Thus Fujiwara's work is a generalization of the work of Taylor–Wiles in the above three cases, and to treat the mixed case, one need to work out Wiles' argument, which works well at least for totally real fields unramified at p.

We should remark that the flat condition (FL) implies that the corresponding local ring on the Hecke side is a local ring of the Hecke algebra of weight 2, of level N (not Np) and with 'Neben' character trivial at p. Thus this case does not occur when the auxiliary level $N = 1$ simply because $h_2(1; \mathcal{O}) = 0$. Therefore if $N = 1$ (and $p \geq 5$) and $k = 2$, we are always in the strict case, because if $\overline{\rho}$ is flat, it has to be associated with a level N weight 2 primitive eigenvalue system λ (a theorem of Mazur–Ribet [R2] Sections 6–7).

Understanding this part requires a good knowledge of Galois cohomology, including the duality theorem mentioned above, which we will cover in Chapter 4. Thus we will give a proof of the Selmer case (SE) in the following subsection assuming the duality theorems.

Although, for simplicity, we have not allowed (in our treatment) ramification outside p and ∞ for our starting deformations (that is, $Q = \emptyset$), in the application to the Shimura–Taniyama conjecture and hence Fermat's last theorem, it is important to include the cases where ramification is allowed outside p. The strategy of Wiles to include ramification outside p is as follows: First treat the case where ramification outside p is minimal, that is, the following cases: If a prime ℓ ramifies in $\overline{\rho}$, the restriction of $\overline{\rho}$ to D_ℓ is supposed to be isomorphic to a representation satisfying one of the following conditions:

(A) $\overline{\rho}$ ramifies at ℓ and $\overline{\rho}|_{D_\ell} \cong \left(\begin{smallmatrix}\chi_1 & * \\ 0 & \chi_2\end{smallmatrix}\right)$ with unramified characters χ_j with $\chi_1\chi_2^{-1} = \omega$, where ω is the Teichmüller character modulo p;
(B) $\overline{\rho}|_{I_\ell} \cong \left(\begin{smallmatrix}\chi_\ell & * \\ 0 & 1\end{smallmatrix}\right)$ with $\chi_\ell \neq 1$;
(C) $H^1(\mathrm{Gal}(\overline{\mathbb{Q}}_\ell/\mathbb{Q}_\ell), Ad(\overline{\rho})) = 0$;
(D) Absolutely irreducible $\overline{\rho}|_{I_p} = \mathrm{Ind}_{\mathbb{Q}_{\ell^2}}^{\mathbb{Q}_\ell} \varphi$ for the unique unramified quadratic extension $\mathbb{Q}_{\ell^2}/\mathbb{Q}_\ell$ with a **ramified** character φ (potentially flat case).

Here we have added one more case: Case (D) to Wiles' classification, because now Case (D) can be dealt with ([Dd1] [CDT]). The above four ℓ-types basically exhaust all possibility of minimal ramification (that is, conductor of $\overline{\rho}|_{D_\ell}$ is minimal up to character twists).

Wiles considers a deformation problem, denoted by $\mathscr{D} = (\bullet, \Sigma, \mathcal{O}, \mathcal{M})$ (depending on the above three cases (A), (B) and (C)) and identifies the universal deformation ring with an appropriate local ring of a Hecke algebra. Here '$\bullet$' indicates a choice of the p-types: (FL), (ST) and (SE) (also Wiles considers one more case (ord) without the weight 2 condition), Σ is a finite set of primes where deformation is allowed to ramify, $\mathcal{O}$ is the base ring, and $\mathcal{M} \subset \Sigma$ are the primes where one of the conditions (A)–(C) is imposed.

In the language of local (admissible) representations of $GL_2(\mathbb{Q}_\ell)$, Case (A) (and Case (C)) corresponds to *'special'*, Case (B) to *'principal'* and Case (D) to *'super-cuspidal'* representations. The definition of the deformation problem $\mathscr{D}$ at ℓ is basically conforming to the classification of the corresponding local representations at ℓ (related to the local Langlands conjecture at the place ℓ: [ARL] Chapter III). We call the problem $\mathcal{D}$ *minimal* if $\Sigma = \{p\} \sqcup \mathcal{M}$. Once Wiles can identify the deformation ring with a local ring of an appropriate Hecke algebra when $\Sigma = \{p\} \cup \mathcal{M}$, analyzing the relation between the local ring and another appropriate one residing in the Hecke algebra with level q added which surjects onto the original local ring, he lets Σ grow allowing arbitrary ramification at $q \in \Sigma - (\mathcal{M} \cup \{p\})$ and identifies the universal deformation ring with the local ring of the Hecke algebra (with level added). Here the prime q is another prime which could be $\ell \in \mathcal{M}$. In other words, if one can prove the identification in a minimal case, one can identify, with an appropriate local ring of a Hecke algebra, the universal deformation ring (parametrizing deformations of one of the p-types: (FL), (ST) and (SE)) ramifying at a given finite set $\Sigma \supset \mathcal{M}$ ([W2] Theorem 2.17).

Recently, potentially flat representations at ℓ and p (that is, ℓ-type is in Case (D)) for weight 2 Hecke algebras, and p-type is flat over some ramified extension of $\mathbb{Q}_p$) have been studied in [CDT], adding one more cases to the p-types. This allows them to prove the Shimura–Taniyama conjecture for rational elliptic curves with conductor not divisible by 27, and it is now announced that the conjecture has been fully proven by Breuil, Conrad, Diamond and Taylor.

3.2.8 Tangential dimensions of deformation rings

We now prove Theorem 3.31 in the Selmer case. We keep the notation introduced in the previous subsections. However, here we write $D_p = \mathrm{Gal}(\overline{\mathbb{Q}}_p/\mathbb{Q}_p)$ and I_p for the inertia group of D_p. We simply write $H^r_p(M)$ (resp. $H^r(\mathbb{Q}_p^{ur}, M)$ and $H^r(\mathbb{Q}_p^{ur}/\mathbb{Q}_p, M^{I_p})$) for the continuous local Galois cohomology group $H^r_{ct}(D_p, M)$ (resp. $H^r_{ct}(I_p, M)$ and $H^r_{ct}(D_p/I_p, M^{I_p})$), where M is a discrete D_p-module (see Subsection 4.3.3 in Chapter 4 for definitions of continuous cohomology).

Recall that we are looking into the Galois representation into $GL_2(h)$ for a local ring h of $h_k(p, \chi; \mathcal{O})$, where χ is a Dirichlet character modulo p. By class field theory, we regard χ as a character of $\mathfrak{G}_p$. Similarly, the norm character N induces as its p-adic avatar the p-adic cyclotomic character $\nu : \mathfrak{G}_p \to \mathbb{Z}_p^\times \subset \mathcal{O}^\times$. The character ν is given by the Galois action on $\mu_{p^\infty} = \bigcup_n \mu_{p^n}$. We call a Galois representation $\rho : \mathfrak{G}_Q \to GL_2(A)$ a deformation (with values in A) of $\overline{\rho}$ if ρ mod $\mathfrak{m}_A = \overline{\rho}$. We consider the functor $\mathcal{F}_Q : CL_{\mathcal{O}} \to SETS$ associating with each A strict equivalence classes of deformations ρ with values in A of $\overline{\rho}$ satisfying the following conditions:

(un$_Q$) *ρ is unramified outside $\{p, \infty\} \sqcup Q$;*

(det) *$det\rho = \phi = \chi\nu^{k-1}$ for a fixed $k \geq 2$ regarding χ as having values in A, where ν is the p-adic cyclotomic character;*

(reg) *We have an exact sequence of D_p-modules:*

$$0 \to V(\varepsilon_\rho) \to V(\rho) \to V(\delta_\rho) \to 0,$$

where δ_ρ is unramified, $V(\delta_\rho) \cong A$ as A-modules and D_p acts on $V(\delta_\rho)$ (resp. $V(\varepsilon_\rho)$) via $\delta_\rho : D_p \to A^\times$ (resp. $\varepsilon_p : D_p \to A^\times$).

Here for any character ξ of a group having values in $\mathcal{O}$, we often regard it as a character with values in $A^\times$ by composing the structure homomorphism: $\mathcal{O} \to A$ of $\mathcal{O}$-algebra structure. In this sense, the Teichmüller character ω (originally having values in $\mathcal{O}^\times$) is considered to be a character with values in $A^\times$ in (reg). Since ϕ is an odd character ramifying only at p, it really ramifies, and hence, the condition (reg) implies the regularity condition (rg$_p$) at p.

Let $\mathcal{Q}_m$ be the set of finite sets of primes q with $q \equiv 1 \bmod p^m$ such that $\overline{\rho}(Frob_q)$ has two distinct eigenvalues. Put $\mathcal{Q} = \bigsqcup_{m>0} \mathcal{Q}_m$. By the argument of Wiles (see Proposition 3.42 below in the text and [W2] page 523), under the absolute irreducibility of $\overline{\rho}$ over $\kappa = \mathbb{Q}(\sqrt{(-1)^{(p-1)/2}p})$: (ai$_\kappa$), there are infinitely many such Q for a given m. For each $Q \in \mathcal{Q}_m$, we write (R_Q, ρ_Q) for the universal couple representing $\mathcal{F}_Q$ (for the terminology 'representable', see Subsection 4.1.2 in Chapter 4; but, (R_Q, ϱ_Q) is just the universal ring for deformations of $\overline{\rho}$ satisfying (un$_Q$), (det) and (reg)). In particular, we put $R = R_\emptyset$. Then $\{R, R_Q, h_Q\}_{Q \in \mathcal{Q}}$ gives a Taylor–Wiles system in the sense of Subsection 3.2.6.

Here we check the following three cases: the Selmer case in detail and the strict and ordinary flat cases briefly, the infinity of the q's as above and the conditions (tw1–3) in Theorem 3.35, which finishes the proof of Theorem 3.31 in these three cases.

We quote the following Poitou–Tate exact sequence from Theorem 4.50 in Chapter 4:

THEOREM 3.37 *Suppose that $p > 2$. Let $\Sigma = \{p\} \sqcup Q$. Let M be a finite discrete $\mathbb{F}[\mathfrak{G}_Q]$-module. Let $B_q \subset H^1_q(M)$ be a submodule for each prime $q \in \Sigma$, and let $B^\perp_q \subset H^1_q(M^*(1))$ be the orthogonal complement of B_q under the pairing of the duality theorems 4.43 and 4.45 in Chapter 4. We write $H^1_B(\mathbb{Q}, M) = \beta(M)^{-1}(\prod_{q\in\Sigma} B_q)$ for the restriction map: $\beta(M) : H^1_{ct}(\mathfrak{G}_Q, M) \to \prod_{q\in\Sigma} H^1_q(M)$ and put $H^1_{B^\perp}(\mathbb{Q}, M^*(1)) = \beta(M^*(1))^{-1}$*

$(\prod_{q\in\Sigma} B_q^\perp)$. *Then we have the following exact sequence:*

$$0 \to H^1_B(\mathbb{Q}, M) \to H^1_{ct}(\mathfrak{G}_Q, M) \to \prod_{q\in\Sigma} H^1_q(M)/B_q \to$$
$$H^1_{B^\perp}(\mathbb{Q}, M^*(1))^* \to H^2_{ct}(\mathfrak{G}_Q, M) \to \prod_{q\in\Sigma} H^2_q(M) \to H^0_{ct}(\mathfrak{G}_Q, M^*(1))^* \to 0.$$

We would like to apply the above theorem to the following $\mathfrak{G}_Q$-module. Let $W \subset M_2(\mathbb{F}) = ad(\overline{\rho})$ be the subspace made of trace 0 matrices. We let $\mathfrak{G}_p$ act on W (and $ad(\overline{\rho})$) by conjugation: $w \mapsto \overline{\rho}(\sigma)w\overline{\rho}(\sigma)^{-1}$. We write this $\mathfrak{G}_p$-module W as $Ad(\overline{\rho})$. We regard the group μ_p of p-th roots of unity as a $\mathfrak{G}_p$-module on which $\mathfrak{G}_p$ acts via the Teichmüller character $\omega : \mathfrak{G}_p \to \mathbb{Z}_p^\times$. For each $\mathbb{F}[\mathfrak{G}_Q]$-module M, we define the arithmetic dual $M^*(1)$ of M by $M^*(1) = \mathrm{Hom}_{\mathbb{Z}}(M, \mu_p) = \mathrm{Hom}_{\mathbb{F}}(M, \mathbb{F}) \otimes_{\mathbb{F}} \mu_p$, on which $\mathfrak{G}_Q$ acts by $\phi \mapsto \sigma \circ \phi \circ \sigma^{-1}$ ($\phi \in M^*(1)$). We suppose that $p > 2$ as in Theorem 3.31. Then the exact sequence:

$$0 \to Ad(\overline{\rho}) \to M_2(\mathbb{F}) \xrightarrow{Tr} \mathbb{F} \to 0$$

is split as $\mathbb{F}[\mathfrak{G}_p]$-modules, and the pairing $(X, Y) \mapsto Tr(XY)$ induces a self duality $Ad(\overline{\rho})^* \cong Ad(\overline{\rho})$ as $\mathbb{F}[\mathfrak{G}_p]$-modules.

We take a base of the representation space $V(\overline{\rho})$ of $\overline{\rho}$ so that $\overline{\rho}|_{D_p} = \left(\begin{smallmatrix} \overline{\varepsilon} & * \\ 0 & \overline{\delta} \end{smallmatrix}\right)$. Then we define a one-dimensional subspace of $Ad(\overline{\rho})$ by

$$W_0 = \left\{ \left(\begin{smallmatrix} 0 & * \\ 0 & 0 \end{smallmatrix}\right) \in Ad(\overline{\rho}) \right\}.$$

We prove the boundedness in the Selmer case here, and after giving the proof of the Selmer case, we give a sketch of the proof in the other two cases: the ordinary-flat and strict cases. Thus, until then, we may assume that

$$\det(\overline{\rho}) \not\equiv \omega \mod \mathfrak{m}_{\mathcal{O}}.$$

We define, as in [W2] pages 460–461,

$$B_p = H^1_{Se}(\mathbb{Q}_p, Ad(\overline{\rho})) = \mathrm{Ker}(H^1_p(Ad(\overline{\rho})) \xrightarrow{\delta} H^1(\mathbb{Q}_p^{ur}, W/W_0)),$$

where δ is the composite of the restriction map with respect to $D_p \supset I_p$ and the morphism induced by the projection: $W \to W/W_0$. For $q \in Q$, we put $B_q^\perp = \{0\}$ and $B_q = H^1_q(Ad(\overline{\rho}))$. We then define $\mathrm{Sel}_Q(Ad(\overline{\rho})) = H^1_{\mathscr{D}^Q}(\mathbb{Q}, Ad(\overline{\rho}))$ by $H^1_B(\mathbb{Q}, Ad(\overline{\rho}))$ (as in Theorem 3.37 for $\Sigma = \{p\} \sqcup Q$). Here $\mathscr{D}^Q$ indicates the deformation problem $\mathscr{F}_Q$. We write the dual as

$$H^1_{\mathscr{D}^{Q*}}(\mathbb{Q}, Ad(\overline{\rho})^*(1)) = H^1_{B^\perp}(\mathbb{Q}, Ad(\overline{\rho})^*(1))$$

as in [W2] page 473. We now claim

Lemma 3.38 *The tangent space* $t_Q = t_{R_Q/\mathcal{O}} = \mathrm{Hom}_{\mathbb{F}}(\mathfrak{m}_{R_Q}/(\mathfrak{m}_{R_Q}^2 + \mathfrak{m}_{\mathcal{O}}), \mathbb{F})$ *is isomorphic to* $H^1_{\mathcal{D}Q}(\mathbb{Q}, Ad(\overline{\rho}))$ *under the isomorphism introduced in Lemma 2.29.*

Proof We recall the isomorphism in Lemma 2.29, using the notation introduced there. We have a canonical isomorphism of t_Q with the space of derivations $Der_{\mathcal{O}}(R_Q, \mathbb{F}) \subset Der_{\mathcal{O}}(\mathcal{R}, \mathbb{F})$, where $\mathcal{R}$ is the universal ring of the deformation problem over $\mathfrak{G}_Q$ of $\overline{\rho}$ without condition. We write $\iota : t_{\mathcal{R}/\mathcal{O}} \cong H^1_{ct}(\mathfrak{G}_Q, Ad(\overline{\rho}))$ for the isomorphism of Lemma 2.29. We now recall the construction of ι. For each derivation $\delta : \mathcal{R} \to \mathbb{F}$, we consider a map $\varphi_\delta : \mathcal{R} \to \mathbb{F}[\varepsilon]$ $(\varepsilon^2 = 0)$ given by $\varphi_\delta(r) = (r \mod \mathfrak{m}_{\mathcal{R}}) + \delta(r)\varepsilon$. Then φ_δ is an $\mathcal{O}$-algebra homomorphism, and all $\mathcal{O}$-algebra homomorphisms of $\mathcal{R}$ into $\mathbb{F}[\varepsilon]$ in $CL_{\mathcal{O}}$ are obtained in this way. Then φ_δ gives rise to a deformation $\rho : \mathfrak{G}_Q \to GL_2(\mathbb{F}[\varepsilon])$ by $\rho = \varphi_\delta \circ \varrho$ for the universal one $\varrho : \mathfrak{G}_Q \to GL_2(\mathcal{R})$. After getting ρ, we construct a 1-cocycle $u = u_\delta : \mathfrak{G}_Q \to ad(\overline{\rho})$ by $\rho(g) = \overline{\rho}(g) + u(g)\overline{\rho}(g)\varepsilon$. Then $\iota(\delta)$ is given by the class of u.

Since $Spec(R_Q)$ is a closed subscheme of $Spec(\mathcal{R})$ (that is, the universal map: $\mathcal{R} \to R_Q$ is surjective), t_Q is a subspace of $t_{\mathcal{R}/\mathcal{O}}$. Thus we study how the extra conditions imposed on $\rho \in \mathcal{F}_Q(\mathbb{F}[\varepsilon])$ give a restriction on tangent vectors. Since we have

$$\det(\overline{\rho}(g) + u(g)\overline{\rho}(g)\varepsilon) = \det \overline{\rho}(g) \det(1 + u(g)\varepsilon) = \phi \cdot (1 + Tr(u(g))\varepsilon),$$

we see that $\det \rho = \phi \iff Tr(u) = 0$. Thus under (det), u has values in $Ad(\overline{\rho})$.

Local conditions at primes in Σ are described by the choice of local subcohomology B_q. Since there is no condition imposed locally at $q \in Q$, our choice of B_q has to be the full local cohomology $H^1_q(Ad(\overline{\rho}))$ for $q \in Q$, and hence $B_q^\perp = 0$. At p, ρ is upper–triangular on D_p if and only if $u(D_p)$ is contained in upper triangular matrices. By the regularity condition (rg$_p$): $\overline{\varepsilon} \neq \overline{\delta}$ on I_p, the upper triangularity of $u(D_p)$ is equivalent to that of $u(I_p)$. Since δ_p is unramified,

$$u(I_p) \subset \left\{ \left(\begin{smallmatrix} * & * \\ 0 & 0 \end{smallmatrix}\right) \in Ad(\overline{\rho}) \right\}.$$

Since $Tr(u) = 0$, the upper left corner of $u(I_p)$ has to be 0, and hence $u(I_p) \subset W_0$. This shows that the right choice of B_p at p is

$$B_p = \mathrm{Ker}(H^1_p(Ad(\overline{\rho})) \to H^1(\mathbb{Q}_p^{ur}, W/W_0)),$$

which completes the proof. □

By Lemma 2.28 in Chapter 2, the minimal number of generators of the ring R_Q over $\mathcal{O}$ is given by the tangential dimension $\dim_{\mathbb{F}} t_Q$. Thus we shall give an estimate of $\dim_{\mathbb{F}} H^1_{\mathcal{D}Q}(\mathbb{Q}, Ad(\overline{\rho}))$. We claim

LEMMA 3.39 *Suppose $p > 2$, (ai$_{\mathbb{Q}}$) for $\overline{\rho}$ and that $\overline{\rho}$ is unramified outside p and ∞. Then $\overline{\rho}$ is absolutely irreducible over $\mathbb{Q}(\sqrt{(-1)^{(p-1)/2}p})$ if and only if $Ad(\overline{\rho})$ is irreducible as an $\mathbb{F}[\mathfrak{G}_p]$-module.*

Proof Suppose that $Ad(\overline{\rho})$ is reducible. Since $Ad(\overline{\rho})$ is semi-simple (because of the simplicity of $\overline{\rho}$), $Ad(\overline{\rho})$ contains a one-dimensional subspace $V(\xi)$ on which $\mathfrak{G}_p$ acts by a character ξ. Let $\phi \in V(\xi)$ be a generator. Since $Ad(\overline{\rho}) \subset ad(\overline{\rho}) = \mathrm{Hom}_{\mathbb{F}}(V(\overline{\rho}), V(\overline{\rho}))$, we may regard $\phi : V(\overline{\rho}) \to V(\overline{\rho})$ as a linear map such that $\phi(\sigma v) = \xi(\sigma)\sigma\phi(v)$. Thus ϕ gives an isomorphism $V(\overline{\rho}) \cong V(\overline{\rho} \otimes \xi)$. If ξ is the identity, ϕ commutes with $\overline{\rho}$, and hence, by Schur's lemma (Proposition 2.5 in Chapter 2), ϕ has to be scalar. This is impossible, because $Tr(\phi) = 0$ and $p > 2$. Thus by Lemma 2.15 in Chapter 2, $\overline{\rho} \cong \mathrm{Ind}_{\mathrm{Ker}(\xi)}^{\mathfrak{G}_p} \eta$ for a representation η of $\mathrm{Ker}(\xi)$. Since $\overline{\rho}$ is two-dimensional, ξ is of order 2. Since $\kappa = \overline{\mathbb{Q}}^{\mathrm{Ker}(\xi)}$ is a quadratic extension of $\mathbb{Q}$ unramified outside p, we know that $\kappa = \mathbb{Q}(\sqrt{(-1)^{(p-1)/2}p})$ for $p > 2$ by the uniqueness of such an extension. Thus $\overline{\rho}$ becomes reducible over κ.

If $\overline{\rho} = \mathrm{Ind}_{\kappa}^{\mathbb{Q}} \eta$, then it is easy to see that $Ad(\overline{\rho}) \cong (\mathrm{Ind}_{\kappa}^{\mathbb{Q}} \eta^{-1}\eta^{\sigma}) \oplus \xi$ for the quadratic character ξ factoring through $\mathrm{Gal}(\kappa/\mathbb{Q})$, where $\eta^{\sigma}(g) = \eta(\sigma g \sigma^{-1})$ for $\sigma \in \mathfrak{G}_p$ inducing the non-trivial automorphism on κ. □

By the lemma, $H^0_{ct}(\mathfrak{G}_Q, Ad(\overline{\rho})^*(1)) = 0$. Thus by Theorem 3.37 combined with the duality theorems 4.43 and 4.45 in Chapter 4, we have

$$\begin{aligned}\dim_{\mathbb{F}} t_Q - \dim_{\mathbb{F}} H^1_{(\mathcal{D}Q)^*}(\mathbb{Q}, Ad(\overline{\rho})^*(1)) &= h_p + \dim_{\mathbb{F}} H^1_{ct}(\mathfrak{G}_Q, Ad(\overline{\rho})) \\ &- \dim_{\mathbb{F}} H^2_{ct}(\mathfrak{G}_Q, Ad(\overline{\rho})) + \sum_{q \in Q} \dim_{\mathbb{F}} H^0_q(Ad(\overline{\rho})^*(1)),\end{aligned} \tag{3.54}$$

where $h_p = \dim_{\mathbb{F}} H^0_p(Ad(\overline{\rho})^*(1)) - \dim_{\mathbb{F}} H^1_p(Ad(\overline{\rho}))/B_p$. By the global Euler characteristic formula (Theorem 4.53 in Chapter 4), we have

$$\begin{aligned}&- \dim H^0_{ct}(\mathfrak{G}_Q, Ad(\overline{\rho})) + \dim H^1_{ct}(\mathfrak{G}_Q, Ad(\overline{\rho})) - \dim_{\mathbb{F}} H^2_{ct}(\mathfrak{G}_Q, Ad(\overline{\rho})) \\ &\qquad = - \dim_{\mathbb{F}} H^0(\mathrm{Gal}(\mathbb{C}/\mathbb{R}), Ad(\overline{\rho})) + \dim_{\mathbb{F}} Ad(\overline{\rho}) = 2.\end{aligned}$$

From this we conclude that

$$d_Q := \dim_{\mathbb{F}} t_Q - \dim_{\mathbb{F}} H^1_{(\mathscr{D}Q)^*}(\mathbb{Q}, Ad(\bar{\rho})^*(1)) \tag{3.55}$$
$$= h_p + 2 + \sum_{q \in Q} H^0_q(Ad(\bar{\rho})^*(1)),$$

since $H^0_{ct}(\mathfrak{G}_Q, Ad(\bar{\rho})) = H^0_{ct}(\mathfrak{G}_Q, Ad(\bar{\rho})^*(1)) = 0$.

We now prove the following inequalities which are key facts for proving that $Q \in \mathscr{Q}_m$ satisfies (tw1–3):

Proposition 3.40 *Suppose we are either in the Selmer case, the strict case or the ordinary-flat case. Then we have*

$$d := \dim_{\mathbb{F}} t_\emptyset \le \dim_{\mathbb{F}} H^1_{\mathscr{D}^*}(\mathfrak{G}_p, Ad(\bar{\rho})^*(1)); \tag{nq1}$$
$$\dim H^0_q(Ad(\bar{\rho})^*(1)) = 1 \text{ if } q \equiv 1 \mod p \text{ and satisfies } (rg_q); \tag{nq2}$$
$$d_Q = d - \dim H^1_{\mathscr{D}^*}(\mathfrak{G}_p, Ad(\bar{\rho})^*(1)) + \sum_{q \in Q} \dim H^0_q(Ad(\bar{\rho})^*(1)) \le |Q|, \tag{nq3}$$

where d is the minimum number of generators of R over $\mathcal{O}$.

Proof We prove the inequalities first in the Selmer case, following [W2] Proposition 1.9. We first prove $h_p+2 \le 0$; then it is immediate to conclude (nq1) and (nq3) from (3.55). We look at the exact sequence associated with the following short exact sequence:

$$0 \to W_0 \to W = V(Ad(\bar{\rho})) \to W/W_0 \to 0,$$

where W_0 is a D_p-stable subspace consisting of upper nilpotent matrices in $V(Ad(\bar{\rho}))$ taking a base of $V(\bar{\rho})$ so that the restriction to D_p is upper triangular. For the image $\mathrm{Im}(u)$ for $u : H^1_p(W) \to H^1_p(W/W_0)$, the exact sequence tells us the exactness of the following sequence (see Corollary 4.28 in Chapter 4):

$$0 \to \mathrm{Im}(u) \to H^1_p(W/W_0) \to H^2_p(W_0) \to H^2_p(W) \to H^2_p(W/W_0) \to 0,$$

since cohomological dimension of the local Galois group at p is equal to 2 (see Theorem 4.43 in Chapter 4). Thus we have (see (4.46) in Chapter 4):

$$\dim_{\mathbb{F}} \mathrm{Im}(u) = \dim_{\mathbb{F}} H^1_p(W/W_0) - \dim_{\mathbb{F}} H^2_p(W_0) \tag{3.56}$$
$$+ \dim_{\mathbb{F}} H^2_p(W) - \dim_{\mathbb{F}} H^2_p(W/W_0).$$

Thus writing the following commutative diagram (with exact lower sequence; see Theorem 4.33 and Corollary 4.27 in Chapter 4):

$$\begin{array}{ccccccc} & & & H^1_p(W) & & & \\ & & & u\downarrow & \searrow\delta & & \\ 0\to H^1_{ct}(\mathbb{Q}_p^{ur}/\mathbb{Q}_p,(W/W_0)^{I_p}) & \to & & H^1_p(W/W_0) & \to & H^1_{ct}(\mathbb{Q}_p^{ur},W/W_0)^{D_p}\to 0, & \end{array}$$

where $I_p = \mathrm{Gal}(\overline{\mathbb{Q}}_p/\mathbb{Q}_p^{ur})$ for the maximal unramified extension $\mathbb{Q}_p^{ur}/\mathbb{Q}_p$ and $D_p = \mathrm{Gal}(\overline{\mathbb{Q}}_p/\mathbb{Q}_p)$. By the commutativity of the triangle diagram as above, the image of δ contains $\mathrm{Im}(u) \mod H^1_{ct}(\mathbb{Q}_p^{ur}/\mathbb{Q}_p,(W/W_0)^{I_p})$. Note here that

$$h_p = \dim_{\mathbb{F}} H^0_p(W^*(1)) - \dim_{\mathbb{F}} \mathrm{Im}(\delta).$$

In any case,

$$\dim_{\mathbb{F}} \mathrm{Im}(\delta) \geq \dim_{\mathbb{F}} \mathrm{Im}(u) - \dim_{\mathbb{F}} H^1_{ct}(\mathbb{Q}_p^{ur}/\mathbb{Q}_p,(W/W_0)^{I_p}).$$

We now compute $\dim_{\mathbb{F}} Z$ for $Z = H^1_{ct}(\mathbb{Q}_p^{ur}/\mathbb{Q}_p,(W/W_0)^{I_p})$: We see from the cyclicity and the torsion-freeness of $D_p/I_p = \langle Frob_p\rangle$ (see (4.16) in Chapter 4) that

$$Z = H^1_{ct}(\mathbb{Q}_p^{ur}/\mathbb{Q}_p,(W/W_0)^{I_p}) \cong (W/W_0)^{I_p}/(Frob_p - 1)(W/W_0)^{I_p}.$$

Then we have the following exact sequence:

$$\begin{aligned} 0 \longrightarrow H^0(\mathbb{Q}_p^{ur}/\mathbb{Q}_p,(W/W_0)^{I_p}) \longrightarrow (W/W_0)^{I_p} &\xrightarrow{Frob_p-1} (W/W_0)^{I_p} \\ &\longrightarrow H^1_{ct}(\mathbb{Q}_p^{ur}/\mathbb{Q}_p,(W/W_0)^{I_p}) \longrightarrow 0. \end{aligned}$$

Thus we have

$$\begin{aligned} \dim_{\mathbb{F}} H^1_{ct}(\mathbb{Q}_p^{ur}/\mathbb{Q}_p,(W/W_0)^{I_p}) &= \dim_{\mathbb{F}} H^0(\mathbb{Q}_p^{ur}/\mathbb{Q}_p,(W/W_0)^{I_p}) \\ &= \dim_{\mathbb{F}} H^0_p(W/W_0). \end{aligned}$$

This shows that $\dim_{\mathbb{F}} Z = \dim_{\mathbb{F}} H^0_p(W/W_0)$. Then we have

$$\begin{aligned} \dim_{\mathbb{F}} \mathrm{Im}(\delta) \geq \dim_{\mathbb{F}} H^1_p(W/W_0) &- \dim_{\mathbb{F}} H^2_p(W_0) + \dim_{\mathbb{F}} H^2_p(W) \\ &- \dim_{\mathbb{F}} H^2_p(W/W_0) - \dim_{\mathbb{F}} H^0_p(W/W_0). \end{aligned}$$

By the local Euler characteristic formula (Theorem 4.52 in Chapter 4), we see

$$\dim_{\mathbb{F}} H^1_p(W/W_0) - \dim_{\mathbb{F}} H^2_p(W/W_0) - \dim_{\mathbb{F}} H^0_p(W/W_0) = \dim_{\mathbb{F}} W/W_0 = 2.$$

We know now

$$\begin{aligned} h_p &= \dim_{\mathbb{F}} H^0_p(W(1)) - \dim_{\mathbb{F}} \mathrm{Im}(\delta) \\ &\leq \dim_{\mathbb{F}} H^0_p(W(1)) - 2 + \dim_{\mathbb{F}} H^2_p(W_0) - \dim_{\mathbb{F}} H^2_p(W) \\ &= -2 + \dim_{\mathbb{F}} H^0_p(W_0^*(1)). \end{aligned} \tag{3.57}$$

The last equality in the above equation follows from the local duality theorem (Theorem 4.43):

$$\dim_{\mathbb{F}} H^q_p(M^*(1)) = \dim_{\mathbb{F}} H^{2-q}_p(M)$$

for finite discrete $\mathbb{Z}_p[D_p]$-modules M (note here $W^* \cong W$) applied to $q = 0$ and 2. Equation (3.57) shows that

$$h_p + 2 \leq \dim_{\mathbb{F}} H^0_p(W_0^*(1)).$$

When $H^0_p(W_0^*(1)) = 0$, we see that $h_p + 2 \leq 0$.

Write $\overline{\rho} = \begin{pmatrix} \omega\eta & * \\ 0 & \overline{\delta} \end{pmatrix}$ on I_p (under the notation of (ord), $\eta\omega = \overline{\delta}$). Since we are in the Selmer case, $\overline{\delta}\eta^{-1} \neq 1$. Then we see $H^0_p(W_0^*(1)) = 0$ and hence $h_p + 2 \leq 0$.

Now we look at the local cohomology at q $\in Q$. We assume $q \equiv 1$ mod p and (rg$_q$), that is, $\overline{\rho}(Frob_q) \cong \begin{pmatrix} \overline{\alpha} & 0 \\ 0 & \overline{\beta} \end{pmatrix}$ with $\overline{\alpha} \neq \overline{\beta}$. We consider the deformation of $\overline{\rho}$ unramified outside $Q \cup \{p, \infty\}$ with ordinarity (ord) and the determinant condition (det). Then we put $h_q = \dim_{\mathbb{F}} H^0_q(W(1))$. Since $q \equiv 1 \mod p$, $W \cong W(1)$ and hence, $h_q = 1$ by the regularity condition $\overline{\alpha} \neq \overline{\beta}$. Then by (3.55), we get

$$\dim_{\mathbb{F}} t_Q - \dim_{\mathbb{F}} H^1_{(\mathscr{D}Q)^*}(\mathfrak{G}_Q, W(1)) = h_p + 2 + \sum_{q \in Q} h_q. \tag{3.58}$$

The assertions (nq1–3) then follows from the fact: $h_p + 2 \leq 0$. □

Write $r = \dim H^1_{\mathscr{D}^*}(\mathfrak{G}_p, Ad(\overline{\rho})(1))$ for $Q = \emptyset$ and $\mathscr{D}^* = \mathscr{D}^{\emptyset^*}$, which is greater than or equal to the tangential dimension $\dim t_\emptyset$ of the functor $\mathscr{F}_\emptyset$ and hence the number d of (minimum) generators of R over $\mathcal{O}$. If we find Q such that $H^1_{(\mathscr{D}Q)^*}(\mathfrak{G}_Q, Ad(\overline{\rho})(1)) = 0$ and $|Q| = r$, then we see from (nq1–2) that this Q satisfies (tw1–3).

We now quote the following classical result from Dickson [LGF] Section 260 on subgroups G of $PGL_2(\mathbb{F}) = GL_2(\mathbb{F})/\mathbb{F}^\times$: We suppose that $p \geq 5$, because we know that $h^{ord}(p, \chi; \mathcal{O}) = 0$ if $p \leq 7$.

- (PL) When $p||G|$ and G is not contained in a Borel subgroup (that is, a conjugate of the group of upper triangular matrices) in $PGL_2(\overline{\mathbb{F}})$, then G is conjugate either to $PSL_2(k)$ or $PGL_2(k)$ for a subfield $k \subset \overline{\mathbb{F}}_p$ in $PGL_2(\overline{\mathbb{F}}_p)$;
- (Sm) When $p \nmid |G|$, if G is neither dihedral nor cyclic, then G is isomorphic to either A_4, S_4 or A_5.

It is easy to see the second assertion, following [Se] Proposition 16: Suppose $p \nmid |G|$. Lift G to a subgroup $\widetilde{G}$ in $GL_2(\mathbb{F})$. We still have $p \nmid |\widetilde{G}|$.

Thus we can lift the representation of $\widetilde{G} \subset GL_2(\mathbb{F})$ to $\widetilde{G} \subset GL_2(K)$ for a p-adic field K with residue field $\mathbb{F}$ (see Corollary 2.7 in Chapter 2). Since $\widetilde{G}$ is a finite group, we may assume after conjugation that $\widetilde{G} \subset GL_2(F)$ for a number field $F \subset K$. Embed $F \subset \mathbb{C}$, and hence $\widetilde{G} \subset GL_2(\mathbb{C})$. Thus $G \subset PGL_2(\mathbb{C})$. Since G is finite, it is in a maximal compact subgroup of $PGL_2(\mathbb{C})$, which is conjugate to the 2×2 special unitary matrices modulo center: $PSU_2(\mathbb{C})$. Thus we may assume that $G \subset PSU_2(\mathbb{C})$. Now we invoke an exotic isomorphism $PSU_2(\mathbb{C}) \cong SO_3(\mathbb{R})$. We consider the symmetric second tensor representation $Sym^2 : SL_2(\mathbb{C})/\{\pm 1\} \to GL_3(\mathbb{C})$. Since Sym^2 preserves a non-degenerate symmetric bilinear form on $\mathbb{C}^3$, suitably choosing conjugation, we have an injective homomorphism $PSU_2(\mathbb{C}) \hookrightarrow SO_3(\mathbb{R})$. By comparing dimension, we conclude that this is a surjective isomorphism. Once this isomorphism is established, G is a group of rotation acting on an Euclidean 3-space. If G is neither dihedral nor cyclic, the shape made by moving around one point off origin by the action of G is a regular polyhedron. The only possibilities are, as is well known, tetrahedron, cube and icosahedron. The symmetric groups of the above polyhedron are isomorphic to A_4, S_4 and A_5 (Exercise 1). As for the first assertion, it is a bit technical (except for when $\mathbb{F} = \mathbb{F}_p$: Exercise 2); so, we just refer the reader to the book by Dickson.

Before showing the existence of infinitely many such sets Q in $\mathcal{Q}_m$, we prepare one more lemma. Write $H^q(F/K, M)$ for $H^q_{ct}(\mathrm{Gal}(F/K), M)$ for a Galois extension F/K and a discrete $\mathrm{Gal}(F/K)$-module M.

Lemma 3.41 (Wiles) *Let* $K = \overline{\mathbb{Q}}^{\mathrm{Ker}(\overline{\rho})}$. *Then we have*

$$H^1(K(\mu_p)/\mathbb{Q}, W(1)) = 0.$$

Proof We may assume that p divides $|\mathrm{Im}(\overline{\rho})|$. We have an exact sequence (from the inflation-restriction sequence: Theorem 4.33):

$$0 \to H^1(K/\mathbb{Q}, W(1)^{\mathrm{Gal}(K(\mu_p)/K)}) \to H^1(K(\mu_p)/\mathbb{Q}, W(1)) \to H^1(K(\mu_p)/K, W(1)) = 0.$$

The vanishing of the last cohomology group follows from the fact that p is prime to $|\mathrm{Gal}(K(\mu_p)/K)|$ (see Proposition 4.21). Since $\mathrm{Gal}(K(\mu_p)/K)$ fixes W element by element,

$$W(1)^{\mathrm{Gal}(K(\mu_p)/K)} = 0 \text{ if } \mathrm{Gal}(K(\mu_p)/K) \neq 1,$$

and hence $H^1(K(\mu_p), W(1)) = 0$. Thus we may suppose that $K(\mu_p) = K$.

Let Z be the center of $G = \mathrm{Gal}(K/\mathbb{Q})$. By absolute irreducibility, Z is in the center of $GL_2(\mathbb{F})$ (Lemma 2.12 in Chapter 2); so, Z acts trivially on W. Then we have another exact sequence:

$$0 \to H^1(G/Z, W(1)^Z) \to H^1(G, W(1)) \to H^1(Z, W(1)) = 0.$$

The vanishing of the last cohomology again follows from the fact that $p \nmid |Z|$. Since Z acts trivially on W, if Z acts non-trivially on μ_p, $W(1)^Z = 0$. Thus we get $H^1(G, W(1)) = 0$, and hence we may suppose that Z acts on μ_p trivially. Then $\mathrm{Gal}(\mathbb{Q}(\mu_p)/\mathbb{Q})$ is the quotient of G/Z, which is isomorphic to either $PGL_2(k)$ or $PSL_2(k)$ for a subfield $k \subset \overline{\mathbb{F}}_p$ by (PL). Since $PSL_2(k)$ for $p \geq 5$ is known to be simple and $(PGL_2(k) : PSL_2(k)) = 2$ if $p \geq 5$, this is impossible, because $[\mathbb{Q}(\mu_p) : \mathbb{Q}] > 2$.

We leave to the reader the remaining case: $p = 3$ (Exercise 3), because the Hecke algebra $h_k^{ord}(p, \chi; \mathcal{O})$ vanishes if $p \leq 7$. □

We now finish the proof of the existence of the Taylor–Wiles system satisfying (tw1–4). We have the following commutative diagram with exact rows:

$$\begin{array}{ccccc}
0 \to \mathrm{Ker}(\beta'_Q) & \longrightarrow & H^1_{ct}(\mathfrak{G}_Q, W^*(1)) & \xrightarrow{b'_Q} & \prod_{q\in Q} H^1(\mathbb{Q}_q^{ur}, W^*(1)) \\
\downarrow & & \downarrow res & & \wr\downarrow \\
0 \to \mathrm{Hom}(\mathfrak{H}, W^*(1)) & \longrightarrow & \mathrm{Hom}(\mathfrak{H}_Q, W^*(1)) & \longrightarrow & \prod_{q\in Q} H^1(\mathbb{Q}_q^{ur}, W^*(1)),
\end{array}$$

where $\mathfrak{H} = \mathrm{Gal}(\mathbb{Q}^{(p)}/K(\mu_p)) \subset \mathfrak{G}_p$ and $\mathfrak{H}_Q = \mathrm{Gal}(\mathbb{Q}^{Q\cup\{p\})}/K(\mu_p)) \subset \mathfrak{G}_Q$. The map *res* as above is injective by Lemma 3.41. By the inflation map, $H^1_{ct}(\mathfrak{G}, W^*(1))$ is embedded into $H^1_{ct}(\mathfrak{G}_Q, W^*(1))$ by Theorem 4.33 in Chapter 4. This tells us that any cohomology class in $H^1_{ct}(\mathfrak{G}_Q, W^*(1))$ unramified at all $q \in Q$ actually comes from $H^1_{ct}(\mathfrak{G}, W^*(1))$. In particular, the kernel of the restriction map: $H^1_{ct}(\mathfrak{G}_Q, W^*(1)) \to \prod_{q\in Q} H^1_q(W^*(1))$ is contained in the kernel $\mathrm{Ker}(b_Q)$ of the following restriction map:

$$b_Q : H^1_{ct}(\mathfrak{G}, W^*(1)) \to \prod_{q\in Q} H^1_q(W^*(1)). \tag{3.59}$$

This fact that we can shift from the larger group $\mathfrak{G}_Q$ to $\mathfrak{G}$ becomes clear from the lower sequence of the diagram.

Since $B_q^\perp = 0$ for all $q \in Q$, enlarging Q, we expect to attain the vanishing: $H^1_{\mathscr{D}_Q^*}(\mathfrak{G}_Q, W^*(1)) = 0$. We would like to prove

Proposition 3.42 *There exist infinitely many finite sets Q such that for a given integer $m > 0$*

(1) $|Q| = r \geq \dim_{\mathbb{F}} \mathrm{Sel}_\Sigma(Ad(\overline{\rho})) = \dim_{\mathbb{F}} t_Q = d_Q$;

(2) $q \equiv 1 \mod p^m$ *for all* $q \in Q$;

(3) b_Q *is injective, and hence* $H^1_{(\mathscr{D}^Q)^*}(\mathfrak{G}_Q, W^*(1)) = 0$.

Proof We give a version of the argument of Wiles proving a similar fact (in [W2] (3.8) in a different setting, his argument is applied to V^*_λ in place of $W^*(1)$ here). By the above lemma (Lemma 3.41) combined with the inflation–restriction sequence (Theorem 4.33 in Chapter 4), we have an exact sequence:

$$0 = H^1(K(\mu_p)/\mathbb{Q}, W(1)) \to H^1_{ct}(\mathbb{Q}^{(p)}/\mathbb{Q}, W(1)) \xrightarrow{\iota} H^1_{ct}(\mathfrak{H}, W(1))^G \cong \mathrm{Hom}_G(\mathfrak{H}, W(1))$$

for $G = \mathrm{Gal}(K(\mu_p)/\mathbb{Q})$. The last isomorphism follows from the fact that a 1-cocycle for the trivial action is a homomorphism (see (4.16) in Chapter 4).

First we show that by adding primes to Q, we can make b_Q injective. Pick a non-zero $x \in \mathrm{Ker}(\mathrm{b}_Q)$, and let $f_x : \mathfrak{H} \to W(1)$ be the image of x under ι. Since we have excluded the case where $\overline{\rho}$ is induced from κ, the image of $\overline{\rho}$ in $\mathrm{PGL}_2(\mathbb{F})$ is either A_4, S_4, S_5 or of order divisible by p. Then we claim ([W2] Lemma 1.10) that there exists $\sigma \in \mathfrak{G}_p$ such that

(1) $\overline{\rho}(\sigma)$ has order ℓ with $\ell \geq 3$ prime to p;

(2) σ fixes $\mathbb{Q}(\mu_{p^m})$ (and hence, $\det\overline{\rho}(\sigma) = 1$).

We now finish the proof of the proposition, accepting the claim, which we will prove in the next lemma. We want to show that we can choose σ as above satisfying one more condition:

(3) $f_x(\sigma^\ell) \neq 0$.

Let L be the minimal Galois extension of $K(\mu_p)$ so that f_x factors through the Galois group $\mathrm{Gal}(L/K(\mu_p))$. Then L is an abelian extension of exponent p. Thus we have an exact sequence:

$$1 \to X = \mathrm{Gal}(L/K(\mu_p)) \to \mathrm{Gal}(L/\mathbb{Q}) \to G = \mathrm{Gal}(K(\mu_p)/\mathbb{Q}) \to 1.$$

Then $\mathrm{Gal}(K(\mu_p)/\mathbb{Q})$ acts on X by conjugation, and $f_x : X \to W(1)$ is a morphism of G-modules. Let σ' be an element satisfying (1)–(2), and write $\overline{\sigma}'$ for the image of σ' in $\mathrm{Gal}(K(\mu_p)/\mathbb{Q})$. By (1)–(2), $Ad(\overline{\rho})(\sigma')$ has three distinct eigenvalues on W, and one of them is equal to 1. Thus we can decompose $X = X[1] \oplus X'$ with $X' \neq 0$ and $X[1] \neq 0$ so that $\overline{\sigma}' - 1$ is an automorphism on X' and $\overline{\sigma}' = 1$ on $X[1]$. Since $Ad(\overline{\rho})$ is absolutely irreducible, if $f_x(X[1]) = 0$, then $f_x(X') = f_x(X)$ is a proper invariant subspace of $W(1)$, which is impossible. Thus we can find $\tau \in X[1]$ such that $f_x(\tau) \neq 0$. We then define

$$\begin{aligned}\tau' = 1 \times \tau \in \mathrm{Gal}(L(\mu_{p^m})/K(\mu_p)) &= \mathrm{Gal}(K(\mu_{p^m})/K(\mu_p)) \times \mathrm{Gal}(L/K(\mu_p)) \\ &\cong \mathrm{Gal}(L(\mu_{p^m})/L) \times \mathrm{Gal}(L(\mu_{p^m})/K(\mu_{p^m})),\end{aligned}$$

because L is linearly disjoint from $K(\mu_{p^m})$ (over $K(\mu_p)$) by our construction ($f_x(X) \subset W$) because $\mathrm{Gal}(K(\mu_p)/\mathbb{Q})$ acts non-trivially on $\mathrm{Gal}(L/L(\mu_p))$ while $\mathrm{Gal}(K(\mu_p)/\mathbb{Q})$ acts trivially on $\mathrm{Gal}(K(\mu_{pm}/K(\mu_p))$. Since $\tau \in X[1]$, τ' commutes with σ' in $\mathrm{Gal}(L(\mu_{p^m})/\mathbb{Q})$. Then $f_x((\tau'\sigma')^\ell) = f_x(\tau'^\ell \sigma'^\ell) = f_x(\tau'^\ell) + f_x(\sigma'^\ell) = \ell f_x(\tau') + f_x(\sigma'^\ell)$ because $\sigma'^\ell \in X$, and $0 \neq \ell f_x(\tau') = f_x((\tau'\sigma')^\ell) - f_x(\sigma'^\ell)$. This shows one of $\tau'\sigma'$ or σ' satisfies (1), (2) and (3), which we write as σ. We then choose a prime q' so that $Frob_{q'} = \sigma$ in $\mathrm{Gal}(L(\mu_{p^m})/\mathbb{Q})$. Then $f_x(Frob_{q'}) \neq 0$ implies $\mathrm{b}_{Q\cup\{q'\}}(x) \neq 0$. By (2), we see that $q' \equiv 1 \mod p^m$. By (1) and (2), the characteristic roots α and β of $\overline{\rho}(\sigma)$ satisfy $\alpha\beta = 1$ and hence primitive ℓ-th roots. Therefore $\ell \geq 3$ (with $p \nmid \ell$) shows that $\alpha \neq \beta$. This provides us with infinitely many disjoint choices of Q, by the Chebotarev density theorem.

As already remarked, the injectivity of b_Q implies the vanishing of $H^1_{(\mathcal{D}Q)^*}(\mathfrak{G}_Q, W^*(1))$ since

$$H^1_{(\mathcal{D}Q)^*}(\mathfrak{G}_Q, W^*(1)) \subset \mathrm{Ker}(\mathrm{b}_Q),$$

which follows from the fact that $B_q^\perp = 0$ for all $q \in Q$ (Theorem 4.43 in Chapter 4).

After proving the injectivity of b_Q, we remove one by one elements of Q keeping injectivity of b_Q. This is possible because $\dim_{\mathbb{F}} H^1_q(W^*(1)) > 0$ (nq2). Then after a finite number of steps, we reach the equality: $|Q| = \dim_{\mathbb{F}} H^1_{\mathcal{D}^*}(\mathfrak{G}_p, W^*(1))$, because in the minimal choice of Q, each q kills a one-dimensional quotient of $H^1_{\mathcal{D}^*}(\mathfrak{G}_p, W^*(1))$. We then put $r = \dim_{\mathbb{F}} H^1_{\mathcal{D}^*}(\mathfrak{G}_p, W^*(1))$, which is independent of Q. We still have infinitely many choices of Q, because we can start this element-removing process from another Q' disjoint from Q. □

We needed to know the following claim ([W2] Lemma 1.10):

LEMMA 3.43 *Suppose that $p > 2$. Suppose that the image of $\overline{\rho}$ in $PGL_2(\mathbb{F})$ is either of order divisible by p or isomorphic to one of the following groups: A_4, S_4 and A_5. Then there exists $\sigma \in \mathfrak{G}_p$ such that*

(1) *$\overline{\rho}(\sigma)$ has order ℓ with $\ell \geq 3$ prime to p;*
(2) *$\det \overline{\rho}(\sigma) = 1$ and σ fixes $\mathbb{Q}(\mu_{p^m})$.*

Proof Let G be the image of $\overline{\rho}$ in $GL_2(\mathbb{F})$ and Z be the center of Z. Since $\overline{\rho}$ is absolutely irreducible, the subgroup Z is made of scalar matrices in G. Thus $G/Z \subset PGL_2(\mathbb{F})$. We need to use the classification of subgroups of $PGL_2(\overline{\mathbb{F}})$ stated in (PL): If $p||G|$, then $G/Z \cong PGL_2(k)$ or $G/Z \cong PSL_2(k)$ for a subfield $k \subset \overline{\mathbb{F}}$. We can then always find an element in the derived group $D(G/Z)$ of order $\ell \geq 3$ prime to p and lift it to the derived group $D(G)$ of G because of surjectivity of the projection $D(G) \twoheadrightarrow D(G/Z)$. We can apply the same proof when G/Z is isomorphic either to A_4, S_4 and A_5 (see (Sm)). □

In the flat case and strict case, what we have not proved is the inequality $h_p + 2 \leq 0$. Once $h_p + 2 \leq 0$ is proved, everything will follow from this (although in our case of auxiliary level 1, the flat case is empty). Here we shall give a sketch of Wiles' proof of this inequality in the two cases separately.

Remark 3.44 Here we give an argument proving $h_p + 2 = 0$ in the strict case: $\overline{\rho}|_{I_p}$ is indecomposable, reducible and $\overline{\delta}\eta^{-1} = 1$, where $\overline{\rho} = \begin{pmatrix} \eta\omega & * \\ 0 & \overline{\delta} \end{pmatrix}$. In this case, $H^0_p(W_0^*(1)) \neq 0$. We use the notation introduced in the proof of Proposition 3.40. We put, similarly to H^1_{Se},

$$B_p = H^1_{str}(\mathbb{Q}_p, W) = \mathrm{Ker}(H^1_p(W) \to H^1_p(W/W_0))$$

and define also $H^1_{\mathscr{D}Q}$ and $H^1_{(\mathscr{D}Q)^*}$ by H^1_B and $H^1_{B^\perp}$, respectively, where $B_q = H^1_q(W)$ and $B_q^\perp = 0$ for all $q \in Q$. Then we have

$$\begin{aligned} h_p &= \dim_{\mathbb{F}} H^0_p(W(1)) - \dim_{\mathbb{F}} \mathrm{Im}(u) \qquad (3.60)\\ &= \dim_{\mathbb{F}} H^0_p(W(1)) - \dim_{\mathbb{F}} H^1_p(W/W_0) + \dim_{\mathbb{F}} H^2_p(W_0)\\ &\quad - \dim_{\mathbb{F}} H^2_p(W) + \dim_{\mathbb{F}} H^2_p(W/W_0)\\ &= \dim_{\mathbb{F}} H^0_p(W(1)) - 2 - \dim_{\mathbb{F}} H^0_p(W/W_0)\\ &\quad + \dim_{\mathbb{F}} H^2_p(W_0) - \dim_{\mathbb{F}} H^2_p(W)\\ &= -2 + \dim_{\mathbb{F}} H^0_p(W_0^*(1)) - \dim_{\mathbb{F}} H^0_p(W/W_0) = -2, \end{aligned}$$

because $\dim_{\mathbb{F}} H^0_p(W/W_0) = \dim_{\mathbb{F}} H^0_p(W_0^*(1)) = 1$ and

$$\dim_{\mathbb{F}} H^2_p(W_0) = \dim_{\mathbb{F}} H^0_p(W_0^*(1)) = \dim_{\mathbb{F}} H^0_p(W(1)) = \dim_{\mathbb{F}} H^2_p(W) = 1.$$

Remark 3.45 As already remarked in the previous subsection, the flat case does not occur if the auxiliary level is 1. However, the formal computation of $h_p + 2$, showing that the quantity is non-positive, still works well (and therefore, when $N > 1$, Taylor and Wiles could prove the universality of the Hecke algebra). We just give a sketch of the argument here for the reader's convenience: In the flat case, we need to use tools, such as Fontaine's theory of p-adic comparison of cohomology groups, Bloch–Kato's finite cohomology groups H^1_f and the theory of the finite flat group scheme over p-adic base, which are a little beyond the scope of this book. So we shall be content with giving only a sketch of the proof of the inequality $h_p + 2 \leq 0$. Thus we may assume that $H^0_p(W_0^*(1)) = 1$ and $\overline{\rho}$ is flat and ordinary at p (then, by the classification theorem of finite flat group schemes, automatically $\det \overline{\rho} = \omega$ on D_p).

We put $B_q = H^1_f(\mathbb{Q}_p, W)$ using a finite cohomology group at p (see below its definition as an extension group). The definition of B_q and $B_q^{\perp}$ are the same for $q \in Q$. Then we put $H^1_{\mathscr{D}\mathscr{Q}} = H^1_B$ and $H^1_{(\mathscr{D}\mathscr{Q})^*} = H^1_{B^{\perp}}$, respectively. Thus $\overline{\rho}|_{D_p} \cong \begin{pmatrix} \eta\omega & 0 \\ 0 & \eta \end{pmatrix}$ for an unramified η. Thus we may assume that $\mathbb{F} = \mathbb{F}_p$, twisting by the unramified character η^{-1}. To compute the finite cohomology $H^1_f(\mathbb{Q}_p, W)$, we need to introduce the category of filtered Dieudonné modules: $\mathscr{MF}$ (cf. [FL]). The category $\mathscr{MF} = \underline{MF}^{f,2}_{tor}$ in [FL] is made of $\mathbb{F}_p$-vector space M with filtration $M^1 \subset M$ and a linear isomorphism $\varphi : M/M^1 \oplus M^1 \cong M$. Then we know from [FL] 9.12 that $H^1_f(\mathbb{Q}_p, ad(\overline{\rho})) \cong \mathrm{Ext}^1_{\mathscr{MF}}(M, M)$ as $\mathbb{F}_p$-vector spaces, where M is the Dieudonné module associated with $\overline{\rho} \cong (\mathbb{Z}/p\mathbb{Z}) \oplus \mu_p$. Then we claim

(1) $\dim_{\mathbb{F}_p} \mathrm{Ext}^1_{\mathscr{MF}}(M, M) = 3$;

(2) The composite $\mathrm{Ext}^1_{\mathscr{MF}}(M, M) \subset H^1_{ct}(\mathbb{Q}_p, ad(\overline{\rho})) \xrightarrow{tr} H^1_{ct}(\mathbb{Q}_p, \mathbb{F})$ has to be non-trivial.

Here is how the computation goes (see [Ra] Section 3 and [TW] page 565): Fix a base $\{e_0, e_1\}$ of M with $e_1 \in M^1$. Then write $\varphi(e_0, 0) = \alpha e_0 + \beta e_1$ and $\varphi(0, e_1) = \gamma e_0 + \delta e_1$. Now put $\Phi = \begin{pmatrix} \alpha & \gamma \\ \beta & \delta \end{pmatrix}$. Then for an $\mathbb{F}_p$-linear map $T : M \to M$ preserving the filtration (so, the matrix t of T is lower triangular), $t\Phi$ is the matrix of $T \circ \varphi$, and $\Phi\widetilde{t}$ is the matrix of $\Phi \circ \widetilde{T}$, where $\widetilde{T}$ is the linear map on $M/M^1 \oplus M^1$ by T and $\widetilde{t}$ is the matrix of $\widetilde{T}$ (thus $\widetilde{t}$ is the diagonal matrix given by the semi-simplification of T). Then by the argument in [TW] page 565, we get

$$\mathrm{Ext}^1_{\mathscr{MF}}(M, M) \cong M_2(\mathbb{F}_p)/X$$

for

$$X = \{t\Phi - \Phi\widetilde{t} \mid t \text{ is the lower triangular in } M_2(\mathbb{F}_p)\} \cong \left\{ \begin{pmatrix} a-a & 0 \\ b & c-c \end{pmatrix} \middle| a, b, c \in \mathbb{F}_p \right\}.$$

Then we see easily that X consists of lower triangular matrices. This shows that $\dim_{\mathbb{F}_p} \mathrm{Ext}^1_{\mathscr{MF}}(M, M) = 3$. The argument proving (2) is exactly the same as in [TW]. Thus we get

$$\dim_{\mathbb{F}} H^1_f(\mathbb{Q}_p, W) = \dim_{\mathbb{F}} H^0_p(W) + 1. \tag{3.61}$$

This shows that

$$h_p = \dim_{\mathbb{F}} H^0_p(W(1)) - \dim_{\mathbb{F}} H^1_p(W) + \dim_{\mathbb{F}} H^0_p(W) + 1 = -2. \tag{3.62}$$

Exercises

(1) Fill in the details in the proof of (Sm).

(2) Prove (PL) when $G \subset PGL_2(\mathbb{F}_p)$.

(3) Finish the proof of Lemma 3.41 in the case where $p = 3$. In this case, we have $PSL_2(\mathbb{F}) \cong A_4$ and $PGL_2(\mathbb{F}_3) \cong S_4$ (see [W2] Proposition 1.11 for an answer).

4
Cohomology Theory of Galois Groups

In this chapter, we first describe the general theory of forming cohomology groups out of an abelian category and a left exact functor. Then we apply the theory to the category of discrete Galois modules and study resulting Galois cohomology groups. We would like to give a full proof of the Tate duality theorems and the Euler characteristic formulas of Galois cohomology groups.

4.1 Categories and Functors

In this section, we describe briefly the theory of categories to provide the basics for our later study of extension groups.

4.1.1 Categories

A *category* $\mathscr{C}$ consists of two data: objects of $\mathscr{C}$ and morphisms of $\mathscr{C}$. For any two objects X and Y of $\mathscr{C}$, we have a set $\mathrm{Hom}_{\mathscr{C}}(X, Y)$ of morphisms satisfying the following three rules:

(ct1) *For three objects X, Y, Z, there is a composition map:*

$$\mathrm{Hom}_{\mathscr{C}}(Y, Z) \times \mathrm{Hom}_{\mathscr{C}}(X, Y) \to \mathrm{Hom}_{\mathscr{C}}(X, Z) : (g, f) \mapsto g \circ f;$$

(ct2) *(Associativity). For three morphisms: $X \xrightarrow{f} Y \xrightarrow{g} Z \xrightarrow{h} W$, we have $h \circ (g \circ f) = (h \circ g) \circ f$;*

(ct3) *For each object X, there is a specific element $1_X \in \mathrm{Hom}_{\mathscr{C}}(X, X)$ such that $1_X \circ f = f$ and $g \circ 1_X = g$ for all $f : Y \to X$ and $g : X \to Z$.*

For two objects X and Y in $\mathscr{C}$, we write $X \cong Y$ if there exist morphisms $f : X \to Y$ and $g : Y \to X$ such that $f \circ g = 1_Y$ and $g \circ f = 1_X$.

EXAMPLE 4.1 Table of some categories used in this book:

Category	Objects	Morphisms
SETS	sets	maps between sets
AB	abelian groups	group homomorphisms
ALG	algebras	algebra homomorphisms
A-ALG	A-algebras	A-algebra homomorphisms
$CL_{\mathcal{O}}$	pro-artinian $A \in \mathcal{O}$-ALG with $A/\mathfrak{m}_A = \mathcal{O}/\mathfrak{m}$	morphisms of local $\mathcal{O}$-algebras
A-MOD	A-modules	A-linear maps
G-$\mathcal{M}OD$	discrete G-modules for a profinite group G	continuous G-linear maps

A category $\mathscr{C}'$ is a *subcategory* of $\mathscr{C}$ if the following two conditions are satisfied:

(i) Each object of $\mathscr{C}'$ is an object of $\mathscr{C}$ and $\mathrm{Hom}_{\mathscr{C}'}(X, Y) \subset \mathrm{Hom}_{\mathscr{C}}(X, Y)$;
(ii) The composition of morphisms is the same in $\mathscr{C}$ and $\mathscr{C}'$.

A subcategory $\mathscr{C}'$ is called a *full* subcategory of $\mathscr{C}$ if $\mathrm{Hom}_{\mathscr{C}'}(X, Y) = \mathrm{Hom}_{\mathscr{C}}(X, Y)$ for any two objects X and Y in $\mathscr{C}'$.

4.1.2 Functors

A *covariant* (resp. *contravariant*) *functor* $F : \mathscr{C} \to \mathscr{C}'$ is a rule associating an object $F(X)$ of $\mathscr{C}'$ and a morphism $F(f) \in \mathrm{Hom}_{\mathscr{C}'}(F(X), F(Y))$ (resp. $F(f) \in \mathrm{Hom}_{\mathscr{C}'}(F(Y), F(X))$) to each object X of $\mathscr{C}$ and each morphism $f \in \mathrm{Hom}_{\mathscr{C}}(X, Y)$ satisfying

$$F(f \circ h) = F(f) \circ F(h) \quad (\text{resp. } F(f \circ h) = F(h) \circ F(f)) \qquad \text{(F)}$$

and

$$F(1_X) = 1_{F(X)}.$$

EXAMPLE 4.2 Let G be a profinite group. Then the category G-$\mathcal{M}OD$ of discrete G-modules consists of discrete modules with continuous G-action and continuous G-linear maps. Then the association of the G-invariant submodule to each object in the category:

$$M \mapsto H^0(G,M) = \{m \in M | gm = m \ \forall g \in G\}$$

is a covariant functor from G-$\mathcal{MOD}$ into AB. Each G-linear homomorphism $\phi : M \to N$ induces $\phi^G : M^G \to N^G$ by G-linearity, which satisfies (F).

A *morphism* f between two contravariant functors $F, G : \mathscr{C} \to \mathscr{C}'$ is a system of morphisms $\{\phi(X) \in \mathrm{Hom}_{\mathscr{C}'}(F(X), G(X))\}_{X\in\mathscr{C}}$ making the following diagram commutative for all $u \in \mathrm{Hom}_{\mathscr{C}}(X,Y)$:

$$\begin{array}{ccc} F(Y) & \xrightarrow{F(u)} & F(X) \\ {\scriptstyle \phi(Y)}\downarrow & & \downarrow{\scriptstyle \phi(X)} \\ G(Y) & \xrightarrow[G(u)]{} & G(X). \end{array} \tag{4.1}$$

Thus we can define the category of contravariant functors $CTF(\mathscr{C},\mathscr{C}')$ using the above definition of morphisms between functors. Similarly, we can define the category $COF(\mathscr{C},\mathscr{C}')$ of covariant functors by reversing the direction of morphisms $F(u)$ and $G(u)$. We write $\mathrm{id}_{\mathscr{C}} : \mathscr{C} \to \mathscr{C}$ for the identity functor taking each object X to X and each morphism ϕ to ϕ. When we have two functors $F : \mathscr{C} \to \mathscr{C}'$ and $G : \mathscr{C}' \to \mathscr{C}$ such that $G \circ F \cong \mathrm{id}_{\mathscr{C}}$ and $F \circ G \cong \mathrm{id}_{\mathscr{C}'}$ (in the categories $COF(\mathscr{C},\mathscr{C})$ and $COF(\mathscr{C}',\mathscr{C}')$, respectively) for each object Y of $\mathscr{C}$ and X of $\mathscr{C}'$, we say that the two categories are equivalent. When a functor $F : \mathscr{C} \to \mathscr{C}'$ gives an equivalence of $\mathscr{C}$ to a full subcategory of $\mathscr{C}'$, we call F *fully faithful*.

4.1.3 Representability

Fix a category $\mathscr{C}$. For each object X in $\mathscr{C}$, we associate a contravariant functor $\underline{X} : \mathscr{C} \to SETS$ by

$$\underline{X}(S) = \mathrm{Hom}_{\mathscr{C}}(S,X).$$

For each morphism $\phi : T \to S$, $\underline{X}(\phi) : \underline{X}(S) \to \underline{X}(T)$ is given by

$$\underline{X}(\phi) : (S \xrightarrow{\eta} X) \mapsto \eta \circ \phi \in \mathrm{Hom}_{\mathscr{C}}(T,X) = \underline{X}(T).$$

If $f : X \to Y$ be a morphism in $\mathscr{C}$, we have $\iota(f) \in \mathrm{Hom}_{CTF}(\underline{X}, \underline{Y})$ given by

$$\iota(f)(S)(\phi : S \to X) = f \circ \phi.$$

We leave the reader with the task of verifying $\iota(f \circ g) = \iota(f) \circ \iota(g)$.

LEMMA 4.3 (UNICITY-LEMMA) *The above functor: $\mathscr{C} \to CTF$ given by $X \mapsto \underline{X}$ is fully faithful.*

Proof We only need to prove $\mathrm{Hom}_{\mathscr{C}}(X, Y) \cong \mathrm{Hom}_{CTF}(\underline{X}, \underline{Y})$ functorially. Here the word '*functorial*' means that the isomorphism commutes with the composition of the morphisms. If this is true, $\underline{X} \cong \underline{Y}$ implies $X \cong Y$, and thus the functor ι gives rise to an equivalence of $\mathscr{C}$ with a full subcategory of CTF. The morphism $\iota : \mathrm{Hom}_{\mathscr{C}}(X, Y) \to \mathrm{Hom}_{CTF}(\underline{X}, \underline{Y})$ is given by $f \mapsto \iota(f)$. We define the inverse $\pi : \mathrm{Hom}_{CTF}(\underline{X}, \underline{Y}) \to \mathrm{Hom}_{\mathscr{C}}(X, Y)$ of ι by

$$F \mapsto F(X)(1_X) \in \underline{Y}(X) = \mathrm{Hom}_{\mathscr{C}}(X, Y),$$

where $F(X) : \underline{X}(X) \to \underline{Y}(X)$ by definition. We compute

$$\pi(\iota(f)) = \iota(f)(X)(1_X) = f \circ 1_X = f.$$

Thus $\pi \circ \iota$ is the identity map.

We shall show that $\iota(\pi(F)) = F$. We put $f = F(X)(1_X) = \pi(F)$. If $S \xrightarrow{\phi} X \in \underline{X}(S)$ is a morphism, then the following diagram is commutative:

$$\begin{array}{ccccccc}
 & & 1_X & \mapsto & F(X)(1_X) & = & f \\
\xi & \in & \underline{X}(X) & \xrightarrow{F(X)} & \underline{Y}(X) & \ni & \eta \\
\downarrow & & \downarrow{\scriptstyle \underline{X}(\phi)} & & {\scriptstyle \underline{Y}(\phi)}\downarrow & & \downarrow \\
\xi \circ \phi & \in & \underline{X}(S) & \xrightarrow[F(S)]{} & \underline{Y}(S) & \ni & \eta \circ \phi.
\end{array} \tag{4.2}$$

Then we have

$$\iota(\pi(F))(S)(\phi) = \iota(f)(S)\circ\phi = f\circ\phi = F(X)(1_X)\circ\phi = F(S)(1_X\circ\phi) = F(S)(\phi).$$

This shows that $\iota \circ \pi$ is the identity map. Since ι is compatible with composition, π has to be compatible with composition. □

Let $\mathscr{C}$ be a category. We consider the functors $\iota : \mathscr{C} \to CTF(\mathscr{C}, SETS)$ and $\iota' : \mathscr{C} \to COF(\mathscr{C}, SETS)$ given by $\iota(X)(S) = \underline{X}(S) = \mathrm{Hom}_{\mathscr{C}}(S, X)$ and $\iota'(X)(S) = \overline{X}(S) = \mathrm{Hom}_{\mathscr{C}}(X, S)$, which can be checked to be fully faithful by the same argument as above (reversing appropriate arrows). If $F \in COF(\mathscr{C}, SETS)$ (resp. $F \in CTF(\mathscr{C}, SETS)$) and we find $X \in \mathscr{C}$ such that $I : \overline{X} \cong F$ (resp. $I : \underline{X} \cong F$), F is called *representable* by X.

Then for $S \xrightarrow{\phi} X \in \underline{X}(S)$ (resp. $X \xrightarrow{\phi} S \in \overline{X}(S)$), the following diagrams are commutative:

$$\begin{array}{ccccccc} & & 1_X & \mapsto & I(X)(1_X) & = & \xi \\ 1_X & \in & \underline{X}(X) & \xrightarrow{I(X)} & F(X) & \ni & \xi \\ \downarrow & & \downarrow \underline{X}(\phi) & & F(\phi)\downarrow & & \downarrow \\ \phi & \in & \underline{X}(S) & \xrightarrow[I(S)]{} & F(S) & \ni & F(\phi)(\xi) \\ & & \phi & \mapsto & F(\phi)(\xi), & & \end{array} \tag{4.3}$$

$$\begin{array}{ccccccc} & & 1_X & \mapsto & I(X)(1_X) & = & \xi \\ 1_X & \in & \overline{X}(X) & \xrightarrow{I(X)} & F(X) & \ni & \xi \\ \downarrow & & \downarrow \overline{X}(\phi) & & F(\phi)\downarrow & & \downarrow \\ \phi & \in & \overline{X}(S) & \xrightarrow[I(S)]{} & F(S) & \ni & F(\phi)(\xi) \\ & & \phi & \mapsto & F(\phi)(\xi). & & \end{array} \tag{4.4}$$

Start from an element $\eta \in F(S)$. The above diagram tells us that there exists a unique ϕ such that η is given by $F(\phi)(\xi)$ for $\xi = I(X)(1_X)$. Therefore each η is a specialization under a unique ϕ of the universal object ξ. If there is another ξ' which is universal in the above sense, then there exist $\phi : X \to X$ such that $F(\phi)(\xi) = \xi'$ and $\phi' : X \to X$ such that $F(\phi')(\xi') = \xi$. Both ϕ and ϕ' are unique under the above requirement. By the uniqueness, $\phi \circ \phi' = \phi' \circ \phi = 1_X$, because, for example, $\alpha = 1_X$ and $\alpha = \phi \circ \phi'$ both satisfy $F(\alpha)(\xi') = \xi'$. Thus ξ is determined up to automorphisms of X.

If $\underline{X} \cong \underline{Y}$ in $CTF(\mathscr{C}, SETS)$, then we have

$$f \in \mathrm{Hom}_{CTF}(X, Y) \cong \mathrm{Hom}_{\mathscr{C}}(X, Y)$$

and

$$g \in \mathrm{Hom}_{CTF}(Y, X) \cong \mathrm{Hom}_{\mathscr{C}}(Y, X)$$

such that $f \circ g = 1_{\underline{Y}}$ and $g \circ f = 1_{\underline{X}}$. This implies $X \cong Y$. That is,

$$X \cong Y \iff \underline{X} \cong \underline{Y}.$$

Similarly we have

$$X \cong Y \iff \overline{X} \cong \overline{Y}.$$

EXAMPLE 4.4 Let $\mathcal{O}$ be a valuation ring finite flat over $\mathbb{Z}_p$. We consider the Galois group $\mathfrak{G} = \mathfrak{G}_p = \mathrm{Gal}(\mathbb{Q}^{(p,\infty)}/\mathbb{Q})$ (unramified outside p and ∞) defined in Subsection 3.2.2. We fix a Galois representation $\overline{\rho} : \mathfrak{G} \to$

$GL_2(\mathbb{F})$ for the residue field $\mathbb{F}$ of $\mathcal{O}$ and define the following covariant functor $\mathcal{F} : CL_{\mathcal{O}} \to SETS$ by

$$\mathcal{F}(A) = \{\rho : \mathfrak{G} \to GL_2(A) | \rho \mod \mathfrak{m}_A = \overline{\rho}\} / \approx, \tag{4.5}$$

where '$\approx$' is the strict equivalence, that is, conjugation by $\widehat{GL}_2(A) = 1_2 + M_2(\mathfrak{m}_A)$. Each element $\rho \in \mathcal{F}(A)$ is of course supposed to be a continuous representation. For any morphism $\alpha \in \mathrm{Hom}_{CL}(A, B)$, we define $\mathcal{F}(\alpha) : \mathcal{F}(A) \to \mathcal{F}(B)$ by $\mathcal{F}(\alpha)(\rho) = \alpha \circ \rho$. In this way, $\mathcal{F}$ forms a covariant functor. Under the absolute irreducibility of $\overline{\rho}$, Mazur's theorem (Theorem 2.26) tells us that there exists a universal couple (R, ϱ) made of $\varrho : \mathfrak{G} \to GL_2(R) \in \mathcal{F}(R)$ with $R \in CL_{\mathcal{O}}$ such that

$$\mathcal{F}(A) \cong \mathrm{Hom}_{CL}(R, A) \text{ via } \rho \mapsto \alpha \iff \alpha \circ \varrho \approx \rho.$$

This shows that the covariant functor $\mathcal{F}$ is representable by $R \in CL_{\mathcal{O}}$, and the universal object in $\mathcal{F}(R)$ is given by ϱ.

We pick a character $\phi : \mathfrak{G} \to \mathcal{O}^{\times}$ such that $\phi \mod \mathfrak{m}_{\mathcal{O}} = \det(\overline{\rho})$. Then we can think of the following three subfunctors of $\mathcal{F}$:

$$\begin{aligned} \mathcal{F}^{\phi}(A) &= \{\rho \in \mathcal{F}(A) | \det(\rho) = \phi\} \\ \mathcal{F}^{ord}(A) &= \{\rho \in \mathcal{F}(A) | \rho \text{ is } p\text{-ordinary}\} \\ \mathcal{F}^{ord,\phi}(A) &= \mathcal{F}^{ord}(A) \cap \mathcal{F}^{\phi}(A), \end{aligned} \tag{4.6}$$

where we regard ϕ as a character with value in $A^{\times}$ projecting the original ϕ down to A by the structure morphism $\mathcal{O} \to A$. Here we recall that ρ is called p-ordinary if $\rho|_{D_p} \cong \left(\begin{smallmatrix} \varepsilon & * \\ 0 & \delta \end{smallmatrix}\right)$ with unramified δ. Of course, the last two functors have meaning only when $\overline{\rho}$ is p-ordinary. Let $CNL_{\mathcal{O}}$ be the subcategory of $CL_{\mathcal{O}}$ made of noetherian local $\mathcal{O}$-algebras. Under the absolute irreducibility of $\overline{\rho}$, $\mathcal{F}^{\phi}$ is representable in $CL_{\mathcal{O}}$ and even in $CNL_{\mathcal{O}}$, and $\mathcal{F}^{ord}$ and $\mathcal{F}^{ord,\phi}$ are also representable in $CNL_{\mathcal{O}}$ if $\overline{\rho}$ is absolutely irreducible and regular (see Proposition 3.30, (ai_F) and (rg_p) in subsection 3.2.4).

4.1.4 Abelian categories

If one can equip $\mathrm{Hom}_{\mathcal{C}}(X, Y)$ with a functorial addition making it into an abelian group, $\mathcal{C}$ is called an *additive* category. An abelian category $\mathcal{C}$ is an additive category which has a (functorial) notion of cokernel, kernel and image. For example, A-MOD, AB, G-$\mathcal{M}OD$ etc. are abelian categories.

Let us recall the formal definition of additive and abelian categories in the following paragraphs. Let $\{0\}$ be a set consisting of a single element '0'. We consider the covariant functor $F_0 : \mathscr{C} \to SETS$ given by $F_0(Y) = \{0\}$ for all $Y \in \mathscr{C}$ and $F_0(\phi) = 1_{\{0\}}$ for any morphism ϕ in $\mathscr{C}$. If F_0 is representable by an object $X_0 \in \mathscr{C}$, X_0 is called an *initial* object. Thus, for each $X \in \mathscr{C}$, $F_0(X) = \{0\} = \mathrm{Hom}_{\mathscr{C}}(X_0, X)$ consists of a unique element i. In other words, for each $X \in \mathscr{C}$, there is a unique morphism $i : X_0 \to X$ such that $F_0(i) = 1_{\{0\}}$.

We can also consider the contravariant functor $F^0 : \mathscr{C} \to SETS$ given by $F^0(Y) = \{0\}$ for all $Y \in \mathscr{C}$ and $F^0(\phi) = 1_{\{0\}}$ for any morphism ϕ in $\mathscr{C}$. If F^0 is representable by $X^0 \in \mathscr{C}$, X^0 is called a *final* object of $\mathscr{C}$. Thus, for each $X \in \mathscr{C}$, there is a unique morphism $p : X \to X^0$ such that $F^0(p) = 1_{\{0\}}$. By definition, we have a unique morphism $e : X_0 \to X^0$ such that $e = p \circ i$.

We consider the following condition:

$\mathscr{C}$ *has an initial and a final object which are isomorphic under e.* (Ab0)

If $\mathscr{C}$ satisfies (Ab0), we identify the initial and the final object by e and call it the zero-object $\mathbf{0} = \mathbf{0}_{\mathscr{C}}$. We write ${}_X\mathbf{0}$ (resp. $\mathbf{0}_X$) for the unique element in $\mathrm{Hom}_{\mathscr{C}}(\mathbf{0}, X)$ (resp. $\mathrm{Hom}_{\mathscr{C}}(X, \mathbf{0})$). We assume that (Ab0) holds. Then we have a unique $\mathbf{0}_{X,Y} \in \mathrm{Hom}_{\mathscr{C}}(X, Y)$ given by $\mathbf{0}_{X,Y} = \mathbf{0}_Y \circ {}_X\mathbf{0}$, which is called the *zero-map*.

For two objects $X, Y \in \mathscr{C}$, we consider the covariant functor $\underline{X \oplus Y} : \mathscr{C} \to SETS$ defined by $Z \mapsto \mathrm{Hom}_{\mathscr{C}}(X, Z) \times \mathrm{Hom}_{\mathscr{C}}(Y, Z)$. If this functor is representable by an object, we call the object the *direct sum* of X and Y, and write it as $X \oplus Y$. In other words, there exists $\iota_X : X \to X \oplus Y$ and $\iota_Y : Y \to X \oplus Y$ such that the map $\mathrm{Hom}_{\mathscr{C}}(X \oplus Y, Z) \ni f \mapsto (f \circ \iota_X, f \circ \iota_Y) \in \mathrm{Hom}_{\mathscr{C}}(X, Z) \times \mathrm{Hom}_{\mathscr{C}}(Y, Z)$ is bijective for all Z, that is, (ι_X, ι_Y) is the universal object. The morphism $\iota_X : X \to X \oplus Y$ is called the *inclusion* of X into $X \oplus Y$.

Similarly we consider the contravariant functor $\overline{X \times Y} : \mathscr{C} \to SETS$ defined by $Z \mapsto \mathrm{Hom}_{\mathscr{C}}(Z, X) \times \mathrm{Hom}_{\mathscr{C}}(Z, Y)$. If this functor is representable by an object, we call the object the *direct product* of X and Y, and write it as $X \times Y$. In other words, there exists $\pi_X : X \times Y \to X$ and $\pi_Y : X \times Y \to Y$ such that the map $\mathrm{Hom}_{\mathscr{C}}(Z, X \times Y) \ni f \mapsto (\pi_X \circ f, \pi_Y \circ f) \in \mathrm{Hom}_{\mathscr{C}}(Z, X) \times \mathrm{Hom}_{\mathscr{C}}(Z, Y)$ is bijective for all Z, that is, (π_X, π_Y) is the universal object. The morphism $\pi_X : X \times Y \to X$ is called the *projection* of $X \times Y$ onto X.

We have a unique morphism $\theta : X \oplus Y \to X \times Y$ such that

$$\pi_X \circ \theta \circ \iota_X = 1_X, \pi_X \circ \theta \circ \iota_Y = \mathbf{0}_{Y,X}, \ \pi_Y \circ \theta \circ \iota_X = \mathbf{0}_{X,Y}, \ \pi_Y \circ \theta \circ \iota_Y = 1_Y.$$

We consider the following condition:

(Ab1) *For any two objects* $X, Y \in \mathscr{C}$, *there exist* $X \oplus Y$ *and* $X \times Y$ *in* $\mathscr{C}$, *and* $\theta : X \oplus Y \cong X \times Y$.

Suppose (Ab1). Let

$$\Delta_X : X \to X \times X = X \oplus X \quad (\text{resp. } \nabla_X : X \times X = X \oplus X \to X)$$

be the morphism corresponding to $(1_X, 1_X)$. For $f, g \in \mathrm{Hom}_{\mathscr{C}}(X, Y)$, let $(f, g) : X \times X \to Y \times Y$ be the corresponding morphism. And we define

$$f + g : X \to Y \in \mathrm{Hom}_{\mathscr{C}}(X, Y) \text{ by } \nabla_Y \circ (f, g) \circ \Delta_X .$$

We then have

$$(f + g) + h = f + (g + h) \quad \text{and} \quad \mathbf{0}_{X,Y} + f = f + \mathbf{0}_{X,Y} = f.$$

We consider the following condition:

$Hom_{\mathscr{C}}(X, Y)$ is an abelian group under '+' with identity $\mathbf{0}_{X,Y}$. (Ab2)

A category satisfying (Ab0–2) is called an *additive* category.

Suppose that $\mathscr{C}$ is an additive category. For the category $\mathscr{C}$ to form an abelian category, $\mathscr{C}$ needs to have 'kernel', 'cokernel' and 'image' of morphisms. Here an object $K \in \mathscr{C}$ with $i : K \to X$ is called a *kernel* of $f : X \to Y$ if

$$0 \to \mathrm{Hom}_{\mathscr{C}}(Z, K) \xrightarrow{\underline{i}} \mathrm{Hom}_{\mathscr{C}}(Z, X) \xrightarrow{\underline{f}} \mathrm{Hom}_{\mathscr{C}}(Z, Y)$$

is an exact sequence in AB for all Z, where $\underline{f}(\phi) = f \circ \phi$. Similarly $p : Y \to C$ is called a *cokernel* of p if

$$0 \to \mathrm{Hom}_{\mathscr{C}}(C, Z) \xrightarrow{\overline{p}} \mathrm{Hom}_{\mathscr{C}}(Y, Z) \xrightarrow{\overline{f}} \mathrm{Hom}_{\mathscr{C}}(X, Z)$$

is exact for all $Z \in \mathscr{C}$, where $\overline{f}(\phi) = \phi \circ f$. If $Y \xrightarrow{p} C$ is a cokernel of $X \xrightarrow{f} Y$, $I = \mathrm{Im}(f)$ is defined to be the kernel of p. Thus we have $j : I \to Y$. Similarly, the cokernel of $K \xrightarrow{i} X$ is defined to be the *coimage* of f, and we write $q : X \to L$ for the canonical map. Looking into the following two exact sequences:

$$0 \to \mathrm{Hom}_{\mathscr{C}}(L, Y) \xrightarrow{\overline{q}} \mathrm{Hom}_{\mathscr{C}}(X, Y) \xrightarrow{\overline{i}} \mathrm{Hom}_{\mathscr{C}}(K, Y) \qquad (4.7)$$

$$0 \to \mathrm{Hom}_{\mathscr{C}}(L, I) \xrightarrow{\underline{j}} \mathrm{Hom}_{\mathscr{C}}(L, Y) \xrightarrow{\underline{p}} \mathrm{Hom}_{\mathscr{C}}(L, C), \qquad (4.8)$$

we claim to have a unique $\tau : L \to I$ such that $f = j \circ \tau \circ q$. In fact, $\underline{f} \circ \underline{i} = \mathbf{0}$ implies $f \circ i = \mathbf{0} \Leftrightarrow f \in \mathrm{Ker}(\overline{i})$ by the unicity-lemma. This shows that $f = g \circ q \in \mathrm{Im}(\overline{q})$ for $g \in \mathrm{Hom}_{\mathscr{C}}(L, Y)$ by the exactness of (4.7). By

$\overline{f} \circ \overline{p} = \mathbf{0}$, we similarly have $p \circ f = \mathbf{0}$ and hence, $\overline{q}(p \circ g) = p \circ g \circ q = \mathbf{0}$. By the injectivity of $\overline{q}$, we have $\underline{p}(g) = p \circ g = \mathbf{0} \Leftrightarrow g \in \mathrm{Ker}(\underline{p}) = \mathrm{Im}(\underline{j})$, and hence $g = j \circ \tau$ for $\tau \in \mathrm{Hom}_{\mathscr{C}}(L, I)$. This proves the claim.

An additive category $\mathscr{C}$ is called *abelian* if the following two conditions are satisfied:

(Ab3) *For every morphism $X \xrightarrow{f} Y$ in $\mathscr{C}$, its kernel and cokernel exist in $\mathscr{C}$;*

(Ab4) *The morphism $\tau : L \to I$ as above is an isomorphism.*

Suppose now that $\mathscr{C}$ is an abelian category. Then $\overline{X}(Y) = \mathrm{Hom}_{\mathscr{C}}(X, Y)$ is an abelian group. That is $\overline{X} \in COF(\mathscr{C}, AB)$. A sequence $F \to G \to H$ of functors in $COF(\mathscr{C}, AB)$ is called exact if $F(X) \to G(X) \to H(X)$ is exact for all X in $\mathscr{C}$. If $X \xrightarrow{\alpha} Y$ is a morphism in $\mathscr{C}$, then $X \xrightarrow{\alpha} Y \to \mathrm{Coker}(\alpha) \to 0$ is exact. Then by definition and the unicity-lemma, we see that

$$0 \to \overline{\mathrm{Coker}(\alpha)} \to \overline{Y} \xrightarrow{\overline{\alpha}} \overline{X}$$

is exact in $COF(\mathscr{C}, AB)$. That is, $\overline{\mathrm{Coker}(\alpha)} = Ker(\overline{\alpha})$. Then we see that

$$\begin{aligned} 0 \to \overline{X} \xrightarrow{\overline{\alpha}} \overline{Y} \xrightarrow{\overline{\beta}} \overline{Z} \text{ is exact} &\iff \overline{\mathrm{Coker}(\beta)} = \mathrm{Ker}(\overline{\beta}) \cong \overline{X} \text{ via } \overline{\alpha} \quad (4.9)\\ &\iff \mathrm{Coker}(\beta) \cong X \text{ via } \alpha \text{ by the unicity-lemma} \\ &\iff Z \xrightarrow{\beta} Y \xrightarrow{\alpha} X \to 0 \text{ is exact.} \end{aligned}$$

A similar assertion also holds for $\underline{X}$.

4.2 Extension of Modules

In this section, we describe the basics of the theory of module extension functors, and we relate it to group cohomology in the following section.

4.2.1 Extension groups

We fix a ring Λ with identity, which may be non-commutative. We consider the category of Λ-modules Λ-MOD. Thus the objects of Λ-MOD are Λ-modules, and $\mathrm{Hom}_{\Lambda}(M, N)$ for two Λ-modules M and N is the abelian group of all Λ-linear maps from M into N.

When Λ is a topological ring, we would prefer to consider only Λ-modules with continuous action, that is, continuous Λ-modules, or we might want to impose further restrictions, like compactness or discreteness, to the Λ-modules we study. The totality of such Λ-modules makes a subcategory of Λ-MOD, whose set of homomorphisms consists of continuous ones. To accommodate such subcategories in an algebraic way without referring to topology, we consider subcategories $\mathscr{C}$ of Λ-MOD satisfying a set of conditions sufficient to define extension functors. First of all, since $\mathscr{C}$ is a subcategory of Λ-MOD:

- Objects of $\mathscr{C}$ consist of a collection $Ob(\mathscr{C})$ of Λ-modules;
- We have a set of $\mathscr{C}$-homomorphisms: $\mathrm{Hom}_{\mathscr{C}}(M,N) \subset \mathrm{Hom}_{\Lambda}(M,N)$ for $M, N \in Ob(\mathscr{C})$;
- The identity map $\mathrm{id}_M : M \to M$ is in $\mathrm{Hom}_{\mathscr{C}}(M,M)$ for each object M;
- $g \circ f \in \mathrm{Hom}_{\mathscr{C}}(M,L)$ for $f \in \mathrm{Hom}_{\mathscr{C}}(M,N)$ and $g \in \mathrm{Hom}_{\mathscr{C}}(N,L)$.

We impose on $\mathscr{C}$ the following four conditions:

(C1) The set $\mathrm{Hom}_{\mathscr{C}}(X,Y) \subset \mathrm{Hom}_{\Lambda}(X,Y)$ is a subgroup;
(C2) If $f : X \to Y$ is a homomorphism in $\mathscr{C}$, $\mathrm{Ker}(f)$ and $\mathrm{Coker}(f)$ are both inside $\mathscr{C}$;
(C3) If X and Y are in $\mathscr{C}$, then the direct product $X \times Y$ is in $\mathscr{C}$;
(C4) The zero module $\{0\}$ is in $\mathscr{C}$.

These conditions guarantee that $\mathscr{C}$ is an abelian category (see Subsection 4.1.4 for a formal definitions of abelian categories). Hereafter we fix such a category $\mathscr{C}$ and work only in $\mathscr{C}$. We call the Λ-linear map: $X \to Y$ for objects X and Y in $\mathscr{C}$ a *$\mathscr{C}$-morphism* if it is in $\mathrm{Hom}_{\mathscr{C}}(X,Y)$. Similarly, an isomorphism which is also a $\mathscr{C}$-morphism is called a $\mathscr{C}$-isomorphism.

For a given pair of Λ-modules M and N in $\mathscr{C}$, we would like to find all Λ-modules E in $\mathscr{C}$ which fit into the following exact sequence in $\mathscr{C}$:

$$0 \to N \xrightarrow{\iota_N} E \xrightarrow{\pi_M} M \to 0.$$

We call such E an extension in $\mathscr{C}$ of Λ-module M by N. Two extensions E and E' are called isomorphic if we have a $\mathscr{C}$-isomorphism $\xi : E \cong E'$ making the following diagram commutative:

$$\begin{array}{ccccc} N & \hookrightarrow & E & \twoheadrightarrow & M \\ \| & & \xi \downarrow & & \| \\ N & \hookrightarrow & E' & \twoheadrightarrow & M. \end{array}$$

We write $E(M,N) = E_{\mathscr{C}}(M,N)$ for the set of all isomorphism classes of extensions of M by N. When $\mathscr{C} = \Lambda$-MOD, we write $E_{\Lambda}(M,N)$ for $E_{\mathscr{C}}(M,N)$. Note that $M \oplus N \in E(M,N)$; so, $E(M,N) \neq \emptyset$. An extension E is called *split*, if we have a $\mathscr{C}$-morphism $\iota_M : M \to E$ such that $\pi_M \circ \iota_M = \mathrm{id}_M$. Then $E \cong M \oplus N$ by $e \mapsto \iota_M(\pi_M(e)) \oplus (e - \iota_M(\pi_M(e)))$. The map ι_M is called a section of π_M. This shows that the class $M \oplus N \in E(M,N)$ is the unique split extension class. If we have a projection $\pi_N : E \to N$ such that $\pi_N \circ \iota_N = \mathrm{id}_N$, then again $E \cong M \oplus N$ by $e \mapsto (e - \iota_N(\pi_N(e))) \oplus \iota_N(\pi_N(e))$, because $\mathrm{Ker}(\pi_N) \cong M$ by π_M in this case.

If $\Lambda = \mathbb{Z}$ and $M = N = \mathbb{Z}/p\mathbb{Z}$ for a prime p, then we have at least two extensions: $(\mathbb{Z}/p\mathbb{Z})^2$ and $\mathbb{Z}/p^2\mathbb{Z}$ in $E_{\mathbb{Z}}(\mathbb{Z}/p\mathbb{Z}, \mathbb{Z}/p\mathbb{Z})$.

Now we would like to study how $E(M,N)$ changes if we change M and N by their homomorphic image (or source). For a given $\mathscr{C}$-morphism $M \xrightarrow{\varphi} X$ and $N \xrightarrow{\phi} X$, the fiber product $T = M \times_X N$ is a Λ-module in $\mathscr{C}$ with the following property:

(FP1) We have two projections

$$\alpha : T \to M \text{ and } \beta : T \to N$$

in $\mathscr{C}$ making the following diagram commutative:

$$\begin{array}{ccc} M \times_X N & \xrightarrow{\alpha} & M \\ {\scriptstyle\beta}\downarrow & & \downarrow{\scriptstyle\varphi} \\ N & \xrightarrow[\phi]{} & X; \end{array}$$

(FP2) If the following diagram in $\mathscr{C}$ is commutative:

$$\begin{array}{ccc} Y & \xrightarrow{\alpha'} & M \\ {\scriptstyle\beta'}\downarrow & & \downarrow{\scriptstyle\varphi} \\ N & \xrightarrow[\phi]{} & X, \end{array}$$

then there exists a unique $\mathscr{C}$-morphism $\gamma : Y \to M \times_X N$ such that $\alpha' = \alpha \circ \gamma$ and $\beta' = \beta \circ \gamma$.

If two fiber products T and T' exist in $\mathscr{C}$, then we have $\gamma : T' \to T$ and $\gamma' : T \to T'$ satisfying (FP2) for $Y = T'$ and $Y = T$, respectively. Then id_T and $\gamma \circ \gamma' : T \to T$ satisfy (PF2) for $Y = T$, and by the uniqueness, $\gamma \circ \gamma' = \mathrm{id}_T$. Similarly, $\gamma' \circ \gamma = \mathrm{id}_{T'}$ and hence $T \cong T'$. Thus the fiber product of M and N is unique in $\mathscr{C}$ up to isomorphisms if it exists. It is easy to see that

$$M \times_X N = \{(m,n) \in M \times N | \varphi(m) = \phi(n)\}$$

satisfies the property (FP1–2) for $\mathscr{C} = \Lambda\text{-}MOD$, and the two projections $\alpha : M \times_X N \to M$ and $\beta : M \times_X N \to N$ take (m,n) to m and n, respectively. For this choice, $\gamma(y)$ is given by $(\alpha'(y), \beta'(y)) \in M \times_X N$. Thus fiber products exist in Λ-MOD. If, further, φ and ϕ are $\mathscr{C}$-morphisms, then by the existence of $M \times N$ in $\mathscr{C}$, the above $M \times_X N$ in Λ-MOD is actually the kernel of $(\alpha - \beta) \circ (\varphi \oplus \phi)$, which is therefore a member of $\mathscr{C}$. This shows the existence of the fiber product in $\mathscr{C}$. In functorial terms, the fiber product represents the functor:

$$Y \mapsto \{(\alpha', \beta') \in \mathrm{Hom}_{\mathscr{C}}(Y, M \times N) | \varphi \circ \alpha' = \phi \circ \beta'\}$$

from $\mathscr{C}$ to $SETS$.

Let $N \hookrightarrow E \twoheadrightarrow M$ be an extension in $\mathscr{C}$. For a $\mathscr{C}$-morphism $\varphi : M' \to M$, we look at the fiber product $E' = E \times_M M'$. Let $\pi' : E' \to M'$ be the projection. Since $\pi : E \to M$ is a surjection, for each $m' \in M'$, we find $e \in E$ such that $\pi(e) = \varphi(m')$. By definition, $\pi'(e, m') = m'$, and π' is a surjection. Then

$$\begin{aligned} \mathrm{Ker}(\pi') &= \{(e, m') \in E \times_M M' | \pi'(m) = 0\} \\ &= \{(e, m') \in E \times M' | \pi(e) = \pi'(m) = 0\} = \mathrm{Ker}(\pi) = \mathrm{Im}(\iota_N) \cong N. \end{aligned}$$

Thus we get an extension $N \hookrightarrow E' \twoheadrightarrow M'$ in $E(M', N)$. Namely we have $E(\varphi, N) : E(M, N) \to E(M', N)$ taking

$$N \hookrightarrow E \twoheadrightarrow M \quad \text{to} \quad N \hookrightarrow E' = E \times_M M' \twoheadrightarrow M'.$$

Note that for two $\mathscr{C}$-morphisms: $M'' \xrightarrow{\varphi'} M' \xrightarrow{\varphi} M$, it is easy to check that

$$E' \times_{M'} M'' = (E \times_M M') \times'_M M'' \cong E \times_{M, \varphi \circ \varphi'} M''.$$

This shows that

$$E(\varphi', N) \circ E(\varphi, N) = E(\varphi \circ \varphi', N);$$

so, the functor $M \mapsto E(M, N)$ for a fixed N is a contravariant functor.

Suppose that we have two $\mathscr{C}$-morphisms $\varphi : X \to M$ and $\phi : X \to N$. We define a fiber sum (or push-out) $S = M \oplus_X N$ under X by the following conditions:

(FS1) We have two inclusions $\alpha : M \to S$ and $\beta : N \to S$ in $\mathscr{C}$ making the following diagram commutative:

$$\begin{array}{ccc} X & \xrightarrow{\varphi} & M \\ {\scriptstyle \phi}\downarrow & & \downarrow{\scriptstyle \alpha} \\ N & \xrightarrow[\beta]{} & S; \end{array}$$

(FS2) If the following diagram in $\mathscr{C}$ is commutative:

$$\begin{array}{ccc} X & \xrightarrow{\varphi} & M \\ \phi\downarrow & & \downarrow\alpha' \\ N & \xrightarrow[\beta']{} & Y, \end{array}$$

then there exists a unique morphism $\gamma : S \to Y$ such that $\alpha' = \gamma \circ \alpha$ and $\beta' = \gamma \circ \beta$.

In the same way as in the case of fiber products, the fiber sum is unique up to isomorphisms if it exists. We define $S = M \oplus_X N$ to be the quotient of $M \times N$ by the Λ-submodule generated by $\varphi(x) - \phi(x)$ for all $x \in X$ (that is, the cokernel of $\varphi - \phi : X \to M \times N$). The inclusions α and β are induced by the inclusions $M \hookrightarrow M \oplus N$ and $N \hookrightarrow M \oplus N$.

If $N \hookrightarrow E \twoheadrightarrow M$ is an extension in $\mathscr{C}$, then for $\phi : N \to N'$, it is easy to check that $N' \backslash (N' \oplus_N E) \cong M$ and $N' \hookrightarrow N' \oplus_N E \twoheadrightarrow M$ is an extension in $\mathscr{C}$. The association $N \hookrightarrow E \twoheadrightarrow M \mapsto N' \hookrightarrow N' \oplus_N E \twoheadrightarrow M$ gives rise to a map $E(M, \phi) : E(M, N) \to E(M, N')$. From the above argument, we get the following fact:

THEOREM 4.5 *The association $(M, N) \mapsto E_{\mathscr{C}}(M, N)$ is a functor from $\mathscr{C} \times \mathscr{C}$ into the category SETS of sets, contravariant with respect to the left variable and covariant with respect to the right variable. This means that for morphisms $M'' \xrightarrow{\varphi'} M' \xrightarrow{\varphi} M$ and $N \xrightarrow{\phi} N' \xrightarrow{\phi'} N''$ in $\mathscr{C}$, $E(\varphi', N) \circ E(\varphi, N) = E(\varphi \circ \varphi', N)$ and $E(M, \phi') \circ E(M, \phi) = E(M, \phi' \circ \phi)$.*

Exercises

(1) Compute $E_{\mathbb{Z}}(\mathbb{Z}/p\mathbb{Z}, \mathbb{Z}/p\mathbb{Z})$ for a prime p.
(2) Compute $E_{\mathbb{Z}}(\mathbb{Z}, \mathbb{Z})$.
(3) Show the existence and the uniqueness of the fiber sum $M \oplus_X N$ in $\mathscr{C}$.
(4) Give a detailed proof of Theorem 4.5.

4.2.2 Extension functors

We would like to find a mechanical way of computing the extension groups. An object I in $\mathscr{C}$ is called *$\mathscr{C}$-injective* if for every $\mathscr{C}$-morphism $\varphi : M \to I$ and every injective $\mathscr{C}$-morphism $i : M \hookrightarrow N$, there exists a

$\mathcal{C}$-linear map $\phi : N \to I$ extending φ, that is, the following diagram is commutative:

$$\begin{array}{ccc} M & \hookrightarrow & N \\ \varphi \downarrow & \swarrow \exists\phi & \\ I. & & \end{array}$$

An *injective presentation* of N is an exact sequence $N \hookrightarrow I \overset{\pi}{\twoheadrightarrow} S$ for a $\mathcal{C}$-injective module I. We always assume

(EI) $\mathcal{C}$ has enough injectives, that is, for a given N in $\mathcal{C}$, an injective presentation exists in $\mathcal{C}$.

Then we apply the covariant functor $* \mapsto \mathrm{Hom}_{\mathcal{C}}(M, *)$ to the above sequence, getting the following exact sequence:

$$0 \to \mathrm{Hom}_{\mathcal{C}}(M,N) \to \mathrm{Hom}_{\mathcal{C}}(M,I) \xrightarrow{\pi_*} \mathrm{Hom}_{\mathcal{C}}(M,S). \qquad (4.10)$$

Then we define $\mathrm{Ext}^1_{\mathcal{C}}(M,N) = \mathrm{Coker}(\pi_*)$.

We claim that the cokernel $\mathrm{Coker}(\pi_*)$ is independent of the choice of the injective presentation. To show this, we pick a $\mathcal{C}$-morphism $\phi : N \to N'$, and let $N' \xrightarrow{i'} I' \xrightarrow{\pi'} S'$ be an injective presentation of N'. Then we have the following commutative diagram:

$$\begin{array}{ccc} N & \hookrightarrow & I \\ i' \circ \phi \downarrow & \swarrow \exists\phi_1 & \\ I'. & & \end{array}$$

The $\mathcal{C}$-injectivity of I' implies that $i' \circ \phi$ extends to $\phi_1 : I \to I'$. We call ϕ_1 a lift of ϕ. Then ϕ_1 induces a $\mathcal{C}$-morphism $\phi_2 : S \to S'$, making the following diagram commutative:

$$\begin{array}{ccccc} N & \longrightarrow & I & \xrightarrow{\pi} & S \\ \phi\downarrow & & \phi_1\downarrow & & \phi_2\downarrow \\ N' & \longrightarrow & I' & \xrightarrow{\pi'} & S'. \end{array}$$

From this, we get another commutative diagram:

$$\begin{array}{ccccc} 0 \to \mathrm{Hom}_{\mathcal{C}}(M,N) & \longrightarrow & \mathrm{Hom}_{\mathcal{C}}(M,I) & \xrightarrow{\pi_*} & \mathrm{Hom}_{\mathcal{C}}(M,S) \\ \phi_*\downarrow & & \phi_{1,*}\downarrow & & \phi_{2,*}\downarrow \\ 0 \to \mathrm{Hom}_{\mathcal{C}}(M,N') & \longrightarrow & \mathrm{Hom}_{\mathcal{C}}(M,I') & \xrightarrow{\pi'_*} & \mathrm{Hom}_{\mathcal{C}}(M,S'). \end{array}$$

Suppose now that we have two lifts $\phi_1, \phi'_1 : I \to I'$ of ϕ. Then $\phi_1 - \phi'_1|_N = \phi - \phi = 0$. Thus $\phi_1 - \phi'_1 = \tau \circ \pi$ for a $\mathcal{C}$-morphism $\tau : S' \to I'$, and hence, $\phi_2 - \phi'_2 = \pi'_* \circ \tau$. This implies that the morphisms of $\mathrm{Coker}(\pi_*)$ into $\mathrm{Coker}(\pi'_*)$ induced by ϕ_2 and ϕ'_2 are equal, which we write as ϕ_*. We apply the above argument to $\phi = \mathrm{id}_N : N = N$ and its inverse ϕ'. Then $\phi_* \circ \phi'_* = \mathrm{id}_{\mathrm{Coker}(\pi'_*)}$ and $\phi'_* \circ \phi_* = \mathrm{id}_{\mathrm{Coker}(\pi_*)}$ showing $\mathrm{Coker}(\pi_*) \cong \mathrm{Coker}(\pi'_*)$ canonically.

We fix an injective presentation $N \hookrightarrow I \xrightarrow{\pi} S$ for each N in $\mathscr{C}$ and define the functor $(M,N) \mapsto \mathrm{Ext}^1_{\mathscr{C}}(M,N) = \mathrm{Coker}(\pi_*)$. This functor is defined on $\mathscr{C}$ and has values in AB. The above argument shows that the association $\mathscr{C} \to AB$ given by $N \mapsto \mathrm{Ext}^1_{\mathscr{C}}(M,N)$ is a covariant functor, that is, $\phi_* = \mathrm{Ext}^1_{\mathscr{C}}(M,\phi) : \mathrm{Ext}^1_{\mathscr{C}}(M,N) \to \mathrm{Ext}^1_{\mathscr{C}}(M,N')$ satisfies

$$\mathrm{Ext}^1_{\mathscr{C}}(M,\phi') \circ \mathrm{Ext}^1_{\mathscr{C}}(M,\phi) = \mathrm{Ext}^1_{\mathscr{C}}(M,\phi' \circ \phi) \tag{4.11}$$

for two $\mathscr{C}$-morphisms $N \xrightarrow{\phi} N' \xrightarrow{\phi'} N''$.

Let $\varphi : M' \to M$ be a $\mathscr{C}$-morphism. This induces

$$\varphi^*_X = \mathrm{Hom}(\varphi, X) : \mathrm{Hom}_{\mathscr{C}}(M,X) \to \mathrm{Hom}_{\mathscr{C}}(M',X)$$

given by $\phi \mapsto \phi \circ \varphi$, and we have the following commutative diagram:

$$\begin{array}{ccccccc}
\mathrm{Hom}(M,N) & \longrightarrow & \mathrm{Hom}(M,I) & \longrightarrow & \mathrm{Hom}(M,S) & \longrightarrow & \mathrm{Ext}^1(M,N) \\
{\scriptstyle \varphi^*_N}\downarrow & & {\scriptstyle \varphi^*_I}\downarrow & & \downarrow{\scriptstyle \varphi^*_S} & & \downarrow{\scriptstyle \mathrm{Ext}(\varphi,N)} \\
\mathrm{Hom}(M',N) & \longrightarrow & \mathrm{Hom}(M',I) & \longrightarrow & \mathrm{Hom}(M',S) & \longrightarrow & \mathrm{Ext}^1(M',N).
\end{array}$$

It is easy to check that

$$\mathrm{Ext}^1(\varphi',N) \circ \mathrm{Ext}^1(\varphi,N) = \mathrm{Ext}^1(\varphi \circ \varphi',N) \tag{4.12}$$

for two $\mathscr{C}$-morphisms $M'' \xrightarrow{\varphi'} M' \xrightarrow{\varphi} M$.

Thus the functor $(M,N) \mapsto \mathrm{Ext}^1_{\mathscr{C}}(M,N)$ is contravariant with respect to M and covariant with respect to N.

We now claim

THEOREM 4.6 *We have an isomorphism $\imath(M,N) : E_{\mathscr{C}}(M,N) \cong \mathrm{Ext}^1_{\mathscr{C}}(M,N)$ such that $\imath(M',N) \circ E(\varphi,N) = \mathrm{Ext}^1_{\mathscr{C}}(\varphi,N) \circ \imath(M,N)$ for each $\mathscr{C}$-morphism $\varphi : M' \to M$ and $\imath(M,N') \circ E_{\mathscr{C}}(M,\phi) = \mathrm{Ext}^1_{\mathscr{C}}(M,\phi) \circ \imath(M,N)$ for each $\mathscr{C}$-morphism $\phi : N \to N'$. In other words, the system of isomorphisms $\imath(M,N)$ gives an isomorphism between two functors $E_{\mathscr{C}}$ and $\mathrm{Ext}^1_{\mathscr{C}}$. In particular, the set $E_{\mathscr{C}}(M,N)$ has a natural structure of an abelian group.*

Proof We pick an extension $N \overset{\alpha}{\hookrightarrow} E \twoheadrightarrow M \in E_{\mathscr{C}}(M,N)$. We look at the following diagram:

$$\begin{array}{ccc} N & \hookrightarrow & E \\ \downarrow & \swarrow \exists\alpha_* & \\ I, & & \end{array}$$

which induces the following commutative diagram:

$$\begin{array}{ccccc} N & \hookrightarrow & E & \twoheadrightarrow & M \\ \| & & \alpha_* \downarrow & & \downarrow \alpha_{*,M} \\ N & \hookrightarrow & I & \overset{\pi}{\twoheadrightarrow} & S. \end{array}$$

Now we associate the class of $[\alpha_{*,M}] \in \mathrm{Coker}(\pi_*)$ with the extension $N \overset{\alpha}{\hookrightarrow} E \twoheadrightarrow M$. If we have another lift $\alpha'_* : E \to I$ making the first diagram commutative, then $\alpha'_* - \alpha_* = 0$ on N, and hence it factors through $E/N = M$. This shows that $[\alpha_{*,M}] = [\alpha'_{*,M}]$, getting $\iota(M,N) : E(M,N) \to \mathrm{Ext}^1_{\mathscr{C}}(M,N)$.

We now construct the inverse of ι. We start from $[\alpha] \in \mathrm{Coker}(\pi_*)$ for $\alpha : M \to S = I/N$. We put $E = I \times_S M$. Then we get an extension $N \hookrightarrow E \twoheadrightarrow M$. If $[\alpha] = [\alpha']$, then there exists $\tau : M \to I$ such that $\pi\tau = \alpha - \alpha'$. We then define $I \times_{S,\alpha} M \cong I \times_{S,\alpha'} M$ by $(i,m) \mapsto (i - \tau(m), m)$. Thus we get a well-defined map $\iota'(M,N) : \mathrm{Ext}^1(M,N) \to E(M,N)$. It is easy to check by following the definition that $\iota(M,N) \circ \iota'(M,N) = \mathrm{id}_{\mathrm{Ext}}$ and $\iota'(M,N) \circ \iota(M,N) = \mathrm{id}_E$. We leave the reader to check the functoriality of ι. □

There is one more way of constructing $E(M,N)$ using projective presentations. A Λ-module P in $\mathscr{C}$ is called *$\mathscr{C}$-projective* if any $\mathscr{C}$-morphism $\alpha : P \to N$ can be extended to $\alpha_E : P \to E$ for each surjective $\mathscr{C}$-morphism $\pi : E \twoheadrightarrow N$ so that $\pi\alpha_E = \alpha$, that is, the following diagram is commutative:

$$\begin{array}{ccc} & & P \\ \exists\alpha_E & \swarrow & \downarrow \alpha \\ E & \twoheadrightarrow & N. \end{array}$$

Thus the notion of $\mathscr{C}$-projective modules is the dual of that of $\mathscr{C}$-injective Λ-modules, in the sense that we have reversed the direction of the arrows in the definition, and injectivity of arrows is replaced by surjectivity. Any Λ-free module is Λ-MOD-projective. A *projective presentation* of Λ-module M in $\mathscr{C}$ is an exact sequence $T \overset{\iota}{\hookrightarrow} P \twoheadrightarrow M$ with $\mathscr{C}$-projective P. For the moment, we assume

(EP) $\mathscr{C}$ has enough projectives, that is, for each $M \in \mathscr{C}$, there exists a $\mathscr{C}$-projective presentation of M.

As we will see, sometimes (EP) may not be satisfied even if (EI) holds for $\mathscr{C}$. Then applying the functor $X \mapsto \mathrm{Hom}_{\mathscr{C}}(X,N)$, we get an exact sequence:

$$0 \to \mathrm{Hom}_{\mathscr{C}}(M,N) \to \mathrm{Hom}_{\mathscr{C}}(P,N) \xrightarrow{\imath^*} \mathrm{Hom}_{\mathscr{C}}(T,N).$$

Then we can show that $\mathrm{Coker}(\imath^*)$ is independent of the choice of the presentation (just reversing the arrows in the proof in the case of injective presentations) and is isomorphic to $E_{\mathscr{C}}(M,N)$. One can find details in [HAL] Chapter III.

Exercises

(1) Show that (4.10) is exact.

(2) Give a detailed proof for (4.11) and (4.12).

(3) Give a detailed proof of Theorem 4.6.

(4) Show that the addition on $E(M,N) = E_{\Lambda-MOD}(M,N)$ is actually given by the following procedure: Let $N \hookrightarrow E \twoheadrightarrow M$ and $N \hookrightarrow E' \twoheadrightarrow M$ be two extensions in $E(M,N)$. Let $\Delta_M \colon M \to M \oplus M$ be the diagonal map ($\Delta_M\,(a) = a \oplus a$) and $\nabla_N : N \oplus N \to N$ be the summation ($\nabla_N(n \oplus n') = n + n'$). Then the sum of the two extensions is given by

$$E(\Delta_M, \nabla_N)(N \oplus N \hookrightarrow E \oplus E' \twoheadrightarrow M \oplus M).$$

(5) Define $\mathrm{Ext}^1_{\mathscr{C}}(M,N)$ by $\mathrm{Coker}(\imath^*)$ using projective presentation $T \overset{\imath}{\hookrightarrow} P \twoheadrightarrow M$, and prove the counterpart of Theorem 4.6 in this setting. Further show that the additive structure of $E(M,N)$ is independent of the choice of either an injective presentation of N or a projective presentation of M.

4.2.3 Cohomology groups of complexes

A *graded module* in $\mathscr{C}$ is an infinite direct sum $M^\bullet = \bigoplus_{j \in \mathbb{Z}} M_j$ of Λ-modules M_j in $\mathscr{C}$. We suppose that either $M_j = 0$ for $j < -N$ or $M_j = 0$ for $j > N$ with sufficiently large N. A $\mathscr{C}$-morphism $f : M^\bullet \to N^\bullet$ of graded modules in $\mathscr{C}$ is called a morphism of *degree* k if $f(M_j) \subset N_{j+k}$ for all j. If there is a $\mathscr{C}$-endomorphism $\partial : M^\bullet \to M^\bullet$ of degree 1 with $\partial \circ \partial = 0$, we call the pair $(M^\bullet, \partial)$ a *complex* in $\mathscr{C}$. A $\mathscr{C}$-*chain map*

$f : (M^\bullet, \partial) \to (N^\bullet, \delta)$ of degree r is a $\mathscr{C}$-morphism of degree r such that $f \circ \partial = \delta \circ f$. For a given complex $(M^\bullet, \partial)$, we define its cohomology group $H^\bullet(M^\bullet, \partial)$ by

$$H^q(M^\bullet, \partial) = \frac{\mathrm{Ker}(\partial : M_q \to M_{q+1})}{\mathrm{Im}(\partial : M_{q-1} \to M_q)}.$$

Any chain map $f : (M^\bullet, \partial) \to (N^\bullet, \delta)$ of degree r induces a linear map $[f] : H^q(M^\bullet, \partial) \to H^{q+r}(N^\bullet, \delta)$.

LEMMA 4.7 *Suppose that the following diagram is commutative with two exact rows made up of* Λ*-modules:*

$$\begin{array}{ccccc} M & \xrightarrow{a} & L & \xrightarrow{b} & N \to 0 \\ {\scriptstyle d}\downarrow & & {\scriptstyle d'}\downarrow & & \downarrow{\scriptstyle d''} \\ 0 \to M' & \xrightarrow{a'} & L' & \xrightarrow{b'} & N'. \end{array}$$

Then there exists a Λ*-linear map* $\delta : \mathrm{Ker}(d'') \to \mathrm{Coker}(d)$, *and the following sequence is exact:*

$$\mathrm{Ker}(d) \to \mathrm{Ker}(d') \to \mathrm{Ker}(d'') \xrightarrow{\delta} \mathrm{Coker}(d) \to \mathrm{Coker}(d') \to \mathrm{Coker}(d'').$$

This lemma is often called the snake lemma.

Proof By the exactness of the first row and commutativity, the maps a and b induce $\mathrm{Ker}(d) \xrightarrow{a} \mathrm{Ker}(d') \xrightarrow{b} \mathrm{Ker}(d'')$. Similarly, a' and b' induce $\mathrm{Coker}(d) \xrightarrow{a'} \mathrm{Coker}(d') \xrightarrow{b'} \mathrm{Coker}(d'')$. It is easy to check that they are exact at the middle terms.

Now let us define $\delta : \mathrm{Ker}(d'') \to \mathrm{Coker}(d)$. Pick $x \in \mathrm{Ker}(d'')$. By the surjectivity of b, we have $y \in L$ such that $a(y) = x$. The choice of y is unique modulo $\mathrm{Im}(a)$. Then we apply d' to y getting $d'(y) \in L'$. Thus $d'(y)$ is unique modulo $\mathrm{Im}(d' \circ a) = \mathrm{Im}(a' \circ d)$. Apply b' to $d'(y)$, getting $b'(d'(y)) = d''(b(y)) = d''(x) = 0$ because $x \in \mathrm{Ker}(d'')$. Thus $b'(d'(y)) \in \mathrm{Im}(a')$; so, we take $z \in M'$ with $a'(z) = b'(d'(y))$. The element $z \in M'$ is unique modulo $a'^{-1}(\mathrm{Im}(a' \circ d)) = \mathrm{Im}(d)$, determining a unique class $[z] \in \mathrm{Coker}(d)$. Then define $\delta(x) = [z]$.

We check the exactness of the sequence at $\mathrm{Ker}(d'')$. By definition, if $x \in \mathrm{Ker}(d'') \cap b(\mathrm{Ker}(d'))$, then $y \in \mathrm{Ker}(d')$; so, $d'(y) = 0$. This shows that $\delta \circ b = 0$. Suppose that $\delta(x) = 0$. Then $z \in \mathrm{Im}(d)$; so, we can choose $t \in M$ so that $z = d(t)$. Since we can change y modulo $\mathrm{Im}(a)$, we replace y by $y' = y - a(t)$. Then $d'(y') = d'(y - a(t)) = d'(y) -$

$d'(a(t)) = d'(y) - a'(d(t)) = 0$. This shows that $y' \in \mathrm{Ker}(d')$ and hence $x = b(y') \in \mathrm{Im}(b : \mathrm{Ker}(d') \to \mathrm{Ker}(d''))$.

We check the exactness at $\mathrm{Coker}(d)$. Since $a'(z) = d'(y)$, $a' \circ \delta = 0$. Suppose that $a'([s]) = 0$ for $s \in M'$. Then $a'(s) = d'(y')$ for $y' \in L$. So for $x' = b(y')$, we see that $d''(x') = d''(b(y')) = b'(d'(y')) = b'(a'(s)) = 0$. Thus $x' \in \mathrm{Ker}(d'')$. Then by definition, $\delta(x') = [s]$. This completes the proof. □

Proposition 4.8 *Let* $0 \to (M^\bullet, d) \xrightarrow{a} (L^\bullet, d') \xrightarrow{b} (N^\bullet, d'') \to 0$ *be an exact sequence of* $\mathscr{C}$*-chain maps of degree* 0. *Then we have a connection map* $\delta_q : H^q(N^\bullet, d'') \to H^{q+1}(M^\bullet, d)$ *for each* q *and a long exact sequence:*

$$H^q(M^\bullet, d) \xrightarrow{[a]_q} H^q(L^\bullet, d') \xrightarrow{[b]_q} H^q(N^\bullet, d'')$$
$$\xrightarrow{\delta_q} H^{q+1}(M^\bullet, d) \xrightarrow{[a]_{q+1}} H^{q+1}(L^\bullet, d') \xrightarrow{[b]_{q+1}} H^{q+1}(N^\bullet, d'').$$

Proof Because of the exactness of the complexes, we have the following commutative diagram with exact rows:

$$\begin{array}{ccccccc} 0 \to M_q & \xrightarrow{a_q} & L_q & \xrightarrow{b_q} & N_q & \longrightarrow & 0 \\ \downarrow d_q & & \downarrow d'_q & & \downarrow d''_q & & \\ 0 \to M_{q+1} & \xrightarrow{a_{q+1}} & L_{q+1} & \xrightarrow{b_{q+1}} & N_{q+1} & \longrightarrow & 0. \end{array}$$

This yields another commutative diagram with exact rows:

$$\begin{array}{ccccc} M_q/\mathrm{Im}(d_{q-1}) & \xrightarrow{[a_q]} & L_q/\mathrm{Im}(d'_{q-1}) & \xrightarrow{[b_q]} & N_q/\mathrm{Im}(d''_{q-1}) \to 0 \\ \downarrow [d_q] & & \downarrow [d'_q] & & \downarrow [d''_q] \\ 0 \to \mathrm{Ker}(d_{q+1}) & \xrightarrow{[a_{q+1}]} & \mathrm{Ker}(d'_{q+1}) & \xrightarrow{[b_{q+1}]} & \mathrm{Ker}(d''_{q+1}). \end{array}$$

The exactness of the first row comes from the snake lemma applied to cokernels of differential maps of the first diagram for degree $q - 1$. The exactness of the second row comes from the snake lemma applied to the kernels of the first diagram at degree $q + 1$. Note that the kernels of the vertical maps of the second diagram are the cohomology groups of degree q, and the cokernels are those of degree $q + 1$. Now applying the snake lemma to the second diagram, we get the long exact sequence and the connection map δ_q. □

We consider the following condition:

(CN) *The connection map δ is a $\mathscr{C}$-morphism as long as the diagram in the lemma is in $\mathscr{C}$.*

We remark that in the above proof, we have not used the condition (CN), since the condition is always valid for the target category AB of the cohomology functors.

4.2.4 Higher extension groups

Two degree r $\mathscr{C}$-chain maps $f, g : (M^\bullet, \partial) \to (N^\bullet, \delta)$ are called *homotopy equivalent* if there exists a $\mathscr{C}$-morphism $\Delta : M^\bullet \to N^\bullet$ of degree $r-1$ such that $f - g = \delta \circ \Delta + \Delta \circ \partial$. We write $f \sim g$ if f and g are homotopy equivalent. This is an equivalence relation, and we have the identity of cohomology maps $[f] = [g]$ if $f \sim g$.

For a given Λ-module M in $\mathscr{C}$, a *$\mathscr{C}$-resolution* of M is an exact sequence in $\mathscr{C}$:

$$0 \to M \xrightarrow{\varepsilon} M_0 \xrightarrow{\partial_0} M_1 \xrightarrow{\partial_1} M_2 \to \cdots \to M_j \xrightarrow{\partial_j} M_{j+1} \to \cdots .$$

Thus we may put $M^\bullet = \bigoplus_{j=0}^{\infty} M_j$ (regarding $M_j = 0$ if $j < 0$), and $\partial : M^\bullet \to M^\bullet$ is a differential map making $M^\bullet$ a $\mathscr{C}$-complex. We sometimes write the resolution as $0 \to M \xrightarrow{\varepsilon} (M^\bullet, \partial)$, and ε is called the *augmentation* map.

An *injective* resolution of M is a resolution $0 \to M \xrightarrow{\varepsilon_M} (M^\bullet, \partial)$ with $\mathscr{C}$-injective M_j for all j. Since $\mathscr{C}$ has enough injectives, we have an injective presentation $\varepsilon_M : M \hookrightarrow M_0$. Suppose we have an exact sequence: $0 \to M \xrightarrow{\varepsilon_M} M_0 \to \cdots \xrightarrow{\partial_{j-1}} M_j$. Then taking an injective presentation $0 \to \mathrm{Coker}(\partial_{j-1}) \xrightarrow{f} M_{j+1}$ and defining $\partial_j : M_j \to M_{j+1}$ by the composite $\partial_j : M_j \twoheadrightarrow \mathrm{Coker}(\partial_{j-1}) \xrightarrow{f} M_{j+1}$, we see that $0 \to M \xrightarrow{\varepsilon_M} M_0 \to \cdots \to M_j \xrightarrow{\partial_j} M_{j+1}$ is exact. Thus under (EI), we always have a $\mathscr{C}$-injective resolution of a given M in $\mathscr{C}$.

For two Λ-modules M and N in $\mathscr{C}$, we take a $\mathscr{C}$-resolution $0 \to M \xrightarrow{\varepsilon} (M^\bullet, \partial)$ and an injective $\mathscr{C}$-resolution $0 \to N \xrightarrow{\varepsilon} (N^\bullet, \delta)$ and consider the group made of homotopy classes of degree r $\mathscr{C}$-chain maps from $(M^\bullet, \partial)$ into $(N^\bullet, \delta)$, which we write $\mathrm{Ext}^r_{\mathscr{C}}(M, N)$.

PROPOSITION 4.9 *The abelian group $\mathrm{Ext}^r_{\mathscr{C}}(M, N)$ does not depend on the choice of the resolution $(M^\bullet, \partial)$ and the injective resolution $(N^\bullet, \delta)$. Moreover, we have $\mathrm{Ext}^0_{\mathscr{C}}(M, N) \cong \mathrm{Hom}_{\mathscr{C}}(M, N)$ and $\mathrm{Ext}^1_{\mathscr{C}}(M, N) \cong E_{\mathscr{C}}(M, N)$.*

Proof Let $\varphi : M \to N$ be a $\mathscr{C}$-morphism. Then we have the following diagram:

$$\begin{array}{ccc} 0 \to M & \longrightarrow & M_0 \\ \varphi \big\downarrow & & \exists\varphi_0 \big\downarrow \\ 0 \to N & \xrightarrow[\varepsilon_N]{} & N_0. \end{array}$$

We claim the existence of a $\mathscr{C}$-morphism φ_0 making the above diagram commutative. This follows from the $\mathscr{C}$-injectivity of N_0 applied to $\varepsilon_N \circ \varphi : M \to N_0$. Then φ_0 induces a $\mathscr{C}$-morphism: $\mathrm{Coker}(\varepsilon_M) \to \mathrm{Coker}(\varepsilon_N)$, which is still written as φ_0. Then we have the following diagram:

$$\begin{array}{ccc} 0 \to \mathrm{Coker}(\varepsilon_M) & \longrightarrow & M_1 \\ \varphi_0 \big\downarrow & & \exists\varphi_1 \big\downarrow \\ 0 \to \mathrm{Coker}(\varepsilon_N) & \xrightarrow[\delta_0]{} & N_1. \end{array}$$

The existence of a $\mathscr{C}$-morphism φ_1 making the above diagram commutative again follows from the injectivity of N_1 applied to $\delta_0 \circ \varphi_0$. Repeating the above process, we get a $\mathscr{C}$-chain map of degree 0: $\varphi^\bullet : M^\bullet \to N^\bullet$, which we call a lift of φ. Suppose that we have two lifts $\varphi^\bullet$ and $\varphi'^\bullet$. Then $\varphi_0 - \varphi_0' = 0$ on $\mathrm{Im}(\varepsilon_M)$, and hence $\varphi_0 - \varphi_0'$ factors through $\mathrm{Coker}(\varepsilon_M) \cong \mathrm{Im}(\partial_0)$. Thus we have the following commutative diagram by the injectivity of N_0:

$$\begin{array}{ccc} \mathrm{Im}(\partial_0) & \hookrightarrow & M_1 \\ \varphi_0 - \varphi_0' \downarrow & \swarrow \exists\Delta_1 & \\ N_0. & & \end{array}$$

We put here $\Delta_0 : M_0 \to N_1 = \{0\}$ as the zero map. Thus we have the homotopy relation:

$$\varphi_0 - \varphi_0' = \Delta_1 \circ \partial_0 + \delta_{-1} \circ \Delta_0.$$

Suppose now by induction on j that we have $\Delta_k : M_k \to N_{k-1}$ for $k \le j$ such that

$$\varphi_{k-1} - \varphi_{k-1}' = \Delta_k \circ \partial_{k-1} + \delta_{k-2} \circ \Delta_{k-1}$$

for all $k \le j$. Then we look at

$$\varphi_j - \varphi_j' - \delta_{j-1} \circ \Delta_j : M_j \to N_j.$$

Note that

$$\begin{aligned}(\varphi_j - \varphi'_j - \delta_{j-1} \circ \Delta_j) \circ \partial_{j-1} &= \delta_j \circ (\varphi_j - \varphi'_j) - \delta_{j-1} \circ \Delta_j \circ \partial_{j-1} \\ &= \delta_j \circ (\varphi_j - \varphi'_j) - \delta_{j-1} \circ (\varphi_j - \varphi'_j - \delta_{j-1} \circ \Delta_j) \\ &= 0.\end{aligned}$$

Thus $\phi = \varphi_j - \varphi'_j - \delta_{j-1} \circ \Delta_j$ factors through $\mathrm{Coker}(\partial_{j-1}) = \mathrm{Im}(\partial_j)$ and we have another commutative diagram by the injectivity of N_j:

$$\begin{array}{ccc} \mathrm{Im}(\partial_j) & \hookrightarrow & M_{j+1} \\ \phi \downarrow & \swarrow \exists\Delta_{j+1} & \\ N_j. & & \end{array}$$

This shows that $\varphi^\bullet - \varphi'^\bullet = \delta \circ \Delta + \Delta \circ \partial$ and $\varphi^\bullet \sim \varphi'^\bullet$. Namely we have a well- defined map: $\mathrm{Hom}_{\mathscr{C}}(M,N) \to \mathrm{Ext}^0_{\mathscr{C}}(M,N)$. Since $M \subset M_0$ and $N \subset N_0$, this map is injective. On the other hand, if $\varphi^\bullet : M^\bullet \to N^\bullet$ is a chain map, φ_0 induces a $\mathscr{C}$-morphism $\varphi : M = \mathrm{Ker}(\partial_0) \to \mathrm{Ker}(\delta_0) = N$, which gives rise to the original $\varphi^\bullet$ via the above construction (up to homotopy). This shows that $\mathrm{Ext}^0_{\mathscr{C}}(M,N) \cong \mathrm{Hom}_{\mathscr{C}}(M,N)$ canonically.

Now we look at $\mathrm{Ext}^r_{\mathscr{C}}(M,N)$, which consists of homotopy equivalence classes of $\mathscr{C}$-chain maps of degree r from $M^\bullet$ into $N^\bullet$. We now define $N[r]^\bullet = \bigoplus_{j=-r}^{\infty} N[r]_j$ for $N[r]_j = N_{j+r}$. Then each degree r chain map: $M^\bullet \to N^\bullet$ can be regarded as a degree 0 chain map: $M^\bullet \to N[r]^\bullet$, and by the same computation above,

$$\mathrm{Ext}^r_{\mathscr{C}}(M,N) = H^r(\mathrm{Hom}_{\mathscr{C}}(M,N^\bullet),\delta_*). \tag{4.13}$$

We repeat the beginning of the argument: We have the following diagram:

$$\begin{array}{ccc} 0 \to M & \xrightarrow{\varepsilon_M} & M_0 \\ \varphi\Big\downarrow & & \Big\downarrow\exists\varphi_0 \\ N_{r-1} = N[r]_1 & \xrightarrow{\delta_{r-1}} & N[r]_0 = N_r. \end{array}$$

Pick $\varphi \in \mathrm{Hom}_{\mathscr{C}}(M,N_r)$ with $\delta_r \circ \varphi = 0$ (because φ has to be a part of the chain map). We start constructing a lift $\varphi_j : M_j \to N[r]_j$. If we change φ by $\varphi + \delta_{r-1} \circ \phi$ for $\phi \in \mathrm{Hom}_{\mathscr{C}}(M,N_{r-1})$, the outcome is homotopy equivalent to the original lift (by definition), and we get

$$\mathrm{Ext}^r_{\mathscr{C}}(M,N) = \frac{\mathrm{Ker}(\delta_{r,*} : \mathrm{Hom}_{\mathscr{C}}(M,N_r) \to \mathrm{Hom}_{\mathscr{C}}(M,N_{r+1}))}{\delta_{r-1,*}(\mathrm{Hom}_{\mathscr{C}}(M,N_{r-1}))}.$$

Thus we need to show that $H^\bullet(\mathrm{Hom}_{\mathscr{C}}(M,N^\bullet),\delta_*)$ is independent of the choice of the resolution $N^\bullet$. We take another resolution $N'^\bullet$. Then applying the above argument, replacing $(M^\bullet,N^\bullet,\varphi : M \to N)$ by

$(N^\bullet, N'^\bullet, \mathrm{id} : N \to N)$, we have a lift $\iota : N^\bullet \to N'^\bullet$ whose homotopy class is uniquely determined. Thus we have a unique map: $[\iota] : H^\bullet(\mathrm{Hom}_{\mathscr{C}}(M, N^\bullet)) \to H^\bullet(\mathrm{Hom}_{\mathscr{C}}(M, N'^\bullet))$. Reversing this operation, we get $[\iota'] : H^\bullet(\mathrm{Hom}_{\mathscr{C}}(M, N'^\bullet)) \to H^\bullet(\mathrm{Hom}_{\mathscr{C}}(M, N^\bullet))$. By the uniqueness of the lift up to homotopy, we find that $[\iota] \circ [\iota'] = [\mathrm{id}_{N'}]$ and $[\iota'] \circ [\iota] = [\mathrm{id}_N]$. This shows that the two cohomology groups are canonically isomorphic.

Since we have shown that $\mathrm{Ext}^r_{\mathscr{C}}(M, N)$ is independent of the choice of $M^\bullet$, we can take the trivial resolution: $0 \to M \overset{\varepsilon_M}{\cong} M \to 0$. Then using this, it is easy to see that $\mathrm{Ext}^1_{\mathscr{C}}(M, N) = \mathrm{Coker}(\delta_{0,*})$ for $\delta_{0,*} : \mathrm{Hom}_{\mathscr{C}}(M, N_0) \to \mathrm{Hom}_{\mathscr{C}}(M, \mathrm{Im}(\delta_0))$. Since $N \hookrightarrow N_0 \twoheadrightarrow \mathrm{Im}(\delta_0)$ is an injective presentation, we see from Theorem 4.6 that $\mathrm{Ext}^1_{\mathscr{C}}(M, N) \cong E_{\mathscr{C}}(M, N)$ canonically. □

The proof of the above proposition shows

Corollary 4.10 *Let $M, N \in \mathscr{C}$ and $0 \to M \to M^\bullet$ (resp. $0 \to N \to N^\bullet$) be a resolution in $\mathscr{C}$ (resp. a $\mathscr{C}$-injective chain complex with augmentation from N). Here $N^\bullet$ may or may not be a resolution. Then for every morphism $\varphi : M \to N$, there is a chain map in $\mathscr{C}^\bullet$ $\varphi^\bullet : M^\bullet \to N^\bullet$ such that φ_0 induces φ. The lift $\varphi^\bullet$ is unique up to homotopy equivalence.*

There is another consequence:

Corollary 4.11 *If N is a $\mathscr{C}$-injective module, then $\mathrm{Ext}^r_{\mathscr{C}}(M, N) = 0$ for all M in $\mathscr{C}$ if $r > 0$.*

This follows from the fact that $0 \to N \xrightarrow{\mathrm{id}} N \to 0$ is a $\mathscr{C}$-injective resolution of N.

Let L be a third Λ-module in $\mathscr{C}$. We take a $\mathscr{C}$-injective resolution $0 \to L \to (L^\bullet, d)$. If $g : N^\bullet \to L^\bullet$ is a $\mathscr{C}$-chain map of degree s, then

$$g \circ (\Delta \circ \partial + \delta \circ \Delta) = g \circ \Delta \circ \partial + g \circ \delta \circ \Delta = (g \circ \Delta) \circ \partial + d \circ (g \circ \Delta).$$

Thus g preserves homotopy equivalence. Thus $g \circ f : M^\bullet \to L^\bullet$ for a chain map $f : M^\bullet \to N^\bullet$ of degree r defines a homotopy class in $\mathrm{Ext}^{r+s}_{\mathscr{C}}(M, L)$, which depends only on classes $[f] \in \mathrm{Ext}^r_{\mathscr{C}}(M, N)$ and $[g] \in \mathrm{Ext}^s_{\mathscr{C}}(N, L)$. Thus we have

Corollary 4.12 *The composition of chain maps induces a bilinear form*

$$\mathrm{Ext}^r_{\mathscr{C}}(M, N) \times \mathrm{Ext}^s_{\mathscr{C}}(N, L) \to \mathrm{Ext}^{r+s}_{\mathscr{C}}(M, L).$$

Proposition 4.13 *Let $0 \to N \xrightarrow{a} E \xrightarrow{b} L \to 0$ be an exact sequence* *in $\mathscr{C}$. Then we have connection maps:* $\mathrm{Ext}^r_{\mathscr{C}}(M,L) \to \mathrm{Ext}^r_{\mathscr{C}}(M,N)$ *and the following long exact sequence:*

$$\begin{aligned}\mathrm{Ext}^r_{\mathscr{C}}(M,N) &\to \mathrm{Ext}^r_{\mathscr{C}}(M,E) \to \mathrm{Ext}^r_{\mathscr{C}}(M,L)\\ &\to \mathrm{Ext}^{r+1}_{\mathscr{C}}(M,N) \to \mathrm{Ext}^{r+1}_{\mathscr{C}}(M,E) \to \mathrm{Ext}^{r+1}_{\mathscr{C}}(M,L).\end{aligned}$$

Proof Let $0 \to (N^\bullet, \delta)$ and $0 \to (L^\bullet, d)$ be $\mathscr{C}$-injective resolutions of N and L respectively. As a graded module, we put $E^\bullet = \bigoplus_{j=0}^{\infty}(N_j \oplus L_j)$ and we want to create a differential $\partial : E^\bullet \to E^\bullet$ so that $0 \to E \to (E^\bullet, \partial)$ is an injective resolution of E. If we can do this, we will have an exact sequence of complexes:

$$0 \to \mathrm{Hom}_{\mathscr{C}}(M,N^\bullet) \to \mathrm{Hom}_{\mathscr{C}}(M,E^\bullet) \to \mathrm{Hom}_{\mathscr{C}}(M,L^\bullet) \to 0,$$

and then by Proposition 4.8, we have the desired long exact sequence from (4.13) in the proof of Proposition 4.9, because $\mathrm{Ext}^r_{\mathscr{C}}(M,N) = H^r(\mathrm{Hom}_{\mathscr{C}}(M,N^\bullet))$.

We start from the following commutative diagram:

$$\begin{array}{ccc} 0 \to N & \hookrightarrow & E \\ \varepsilon_N \downarrow & \swarrow \exists \varepsilon' & \\ N_0. & & \end{array}$$

The existence of $\varepsilon' : E \to N_0$ follows from the $\mathscr{C}$-injectivity of N_0. Then we define $\varepsilon_E : E \to E_0 = N_0 \oplus L_0$ by $\varepsilon_E(e) = \varepsilon'(e) \oplus \varepsilon_L(b(e))$. If $\varepsilon_E(e) = 0$, then $\varepsilon_L(b(e)) = 0$, and hence $e \in \mathrm{Im}(a)$, because $\varepsilon_L : L \to L_0$ is injective. Writing $e = a(n)$, we then see that $0 = \varepsilon'(e) = \varepsilon'(a(n)) = \varepsilon_N(n)$ and hence $n = 0$ by the injectivity of ε_N. This shows $e = a(n) = 0$. Thus ε_E is injective. We suppose that we have constructed an exact sequence:

$$0 \to E \xrightarrow{\varepsilon_E} E_0 \xrightarrow{\partial_0} E_1 \to \ldots \to E_j$$

so that the following diagram is commutative up to $k \le j$:

$$\begin{array}{ccccc} N_{k-1} & \hookrightarrow & E_{k-1} = N_{k-1} \oplus L_{k-1} & \twoheadrightarrow & L_{k-1} \\ \delta_{k-1} \downarrow & & \partial_{k-1} \downarrow & & \downarrow d_{k-1} \\ N_k & \hookrightarrow & E_k = N_k \oplus L_k & \twoheadrightarrow & L_k. \end{array} \tag{4.14}$$

Then we have the following commutative diagram:

$$\begin{array}{ccc} 0 \to \mathrm{Coker}(\delta_{j-1}) & \hookrightarrow & \mathrm{Coker}(\partial_{j-1}) \\ \delta_j \downarrow & \exists \partial' \swarrow & \\ N_{j+1}. & & \end{array}$$

We claim that the first row is exact. To see this, we apply the snake lemma to (4.14) for $k = j-1$, which shows that the natural map: $\mathrm{Im}(\partial_{j-1}) = \mathrm{Coker}(\partial_{j-2}) \to \mathrm{Coker}(d_{j-2}) = \mathrm{Im}(d_{j-1})$ is surjective. Again applying the snake lemma to:

$$\begin{array}{ccccccc} \mathrm{Im}(\delta_{j-1}) & \longrightarrow & \mathrm{Im}(\partial_{j-1}) & \longrightarrow & \mathrm{Im}(d_{j-1}) & \longrightarrow & 0 \\ \cap\downarrow & & \cap\downarrow & & \cap\downarrow & & \\ N_j & \longrightarrow & E_j & \longrightarrow & L_j, & & \end{array}$$

we get the desired injectivity. Then the existence of ∂' follows from the $\mathscr{C}$-injectivity of N_{j+1}, and as before, we define $\partial_j : E_j \twoheadrightarrow \mathrm{Coker}(\partial_{j-1}) \to E_{j+1}$ by $\partial_j(x) = \partial'(x) \oplus d_j(b_j(x))$ for $b_j : E_j \to L_j$. We can check similarly as in the case of ε_L, that ∂_j as a map from $\mathrm{Coker}(\partial_{j-1})$ to E_{j+1} is injective. This shows that $E_{j-1} \xrightarrow{\partial_{j-1}} E_j \xrightarrow{\partial_j} E_{j+1}$ is exact. Thus by induction on j, we get the desired $\mathscr{C}$-injective resolution $0 \to E \to E^\bullet$. □

REMARK 4.14 In the above proof of Corollary 4.12, we have lifted a given exact sequence $0 \to N \to E \xrightarrow{b} L \to 0$ in $\mathscr{C}$ to an exact sequence of injective resolutions: $0 \to N^\bullet \to E^\bullet \to L^\bullet \to 0$. Although we have used the surjectivity of b to do that, in fact we can lift an exact sequence $0 \to N \to E \xrightarrow{b} L$ to an exact sequence of complexes: $0 \to N^\bullet \to E^\bullet \to L^\bullet$ in the following way: The above proof applied to $0 \to \mathrm{Im}(b) \to L \to L/\mathrm{Im}(b) \to 0$ tells us that we can choose an injective resolution $L^\bullet$ so that any given injective resolution $\mathrm{Im}(b)^\bullet$ of $\mathrm{Im}(b)$ is embedded into $L^\bullet$. Then applying again the above proof to $0 \to N \to E \to \mathrm{Im}(b) \to 0$, we get an exact sequence $0 \to N^\bullet \to E^\bullet \to \mathrm{Im}(b)^\bullet \to 0$. Combining this with $\mathrm{Im}(b)^\bullet \hookrightarrow L^\bullet$, we get the desired exact sequence: $0 \to N^\bullet \to E^\bullet \to L^\bullet$.

We have used an injective resolution of N to define $\mathrm{Ext}^r_{\mathscr{C}}(M,N)$, and then $\mathrm{Ext}^r_{\mathscr{C}}(M,N) = H^r(\mathrm{Hom}_{\mathscr{C}}(M,N^\bullet),\delta_*)$. We can, instead, use a dual version (basically reversing all arrows in the above construction and using a projective resolution of M). Suppose here that $\mathscr{C}$ has enough projectives. A projective resolution of M is an exact sequence of $\mathscr{C}$-projective modules M_j:

$$\cdots \to M_j \xrightarrow{\partial_j} M_{j-1} \to \cdots \to M_0 \xrightarrow{\pi_M} M \to 0.$$

Then we consider the reversed complex:

$$\mathrm{Hom}_{\mathscr{C}}(M^\bullet, N) = \bigoplus_{j=0}^{\infty} \mathrm{Hom}_{\mathscr{C}}(M_j, N)$$

with differentials $\partial_j^* : \mathrm{Hom}_{\mathscr{C}}(M_{j-1}, N) \to \mathrm{Hom}_{\mathscr{C}}(M_j, N)$ given by sending a map $(\phi : M_{j-1} \to N)$ to $(\phi\partial_j : M_j \to N)$. Then it turns out that

$$\mathrm{Ext}^r_{\mathscr{C}}(M,N) \cong H^r(\mathrm{Hom}_{\mathscr{C}}(M^\bullet, N), \partial^*). \qquad \text{(DF)}$$

In fact this definition of the extension modules is more standard (see [HAL] IV.7-8, for example). The extension module $\mathrm{Ext}^1_{\mathscr{C}}(M,N)$ is also related to the classification of r-extensions, where an r-extension is an exact sequence:

$$0 \to N \to E_r \to E_{r-1} \to \cdots \to E_1 \to M \to 0$$

in $\mathscr{C}$, but the description is not as straightforward as in the case of $r = 0, 1$ (see [HAL] IV.9).

The dual version of Proposition 4.13 is given as follows:

PROPOSITION 4.15 *We suppose* (EP). *Let*

$$0 \to M \xrightarrow{a} E \xrightarrow{b} L \to 0$$

be an exact sequence in $\mathscr{C}$. *Then we have connection maps:* $\mathrm{Ext}^r_{\mathscr{C}}(M,N) \to \mathrm{Ext}^r_{\mathscr{C}}(L,N)$ *and the following long exact sequence:*

$$\mathrm{Ext}^r_{\mathscr{C}}(L,N) \to \mathrm{Ext}^r_{\mathscr{C}}(E,N) \to \mathrm{Ext}^r_{\mathscr{C}}(M,N) \\ \to \mathrm{Ext}^{r+1}_{\mathscr{C}}(L,N) \to \mathrm{Ext}^{r+1}_{\mathscr{C}}(E,N) \to \mathrm{Ext}^{r+1}_{\mathscr{C}}(M,N).$$

The proof is just a reverse (dual) of that of Proposition 4.13; so we leave it to the reader (Exercise 1).

The last remark in this section is that we can realize $\mathrm{Ext}^r_{\mathscr{C}}(M,N)$ as a set of r-extension classes as already remarked. This construction extends to any category $\mathscr{C}$ satisfying (C1-4) and the existence of long exact sequence is known in this general case (cf. [HAL] IV.9). Thus actually, the condition (EI) (or (EP)) is not necessary for a theory of $\mathrm{Ext}_{\mathscr{C}}$.

Exercise

(1) Give a detailed proof of Proposition 4.15.

4.3 Group Cohomology Theory

In this section, we study the basic properties of group cohomology theory.

4.3.1 Cohomology of finite groups

Let G be a group. For any given commutative ring A with identity, we define the group algebra $A[G]$ by the set of all formal A-linear combinations $\sum_{g\in G} a_g g$ of group elements $g \in G$. The product of the two elements in $A[G]$ is given by:

$$\sum_{g\in G} a_g g \cdot \sum_{h\in G} b_h h = \sum_{g,h\in G} a_g b_h gh.$$

Then $A[G]$ is an A-algebra, whose identity is given by the identity of G. When G is finite, for each $A[G]$-module M, we define $H^r_A(G,M) = \mathrm{Ext}^r_{A[G]}(A,M)$, where on A, G acts trivially, that is, $ga = a$ for all $a \in A$ and $g \in G$. Later we will extend this definition to infinite G in different ways. For the moment, we assume that G is finite.

To compute the cohomology group explicitly, we construct a standard $A[G]$-projective resolution of A: Let $A_n = A[\overbrace{G \times G \times \cdots \times G}^{n+1}]$ and regard A_n as a $A[G]$-module by $g(g_0,\ldots,g_n) = (gg_0,\ldots,gg_n)$. Then we define $\partial_n : A_n \to A_{n-1}$ by $\partial_n(g_0,\ldots,g_n) = \sum_{j=0}^n(-1)^j(g_0,\ldots,\widehat{g}_j,\ldots,g_n)$, where $\widehat{g}_j$ indicates that g_j is to be removed. Since any element of A_n is a formal linear combination of $(g_0,\ldots,g_n)$, we extend ∂_n to the whole A_n linearly. By definition, it is obvious that $\partial_{n-1} \circ \partial_n = 0$. Since A_n is an $A[G]$-free module with base $(1,g_1,\ldots,g_n)$, A_n is $A[G]$-projective.

We define $\pi_A : A_0 \to A$ by $\pi_A(\sum_g a_g g) = \sum_g a_g$. We claim now that $\cdots \to A_n \to \cdots \to A_0 \xrightarrow{\pi_A} A \to 0$ is exact. To show this, we now define A-linear maps $\Delta_n : A_n \to A_{n+1}$ by $\Delta_n(g_0,\ldots,g_n) = (1,g_0,\ldots,g_n)$. Then it is a matter of computation that

$$\partial\Delta + \Delta\partial = \mathrm{id}_{A^\bullet}, \tag{4.15}$$

for the complex $A^\bullet = \bigoplus_{n=0}^\infty A_n$. Thus the identity map is homotopy equivalent to the 0 map in the category A-$MOD^\bullet$ (not in $A[G]$-$MOD^\bullet$), and hence the identity map of $H^r(A^\bullet,\partial)$ is the zero-map if $r > 0$. This shows that

$$H^0(A^\bullet,\partial) = A \quad \text{and} \quad H^r(A^\bullet,\partial) = 0 \quad \text{if } r > 0$$

and that $A^\bullet \to A \to 0$ is a $A[G]$-projective resolution.

We can always regard M as a $\mathbb{Z}[G]$-module. Thus we also have $H^r_{\mathbb{Z}}(G,M)$.

PROPOSITION 4.16 *For each $A[G]$-module M, we have a canonical isomorphism:*

$$H^r_A(G,M) \cong H^r_{\mathbb{Z}}(G,M).$$

Proof By (DF) in Subsection 4.2.4, we have

$$H^r_A(G,M) = H^r(\mathrm{Hom}_{A[G]^\bullet}(A^\bullet, M), \partial^*).$$

Since $A^\bullet = \mathbb{Z}^\bullet \otimes_{\mathbb{Z}} A$ by definition, we have (see Exercises 2–3 below)

$$\mathrm{Hom}_{\mathbb{Z}[G]^\bullet}(\mathbb{Z}^\bullet, M) \cong \mathrm{Hom}_{A[G]}(\mathbb{Z}^\bullet \otimes_{\mathbb{Z}} A, M \otimes_A A) = \mathrm{Hom}_{A[G]}(A^\bullet, M).$$

The first isomorphism is given by associating with $f : \mathbb{Z}_n \to M$ a linear map $f_A : \mathbb{Z}_n \otimes_A A \to M \otimes_A A = M$ such that $f_A(z \otimes a) = af(z)$. Using the $\mathbb{Z}[G]$-freeness of $\mathbb{Z}_n$, it is easy to check that $f \mapsto f_A$ is an isomorphism. This shows that

$$H^r_A(G,M) = H^r(\mathrm{Hom}_{A[G]^\bullet}(A^\bullet, M)) = H^r(\mathrm{Hom}_{\mathbb{Z}[G]^\bullet}(\mathbb{Z}^\bullet, M)) = H^r_{\mathbb{Z}}(G,M),$$

as desired. □

Thus, there is no need to specify the base ring A in the definition of $H^r_A(G,M)$; so, we write simply $H^r(G,M)$ for this module (although we need to keep in mind that the definition $H^r(G,M) = \mathrm{Ext}^r_{A[G]}(A,M)$ works only for $A[G]$-modules M).

COROLLARY 4.17 *If G is a finite group and M is a finite G-module, then $H^r(G,M)$ is a finite module.*

Proof Let A be the center of $\mathrm{End}_{\mathbb{Z}[G]}(M)$, which is a finite ring. Then M is an $A[G]$-module. The standard resolution $A^\bullet \to A \to 0$ consists of finite modules. Then each component of the complex $\mathrm{Hom}_{A[G]^\bullet}(A^\bullet, M)$ is again a finite module. Thus $H^r(G,M) = H^r(\mathrm{Hom}_{A[G]^\bullet}(A^\bullet, M), \partial^*)$ is a finite module. □

There is another standard projective resolution $\underline{A}^\bullet$ of A in $A[G]$-MOD (which is called the standard inhomogeneous resolution) given as follows: We put $\underline{A}_n = A_n$ but let G act (inhomogeneously) on $\underline{A}_n$, which is a little different from the homogeneous action on A_n, by

$$g(g_0, \ldots, g_n) = (gg_0, g_1, \ldots, g_n).$$

Then again $\underline{A}_n$ is an $A[G]$ free module with basis $[g_1, \ldots, g_n] = (1, g_1, \ldots, g_n)$.

In particular, $\underline{A}_0$ is the rank one free $A[G]$ module generated by $[\,] = (1)$. Define the differential $\underline{\partial}_n : \underline{A}_n \to \underline{A}_{n-1}$ by

$$\underline{\partial}_n([g_1,\ldots,g_n]) = g_1[g_2,\ldots,g_n] + \sum_{j=1}^{n-1}(-1)^j[g_1,\ldots,\overset{j}{g_jg_{j+1}},\ldots,g_n] + (-1)^n[g_1,\ldots,g_{n-1}].$$

The augmentation: $\underline{A}_0 \to A$ is given by the degree map π_A as in the case of a homogeneous chain complex (this is all right, because $\underline{A}_0 = A_0$ as an $A[G]$-module, that is, the homogeneous and inhomogeneous actions are the same at degree 0). Again we have $\mathrm{id}_{\underline{A}^\bullet} = \underline{\partial}\underline{\Delta} + \underline{\Delta}\underline{\partial}$ for the following A-linear maps:

$$\underline{\Delta}_{-1}(1) = [\,], \quad \underline{\Delta}_n((x_0, x_1, \ldots, x_n)) = [x_0, x_1, \ldots, x_n].$$

This shows that $\underline{A}^\bullet$ is a projective resolution of A in $A[G]$-MOD.

Since a homomorphism $\phi \in \mathrm{Hom}_{A[G]}(\underline{A}_n, M)$ for an $A[G]$-module M is determined by its values at the base $\{[x_1,\ldots,x_n]\}$, we can identify $\mathrm{Hom}_{A[G]}(\underline{A}_n, M)$ with the space $C_n(M)$ of all functions $\phi : \overbrace{G \times G \times \cdots \times G}^{n} \to M$. Thus by evaluation at the standard base,

$$(\mathrm{Hom}_{A[G]^\bullet}(\underline{A}^\bullet, M), \underline{\partial}^*) \cong (C^\bullet(M) = \bigoplus_{n\geq 0} C_n(M), \delta),$$

where

$$\delta_{n-1}(\phi)(x_1,\ldots,x_n) = x_1\phi(x_2,\ldots,x_n) + \sum_{j=1}^{n-1}(-1)^j\phi(x_1,\ldots,x_jx_{j+1},\ldots,x_n + (-1)^n\phi(x_1,\ldots,x_{n-1}).$$

We call a function ϕ an *n-cocycle* (resp. *n-coboundary*) if $\delta_n(\phi) = 0$ (resp. $\phi = \delta_{n-1}(\psi)$). From this, we find

$$u \in \mathrm{Ker}(\delta_1) \iff u(gh) = gu(h) + u(g)$$

for all $g, h \in G$ and

$$\delta_0(m)(g) = (g-1)m$$

for some $m \in M$. Thus for a finite group G,

$$H^0(G, M) = \{m \in M | gm = m\ \forall g \in G\},$$
$$H^1(G, M) = \frac{\{u : G \to M | u(gh) = gu(h) + u(g)\}}{\{g \mapsto (g-1)m | m \in M\}}.$$

In particular, for A with the trivial G-action,

$$H^1(G, A) = \mathrm{Hom}_{gp}(G, A).$$

When G is a finite cyclic group generated by g, we have a very simple projective resolution of $\mathbb{Z}$: We define $C_n = \mathbb{Z}[G]$ for all n. Then we define $\partial_{2n} : C_{2n+1} \to C_{2n}$ by $\partial_{2n}(x) = (g-1)x$ and $\partial_{2n-1} : C_{2n} \to C_{2n-1}$ by $\partial_{2n-1}(x) = N_G x = \sum_{h \in G} hx$. We leave the reader to check that $C^\bullet$ with augmentation $C_0 \to \mathbb{Z}$ given by $\sum_h a_h h \mapsto \sum_h a_h$ is a resolution. From this, we get

Proposition 4.18 *Suppose that G is a finite cyclic group generated by g. Then*

$$H^{2n}(G, M) \cong M^G/N_G M \text{ and } H^{2n-1}(G, M) = \mathrm{Ker}(N_G : M \to M)/(g-1)M$$

for all $n > 0$.

Remark 4.19 If G is a topological (possibly infinite) group, that is, the multiplication: $(g, h) \mapsto gh$ and the inverse $g \mapsto g^{-1}$ are supposed to be continuous, we have continuous G-modules M. In other words, for a G-module M with a topology given, we require the action $G \times M \to M$ given by $(g, m) \mapsto gm$ to be continuous. Then we can define a subcomplex $C^\bullet_{ct}(M) \subset C^\bullet(M)$ by requiring continuity to elements in $C^\bullet(M)$. Obviously the differential δ sends $C^\bullet_{ct}(M)$ into itself, giving rise to a differential of $C^\bullet_{ct}(M)$. The continuous cohomology $H^\bullet_{ct}(G, M)$ is defined by $H^\bullet(C^\bullet_{ct}(M), \delta)$. By definition, the inclusion $C^\bullet_{ct}(M) \hookrightarrow C^\bullet(M)$ induces a canonical map $H^\bullet_{ct}(G, M) \to H^\bullet(G, M)$. Here, note that this map of the two cohomology groups may not be injective and may not be surjective either. Suppose that G is a profinite group topologically generated by an element g of infinite order and that M is a discrete G-module. Then we claim that

$$H^1_{ct}(G, M) \cong M/(g-1)M. \tag{4.16}$$

Let us prove the claim. For each continuous 1-cocycle $u : G \to M$, we easily see that $u(g^n) = (1 + g + \cdots + g^{n-1})u(g)$ for $0 < n \in \mathbb{Z}$ and $u(g^{-1}) = -g^{-1}u(g)$. Thus for a given $x \in M$, we define a map $u : H = \{g^n | n \in \mathbb{Z}\} \to M$ by the above formula. We can easily check that u is a 1-cocycle of H and is continuous under the topology on H induced by G, if M is discrete. Thus by continuity, u extends to a 1-cocycle on G. Then $u \mapsto u(g)$ induces the isomorphism (4.16). We will see later that

$H^2_{ct}(G, M) = 0$ if G is a profinite group topologically generated by an element g of infinite order and M is a discrete G-module.

Let U be a subgroup of G of finite index. We consider the following isomorphism for the $\mathbb{Z}[U]$-module M:

$$\eta : \mathrm{Hom}_{\mathbb{Z}[G]^\bullet}(\mathbb{Z}^\bullet, \mathrm{Hom}_{\mathbb{Z}[U]}(\mathbb{Z}[G], M)) \cong \mathrm{Hom}_{\mathbb{Z}[U]^\bullet}(\mathbb{Z}^\bullet, M). \quad (4.17)$$

The isomorphism η is given by the equation $\eta(\varphi)(x) = (\varphi(x))(1)$ (Exercise 5). The module $\mathrm{Hom}_{\mathbb{Z}[U]}(\mathbb{Z}[G], M)$ is considered to be a $\mathbb{Z}[G]$-module by $g\varphi(x) = \varphi(xg)$. Then $\mathbb{Z}^\bullet$ is also a $\mathbb{Z}[U]$-projective resolution of $\mathbb{Z}$, because $\mathbb{Z}[G]$ is a $\mathbb{Z}[U]$-free module. We have

Lemma 4.20

$$H^\bullet(G, \mathrm{Hom}_{\mathbb{Z}[U]}(\mathbb{Z}[G], M)) \cong H^\bullet(\mathrm{Hom}_{\mathbb{Z}[U]^\bullet}(\mathbb{Z}^\bullet, M)) = H^\bullet(U, M).$$

Choosing a coset decomposition $G = \bigsqcup_{\xi \in \Xi} U\xi$, we see that $\mathbb{Z}[G]$ is a $\mathbb{Z}[U]$-free module with basis Ξ. We consider a linear map θ : $\mathrm{Hom}_{\mathbb{Z}[U]}(\mathbb{Z}[G], M) \to M$ given by $\varphi \mapsto \sum_{\xi \in \Xi} \xi^{-1}\varphi(\xi)$. The map θ is well defined and independent of the choice of Ξ because $(u\xi)^{-1}\varphi(u\xi) = \xi^{-1}u^{-1}\varphi(u\xi) = \xi^{-1}\varphi(\xi)$ for all $u \in U$. For each $g \in G$, $\xi g = u_g\xi_g$ for $u_g \in U$ and $\xi_g \in \Xi$. The map: $\xi \mapsto \xi_g$ is a permutation on Ξ. Then we have

$$\begin{aligned}\theta(g\varphi) &= \sum_\xi \xi^{-1}\varphi(\xi g) = \sum_\xi \xi^{-1}u_g\varphi(\xi_g) \\ &= \sum_\xi (u_g^{-1}\xi)^{-1}\varphi(\xi_g) = \sum_\xi (\xi_g g^{-1})^{-1}\varphi(\xi_g) = g\theta(\varphi).\end{aligned}$$

This shows that θ is a morphism of the $\mathbb{Z}[G]$-module. In particular, we have a linear map:

$$trf_{G/U} : H^\bullet(U, M) = H^\bullet(G, \mathrm{Hom}_{\mathbb{Z}[U]}(\mathbb{Z}[G], M)) \xrightarrow{H^\bullet(\theta)} H^\bullet(G, M), \quad (4.18)$$

which is called the *transfer* map. We can define the restriction map $res_{G/U} : H^\bullet(G, M) \to H^\bullet(U, M)$ by restricting the G-cocycle to H.

Proposition 4.21 *We have $trf_{G/U} \circ res_{G/U}(x) = [G : U]x$. In particular, we have:*

(1) *If M is finite and $|G|$ and $|M|$ are mutually prime, then $H^q(G, M) = 0$ for $q > 0$.*
(2) *If $res_{G/U}(x) = 0$ for p-Sylow subgroups U for each prime factor p of $|G|$, then $x = 0$ in $H^q(G, M)$ $(q > 0)$.*

Proof The identity that $trf_{G/U} \circ res_{G/U}(x) = [G : U]x$ follows easily from the fact that $\theta \circ res$ is the scalar multiplication by $[G : U]$. We write $g = |G| = \prod_p p^{e(p)}$. Then for $g^{(p)} = g/p^{e(p)}$, $g^{(p)}x = trf_{G/U} \circ res_{G/U}(x) = 0$ for a p-Sylow subgroup U. Since the greatest common divisor of $\{g^{(p)}\}_p$ is 1, we can find integers m_p such that $\sum_p m_p g^{(p)} = 1$, and hence $x = \sum_p m_p g^{(p)} x = 0$, which shows (2). As for (1), take U to be the trivial subgroup made up of the identity. Then the multiplication by $|G|$ is an automorphism on cocycles, since $|G|$ is prime to $|M|$. The cohomology of trivial group vanishes for positive degree, because $0 \to \mathbb{Z} \to \mathbb{Z} \to 0$ is a projective resolution of $\mathbb{Z}$. Thus $res_{G/U}(x) = 0$, and hence $0 = trf_{G/U} \circ res_{G/U}(x) = |G|x$ implies $x = 0$. This shows (1). □

Exercises

(1) Give a detailed proof of (4.15).

(2) For any A-module M, show that $M \otimes_A A \cong M$ canonically.

(3) For any $\mathbb{Z}[G]$-free module X and any $A[G]$-module M, show that

$$\mathrm{Hom}_{\mathbb{Z}[G]}(X, M) \otimes_A A = \mathrm{Hom}_{A[G]}(X \otimes_{\mathbb{Z}} A, M).$$

(4) Define $\varphi^\bullet : A^\bullet \to \underline{A}^\bullet$ and $\psi^\bullet : \underline{A}^\bullet \to A^\bullet$ by

$$\varphi_n((g_1, \ldots, g_n)) = [g_1, g_1^{-1} g_2, \ldots, g_{n-1}^{-1} g_n]$$

and

$$\psi_n([g_1, \ldots, g_n]) = (g_1, g_1 g_2, \ldots, g_1 g_2 \cdots g_n).$$

Show that they are chain maps, $\varphi^\bullet \circ \psi^\bullet = \mathrm{id}_{\underline{A}^\bullet}$ and $\psi^\bullet \circ \varphi^\bullet = \mathrm{id}_{A^\bullet}$.

(5) Prove (4.17).

4.3.2 Tate cohomology groups

Now we would like to define the Tate cohomology group for finite G, which is a modification of the usual cohomology groups but allows negative degree. We choose a $\mathbb{Z}[G]$-free (so projective) resolution $(C^\bullet, \partial) \xrightarrow{\varepsilon} \mathbb{Z}$. For example $C^\bullet = \mathbb{Z}^\bullet$ satisfies this condition. We suppose that C_n is free of finite rank over $\mathbb{Z}[G]$ (and hence is free of finite rank over $\mathbb{Z}$). We write $\widehat{C}_n = \mathrm{Hom}_{\mathbb{Z}}(C_n, \mathbb{Z})$ and make it into a G-module by $g\phi(x) = \phi(g^{-1}x)$. Since C_n is $\mathbb{Z}[G]$-free, take a base $u_1, \ldots, u_r$ over $\mathbb{Z}[G]$. Then $\{gu_j | g \in G\}$ is a $\mathbb{Z}$-base of C_n. We take the dual base $\widehat{gu}_j$ of $\widehat{C}_n$ so that $\widehat{gu}_i(hu_j)$ is 1 or 0 according as $g = h$, $i = j$ or not. This shows that $\widehat{C}_n$ is $\mathbb{Z}[G]$-free of finite rank. We have the transpose $\partial_p^t : \widehat{C}_{p-1} \to \widehat{C}_p$ of the differential

$\partial_p : C_p \to C_{p-1}$. We now connect $(C^\bullet, \partial) \xrightarrow{\varepsilon} \mathbb{Z} \xrightarrow{{}^t\varepsilon} (\widehat{C}^\bullet, \partial^t)$ as follows: Define

$$(\widetilde{C}_p, \delta_p) = \begin{cases} (C_p, \partial_p) & \text{if } p > 0 \\ (C_0, {}^t\varepsilon \circ \varepsilon) & \text{if } p = 0 \\ (\widehat{C}_{p+1}, \partial^t_{-p}) & \text{if } p < 0. \end{cases}$$

It is easy to check that $(\widetilde{C}^\bullet, \delta)$ gives a long exact sequence of $\mathbb{Z}[G]$-free modules (Exercise 1). We then define

$$H^\bullet_T(G, M) = H^\bullet((\mathrm{Hom}_{\mathbb{Z}[G]^\bullet}(\widetilde{C}^\bullet, M), \delta^*)). \tag{4.19}$$

The above cohomology group is independent of the choice of the free resolution by the argument (based on $\mathbb{Z}[G]$-projectivity of $\widetilde{C}_n$ for all n) given in the previous section.

THEOREM 4.22 *We have*

$$H^q_T(G, M) = \begin{cases} H^q(G, M) & \textit{if } q > 0 \\ M^G/N_G M & \textit{if } q = 0 \\ \mathrm{Ker}(N_G)/D_G M & \textit{if } q = -1, \end{cases}$$

where the norm map $N_G : M \to M$ is given by $N_G(m) = \sum_{g \in G} gm$ and $D_G M = \sum_{g \in G}(g-1)M$. Moreover $H^{-2}_T(G, \mathbb{Z}) = G^{ab}$, where G^{ab} is the maximal abelian quotient of G.

Proof The assertion for $q > 0$ follows from the definition, because $(\widetilde{C}_n, \delta_n) = (C_n, \partial_n)$ for $n > 0$. Since $0 \to \mathbb{Z} \xrightarrow{\varepsilon} C_0 \xrightarrow{\partial_0} C_1$ is exact, we have another exact sequence:

$$\widetilde{C}_{-2} \xrightarrow{\delta_{-2}} \widetilde{C}_{-1} \xrightarrow{{}^t\varepsilon} \mathbb{Z} \to 0.$$

Note that $M \otimes_{\mathbb{Z}[G]} C_n \cong \mathrm{Hom}_{\mathbb{Z}[G]}(\widehat{C}_n, M)$ by $\nu : m \otimes c \mapsto (\phi \mapsto \sum_{g \in G} \phi(gc)gm)$ for $\phi \in \widehat{C}_n = \mathrm{Hom}_{\mathbb{Z}}(C_n, \mathbb{Z})$ (Exercise 2). Then we have the following commutative diagram:

$$\begin{array}{ccccccc} \xrightarrow{\delta_{-2}} \mathrm{Hom}_{\mathbb{Z}[G]}(\widehat{C}_0, M) & \xrightarrow{{}^t\varepsilon^*} & \mathrm{Hom}_{\mathbb{Z}[G]}(\mathbb{Z}, M) & \xrightarrow{\varepsilon^*} & \mathrm{Hom}_{\mathbb{Z}[G]}(C_0, M) \xrightarrow{\delta_0} \\ \wr\uparrow & & \nu\uparrow & & \uparrow 0 \\ \xrightarrow{\mathrm{id}\otimes\partial_1} M \otimes_{\mathbb{Z}[G]} C_0 & \xrightarrow[\mathrm{id}\otimes\varepsilon]{} & M \otimes_{\mathbb{Z}[G]} \mathbb{Z} & \longrightarrow & 0, \end{array}$$

where the lower sequence is exact and $\nu(m \otimes n) = \sum_{g \in G} ngm$ is the one defined above. Note that $M \otimes_{\mathbb{Z}[G]} \mathbb{Z} = M/D_G M$ and $\mathrm{Hom}_{\mathbb{Z}[G]}(\mathbb{Z}, M) = M^G$.

This shows that $H_T^0(G,M) = M^G/N_GM$ and $H_T^{-1}(G,M) = \mathrm{Ker}(N_G)/D_GM$. To compute $H_T^{-2}(G,M)$, we use the inhomogeneous standard projective resolution $\underline{\mathbb{Z}}^\bullet \twoheadrightarrow \mathbb{Z}$. Note that $\mathrm{id} \otimes \partial_0(m \otimes g) = (g-1)m$. Thus if $M = \mathbb{Z}$, $\mathrm{id} \otimes \partial_0$ is the zero map. Then

$$H_T^{-2}(G,\mathbb{Z}) = \frac{\mathrm{Ker}(\mathrm{id} \otimes \partial_0 : \mathbb{Z} \otimes_{\mathbb{Z}[G]} \mathbb{Z}[G^2] \to \mathbb{Z})}{\langle [g] - [hg] + [h] \rangle_{g,h \in G}} = \frac{\mathbb{Z}[G]}{\langle [g] - [hg] + [h] \rangle} \cong G^{ab}$$

by $\sum_{g \in G} n_g[g] \mapsto \prod_{g \in G} g^{n_g} \mod (G,G)$, because $[g] - [hg] + [h]$ is sent to the commutator (g,h) by this map. □

Since the Tate cohomology is computed by using any free (finite rank) resolution, when G is cyclic, we can take the special resolution in Proposition 4.18. Then one verifies that Proposition 4.18 is still valid for $H_T^q(G,M)$ (including negative q) in place of $H^q(G,M)$.

THEOREM 4.23 *Let M be a G-module for a finite group G. Suppose that there exists an index $k \in \mathbb{Z}$ such that $H_T^k(U,M) = H_T^{k+1}(U,M) = 0$ for all subgroups U of G. Then $H_T^q(U,M) = 0$ for all subgroups U and q.*

Proof First suppose that $k = 1$. By Proposition 4.21, we may assume that G has a prime power order. Thus G is a nilpotent group. We have a proper normal subgroup H of G such that G/H is cyclic. We proceed by induction on the order of G. By assumption, $H^q(H,M) = 0$ for $q = 1,2$. We have the restriction–inflation exact sequence (Theorem 4.33):

$$0 \to H^q(G/H, M^H) \to H^q(G,M) \to H^q(H,M)$$

for $q = 1,2$. By induction of the order of G, the vanishing of H^q ($q = 1,2$) for subgroups of G is equivalent to that for subgroups of the cyclic G/H. Thus we may assume that G is cyclic. Then the assertion for $q = 1,2$ follows from Proposition 4.18. To treat the general case, take an injective presentation $M \hookrightarrow I \twoheadrightarrow S$ of $\mathbb{Z}[G]$-modules. As we will see later, we may assume that I is $\mathbb{Z}[U]$-injective for all U. Then by the long exact sequence and the vanishing $H^q(U,I) = 0$ for all $q > 0$, we have $H^{q-1}(U,S) \cong H^q(U,M)$. Thus we can bring the higher 'q' case down to $q = 1,2$ replacing M by S. This domenstrates the case $k > 0$. We can instead use the projective presentation: $T \hookrightarrow P \twoheadrightarrow M$. Then $H_T^q(U,M) \cong H_T^{q+1}(U,T)$. Since the Tate cohomology is defined by using projective resolution, this shift of degree is valid for all degree q including negative ones. By this, again we can bring the '$q \leq 0$' case into the '$q = 1,2$' case. □

THEOREM 4.24 (J. TATE) *Suppose that G is a finite group. If for a G-module C and for all subgroups U of G,*

(1) $H^1(U,C)=0$ *and*
(2) $H^2(U,C)$ *is a cyclic group of order* $|U|$,

then for all subgroups U of G, $H^k(U,\mathbb{Z})\cong H^{k+2}(U,C)$ for all $k>0$.

In fact, this theorem can be generalized to the torsion-free G-module M under the assumption of the theorem as follows:

$$H^k(U,M)\cong H^{k+2}(U,M\otimes_{\mathbb{Z}}C)$$

for all $k>0$. For the proof of this generalized version, see [CLC] IX.8.

Proof We first assume, shifting the degree by 1, that $H_T^0(U,M)=0$ for all subgroups U of G and $H^1(U,M)$ is a cyclic group of order $|U|$. Let $z\in H^1(G,M)=\mathrm{Ext}^1_{\mathbb{Z}[G]}(\mathbb{Z},M)$ be a generator. Thus we have the corresponding extension: $M\hookrightarrow\overline{M}\twoheadrightarrow\mathbb{Z}$ of $\mathbb{Z}[G]$-modules. The associated long exact sequence is: $\cdots\to H_T^0(U,M)\to H_T^0(U,\overline{M})\to H_T^0(U,\mathbb{Z})\to\cdots$. Note that $H_T^0(U,\mathbb{Z})=\mathbb{Z}/|U|\mathbb{Z}$. We write its generator as 1_U. By the definition of $\overline{M}$, $\delta(1_U)=z_U$ is a generator of $H^1(U,M)$. Thus the connecting map $\delta:H_T^0(U,\mathbb{Z})\to H^1(U,M)$ is an isomorphism. Since $H^1(U,\mathbb{Z})=\mathrm{Hom}_{\mathbb{Z}}(U,\mathbb{Z})=0$ because $|U|<\infty$, we have $H^1(U,\overline{M})=H_T^0(U,\overline{M})=0$. This shows, by the previous theorem, that $H_T^q(U,\overline{M})=0$ for all q. Thus $H^k(U,\mathbb{Z})\cong H^{k+1}(U,M)$.

To prove the theorem, we take an injective presentation $C\hookrightarrow I\twoheadrightarrow S$. We may assume that I remains $\mathbb{Z}[U]$-injective. Since $H_T^q(U,I)=\mathrm{Ext}^q_{\mathbb{Z}[U]}(\mathbb{Z},I)=0$, by the cohomology exact sequence: $H_T^q(U,S)\cong H^{q+1}(U,C)$. Thus by the assumption, $H_T^0(U,S)=0$ for all subgroups $U\subset G$, and $H^1(U,S)$ is cyclic of order $|U|$. Applying the above argument to $M=S$, we get

$$H^k(U,\mathbb{Z})\cong H^{k+1}(U,S)\cong H^{k+2}(U,C),$$

which is the required result. □

Exercises

(1) Prove $\widetilde{C}_{n-1}\xrightarrow{\delta_{n-1}}\widetilde{C}_n\xrightarrow{\delta_n}\widetilde{C}_{n+1}$ is exact for all $n\in\mathbb{Z}$ if $(C^\bullet,\partial)\twoheadrightarrow\mathbb{Z}$ is a $\mathbb{Z}[G]$-free resolution such that $\mathrm{rank}_{\mathbb{Z}}C_n<\infty$ for all $n\geq 0$.
(2) Show that $M\otimes_{\mathbb{Z}[G]}C_n\cong\mathrm{Hom}_{\mathbb{Z}[G]}(\widehat{C}_n,M)$ by $v:m\otimes c\mapsto(\phi\mapsto\sum_{g\in G}\phi(gc)gm)$ for $\phi\in\widehat{C}_n=\mathrm{Hom}_{\mathbb{Z}}(C_n,\mathbb{Z})$.

4.3.3 Continuous cohomology for profinite groups

We fix a profinite group G. Let $\mathscr{C}$ be the category of discrete G-modules. By discrete G-modules M, we mean that M has a discrete topology and that it has a continuous action of G. The morphisms of the category $\mathscr{C}$ are homomorphisms of $\mathbb{Z}[G]$-modules, because continuity under discrete topology does not impose any restrictions. Thus $\mathrm{Coker}(f)$ and $\mathrm{Ker}(f)$ for a morphism $f : M \to N$ in $\mathscr{C}$ are again objects in $\mathscr{C}$ (see Exercise 1). Therefore $\mathscr{C}$ is an abelian category.

Let $\mathscr{U}$ be the system of neighborhoods of $1 \in G$ consisting of normal open subgroups. Then it is easy to check:

$$M \textit{ is a discrete } G\textit{-module} \iff M = \bigcup_{U \in \mathscr{U}} M^U, \tag{4.20}$$

where $M^U = \{m \in M | gm = m \ \forall g \in U\}$. From this, $A[G]$ is not an object of $\mathscr{C}$ (see Exercise 3) if G is infinite, hence there is no $A[G]$-free modules in $\mathscr{C}$ if G is infinite. This shows basically that there are no $A[G]$-projective modules in $\mathscr{C}$. However, for the space $C(G/U, I)$ of all functions $\phi : G/U \to I$, $I_G = \bigcup_{U \in \mathscr{U}} C(G/U, I)$ for any injective $\mathbb{Z}$-module I is actually $\mathscr{C}$-injective (see Proposition 4.25). Here we let G act on I_G by $g\phi(h) = \phi(hg)$. In this way, one can prove that $\mathscr{C}$ has enough injectives (see Proposition 4.25). We define $H^q_{\mathscr{C}}(G, M) = \mathrm{Ext}^q_{\mathscr{C}}(\mathbb{Z}, M)$. By definition,

$$H^0_{\mathscr{C}}(G, M) = \mathrm{Ext}^0_{\mathscr{C}}(\mathbb{Z}, M) = \mathrm{Hom}_{\mathbb{Z}[G]}(\mathbb{Z}, M) \cong M^G \tag{4.21}$$

by $\mathrm{Hom}_{\mathbb{Z}[G]}(\mathbb{Z}, M) \ni \phi \mapsto \phi(1) \in M^G$.

The above fact (4.21) and (4.13) tell us that, for a $\mathscr{C}$-injective resolution $M \hookrightarrow (M^\bullet, \partial)$,

$$H^\bullet(G, M) = H^\bullet(H^0(G, M^\bullet), \partial). \tag{4.22}$$

Since each member M_j of $M^\bullet$ is an object of $\mathscr{C}$, for open normal subgroups $U \subset G$, $M^\bullet = \bigcup_{U \in \mathscr{U}} (M^\bullet)^U$.

Let I be a $\mathscr{C}$-injective module. Then for each G/U-modules $M \subset N$, regarding them as G-modules via the projection: $G \to G/U$, any $\mathscr{C}$-morphism $i : M \to I$ can be extended to a $\mathscr{C}$-morphism $j : N \to I$. Since $j(n) = j(un) = uj(n)$ for $\forall n \in N$ and $\forall u \in U$, j has values in I^U. Similarly i has values in I^U. Thus we have found that any morphism $i : M \to I^U$ can be extended to $j : N \to I^U$. Thus we have

$$I^U \textit{ is } \mathbb{Z}[G/U]\textit{-injective if } I \textit{ is } \mathscr{C}\textit{-injective.} \tag{4.23}$$

Now start from an injective system $(I_U, \iota_{U,V} : I_U \to I_V)_{U,V \in \mathscr{U}}$ of $\mathbb{Z}[G/U]$-

injective modules I_U. Then we claim

$$I = \varinjlim_{U \in \mathscr{U}} I_U \text{ is } \mathscr{C}\text{-injective if the maps } \iota_{U,V} \text{ are all injective.} \quad (4.24)$$

Let us prove this: Let $M \subset N$ be an inclusion in $\mathscr{C}$ with a $\mathscr{C}$-morphism $\alpha : M \to I$. Then $M^U \subset N^U$ and $\alpha_U : M^U \to I$. First suppose that M^U is finitely generated as a $\mathbb{Z}[G]$-module. Then $\mathrm{Im}(\alpha_U) \subset I_{V(U)}$ for sufficiently small $V = V(U) \in \mathscr{U}$. We may assume that $V(U) \subset U$. Then by the $\mathbb{Z}[G/V]$-injectivity of I_V, α_U extends to $\beta_U : N^U \to I_V$. If $U' \subset U$, we have the corresponding $V' = V(U') \subset V = V(U)$ as above. Then the following diagram is commutative:

$$\begin{array}{ccccc} N^U & \xrightarrow{\subset} & N^{U'} & \xrightarrow{\beta_{U'}} & I_{V'} \\ \cup\uparrow & & \cup\uparrow & & \uparrow\| \\ M^U & \xrightarrow[\subset]{} & M^{U'} & \xrightarrow{\alpha_{U'}} & I_{V'}. \end{array}$$

Applying the above argument to $N^{U'}$, we may assume that $\beta_U : N^U \to I_{V(U)}$ satisfies $\iota_{V(U),V(U')}\beta_U = \beta_{U'}$. Then taking an injective limit of β_U, we get an extension $\beta : N \to I$ of α. In general, the module M can be written as a union of $M = \bigcup_j M_j$ such that M_j^U is a $\mathbb{Z}[G/U]$-module of finite type. First we apply the above argument to each M_j, and then taking the limit of extensions of α to N, we can extend α to N. This proves the claim. □

Let J be a $\mathbb{Z}$-module and Λ be a ring (possibly non-commutative). Write ? for an indeterminate object in Λ-*MOD*. Then we have an isomorphism of functors $\eta_? : \mathrm{Hom}_\Lambda(?, \mathrm{Hom}_\mathbb{Z}(\Lambda, J)) \to \mathrm{Hom}_\mathbb{Z}(?, J)$ given by $\eta_A(\varphi)(a) = \varphi(a)(1)$ (Exercise 4). Suppose that J is $\mathbb{Z}$-injective (any divisible module is $\mathbb{Z}$-injective). We would like to prove that $I = \mathrm{Hom}_\mathbb{Z}(\Lambda, J)$ is Λ-injective, where $\lambda \in \Lambda$ acts on $\phi : \Lambda \to J$ by $\lambda\phi(x) = \phi(x\lambda)$. Let $M \subset N$ be Λ-modules. For all $\alpha : M \to I$, $\alpha \in \mathrm{Hom}_\Lambda(M, \mathrm{Hom}_\mathbb{Z}(\Lambda, J)) \cong \mathrm{Hom}_\mathbb{Z}(M, J)$. By the $\mathbb{Z}$-injectivity, we can extend $\eta_M(\alpha) : M \to J$ to $\beta' : N \to J$. Then for β such that $\eta_N(\beta) = \beta'$, $\beta : N \to I$ is an extension of α.

We take $J = \mathbb{Q}/\mathbb{Z}$. Then any cyclic $\mathbb{Z}$-module $\langle a \rangle$ generated by a has a homomorphism $\phi_a : \langle a \rangle \to J$ such that $\phi_a(a) \neq 0$. For example, if the order of a is an integer N, we simply define $\phi(ma) = \frac{m}{N} \bmod \mathbb{Z}$. If the order of a is infinite, we take any $x \neq 0$ in J and put $\phi(ma) = mx$. For any given Λ-module M and $m \in M$, we have a non-zero map $\phi_m : \langle m \rangle \to J$, which extends to a linear map $\Phi_m : M \to J$ by the $\mathbb{Z}$-injectivity of J. Then $\eta_M^{-1}(\Phi_m) = \varphi_m : M \to I$ is a Λ-linear map with $\varphi_m(m) \neq 0$. We define $\iota_M : M \to I_M = \prod_{0 \neq m \in M} I_m$ by $\iota_M(x) = \prod_m \varphi_m(x)$, where $I_m = I$. Obviously the product $\prod_{0 \neq m \in M} I_m$ is an injective module, and hence we get an injective presentation of M in Λ-MOD.

PROPOSITION 4.25 *The category Λ-MOD has enough injectives.*

Let $M \in \mathscr{C}$. We apply the above argument to $\Lambda_U = \mathbb{Z}[G/U]$ for M^U. We have an injective presentation of the Λ_U-module

$$0 \to M^U \hookrightarrow I_{M^U} = \prod_{m \in M - \{0\}} \mathrm{Hom}_{\mathbb{Z}}(\mathbb{Z}[G/U], J).$$

If $V \subset U$, then we have a projection $\pi_{V,U} : G/V \to G/U$ which induces a ring homomorphism $\pi_{V,U} : \mathbb{Z}[G/V] \to \mathbb{Z}[G/U]$, which is a surjection. Then by pull-back, we have an inclusion $i_{U,V} = \pi^*_{V,U} : \mathrm{Hom}_{\mathbb{Z}}(\mathbb{Z}[G/U], J) \to \mathrm{Hom}_{\mathbb{Z}}(\mathbb{Z}[G/V], J)$, which is an injection. We can define $\iota_{U,V} : I_{M^U} \hookrightarrow I_{M^V}$ so that $\iota_{U,V}$ coincides with $i_{U,V}$ on the component $\mathrm{Hom}_{\mathbb{Z}}(\mathbb{Z}[G/U], J)$ indexed by $m \in M^U$ and outside such components, the map is just a zero map. Plainly, we have a commutative diagram:

$$\begin{array}{ccc} M^U & \longrightarrow & I_{M^U} \\ \downarrow & & \downarrow \\ M^V & \longrightarrow & I_{M^V}. \end{array}$$

Taking the injective limit with respect to $U \in \mathscr{U}$, we get an injective presentation:

$$\iota_M = \varinjlim_U \iota_{M^U} : M \hookrightarrow \varinjlim_U I_{M^U} = I_M.$$

We note here that the above module I_M is divisible by definition.

COROLLARY 4.26 *The category of discrete G-modules $\mathscr{C}$ has enough injectives. Moreover, for each object $M \in \mathscr{C}$, we have an injective resolution $M \hookrightarrow M^\bullet$ such that $M^\bullet$ is divisible and is given by $\bigcup_{U \in \mathscr{U}} M^\bullet_U$ for an injective resolution $M^\bullet_U$ of M^U in $\mathbb{Z}[G/U]$-MOD. In particular, this shows that*

$$H^\bullet_{\mathscr{C}}(G, M) \cong \varinjlim_{U \in \mathscr{U}} H^\bullet(G/U, M^U) \cong H^\bullet_{ct}(G, M).$$

Proof We first prove the first identity. Pick a $\mathbb{Z}[G/U]$-injective resolution $M^U \hookrightarrow I^\bullet_U$ of M^U for each $U \in \mathscr{U}$. If $V \subset U$, we have a lift $i^\bullet_{U,V} : I^\bullet_U \to I^\bullet_V$ of the inclusion: $M^U \hookrightarrow M^V$, which is unique modulo homotopy equivalence. Thus $i^\bullet_{U,V}$ induces a unique map of cohomology groups, called the *inflation map* $inf_{U/V} : H^\bullet(G/U, M^U) \to H^\bullet(G/V, M^V)$ independently of the resolution. Thus $\varinjlim_{U\in\mathcal{U}} H^\bullet(G/U, M^U)$ is well defined. In particular, the explicit resolution constructed just above the corollary, called the standard injective resolution, identifies it with $H^\bullet_{\mathscr{C}}(G, M)$.

At finite group level, we can compute $H^\bullet(G/U, M^U)$ by using the inhomogeneous standard projective resolution of $\underline{\mathbb{Z}}^\bullet_U$ in $\mathbb{Z}[G/U]$-*MOD*. Again for $V \subset U$, the projection $G/V \twoheadrightarrow G/U$ induces a surjection $\underline{\mathbb{Z}}^\bullet_V \twoheadrightarrow \underline{\mathbb{Z}}^\bullet_U$. This shows that the injective system $(H^\bullet(G/U, M^U))_{U\in\mathcal{U}}$ induced by this projection of the projective resolutions coincides with that induced by the standard injective resolutions. For any continuous map $\phi : G^n \to M$, $G^n = \bigcup_{m\in M} \phi^{-1}(m)$. Since M is discrete, $\phi^{-1}(m)$ is an open set. By the compactness of G^n, there are finitely many $m_i \in M$ such that $G^n = \bigcup_{m_i} \phi^{-1}(m_i)$. Thus ϕ is locally constant, and therefore, there exists a small $U \in \mathcal{U}$ such that ϕ factors through $(G/U)^n$. So any continuous cocycle or coboundary can be regarded as a cocycle or coboundary of G/U for sufficiently small U. This shows that $H^\bullet_{ct}(G, M) = \varinjlim_{U\in\mathcal{U}} H^\bullet(G/U, M^U)$. □

Corollary 4.27 *Let G be a profinite group topologically generated by an element g of infinite order. Let M be a torsion discrete G-module. Then we have* $H^2_{ct}(G, M) = 0$.

Proof By the assumption, $G = \varprojlim_n G_n$ for cyclic groups G_n with $|G_n| \to \infty$ as $n \to \infty$. Let K_n be the kernel of the projection $G \twoheadrightarrow G_n$. By Proposition 4.18, we have $H^2(G_n, M) \cong M^{G_n}/N_{G_n}M$. We can easily check from the proof of Proposition 4.18 that for $m > n$,

$$inf_{m,n} : H^2(G_n, M^{K_n}) = M^G/N_{G_n}M^{K_n} \to M^G/N_{G_m}M^{K_m} = H^2(G_m, M^{K_m})$$

is induced by the norm map given by $N_{K_n/K_m}(x) = \sum_{\gamma\in K_n/K_m} \gamma x = (K_n : K_m)x$ for $x \in M^G$. Since G is torsion-free, if a prime p divides $|G_n|$ for some n, then for any given $N > 0$, we can find $M \gg 0$ such that $p^N|(K_n : K_M)$. This shows that for a given n and $x \in M^G$, $N_{K_n/K_m}(x) = 0$ for sufficiently large $m > n$, since M is a torsion module. Thus $H^2_{ct}(G, M) = \varinjlim_n H^2(G_n, M^{K_n}) = 0$. □

We have from Proposition 4.13

Corollary 4.28 *Let* $0 \to M \to N \to L \to 0$ *be an exact sequence of discrete G-modules. Then we have a long exact sequence:*

$$H^q_{ct}(G,M) \to H^q_{ct}(G,N) \to H^q_{ct}(G,L) \\ \xrightarrow{\delta} H^{q+1}_{ct}(G,M) \to H^{q+1}_{ct}(G,N) \to H^{q+1}_{ct}(G,L).$$

REMARK 4.29 Corollary 4.26 shows that

$$\mathrm{Ext}^\bullet_{\mathscr{C}}(\mathbb{Z},M) = H^\bullet_{\mathscr{C}}(G,M) \cong \varinjlim_{U\in\mathscr{U}} H^\bullet(G/U,M^U) \\ = \varinjlim_{U\in\mathscr{U}} \mathrm{Ext}^\bullet_{\mathbb{Z}[G/U]}(\mathbb{Z},M^U).$$

Exactly the same argument gives a slightly more general result:

$$\mathrm{Ext}_{\mathscr{C}}(N,M) \cong \varprojlim_{U\in\mathscr{U}} \mathrm{Ext}_{\mathbb{Z}[G/U]}(N,M^U)$$

for any discrete G-module N of finite type. Here U runs over all open normal subgroups of G fixing N element by element. Since N is of finite type, there exists an open normal subgroup of G fixing N.

For open subgroups $V \subset U$ of G, we can think of the category of discrete U-modules $\mathscr{C}_U$. Each U-module M can be regarded naturally as a V-module. Let $M \hookrightarrow M^\bullet_V$ be a $\mathscr{C}_V$-injective resolution. We have a lift $i^\bullet : M^\bullet_U \to M^\bullet_V$ of the identity $\mathrm{id} : M \to M$, which is unique up to homotopy. The induced map of cohomology group $H^\bullet(\mathrm{id}) : H^\bullet_{ct}(U,M) \to H^\bullet_{ct}(V,M)$ is called the restriction map and written as $res_{U/V}$. As usual, $res_{U/V}$ does not depend on the choice of the resolution, coincides with the restriction map defined in the previous subsection and satisfies $res_{V/W} \circ res_{U/V} = res_{U/W}$ (Exercise 5).

If M and N are G-modules, then $\mathrm{Hom}_{\mathbb{Z}}(M,N)$ is again a G-module by $g\phi(m) = g\phi(g^{-1}m)$. Then by definition, $\mathrm{Hom}_{\mathbb{Z}}(M,N)^G = \mathrm{Hom}_{\mathbb{Z}[G]}(M,N)$. Even if $M, N \in \mathscr{C}$, $\mathrm{Hom}(M,N)$ may not be in $\mathscr{C}$. We remedy this by defining

$$Hom(M,N) = \bigcup_{U\in\mathscr{U}} \mathrm{Hom}_{\mathbb{Z}}(M,N)^U,$$

which is a discrete G-module by (4.20). When M is finitely generated over $\mathbb{Z}$, the image of its generator under $\phi \in \mathrm{Hom}_{\mathbb{Z}}(M,N)$ falls in N^U for some small $U \in \mathscr{U}$. Therefore in this case, $Hom(M,N) = \mathrm{Hom}_{\mathbb{Z}}(M,N)$.

PROPOSITION 4.30 *Let* $\langle\ ,\ \rangle : M \times N \to P$ *be a bilinear pairing of discrete G-modules. Suppose that* $\langle gm, gn\rangle = g\langle m,n\rangle$ *for all* $g \in G$ *and that* N *is finitely generated as a* $\mathbb{Z}[G]$*-module. There are canonical morphisms:*

$$H^r_{ct}(G,M) \to H^r_{ct}(G,Hom(N,P)) \to \mathrm{Ext}^r_{\mathscr{C}}(N,P).$$

If one of the following three conditions is satisfied:

(1) *N is $\mathbb{Z}$-free of finite rank,*
(2) *P is divisible,*
(3) *P is a p-divisible (discrete) $\mathbb{Z}_p[G]$-module for a prime p,*

then

$$H^r_{ct}(G, Hom(N,P)) \cong \mathrm{Ext}^r_{\mathscr{C}}(N,P).$$

Proof The first morphism is induced by the morphism of G-modules: $M \to \mathrm{Hom}(N,P)$ which is in turn induced by the pairing. To see the second, we first assume that N is $\mathbb{Z}$-free of finite rank. For any $\mathscr{C}$-injective I, we claim that $Hom(N,I)$ is $\mathscr{C}$-injective, as long as N is $\mathbb{Z}$-free of finite rank. By our assumption on N, we have $\mathrm{Hom}_{\mathbb{Z}}(N,I) = Hom(N,I)$. We have a canonical isomorphism (Exercise 4):

$$\mathrm{Hom}_{\mathbb{Z}[G]}(M, Hom(N,I)) \cong \mathrm{Hom}_{\mathbb{Z}[G]}(M \otimes_{\mathbb{Z}} N, I).$$

Since N is torsion-free, it is $\mathbb{Z}$-flat, and hence $M \mapsto M \otimes_{\mathbb{Z}} N$ preserves exact sequence. Note that I is $\mathscr{C}$-injective if and only if $M \mapsto \mathrm{Hom}_{\mathbb{Z}[G]}(M,I) = \mathrm{Hom}_{\mathscr{C}}(M,I)$ preserves exact sequences. Thus the composite functor

$$M \mapsto M \otimes_{\mathbb{Z}} N \mapsto \mathrm{Hom}_{\mathscr{C}}(M \otimes_{\mathbb{Z}} N, I) = \mathrm{Hom}_{\mathbb{Z}[G]}(M, Hom(N,I))$$

preserves exact sequence. This shows that $Hom(N,I)$ is $\mathscr{C}$-injective, as long as N is $\mathbb{Z}$-free of finite rank.

We continue to assume that N is $\mathbb{Z}$-free of finite rank. Thus we may assume that $N \cong \mathbb{Z}^n$. Then $Hom(N,I) = I^n$. Thus for an injective resolution $P \hookrightarrow P^\bullet$, $Hom(N,P^\bullet) \cong (P^\bullet)^n$ and hence $Hom(N,P) \hookrightarrow Hom(N,P^\bullet)$ is an injective resolution of $Hom(N,P)$. We choose another $\mathscr{C}$-injective resolution: $Hom(N,P) \hookrightarrow Hom(N,P)^\bullet$, which could be different from $Hom(N,P^\bullet)$. Since P_j is $\mathscr{C}$-injective, $Hom(N,P_j)$ is $\mathscr{C}$-injective. Thus we have lifts

$$i^\bullet : Hom(N,P)^\bullet \to Hom(N,P^\bullet) \quad \text{and} \quad j^\bullet : Hom(N,P^\bullet) \to Hom(N,P)^\bullet$$

of id : $Hom(N,P) \to Hom(N,P)$, which are unique up to homotopy and are mutually an inverse of each other. Then we have

$$\begin{aligned} H^\bullet(i) : H^\bullet(G, Hom(N,P)) &= H^\bullet(\mathrm{Hom}_{\mathscr{C}}(\mathbb{Z}, Hom(N,P^\bullet))) \\ &\cong H^\bullet(Hom_{\mathscr{C}}(N,P^\bullet)) = \mathrm{Ext}^\bullet_{\mathscr{C}}(N,P), \end{aligned}$$

which is the desired isomorphism.

For a general N, to create the desired map: $H^q_{ct}(G, Hom(N,P)) \to$

$\mathrm{Ext}^q_{\mathscr{C}}(N,P)$, we need to study double complexes, which yield an appropriate spectral sequence giving rise to the map (see [ADT] 0.3). Here we avoid the use of the spectral sequence, and instead we consider the category $\mathscr{C}^\bullet$ of complexes made up of objects in $\mathscr{C}$. Since $\mathscr{C}$ is an abelian category with enough injectives, so is $\mathscr{C}^\bullet$. For any object X in $\mathscr{C}$, we take a $\mathscr{C}^\bullet$-injective resolution $Hom(X,P^\bullet)^\bullet$ of the complex $Hom(X,P^\bullet)$ in $\mathscr{C}^\bullet$: Write ∂ for the differential of $Hom(X,P^\bullet)^\bullet$ coming from inner $P^\bullet$ and δ for the outer one. We then define a new complex $HOM(X,P)^\bullet$ by putting $HOM(X,P)_\ell = \bigoplus_{j+k=\ell} Hom(X,P_j)_k$. The new differential Δ_ℓ is given by $\delta_k + (-1)^k\partial_j$ on $Hom(X,P_j)_k$. It is easy to check that $\Delta \circ \Delta = 0$. Then the projection $p : HOM(X,P)^\bullet = Hom(X,P^\bullet)^\bullet \twoheadrightarrow Hom(X,P^\bullet)$ is a morphism of complexes: $HOM(X,P)^\bullet \to Hom(X,P^\bullet)$ up to sign, and inclusion $i : Hom(X,P)^\bullet \hookrightarrow HOM(X,P)^\bullet$ is also a morphism in $\mathscr{C}^\bullet$. Then the composite $p \circ i$ for $X = N$ is the desired map.

To show that the above construction is compatible with the one we gave for $\mathbb{Z}$-free N, we take a presentation $N_1 \hookrightarrow N_0 \twoheadrightarrow N$ of $\mathbb{Z}[G]$-modules with $\mathbb{Z}$-free N_j $(j = 0,1)$ (of finite rank). This is possible if N is of finite type over $\mathbb{Z}[G]$ because generators of N are fixed by an open normal subgroup U. Then we look at the exact sequence in $\mathscr{C}^\bullet$:

$$0 \to Hom(N,P^\bullet) \to Hom(N_0,P^\bullet) \to Hom(N_1,P^\bullet).$$

We take an injective resolution in $\mathscr{C}^\bullet$ of each term of the above sequence so that the following diagram is commutative:

$$\begin{array}{ccccc} 0 \longrightarrow Hom(N,P^\bullet)^\bullet & \longrightarrow & Hom(N_0,P^\bullet)^\bullet & \longrightarrow & Hom(N_1,P^\bullet)^\bullet \\ \downarrow & & \downarrow & & \downarrow \\ 0 \longrightarrow Hom(N,P^\bullet) & \longrightarrow & Hom(N_0,P^\bullet) & \longrightarrow & Hom(N_1,P^\bullet). \end{array}$$

The argument in Remark 4.14 applied to $\mathscr{C}^\bullet$ (in place of $\mathscr{C}$ there) shows the existence of such $\mathscr{C}^\bullet$-injective resolutions. Since homotopy equivalence with respect to δ (or ∂) gives rise to that with respect to Δ, we have the following commutative diagram (unique up to homotopy equivalence):

$$\begin{array}{ccccc} 0 \longrightarrow Hom(N,P)^\bullet & \longrightarrow & Hom(N_0,P)^\bullet & \longrightarrow & Hom(N_1,P)^\bullet \\ \downarrow & & \downarrow & & \downarrow \\ 0 \longrightarrow HOM(N,P)^\bullet & \longrightarrow & HOM(N_0,P)^\bullet & \longrightarrow & HOM(N_1,P)^\bullet \\ \downarrow & & \downarrow & & \downarrow \\ 0 \longrightarrow Hom(N,P^\bullet) & \longrightarrow & Hom(N_0,P^\bullet) & \longrightarrow & Hom(N_1,P^\bullet). \end{array}$$

Removing the middle sequence, we know that the two right vertical maps are lifts of the identities of $Hom(N_j, P)$. Thus the map at the extreme left is induced by the lifts of the identity for $\mathbb{Z}$-free N_j. Thus our construction is compatible with the one for $\mathbb{Z}$-free G-modules, and the map is unique up to homotopy.

Suppose that

$$0 \to Hom(N, P) \to Hom(N_0, P) \to Hom(N_1, P) \to 0$$

is exact. Thus, writing $H^\bullet(X)$ (resp. $E^\bullet(X)$) for $H^\bullet(G, Hom(X, P))$ (resp. $\mathrm{Ext}^\bullet_{\mathscr{C}}(X, P)$), we have the following commutative diagram of long exact sequences:

$$\begin{array}{ccccccccc} H^{r-1}(N_0) & \longrightarrow & H^{r-1}(N_1) & \longrightarrow & H^r(N) & \longrightarrow & H^r(N_0) & \longrightarrow & H^r(N_1) \\ \wr\downarrow & & \wr\downarrow & & \downarrow & & \wr\downarrow & & \wr\downarrow \\ E^{r-1}(N_0) & \longrightarrow & E^{r-1}(N_1) & \longrightarrow & E^r(N) & \longrightarrow & E^r(N_0) & \longrightarrow & E^r(N_1). \end{array}$$

Then by the five lemma (see [BCM] Exercise I.1.4.b), $H^r(G, Hom(N, P))$ is isomorphic to $\mathrm{Ext}^r_{\mathscr{C}}(N, P)$. In particular, under the assumptions (2) or (3), we have the desired isomorphism. □

We record here a by-product of the above proof:

LEMMA 4.31 *Let N be a discrete $\mathbb{Z}[G]$-module, which is free of finite rank over $\mathbb{Z}$. Then for a $\mathscr{C}$-injective module I, $Hom(N, I) = Hom_{\mathbb{Z}}(N, I)$ is $\mathscr{C}$-injective, and for an arbitrary discrete $\mathbb{Z}[G]$-module P and its $\mathscr{C}$-injective resolution $P \hookrightarrow P^\bullet$, $Hom(N, P) \hookrightarrow Hom(N, P^\bullet)$ is a $\mathscr{C}$-injective resolution of* $\mathrm{Hom}_{\mathbb{Z}}(N, P) = Hom(N, P)$.

Here is a definition–proposition of the cup product pairing:

COROLLARY 4.32 *Let the notation and the assumption be as in the proposition. Then we have a pairing*

$$\langle\ ,\ \rangle : H^r_{ct}(G, M) \times H^s_{ct}(G, N) \to H^{r+s}_{ct}(G, P)$$

induced by the pairing: $M \times N \to P$.

Proof We consider the following diagram:

$$\begin{array}{ccccc} H^r_{ct}(G, M) & \times & H^s_{ct}(G, N) & \longrightarrow & H^{r+s}_{ct}(G, P) \\ \downarrow & & \wr\downarrow & & \wr\downarrow \\ \mathrm{Ext}^r_{\mathscr{C}}(N, P) & \times & \mathrm{Ext}^s_{\mathscr{C}}(\mathbb{Z}, N) & \longrightarrow & \mathrm{Ext}^{r+s}_{\mathscr{C}}(\mathbb{Z}, P). \end{array}$$

The bottom row is the extension pairing in Corollary 4.12. Then we define the desired pairing simply insisting on making the above diagram commutative. □

The pairing in Corollary 4.32 is called the cup product pairing.

Exercises

(1) For an exact sequence of $\mathbb{Z}[G]$-modules: $0 \to M \to N \to L \to 0$, if two left (or right) terms of the sequence are in $\mathscr{C}$, show that the remaining term is again in $\mathscr{C}$. Find an example of an exact sequence as above in which M and L are discrete modules, but N is not discrete (the maps have to be continuous).

(2) Prove (4.20).

(3) Show that $A[G]$ is not in $\mathscr{C}$ if G is infinite.

(4) Define a morphism of functors $\eta_? : \mathrm{Hom}_\Lambda(?, \mathrm{Hom}_\mathbb{Z}(\Lambda, J)) \to \mathrm{Hom}_\mathbb{Z}(?, J)$ by $\eta_A(\varphi)(a) = \varphi(a)(1)$. Show that this is an isomorphism.

(5) Show that $res_{U/V}$ is independent of the choice of resolutions and $res_{V/W} \circ res_{U/V} = res_{U/W}$.

(6) Suppose that N is $\mathbb{Z}$-free of finite rank. Write down the map of Proposition 4.30: $H^1(G, Hom(N, P)) \to \mathrm{Ext}^1_{\mathscr{C}}(N, P)$ explicitly, associating with an inhomogeneous 1-cocycle $c : G \to Hom(N, P)$ an extension $P \hookrightarrow M \twoheadrightarrow N$ in $\mathscr{C}$.

4.3.4 Inflation and restriction sequences

Let U be a closed normal subgroup of G. For an inhomogeneous q-cocycle $u : U^q \to M$ and $g \in G$, ${}^g u : (g_1, \ldots, g_q) \mapsto g u(g^{-1} g_1 g, \ldots, g^{-1} g_q g)$ is again a q-cocycle of U, and the cohomology class of ${}^g u$ is equal to that of u if $g \in U$, as easily verified by computation. Thus the quotient group G/U acts on $H^q(U, M)$ by $[u] \mapsto [{}^g u]$. We now prove

THEOREM 4.33 *Let U be a closed normal subgroup of G, and suppose that $H^q_{ct}(U, M) = 0$ for all $q = 1, 2, \ldots, p-1$. Then the following sequence is exact:*

$$0 \to H^p_{ct}(G/U, M^U) \xrightarrow{inf_{G/U}} H^p_{ct}(G, M) \xrightarrow{res_{G/U}} H^0(G/U, H^p_{ct}(U, M)) \xrightarrow{trans_{G/U}} H^{p+1}_{ct}(G/U, M^U).$$

We shall give a definition of the *transgression* $trans_{G/U}$, due to Hochshild and Serre [HS], in the following proof of the theorem.

Proof We only prove the theorem for open subgroups U. The general case is left to the reader (who needs to check continuity of cocycles in the proof below applied to a closed subgroup U in place of open U). When $p = 1$, we use the inhomogeneous cochains. Thus for a profinite group X and a discrete X-module N

$$H^1_{ct}(X, N) = \frac{\{c : X \to N : \text{continuous} | c(gh) = gc(h) + c(g)\ \forall g, h \in X\}}{\{g \mapsto (g-1)x | x \in N\}}. \tag{4.25}$$

For the projection $\pi : G \to G/U$, $inf_{G/U}(c) = c \circ \pi$ and $res_{G/U}c = c|_U$. For these two maps, it is easy to show the exactness by a simple computation (Exercise 1).

We now prove the exactness at $H^0(G/U, H^1_{ct}(U, M))$. Let $c : U \to M$ be a cocycle representing a class $[c]$ in $H^0(G/U, H^1_{ct}(U, M))$. Then $gc(g^{-1}ug) - c(u) = (u-1)a(g)$ for a function $a : G \to M$, because $g[c] = [c]$. If $g \in U$, by the cocycle relation, we see that

$$gc(g^{-1}ug) - c(u) = c(ug) + gc(g^{-1}) - c(u) = uc(g) - c(g) = (u-1)c(g).$$

Thus we may take the function a to be c on U and hence may assume that $a(u) = c(u)$ for all $u \in U$. Then we have

$$ga(g^{-1}ug) - a(u) = (u-1)a(g).$$

Let F be the space of continuous functions $f : U \to M$. Then we make F into a G-module by the following G-action: $gf(u) = gf(g^{-1}ug)$. Note that $(g-1)f(u) = gf(g^{-1}ug) - f(u)$. But $\delta(g \mapsto (g-1)f) = 0$ for the differential $\delta : C_1(G, F) \to C_2(G, F)$ of inhomogeneous cochains, and by applying δ to $ga(g^{-1}ug) - a(u) = (u-1)a(g)$, we have

$$\begin{aligned} 0 &= \delta(x \mapsto (x-1)a(u)) = \delta(x \mapsto (u-1)a(x))(g, h) \\ &= g(g^{-1}ug - 1)a(h) - (u-1)a(gh) + (u-1)a(g) \\ &= (u-1)(ga(h) - a(gh) + a(g)). \end{aligned}$$

Now we put $b(g, h) = \delta_1(a)(g, h) = ga(h) - a(gh) + a(g)$. Then the above equation becomes:

$$(u-1)b(g, h) = 0.$$

Thus the 2-cocycle $b : G \times G \to M$ actually has values in M^U.

Note that

$$\begin{aligned} (u-1)(ua(g) + a(u)) &= u(ga(g^{-1}ug) - a(u)) + ua(u) - a(u) \\ &= uga(g^{-1}ug) - a(u) = (u-1)a(ug). \end{aligned}$$

Thus fixing a complete representative set R for $U\backslash G$ so that $1 \in R$, we may normalize a so that $a(ug) = ua(g) + a(u)$ for all $u \in U$ and all $g \in R$. Since $a|_U$ is a 1-cocycle, by computation, we conclude that $a(ug) = ua(g) + a(u)$ for all $u \in U$ and all $g \in G$ (not just in R). Then for all $u \in U$ and $g, h \in G$, we see that $b(u, g) = 0$, and the 2-cocycle relation is

$$ub(g,h) - b(ug,h) + b(u,gh) - b(u,g) = 0.$$

This shows that $b(g,h) = ub(g,h) = b(ug,h)$. Similarly, we can show that $b(g,uh) = b(g,h)$. Thus b factors through G/U.

If $a' : G \to M$ satisfies the same properties as a, that is, $ga'(g^{-1}ug) - a'(u) = (u-1)a'(g)$ and $a' = c$ on U, then

$$(u-1)(a(g) - a'(g)) = ga(g^{-1}ug) - a(u) - (ga'(g^{-1}ug) - a'(u)) = 0,$$

because $a = c = a'$ on U. This shows that $d(g) = a(g) - a'(g) \in M^U$. Then $b - b' = \delta(d) \in \mathrm{Im}(C_1(G/U, M^U) \xrightarrow{\delta} C_2(G/U, M^U))$, and hence we have the identity of the cohomology classes:

$$[b] = [b'] \in H^2(G/U, M^U)$$

for $b' = \delta(a')$. We then define $trans_{G/U}([c])$ by the cohomology class of $[b]$ in $H^2(G/U, M^U)$.

Suppose that $trans_{G/U}([c]) = 0$. Then choosing a 1-cochain $d : G/U \to M^U$ such that $\delta(d) = b$, we see that $a' = a - d$ agrees with c on U and $\delta(a') = 0$; so a' is a 1-cocycle of G inducing c. This shows that

$$\mathrm{Ker}(trans_{G/U}) \supset \mathrm{Im}(res_{G/U}).$$

By definition, if $c \in \mathrm{Im}(res_{G/U})$, we take a to be the 1-cocycle of G restricting c on U. Thus, $\mathrm{Im}(res_{G/U}) \supset \mathrm{Ker}(trans_{G/U})$. This proves the desired exactness for degree 1 cohomology groups.

We now prove the result in general by induction on p. We take a $\mathscr{C}$-injective presentation:

$$0 \to M \to I \to S \to 0.$$

Then by the long exact sequence of cohomology groups, the following sequence:

$$\begin{aligned} H^{p-1}_{\mathscr{C}}(G,M) &\to H^{p-1}_{\mathscr{C}}(G,I) \to H^{p-1}_{\mathscr{C}}(G,S) \\ &\to H^{p}_{\mathscr{C}}(G,M) \to H^{p}_{\mathscr{C}}(G,I) \to H^{p}_{\mathscr{C}}(G,S) \end{aligned}$$

is exact. Since I is $\mathscr{C}$-injective, $H^j_{\mathscr{C}}(G,I) = \mathrm{Ext}^j_{\mathscr{C}}(\mathbb{Z},I) = 0$. This gives the isomorphism: $H^{p-1}_{\mathscr{C}}(G,S) \cong H^p_{\mathscr{C}}(G,M)$. We may take

$$I = I_M = \prod_{m \in M - \{0\}} Hom(\mathbb{Z}[G], \mathbb{Q}/\mathbb{Z}).$$

Since $\mathbb{Z}[U]^r \cong \mathbb{Z}[G]$ as right U-modules by using a coset decomposition $G = \bigsqcup g_i U$, I_M is again $\mathscr{C}_U$-injective. Thus again using the long exact sequence:

$$H^{p-1}_{\mathscr{C}}(U,S) \cong H^p_{\mathscr{C}}(U,M)$$

as G/U-modules. Since $H^q_{\mathscr{C}}(U,M) = 0$ if $0 < q < p$, $H^q_{\mathscr{C}}(U,S) = 0$ if $0 < q < p-1$. By taking U-invariant, we get another exact sequence:

$$0 \to M^U \to I^U \to S^U \to H^1_{\mathscr{C}}(U,M) = 0.$$

Note that I^U is $\mathbb{Z}[G/U]$-injective (4.23). Thus again by the long exact sequence, we get

$$H^{p-1}(G/U, S^U) \cong H^p(G/U, M^U).$$

Then applying an induction hypothesis to S in place of M, we get an exact sequence:

$$\begin{aligned} 0 &\to H^{p-1}(G/U, S^U) \to H^{p-1}_{ct}(G,S) \\ &\to H^0(G/U, H^{p-1}_{ct}(U,S)) \to H^p(G/U, S^U), \end{aligned}$$

which gives rise to the desired sequence by the three isomorphisms as above. □

We can prove a similar theorem for extension groups:

THEOREM 4.34 *Let U be a closed normal subgroup of G. We write $\mathscr{C}_G$ for the category of discrete G-modules. Let M be a discrete G/U-module and N be a discrete G-module. If $H^q_{ct}(U,N) = 0$ for all integers q with $1 \le q \le p$, then we have for all q as above*

$$\mathrm{Ext}^q_{\mathscr{C}_G}(M,N) \cong \mathrm{Ext}^q_{\mathscr{C}_{G/U}}(M, N^U).$$

If M is finite of order a prime power ℓ^r and $p > 1$, we can ease the requirement to have the above equality of extension groups to the following weaker condition: $H^q_{ct}(U,N) = 0$ for all integers q with $1 \le q \le p-1$ and $H^p_{ct}(U,N)[\ell^\infty] = 0$, where '$[\ell^\infty]$' indicates the ℓ-torsion part.

Proof We claim to have the following exact sequence:

$$0 \to \mathrm{Ext}^1_{\mathscr{C}_{G/U}}(M, N^U) \xrightarrow{inf} \mathrm{Ext}^1_{\mathscr{C}_G}(M, N) \xrightarrow{\delta} \mathrm{Hom}_{\mathscr{C}_G}(M, H^1(U, N)). \quad (4.26)$$

To see this, we consider the extension class $e : N^U \hookrightarrow Y_U \twoheadrightarrow M$ of G/U-modules. Then the class of e can be extended to $inf(e) : N \hookrightarrow (N \oplus_{N^U} Y_U) \twoheadrightarrow M$, which is an extension of M by N, where $N \oplus_{N^U} Y_U$ is the fiber sum of $N^U \hookrightarrow N$ and Y_U over N (see Theorem 4.5). Since e is a subsequence of $inf(e)$, if $inf(e) = 0$, we have $e = 0$. Thus inf is an injection. For each extension class $E : N \hookrightarrow Y \twoheadrightarrow M$ in $\mathrm{Ext}^1_{\mathscr{C}_G}(M, N)$, we have a long exact sequences of cohomology groups:

$$0 \to N^U \to Y^U \to M \xrightarrow{\delta_E} H^1_{ct}(U, N).$$

We assign to E the map δ_E. By definition, $\delta_E = 0 \iff E = inf(e)$ for $e : N^U \hookrightarrow Y^U \twoheadrightarrow M$. This shows the exactness of the claimed sequence. Thus we get the desired identity for $p = 1$.

We now prove the general case by induction on $p > 1$. We now take an injective presentation: $N \hookrightarrow I \twoheadrightarrow S$ in $\mathscr{C}_G$. As explained above, we may assume that

$$0 \to N^U \to I^U \to S^U \to H^1(U, N) = 0$$

is a $\mathscr{C}_{G/U}$-injective presentation. Then via a long exact sequence of extension groups and group cohomology, we have the following identity:

$$\mathrm{Ext}^q_{\mathscr{C}_G}(M, N) \cong \mathrm{Ext}^{q-1}_{\mathscr{C}_G}(M, S), \quad \mathrm{Ext}^q_{\mathscr{C}_{G/U}}(M, N^U) \cong \mathrm{Ext}^{q-1}_{\mathscr{C}_{G/U}}(M, S^U)$$

and

$$H^p_{ct}(U, N) = H^{p-1}_{ct}(U, S).$$

From this and the induction assumption applied to S, the desired result follows.

Now assume $|M| = \ell^r$ for a prime ℓ. We first check the last assertion when $p = 2$. We have the following exact sequence:

$$\begin{aligned} 0 &\longrightarrow \mathrm{Ext}^1_{\mathscr{C}_{G/U}}(M, S^U) \longrightarrow \mathrm{Ext}^1_{\mathscr{C}_G}(M, S) \\ &\longrightarrow \mathrm{Hom}_{\mathscr{C}_G}(M, H^1_{ct}(U, S)) = \mathrm{Hom}_{\mathscr{C}_G}(M, H^1_{ct}(U, S)[\ell^\infty]) = 0, \end{aligned}$$

and also we have

$$\mathrm{Ext}^1_{\mathscr{C}_{G/U}}(M, S^U) \cong \mathrm{Ext}^2_{\mathscr{C}_{G/U}}(M, N^U) \text{ and } \mathrm{Ext}^1_{\mathscr{C}_G}(M, S^U) \cong \mathrm{Ext}^2_{\mathscr{C}_G}(M, N).$$

This proves the assertion when $p = 2$. Then we apply this to the initial step of the induction on p. By using the injective presentation: $N \hookrightarrow I \twoheadrightarrow S$, we can reduce the validity of the assertion to the case where $p = 2$, because the top degree vanishing of the ℓ-torsion part reduces to degree 2 and the total vanishing of degree $< p$ assures the vanishing of H^1. The details are left to the reader. □

Exercises

(1) Show the exactness of

$$0 \to H^1(G/U, M^U) \xrightarrow{inf_{G/U}} H^1_{ct}(G, M) \xrightarrow{res_{G/U}} H^0(G/U, H^1_{ct}(U, M)).$$

(2) Show that each closed subgroup of a profinite group G is an intersection of open normal subgroups containing the subgroup.

(3) Give a detailed proof of the last assertion of Theorem 4.34.

4.3.5 Applications to representation theory

Group cohomology can be used to measure obstruction of extending a representation of a subgroup to the entire group. The theory is a version of Schur's theory of projective representations [MRT] Section 11E. The material presented in this subsection will not be used until Section 5.4 in Chapter 5.

Let G be a profinite group with a normal open subgroup H of finite index. We put $\Delta = G/H$. Fix a complete noetherian local $\mathbb{Z}_p$-algebra $\mathcal{O}$ with residue field $\mathbb{F}$. Any algebra A in this section will be assumed to be an object of $CL_{\mathcal{O}}$. For each continuous representation $\rho : H \to GL_n(A)$ and $\sigma \in G$, we define $\rho^\sigma(g) = \rho(\sigma g \sigma^{-1})$.

We take a representation $\pi : H \to GL_n(A)$ for an artinian local $\mathcal{O}$-algebra A with residue field $\mathbb{F}$. We assume the following condition:

$$\overline{\rho} = \pi \mod \mathfrak{m}_A \text{ is absolutely irreducible.} \qquad (\mathrm{AI}_H)$$

For the moment, we assume another condition:

$$\pi = c(\sigma)^{-1}\pi^\sigma c(\sigma) \text{ with some } c(\sigma) \in GL_n(A) \text{ for each } \sigma \in G. \qquad (\mathrm{C})$$

If we find another $c'(\sigma) \in GL_n(A)$ satisfying $\pi = c'(\sigma)^{-1}\pi^\sigma c'(\sigma)$, we have

$$\pi = c'(\sigma)^{-1}c(\sigma)\pi c(\sigma)^{-1}c'(\sigma),$$

and hence by Lemma 2.12 in Chapter 2, $c(\sigma)^{-1}c'(\sigma)$ is a scalar. In particular, for $\sigma, \tau \in G$,

$$c(\sigma\tau)^{-1}\pi^{\sigma\tau}c(\sigma\tau) = \pi = c(\tau)^{-1}\pi^\tau c(\tau) = c(\tau)^{-1}c(\sigma)^{-1}\pi^{\sigma\tau}c(\sigma)c(\tau),$$

and hence, $b(\sigma,\tau) = c(\sigma)c(\tau)c(\sigma\tau)^{-1} \in A^\times$. Thus $c(\sigma)c(\tau) = b(\sigma,\tau)c(\sigma\tau)$. This shows by the associativity of the matrix multiplication that

$$(c(\sigma)c(\tau))c(\rho) = b(\sigma,\tau)c(\sigma\tau)c(\rho) = b(\sigma,\tau)b(\sigma\tau,\rho)c(\sigma\tau\rho)$$

and

$$c(\sigma)(c(\tau)c(\rho)) = c(\sigma)b(\tau,\rho)c(\tau\rho) = b(\tau,\rho)b(\sigma,\tau\rho)c(\sigma\tau\rho),$$

and hence $b(\sigma,\tau)$ is a 2-cocycle of G. If $h \in H$, then

$$\begin{aligned}\pi(g) &= c(h\tau)^{-1}\pi(h\tau g\tau^{-1}h^{-1})c(h\tau)\\ &= c(h\tau)^{-1}\pi(h)c(\tau)\pi(g)c(\tau)^{-1}\pi(h)^{-1}c(h\tau).\end{aligned}$$

Thus $c(h\tau)^{-1}\pi(h)c(\tau) \in A^\times$.

If we let $h \in G$ act on the space $C(G; M_n(A))$ of continuous functions $f : G \to M_n(A)$ by $f|h(g) = \pi(h)^{-1}f(hg)$, then c is an eigenfunction belonging to a character $\xi : H \to A^\times$. Now we take a function $\eta : G \to A^\times$ such that $\eta(h\tau) = \xi^{-1}(h)\eta(\tau)$ for all $h \in H$. For example, writing $G = \bigsqcup_{\tau\in R} H\tau$ (disjoint), we may define $\eta(h\tau) = \xi^{-1}(h)$. We replace c by ηc. Then c satisfies $c(h\tau) = \pi(h)c(\tau)$ for all $h \in H$ and $\tau \in R$. Since $c(hh'\tau) = \pi(hh')c(\tau) = \pi(h)c(h'\tau)$, in fact c satisfies

$$c(h\tau) = \pi(h)c(\tau) \text{ for all } h \in H \text{ and all } \tau \in G. \qquad (\pi)$$

Since $c(1)$ commutes with $\mathrm{Im}(\pi)$, $c(1)$ is scalar. Thus we may also assume that

$$c(1) = 1. \qquad \text{(id)}$$

Note that for h, $h' \in H$,

$$\begin{aligned}b(h\sigma, h'\tau) &= c(h\sigma)c(h'\tau)c(h\sigma h'\tau)^{-1}\\ &= \pi(h)c(\sigma)\pi(h')c(\tau)c(\sigma\tau)^{-1}\pi(h\sigma h'\sigma^{-1})^{-1}\\ &= \pi(h)\pi^\sigma(h')b(\sigma,\tau)\pi(h\sigma h'\sigma^{-1})^{-1} = b(\sigma,\tau).\end{aligned}$$

Thus b is a 2-cocycle factoring through Δ.

If we replace c by c', then by (C), $c'(\sigma) = c(\sigma)\zeta(\sigma)$ for $\zeta(\sigma) \in A^\times$. Thus we see from $c(\sigma)c(\tau) = b(\sigma,\tau)c(\sigma\tau)$ that $c'(\sigma)c'(\tau) = b(\sigma,\tau)\zeta(\sigma)\zeta(\tau)c'(\sigma\tau)\zeta(\sigma\tau)^{-1}$. Thus the 2-cocycle b' made out of c' is cohomologous to b, and the cohomology class $[b] = [\pi] \in H^2(\Delta, A^\times)$ is uniquely determined by π.

If $b(\sigma,\tau) = \zeta(\sigma)\zeta(\tau)\zeta(\sigma\tau)^{-1}$ is further a coboundary of $\zeta : \Delta \to A^\times$, we modify c by $\zeta^{-1}c$. Since ζ factors through Δ, this modification does not destroy the property (π). Then $c(\sigma\tau) = c(\sigma)c(\tau)$ and $c(h\tau) = \pi(h)c(\tau)$

for $h \in H$. Thus c is a representation of G and extends π to G. Let d be another extension of π. Then $\chi(\sigma) = c(\sigma)d(\sigma)^{-1} \in A^\times$ is a character of G, because χ commutes with π. Thus $c = d \otimes \chi$.

We consider another condition

$$\mathrm{Tr}(\pi) = \mathrm{Tr}(\pi^\sigma) \text{ for all } \sigma \in G. \tag{inv}$$

Under (AI_H), it has been proved by Carayol and Serre (Proposition 2.13) that (inv) is actually equivalent to (C). Thus we have

Theorem 4.35 *Let $\pi : H \to GL_n(A)$ be a continuous representation for a p-adic artinian local ring A. Assume (AI_H) and (inv).*

(1) *We can choose c satisfying (π);*
(2) *Choosing c as above, $b(\sigma,\tau) = c(\sigma)c(\tau)c(\sigma\tau)^{-1}$ is a 2-cocycle of Δ with values in $A^\times$;*
(3) *The cohomology class $[b] = [\pi]$ (called the obstruction class of π) of the above b only depends on π and not on the choice of c, etc. There exists a continuous representation π_E of G into $GL_n(A)$ extending π if and only if $[\pi] = 0$ in $H^2(\Delta, A^\times)$;*
(4) *All other extensions of π to G are of the form $\pi_E \otimes \chi$ for a character χ of Δ with values in $A^\times$;*
(5) *If $H^2(\Delta, A^\times) = 0$, then any representation π satisfying (AI_H) and (inv) can be extended to G.*

Corollary 4.36 *If Δ is a p-group, then any representation π with values in $GL_n(\mathbb{F})$ for a finite field $\mathbb{F}$ of characteristic p satisfying* (AI_H) *and* (inv) *can be extended to G.*

Proof This follows from the fact that $|\mathbb{F}^\times|$ is prime to p. Hence $H^2(\Delta, \mathbb{F}^\times) = 0$. □

When Δ is cyclic, $H^2(\Delta, A^\times) \cong A^\times/(A^\times)^d$ for $d = |\Delta|$. If for a generator σ of G, $\xi = c(\sigma^d)\pi(\sigma^d)^{-1} \in (A^\times)^d$, then b is a coboundary of $\zeta(\sigma^j) = \xi^{j/d}$. By extending the scalar to $B = A[X]/(X^d - \xi)$, in $H^2(G, B^\times)$, the class of b vanishes. Thus we have

Corollary 4.37 *Suppose (AI_H) and (inv). If Δ is a cyclic group of order d, then π can be extended to a representation of G into $GL_n(B)$ for a local A-algebra B which is A-free of rank at most $d = |\Delta|$.*

Let $\overline{\rho} = \pi \mod \mathfrak{m}_A$. We suppose that $\overline{\rho}$ can be extended to G. Then we may assume that the cohomology class of $b(\sigma,\tau) \mod \mathfrak{m}_A$ vanishes in $H^2(G,\mathbb{F}^\times)$. Thus we can find $\zeta : G \to A^\times$ such that

$$a(\sigma,\tau) = b(\sigma,\tau)\zeta(\sigma)\zeta(\tau)\zeta(\sigma\tau)^{-1} \mod \mathfrak{m}_A \equiv 1.$$

Then a has values in $\widehat{\mathbf{G}}_m(A) = 1 + \mathfrak{m}_A$. In particular, if the Sylow p-subgroup S of G is cyclic, we have $H^2(S,\widehat{\mathbf{G}}_m(A)) \cong \widehat{\mathbf{G}}_m(A)/\widehat{\mathbf{G}}_m(A)^{|S|}$. Write ξ for the element in $\widehat{\mathbf{G}}_m(A)$ corresponding to a. Then for $B = A[X]/(X^{|S|} - \xi)$, the cohomology class of a vanishes in $H^2(S,\widehat{\mathbf{G}}_m(B))$. This implies that in $H^2(S,B^\times)$, the cohomology class of b vanishes.

COROLLARY 4.38 *Suppose* (AI_H) *and* (*inv*). *Suppose* Δ *has a cyclic Sylow* p*-subgroup of order* q. *If* $\overline{\rho}$ *can be extended to* G, *then* π *can be extended to a representation of* G *into* $GL_n(B)$ *for a local* A*-algebra* B *which is* A*-free of rank at most* q.

We now prove the following fact:

(AI) When Δ is cyclic of odd order and $n = 2$, the condition (AI_H) is equivalent to (AI_G).

Proof Let ρ be an absolutely irreducible representation of G into $GL_2(K)$ for a field K. We assume that Δ is cyclic of odd order. We prove that ρ cannot contain a character of H as a representation of H, which shows the equivalence, since ρ is two-dimensional. Suppose by absurdity that ρ restricted to H contains a character χ. Let $H' = \{g \in G | \chi(ghg^{-1}) = \chi\}$. Then χ can be extended to a character of H' (Corollary 4.37). We pick one extension $\widetilde{\chi} : H' \to B^\times$ for a finite flat extension B/A in CL. Let $\rho' = \rho|_{H'}$. By Shapiro's lemma (4.20), we have

$$\begin{aligned} \mathrm{Hom}_{\mathbb{Z}[H']}(\rho', \mathrm{Ind}_H^{H'} \chi) &= H^0_{ct}(H', \mathrm{Hom}_{\mathbb{Z}}(\rho', \mathrm{Ind}_H^{H'} \chi)) \\ &\cong H^0_{ct}(H, \mathrm{Hom}_{\mathbb{Z}}(\rho'|_H, \chi)) = \mathrm{Hom}_{\mathbb{Z}[H]}(\rho'|_H, \chi), \end{aligned} \quad (4.27)$$

where, by definition, $\mathrm{Ind}_G^H M = \mathrm{Hom}_{\mathbb{Z}[G]}(\mathbb{Z}[H], M)$ and we let $g \in H'$ act on $\phi \in \mathrm{Hom}_{\mathbb{Z}}(M,N)$ by $(g\phi)(x) = g(\phi(g^{-1}x))$ for two H'-modules M and N. If $\rho' = \rho|_{H'}$ remains irreducible, this shows that $\rho' \subset \mathrm{Ind}_H^{H'} \chi$. It is easy to check from the definition that

$$\mathrm{Ind}_H^{H'} \chi \cong \oplus_\xi \widetilde{\chi}\xi,$$

ξ running through all the characters of the cyclic group H'/H. Thus ρ' cannot be irreducible, and we may assume that $H = H'$. Then conjugates of χ under Δ are all distinct. Since, by Shapiro's lemma again, $\rho \subset \mathrm{Ind}_H^G \chi$ and $\rho \cong \rho^\sigma \subset \mathrm{Ind}_{H'}^G \chi'^\sigma$. Therefore $\rho|_{H'}$ contains all conjugates of χ' with the equal multiplicity. Thus $(G : H')|2$, which is absurd because $(G : H)$ is odd. □

In fact we can generalize the above fact in the following form:

LEMMA 4.39 *Let ρ be an absolutely irreducible representation of G into the group $GL_n(K)$ for a field K. Suppose that H is a closed normal subgroup of a p-power index for a prime p. Then absolute irreducibility of ρ over H is equivalent to that over G if $p \nmid n$.*

For the proof, see [H99b].

4.4 Duality in Galois Cohomology

In this section, we describe in detail the duality theory of the Galois cohomology group due to J. Tate and G. Poitou. Our exposition follows basically that of Milne [ADT] Chapter I, but in many places, it is more elementary.

4.4.1 Class formation and duality of cohomology groups

Consider a profinite group G, a discrete G-module C and a family of isomorphisms $inv_U : H^2_{ct}(U, C) \xrightarrow{\cong} \mathbb{Q}/\mathbb{Z}$ indexed by open subgroups U of G. Such a system is called a *class formation* if

(CF1) $H^1_{ct}(U, C) = 0$ for all open subgroups U of G;
(CF2) For all pairs of open subgroups $V \subset U \subset G$, the diagram

$$\begin{array}{ccc} H^2_{ct}(U,C) & \xrightarrow{res_{U/V}} & H^2_{ct}(V,C) \\ {\scriptstyle inv_U}\downarrow & & \downarrow{\scriptstyle inv_V} \\ \mathbb{Q}/\mathbb{Z} & \xrightarrow{x \mapsto nx} & \mathbb{Q}/\mathbb{Z} \end{array}$$

commutes, where $n = (U : V)$.

The map inv_U is called the 'invariant' with respect to U.

THEOREM 4.40 *Let (G, C) be a class formation and $G^{ab} = \varprojlim_{U \in \mathcal{U}} (G/U)^{ab}$ be the maximal continuous abelian quotient of G. Then there exists a canonical map $rec_G : C^G \to G^{ab}$ (called the reciprocity map), whose image is dense in G^{ab} and whose kernel is the intersection $\bigcap_{U \in \mathcal{U}} N_{G/U} C^U$, where $N_{G/U} : C^U \to C^G$ is the map given by $N_{G/U}(x) = \sum_{g \in G/U} gx$.*

Proof Let $V \subset U$ with $n = (U : V)$ be two members of $\mathscr{U}$. We have the following commutative diagram:

$$\begin{array}{ccccccc}
0 \to H^2_{ct}(U/V, C^V) & \longrightarrow & H^2_{ct}(U, C) & \xrightarrow{res_{U/V}} & H^2_{ct}(V, C) & & \\
\Big\downarrow inv_{U/V} & & \wr\Big\downarrow inv_U & & \wr\Big\downarrow inv_V & & \\
0 \to \frac{1}{n}\mathbb{Z}/\mathbb{Z} & \longrightarrow & \mathbb{Q}/\mathbb{Z} & \xrightarrow[n]{} & \mathbb{Q}/\mathbb{Z} & \longrightarrow & 0.
\end{array}$$

Since the last two vertical arrows are isomorphisms, the first one is an isomorphism:

$$inv_{U/V} : H^2(U/V, C^V) \cong \frac{1}{(U:V)}\mathbb{Z}/\mathbb{Z}. \tag{4.28}$$

We have the inflation–restriction exact sequence:

$$0 \to H^1(U/V, C^V) \to H^1_{ct}(U, C) \to H^1_{ct}(V, C),$$

which tells us that $H^1(U/V, C^V) = 0$. Then applying Theorem 4.19 and Theorem 4.24 to G/U for $k = -2$, we get the following commutative diagram:

$$\begin{array}{ccc}
H_T^{-2}(G/U, \mathbb{Z}) & \xrightarrow{\simeq} & H_T^0(G/U, C^U) \\
\wr\Big\downarrow & & \wr\Big\downarrow \\
(G/U)^{ab} & \xrightarrow[rec_{G/U}^{-1}]{} & C^G/N_{G/U}C^U.
\end{array}$$

Taking the projective limit with respect to $U \in \mathscr{U}$, we get an injective map

$$rec_G : C^G/\bigcap_{U\in\mathscr{U}} N_{G/U}(C^U) \hookrightarrow \varprojlim_U C^G/N_{G/U}(C^U) \cong \varprojlim_U (G/U)^{ab} = G^{ab}$$

as desired. □

Let M be a (discrete) G-module which is of finite type as a $\mathbb{Z}$-module. Then by Proposition 4.30, we have a natural map

$$H^r_{ct}(G, \mathrm{Hom}_{\mathbb{Z}}(M, C)) \longrightarrow \mathrm{Ext}^r_{\mathscr{C}}(M, C).$$

This is an isomorphism if M is $\mathbb{Z}$-torsion-free or C is divisible. We also have the cup product pairing:

$$\begin{array}{ccccc}
H^r_{ct}(G, \mathrm{Hom}_{\mathbb{Z}}(M, C)) & \times & H^{2-r}_{ct}(G, M) & \longrightarrow & H^2_{ct}(G, C) \cong \mathbb{Q}/\mathbb{Z} \\
\downarrow & & \wr\downarrow & & \Vert \\
\mathrm{Ext}^r_{\mathscr{C}}(M, C) & \times & \mathrm{Ext}^{2-r}_{\mathscr{C}}(\mathbb{Z}, M) & \longrightarrow & \mathrm{Ext}^2_{\mathscr{C}}(\mathbb{Z}, C),
\end{array}$$

where we agree to define the negative degree cohomology groups and the negative degree extension groups as zero. For any discrete module M, we put $M^* = \mathrm{Hom}_{\mathbb{Z}}(M, \mathbb{Q}/\mathbb{Z})$, which is called the Pontryagin dual module. We have $(M^*)^* \cong M$ canonically. When M is a finite module, $M^* \cong M$ as abelian groups (but may not be as G-modules). By the above pairing, we have the following canonical morphisms:

$$\alpha^r_G(M) : \mathrm{Ext}^r_{\mathscr{C}}(M, C) \to H^{2-r}_{ct}(G, M)^*. \tag{4.29}$$

Here negative degree cohomology groups and extension groups are defined as 0. It is known that

$$i : H^2_{ct}(G, \mathbb{Z}) \cong \mathrm{Hom}_{ct}(G, \mathbb{Q}/\mathbb{Z}) \tag{4.30}$$

in the following way: the cup product pairing associated with $C \otimes_{\mathbb{Z}} \mathbb{Z} \cong C$:

$$\langle\ ,\ \rangle : H^0_{ct}(G, C) \times H^2_{ct}(G, \mathbb{Z}) \to H^2_{ct}(G, C) = \mathbb{Q}/\mathbb{Z}$$

is given by $\langle c, x\rangle = i(x)(rec_G(c))$ (see [CLC] XI.3). Then, in particular, $\alpha^0_G(\mathbb{Z})$ is the reciprocity map rec_G.

THEOREM 4.41 (J. TATE) *Let (G, C) be a class formation and M be a discrete $\mathbb{Z}[G]$-module of finite type. Then we have:*

(1) *The map $\alpha^r_G(M)$ is bijective for all $r \geq 2$, $\alpha^1_G(M)$ is bijective if M is torsion-free, and $\mathrm{Ext}^r_{\mathscr{C}}(M, C) = 0$ if $r \geq 3$;*
(2) *The map $\alpha^1_G(M)$ is bijective if $\alpha^1_U(\mathbb{Z}/m\mathbb{Z})$ is bijective for all $U \in \mathscr{U}$ and all $m \in \mathbb{Z}$ (including $m = 0$);*
(3) *The map $\alpha^0_G(M)$ is bijective for all finite M if $\alpha^0_U(\mathbb{Z}/m\mathbb{Z})$ is bijective for all $U \in \mathscr{U}$ and $m \in \mathbb{Z}$ (including $m = 0$).*

Proof We first claim

$$\mathrm{Ext}^r_{\mathscr{C}}(M, C) = 0 \text{ for } r \geq 4, \text{ and if } M \text{ is torsion-free } \mathrm{Ext}^3_{\mathscr{C}}(M, C) = 0. \tag{4.31}$$

We take a generator $u_1, \ldots, u_r$ of M over $\mathbb{Z}[G]$. Since $M \in \mathscr{C}$, $M = \bigcup_{U \in \mathscr{U}} M^U$ and hence, we find a small open normal subgroup U such that $u_j \in M^U$ for all j, that is, M is a G/U-module. Then we consider the $\mathbb{Z}[G]$-linear map $\pi : P = \mathbb{Z}[G/U]^r \twoheadrightarrow M$ given by $(x_1, \ldots, x_r) \mapsto \sum_{j=1}^r x_j u_j$. Since G/U is a finite group, P is a $\mathbb{Z}$-module of finite rank. Thus M is a $\mathbb{Z}$-module of finite type, and $Q = \mathrm{Ker}(\pi)$ is a $\mathbb{Z}$-free module of finite type. If we know the claim for $\mathbb{Z}$-free $\mathbb{Z}[G]$-modules of finite type, writing down the long exact sequence attached to $Q \hookrightarrow P \twoheadrightarrow M$:

$$\mathrm{Ext}^{r-1}_{\mathscr{C}}(Q, C) \to \mathrm{Ext}^r_{\mathscr{C}}(M, C) \to \mathrm{Ext}^r_{\mathscr{C}}(P, C),$$

we see that $\mathrm{Ext}^r(M,C)=0$ if $r\geq 4$. Thus we may and will assume that M is $\mathbb{Z}$-free.

Let $\widehat{M}=\mathrm{Hom}_{\mathbb{Z}}(M,\mathbb{Z})$. Then by $\mathbb{Z}$-freeness, $\widehat{M}\otimes_{\mathbb{Z}}C\cong \mathrm{Hom}_{\mathbb{Z}}(M,C)$ via $\phi\otimes c\mapsto (m\mapsto \phi(m)c)$. Then by Proposition 4.30, we have

$$\mathrm{Ext}^r_{\mathscr{C}}(M,C)\cong H^r_{ct}(G,\mathrm{Hom}_{\mathbb{Z}}(M,C))\cong H^r_{ct}(G,\widehat{M}\otimes_{\mathbb{Z}}C).$$

We have from Corollary 4.26 that

$$H^r_{ct}(G,\widehat{M}\otimes_{\mathbb{Z}}C)=\varprojlim_{U\in\mathscr{U},\widehat{M}^U=\widehat{M}}H^r(G/U,M\otimes_{\mathbb{Z}}C^U).$$

By the remark after Theorem 4.24, we have

$$H^{r-2}(G/U,\widehat{M})\cong H^r(G/U,\widehat{M}\otimes_{\mathbb{Z}}C^U)$$

for all $r\geq 3$. Write $u_{U/V}$ for the generator of $H^2(U/V,C^V)$ with $inv_{U/V}(u_{U/V})=\frac{1}{(U:V)}$. Then we see that $inf_{U/V}(u_{G/U})=(U:V)u_{G/V}$ by (CF2). Since the cup product commutes with the inflation map (by definition) and the horizontal map of the following diagram sends $x\in H^{r-2}(G/U,\widehat{M})$ to the cup product $x\cup u_{G/U}$, we can check by (CF2) that the following diagram commutes:

$$\begin{array}{ccc} H^{r-2}(G/U,\widehat{M}) & \xrightarrow{\simeq} & H^r(G/U,\widehat{M}\otimes_{\mathbb{Z}}C^U) \\ {\scriptstyle (U:V)inf_{U/V}}\downarrow & & \downarrow{\scriptstyle inf_{U/V}} \\ H^{r-2}(G/V,\widehat{M}) & \xrightarrow{\simeq} & H^r(G/V,\widehat{M}\otimes_{\mathbb{Z}}C^V). \end{array}$$

Since $H^r(\{1_G\},X)=0$ for all $r>0$, by Proposition 4.21 applied to $\{1_{G/U}\}\subset G/U$, we see that the index $(G:U)$ kills $H^r(G/U,X)$ for all G/U-modules X. In particular, $H^{r-2}(G/U,\widehat{M})$ is a torsion module if $r>2$. Thus $\varinjlim_{U\in\mathscr{U}}H^{r-2}(G/U,\widehat{M})$ with respect to $(U:V)inf_{U/V}$ vanishes if $r>2$. This shows the vanishing of $H^r_{ct}(G,\widehat{M}\otimes_{\mathbb{Z}}C)=\mathrm{Ext}^r_{\mathscr{C}}(M,C)$ for $r>2$ as claimed.

Now we suppose that $M=\mathbb{Z}$. Then $\alpha^2_G(\mathbb{Z})$ is the invariant map inv_G by definition, because $H^0_{ct}(G,\mathbb{Z})^*=\mathrm{Hom}_{\mathbb{Z}}(\mathbb{Z},\mathbb{Q}/\mathbb{Z})=\mathbb{Q}/\mathbb{Z}$. To study α^0, we look at the following exact sequence: $0\to\mathbb{Z}\to\mathbb{Q}\to\mathbb{Q}/\mathbb{Z}\to 0$. Since $H^q(G/U,V)$ is killed by $(G:U)$, if V is a vector space over $\mathbb{Q}$, $H^q(G/U,V)=0$. Thus by the cohomology long exact sequence, we get

$$\mathrm{Hom}_{ct}(G,\mathbb{Q}/\mathbb{Z})=H^1_{ct}(G,\mathbb{Q}/\mathbb{Z})\cong H^2_{ct}(G,\mathbb{Z}).$$

Then $\alpha^0_G(\mathbb{Z}):C^G=\mathrm{Hom}_{\mathbb{Z}[G]}(\mathbb{Z},C)\to\mathrm{Hom}_{ct}(G,\mathbb{Q}/\mathbb{Z})^*=G^{ab}$ is the reciprocity map. As for $\alpha^1_G(\mathbb{Z}):\mathrm{Ext}^1_{\mathscr{C}}(\mathbb{Z},C)=H^1_{ct}(G,C)\to H^1_{ct}(G,\mathbb{Z})^*$, by our assumption $H^1_{ct}(G,C)=0$. We need to show that $H^1_{ct}(G,\mathbb{Z})=$

$\mathrm{Hom}_{ct}(G,\mathbb{Z}) = 0$. Let $\phi : G \to \mathbb{Z}$ be a continuous homomorphism. Since G is compact, $\phi(G)$ is a compact subgroup of $\mathbb{Z}$. Since $\mathbb{Z}$ is discrete and torsion-free, it has only one compact subgroup, that is, $\{0\}$.

Next we suppose that $M = \mathbb{Z}/m\mathbb{Z}$. For any $\mathbb{Z}$-module X, we write $X[m] = \{x \in X | mx = 0\}$. Note that $H^1_{ct}(G,\mathbb{Z}) = \mathrm{Hom}_{ct}(G,\mathbb{Z}) = \{0\}$ because $\mathbb{Z}$ has only one compact subgroup $\{0\}$. We get from the long exact sequence attached to $0 \to \mathbb{Z} \xrightarrow{x \mapsto mx} \mathbb{Z} \to \mathbb{Z}/m\mathbb{Z} \to 0$ a short exact sequence:

$$0 = H^1_{ct}(G,\mathbb{Z}) \to H^1_{ct}(G,\mathbb{Z}/m\mathbb{Z}) \to H^2_{ct}(G,\mathbb{Z})[m] = \mathrm{Hom}_{\mathbb{Z}}(G^{ab},\mathbb{Z}/m\mathbb{Z}) \to 0.$$

This shows that $H^1_{ct}(G,\mathbb{Z}/m\mathbb{Z})^* = G^{ab}/mG^{ab}$. Similarly, we get from Proposition 4.15 the following exact sequence:

$$0 \to \mathrm{Hom}_{\mathscr{C}}(\mathbb{Z},C) \otimes_{\mathbb{Z}} \mathbb{Z}/m\mathbb{Z} \to \mathrm{Ext}^1(\mathbb{Z}/m\mathbb{Z},C) \to \mathrm{Ext}^1_{\mathscr{C}}(\mathbb{Z},C)[m] \to 0.$$

Note that $\mathrm{Hom}_{\mathscr{C}}(\mathbb{Z},C) \cong C^G$ and $\mathrm{Ext}^1_{\mathscr{C}}(\mathbb{Z},C) = H^1_{ct}(G,C) = 0$ by (CF1). This shows that $\mathrm{Ext}^1_{\mathscr{C}}(\mathbb{Z}/m\mathbb{Z},C) = C^G/mC^G$. Then

$$\alpha^1_G(\mathbb{Z}/m\mathbb{Z}) : C^G/mC^G \to \mathrm{Hom}_{\mathbb{Z}}(G,\mathbb{Z}/m\mathbb{Z})^* = G^{ab}/mG^{ab}$$

is induced by rec_G. Since $\mathrm{Ext}^1_{\mathscr{C}}(\mathbb{Z},C) = 0$, again by the long exact sequence:

$$\begin{aligned} 0 &= \mathrm{Ext}^1_{\mathscr{C}}(\mathbb{Z},C) \to \mathrm{Ext}^2_{\mathscr{C}}(\mathbb{Z}/m\mathbb{Z},C) \to \mathrm{Ext}^2_{\mathscr{C}}(\mathbb{Z},C)[m] \\ &= H^2_{ct}(G,C)[m] = \mathbb{Q}/\mathbb{Z}[m], \end{aligned}$$

the map $\alpha^2_G(\mathbb{Z}/m\mathbb{Z}) : \mathbb{Q}/\mathbb{Z}[m] \to H^0_{ct}(G,\mathbb{Z}/m\mathbb{Z})^* = \frac{1}{m}\mathbb{Z}/\mathbb{Z}$ is an isomorphism induced by inv_G. As we have seen, $\mathrm{Ext}^3_{\mathscr{C}}(\mathbb{Z},M) = 0$. Then by the long exact sequence, we get a short exact sequence:

$$0 \to \mathrm{Ext}^2_{\mathscr{C}}(\mathbb{Z},C) \otimes_{\mathbb{Z}} \mathbb{Z}/m\mathbb{Z} \to \mathrm{Ext}^3_{\mathscr{C}}(\mathbb{Z}/m\mathbb{Z},C) \to \mathrm{Ext}^3_{\mathscr{C}}(\mathbb{Z},C) = 0.$$

Since $\mathrm{Ext}^2_{\mathscr{C}}(\mathbb{Z},C) = H^2_{ct}(G,C) = \mathbb{Q}/\mathbb{Z}$, $\mathrm{Ext}^2_{\mathscr{C}}(\mathbb{Z},C) \otimes_{\mathbb{Z}} \mathbb{Z}/m\mathbb{Z} = 0$, and hence $\mathrm{Ext}_{\mathscr{C}}(\mathbb{Z}/m\mathbb{Z},C) = 0$. Thus the theorem is valid for any M on which G acts trivially.

Now we treat general M. We take $U \in \mathscr{U}$ such that $M^U = M$. Then M is a module over the finite group G/U. Let $\pi : \mathbb{Z}[G/U] \to \mathbb{Z}$ be the augmentation: $\pi(\sum_{g \in G/U} a_g g) = \sum_{g \in G/U} a_g$. Then $0 \to M = \mathrm{Hom}_{\mathbb{Z}}(\mathbb{Z},M) \xrightarrow{\pi^*} \mathrm{Hom}_{\mathbb{Z}}(\mathbb{Z}[G/U],M) = I$ is an injection with cokernel S. For an injective resolution $M \hookrightarrow M^\bullet$, we see that $\mathrm{Hom}_{\mathbb{Z}}(\mathbb{Z}[G/U],M^\bullet)$ is a $\mathscr{C}$-injective resolution of $\mathrm{Hom}_{\mathbb{Z}}(\mathbb{Z}[G/U],M)$ (Lemma 4.31). Then we have the following version of Shapiro's lemma (see Exercise 5 of Subsection 4.3.1):

$$\mathrm{Hom}_{\mathbb{Z}[G]}(\mathbb{Z}, \mathrm{Hom}_{\mathbb{Z}}(\mathbb{Z}[G/U], M^\bullet)) = \mathrm{Hom}_{\mathbb{Z}[U]}(\mathbb{Z}, M^\bullet).$$

This shows that

$$H^\bullet_{ct}(G, I) = H^\bullet_{ct}(U, M).$$

Since $\mathbb{Z}[G/U] \cong \mathrm{Hom}_{\mathbb{Z}}(\mathbb{Z}[G/U], \mathbb{Z})$ as $\mathbb{Z}[G]$-modules, we have $\mathrm{Hom}_{\mathbb{Z}}(\mathbb{Z}[G/U], M) \cong \mathbb{Z}[G/U] \otimes_{\mathbb{Z}} M$. Then, taking a $\mathscr{C}$-injective resolution $C \hookrightarrow C^\bullet$, we have

$$\begin{aligned} \mathrm{Hom}_{\mathbb{Z}[G]}(\mathrm{Hom}_{\mathbb{Z}}(\mathbb{Z}[G/U], M), C^\bullet) &\cong \mathrm{Hom}_{\mathbb{Z}[G]}(\mathbb{Z}[G/U] \otimes_{\mathbb{Z}} M, C^\bullet) \\ &= \mathrm{Hom}_{\mathbb{Z}[U]}(M, C^\bullet), \end{aligned} \tag{4.32}$$

which is induced by $\phi \mapsto (m \mapsto \phi(1 \otimes m))$. Write $\mathscr{C}_U$ for the category of discrete U-modules. Then the $\mathscr{C}$-injective resolution of C constructed in Subsection 4.3.3 is also a $\mathscr{C}_U$-injective resolution, and we get

$$\begin{aligned} \mathrm{Ext}^\bullet_{\mathscr{C}}(I, C) &= H^\bullet(\mathrm{Hom}_{\mathbb{Z}[G]}(\mathrm{Hom}_{\mathbb{Z}}(\mathbb{Z}[G/U], M), C^\bullet)) \\ &\cong H^\bullet(\mathrm{Hom}_{\mathbb{Z}[U]}(M, C^\bullet)) = \mathrm{Ext}^\bullet_{\mathscr{C}_U}(M, C). \end{aligned}$$

Now we get, writing $E^q_U(X)$ for $\mathrm{Ext}_{\mathscr{C}_U}(X, C)$ and $H^q_U(X)$ for $H^q_{ct}(U, X)$, the following commutative diagram:

$$\begin{array}{ccccccc} \to E^r_G(S) & \longrightarrow & E^r_U(M) & \longrightarrow & E^r_G(M) & \longrightarrow & E^{r+1}_G(S) \to \\ {\scriptstyle \alpha^r_G(S)}\downarrow & & {\scriptstyle \alpha^r_U(M)}\downarrow & & {\scriptstyle \alpha^r_G(M)}\downarrow & & \downarrow{\scriptstyle \alpha^{r+1}_G(S)} \\ \to H^{2-r}_G(S) & \longrightarrow & H^{2-r}_U(M) & \longrightarrow & H^{2-r}_G(M) & \longrightarrow & H^{1-r}_G(S) \to . \end{array}$$

We first apply this to $r = 3$ and we get from (4.31):

$$\begin{array}{ccccc} E^3_U(M) & \longrightarrow & E^3_G(M) & \longrightarrow & E^4_G(S) \\ \wr\downarrow{\scriptstyle \alpha^3_U(M)} & & {\scriptstyle \alpha^3_G(M)}\downarrow & & \wr\downarrow{\scriptstyle \alpha^4_U(S)} \\ H^{-1}_U(M) = 0 & \longrightarrow & H^{-1}_G(M) = 0 & \longrightarrow & H^{-2}_G(S) = 0. \end{array}$$

This shows that $\alpha^3_G(M)$ is the zero map as desired. Now we apply the diagram to $r = 2$.

$$\begin{array}{ccccccccc} \to E^2_U(M) & \longrightarrow & E^2_G(M) & \longrightarrow & E^3_G(S) & \longrightarrow & E^3_U(M) & \longrightarrow & 0 \\ \wr\downarrow{\scriptstyle \alpha^2_U(M)} & & {\scriptstyle \alpha^2_G(M)}\downarrow & & \wr\downarrow{\scriptstyle \alpha^3_G(S)} & & \wr\downarrow{\scriptstyle \alpha^3_U(M)} & & \\ \to H^0_U(M) & \longrightarrow & H^0_G(M) & \longrightarrow & H^{-1}_G(S) = 0 & \longrightarrow & H^{-1}_U(M) = 0 & \longrightarrow & 0. \end{array}$$

Then by the five lemma (see [BCM] Exercise I.1.4.b), $\alpha^2_G(M)$ is surjective. Applying this to S in place of M, we see that $\alpha^2_G(S)$ is surjective, getting the following commutative diagram:

$$\begin{array}{ccccccccc} \to E^2_G(S) & \longrightarrow & E^2_U(M) & \longrightarrow & E^2_G(M) & \longrightarrow & E^3_G(S) & \longrightarrow & 0 \\ \downarrow{\scriptstyle \alpha^2_G(S)} & & \wr\downarrow{\scriptstyle \alpha^2_U(M)} & & \downarrow{\scriptstyle \alpha^2_G(M)} & & \wr\downarrow{\scriptstyle \alpha^3_G(S)} & & \\ \to H^0_G(S) & \longrightarrow & H^0_U(M) & \longrightarrow & H^0_G(M) & \longrightarrow & H^{-1}_G(S) = 0. & & \end{array}$$

Again by the five lemma, $\alpha^2_G(M)$ is an isomorphism. If M is torsion-free, we see that I and S are both torsion-free. Then the same argument for $r = 1$ shows that $\alpha^1_G(M)$ is an isomorphism. This shows (1). As for (2), under the assumption, we do not need to assume that M is torsion-free for $\alpha^1_G(M)$ to be an isomorphism. The same reasoning is valid for $\alpha^0_G(M)$ under the assumption of (3). □

EXAMPLE 4.42 Let K be a finite extension of $\mathbb{Q}_p$. We fix an algebraic closure $\overline{K}$ of K and put $G = \mathrm{Gal}(\overline{K}/K)$. Then G is a profinite group. Let $C = \overline{K}^\times$. Then by the local class field theory and Hilbert's theorem 90 (Exercises 3–4), it is known that (G, C) is a class formation. Since C is divisible (that is, for any $a \in \overline{K}^\times$ and $0 < n \in \mathbb{Z}$ $\sqrt[n]{a} \in \overline{K}^\times$), by Proposition 4.30,

$$\mathrm{Ext}^r_{\mathscr{C}}(M, C) \cong H^r_{ct}(G, Hom(M, C)).$$

Thus the cup product pairing:

$$H^r_{ct}(G, Hom(M, C)) \times H^{2-r}_{ct}(G, M) \to H^2_{ct}(G, C) \cong \mathbb{Q}/\mathbb{Z}$$

gives the duality. In this case, the exact statement of the duality is as follows:

THEOREM 4.43 (J. TATE) *Let M be a finitely generated discrete $\mathbb{Z}[G]$-module. Then $\alpha^r_G(M) : H^r_{ct}(G, Hom(M, \overline{K}^\times)) \to H^{2-r}_{ct}(G, M)^*$ is an isomorphism for all $r \geq 1$. If M is finite, then $\alpha^0_G(M) : \mathrm{Hom}_{\mathbb{Z}[G]}(M, \overline{K}^\times) \cong H^2_{ct}(G, M)^*$. If M is finite, all cohomology groups introduced above are finite and vanish except for the degrees $0, 1, 2$.*

Proof Let R be the p-adic integer ring of K. Let I be the inertia subgroup of G. Let I_{ab} be the image of I in G^{ab}. By the local class field theory, we have the following commutative diagram with exact rows:

$$\begin{array}{ccccccc} 1 \to R^\times & \longrightarrow & K^\times & \xrightarrow{v} & \mathbb{Z} & \longrightarrow & 0 \\ \wr\downarrow & & \downarrow rec_G & & \downarrow\cap & & \\ 1 \to I_{ab} & \longrightarrow & G^{ab} & \longrightarrow & \widehat{\mathbb{Z}} & \longrightarrow & 0. \end{array}$$

The quotient $G^{ab}/I_{ab} = \mathrm{Gal}(\overline{\mathbb{F}}_p/k)$ for the residue field k of K, which is generated by the Frobenius automorphism and is isomorphic to $\widehat{\mathbb{Z}}$. We already have $\alpha_G^0(\mathbb{Z}/m\mathbb{Z}) : \mu_m(K) \to G^{ab}[m]$ and $\alpha_G^1(\mathbb{Z}/m\mathbb{Z}) : K^\times/(K^\times)^m \cong G^{ab}/(G^{ab})^m$ for all positive integers m, where $\mu_m(K)$ is the subgroup of the m-th roots of unity in K. Since $\widehat{\mathbb{Z}}$ is $\mathbb{Z}$-flat, after tensoring $\mathbb{Z}/m\mathbb{Z}$ over $\mathbb{Z}$, the two rows are still exact. This combined with the snake lemma shows that $\alpha_G^1(\mathbb{Z}/m\mathbb{Z})$ is an isomorphism. Similarly, by applying $\mathrm{Hom}_{\mathbb{Z}}(\mathbb{Z}/m\mathbb{Z}, *)$ to the above diagram, we see that $\alpha_G^0(\mathbb{Z}/m\mathbb{Z})$ is an isomorphism. Thus from Theorem 4.41, we know that $\alpha_G^j(M)$ is an isomorphism as described above. When M is finite, as a result of what we have proved, $H_{ct}^r(G, M) = 0$ if $r > 2$. By the duality we proved, we only need to prove the finiteness of H_{ct}^0 and H_{ct}^1. The finiteness of H_{ct}^0 is obvious. To see the finiteness for H_{ct}^1, we use the following exact sequence:

$$1 \to \mu_m \to \overline{K}^\times \xrightarrow{x \mapsto x^m} \overline{K}^\times \to 1.$$

From this we get the following exact sequence:

$$1 \to \mu_m(K) \to K^\times \xrightarrow{x \mapsto x^m} K^\times \to H_{ct}^1(G, \mu_m) \to H_{ct}^1(G, \overline{K}^\times) = 0.$$

The vanishing of $H_{ct}^1(G, \overline{K}^\times)$ follows from the Hilbert theorem 90 (Exercises 3–4). This shows that

$$H_{ct}^1(G, \mu_m) \cong K^\times/(K^\times)^m,$$

which is finite. In general, we pick m so that $mM = 0$. Then we take a finite Galois extension L of K such that $\mu_m(L) = \mu_m$ and $M^U = M$ for $U = \mathrm{Gal}(\overline{K}/L)$. Then M is a finite product of copies of μ_n for $n|m$ as U-modules, and hence $H_{ct}^1(U, M)$ is finite. By the inflation-restriction sequence, we get an exact sequence:

$$0 \to H^1(G/U, M) \to H_{ct}^1(G, M) \to H_{ct}^1(U, M).$$

Since $H^1(G/U, M)$ is finite (Corollary 4.17), $H_{ct}^1(G, M)$ is finite. □

Example 4.44 We have a result analogous to Theorem 4.43 for archimedean local fields: Since $\mathrm{Gal}(\mathbb{C}/\mathbb{R}) \cong \mathbb{Z}/2\mathbb{Z}$, we can directly compute

$H^1(\mathrm{Gal}(\mathbb{C}/\mathbb{R}), M)$ for finite modules M over $\mathrm{Gal}(\mathbb{C}/\mathbb{R})$, and we leave the reader to prove the following theorem (Exercise 2):

THEOREM 4.45 *Let $G = \mathrm{Gal}(\mathbb{C}/\mathbb{R})$, and let M be a finite G-module. Then we have*

(1) *The cup product defines a perfect Pontryagin pairing:*

$$H_T^r(G, \mathrm{Hom}_{\mathbb{Z}}(M, \mathbb{C}^\times)) \times H_T^{2-r}(G, M) \cong \frac{1}{2}\mathbb{Z}/\mathbb{Z}$$

for all $r \in \mathbb{Z}$;

(2) *Let $K = \mathbb{R}$ or $\mathbb{C}$, and put $H = \mathrm{Gal}(\mathbb{C}/K)$. When $K = \mathbb{R}$, we put $|M|_K = |M|$ (the order of M) and when $K = \mathbb{C}$, we write $|M|_K$ for $|M|^2$. Then we have*

$$\frac{|H^0(H, M)| \cdot |H^0(H, \mathrm{Hom}_{\mathbb{Z}}(M, \mathbb{C}^\times))|}{|H^1(H, M)|} = |M|_K.$$

Exercises

(1) Prove (4.32).
(2) Prove Theorem 4.45.
(3) Prove for a finite Galois extension K/F, $H^1(\mathrm{Gal}(K/F), K^\times) = 0$, where $K^\times$ is the multiplicative group of the field K (Hilbert's theorem 90).
(4) Prove for the separable algebraic closure K_s of K,

$$H_{ct}^1(\mathrm{Gal}(K_s/K), K_s^\times) = 0.$$

(5) Prove for a finite Galois extension K/F, $H^q(\mathrm{Gal}(K/F), K) = 0$ for all $q > 0$, where K is the additive group of the field K. Hint: Show first the existence of a normal base of K/F.

4.4.2 Global duality theorems

We now look into the global case. Let $K/\mathbb{Q}$ be a finite extension. If we put $G = \mathrm{Gal}(\overline{\mathbb{Q}}/K)$ and $C = C_K = \varinjlim_{L/K} L_{\mathbb{A}}^\times/L^\times$, then by global class field theory, (G, C) is a class formation. However, the modules C and G are too big to compute cohomology. Thus in this case, we need to restrict ramification to a finite set to get a reasonable theory. This can be done as follows: Let S be a finite set of places of $\mathbb{Q}$ including the archimedean place and Σ be the set of places of K above S. Let K^S/K be the maximal algebraic extension unramified outside S. We write $\mathfrak{G} = \mathfrak{G}_S = \mathrm{Gal}(K^S/K)$. We write $O^S = O_K^S$ for the ring of S-integers; in other words, for each prime ideal $\mathfrak{p} \in \Sigma$, we take a power $\mathfrak{p}^h$ for the class number h of K and write its generator as $\omega_{\mathfrak{p}}$: $\mathfrak{p}^h = (\omega_{\mathfrak{p}})$. Then

$$O_K^S = O_K[\frac{1}{\omega_{\mathfrak{p}}}]_{\mathfrak{p}\in\Sigma}.$$

We can rewrite this ring as

$$O_K^S = K \cap \bigcap_{\mathfrak{p}\notin\Sigma} O_{K,\mathfrak{p}} = \Big\{\alpha \in O_K \Big| \tag{4.33}$$

$$(\alpha) = \frac{\mathfrak{a}}{\mathfrak{b}} \text{ with } \mathfrak{a} \subset O_K \text{ and } \mathfrak{b} \text{ is a product of prime ideals in } \Sigma\Big\},$$

where O_K is the integer ring of K and $O_{K,\mathfrak{p}}$ is the $\mathfrak{p}$-adic completion of O_K.

We ease slightly the definition of class formation as follows: Consider a profinite group G, a discrete G-module C and a family of injections $inv_U : H^2_{ct}(U, C) \hookrightarrow \mathbb{Q}/\mathbb{Z}$ indexed by open subgroups U of G. Such a system is called an S-*class formation* if the following three conditions are satisfied:

(cf1) $H^1_{ct}(U, C) = 0$ for all open subgroups U of G;

(cf2) For all pairs of open subgroups $V \subset U \subset G$, the diagram

$$\begin{array}{ccc} H^2_{ct}(U,C) & \xrightarrow{res_{U/V}} & H^2_{ct}(V,C) \\ {\scriptstyle inv_U}\downarrow & & \downarrow{\scriptstyle inv_V} \\ \mathbb{Q}/\mathbb{Z} & \xrightarrow{x\mapsto nx} & \mathbb{Q}/\mathbb{Z} \end{array}$$

commutes, where $n = (U : V)$.

(cf3) For all pairs of open subgroups $V \subset U \subset G$ with V normal in U, the induced map: $inv_{U/V} : H^2_{ct}(U/V, C^V) \to \frac{1}{n}\mathbb{Z}/\mathbb{Z}$ is a surjective isomorphism for $n = (U : V)$.

Simply restrict ourselves to the ℓ-primary parts of the proof of Theorem 4.41, we immediately get

THEOREM 4.46 *Let (G, C) be an S-class formation and M be a discrete $\mathbb{Z}[G]$-module of finite type. Let $\ell \in S$. Then we have*

(1) *The map $\alpha^r_G(M)[\ell] : \mathrm{Ext}^r_{\mathscr{C}}(M, C)[\ell^\infty] \to H^{2-r}_{ct}(G, M)^*[\ell^\infty]$ is bijective for all $r \geq 2$, $\alpha^1_G(M)[\ell]$ is bijective if M is torsion-free, and $\mathrm{Ext}^r_{\mathscr{C}}(M, C) = 0$ if $r \geq 3$;*

(2) *The map* $\alpha^1_G(M)[\ell]$ *is bijective if* $\alpha^1_U(\mathbb{Z}/\ell^m\mathbb{Z})$ *is bijective for all* $U \in \mathcal{U}$ *and all* $m \in \mathbb{Z}$;

(3) *The map* $\alpha^0_G(M)$ *is bijective for all finite* ℓ*-primary* M *if* $\alpha^0_U(\mathbb{Z}/\ell^m\mathbb{Z})$ *is bijective for all* $U \in \mathcal{U}$ *and* $m \in \mathbb{Z}$.

We write for any finite extension F/K inside K^S

$$F^\times_S = \prod_{\ell \in S} F^\times_\ell = \{a \in F^\times_{\mathbb{A}} | a_\ell = 1 \ \forall \ell \notin S\},$$

where $F_{\mathbb{A}} = F \otimes_{\mathbb{Q}} \mathbb{A}$ and $F_\ell = F \otimes_{\mathbb{Q}} \mathbb{Q}_\ell$. We also put $E_{F,S} = (O^S_F)^\times$, $C_{F,S} = F^\times_S/E_{F,S}$ and $C_S = \varinjlim_F C_{F,S}$, where F runs over all finite extensions of K inside K^S. Then naturally C_S is a discrete $\mathfrak{G}$-module. We want to prove

THEOREM 4.47 (TATE) *The couple* $(\mathfrak{G}_S, C_S)$ *forms an S-class formation, and* $C^{\mathfrak{G}}_S \cong K^\times_{\mathbb{A}}/U_{K,S}K^\times$, *where* $U_{K,S} = \prod_{\ell \notin S} O^\times_{K,\ell}$ *for* $O_{K,\ell} = O_K \otimes_{\mathbb{Z}} \mathbb{Z}_\ell$.

If $\ell \in S$, K^S contains $K(\mu_{\ell^\infty})$, and hence its maximal ℓ-profinite quotient is an infinite group. This shows, by (cf2–3), that inv_U induces a surjective isomorphism of ℓ-primary parts for $\ell \in S$: $H^2_{ct}(U, C) \cong \mathbb{Q}_\ell/\mathbb{Z}_\ell$ for all open subgroups $U \subset \mathfrak{G}$.

Proof We follow the treatment of Milne in [ADT] I.4. We know from class field theory ([T1] Section 11) that (G, C) is a class formation for $G = \mathrm{Gal}(\overline{\mathbb{Q}}/K)$ and $C = C_K = \varinjlim_L L^\times_{\mathbb{A}}/L^\times$, where L runs over all finite extensions of K inside $\overline{\mathbb{Q}}$. Let $\mathfrak{H} = \mathrm{Gal}(\overline{\mathbb{Q}}/K^S)$ and $\pi : G \to \mathfrak{G}$ be the projection (thus $\mathfrak{H} = \mathrm{Ker}(\pi)$). By the inflation-restriction sequence (Theorem 4.33) with respect to $\mathfrak{H}$ and $\mathfrak{G}$, we verify easily (cf1,2,3) for $(\mathfrak{G}, C_S)$ if we can prove $H^r(\mathfrak{G}, C_S) \cong H^r(\mathfrak{G}, C^{\mathfrak{H}})$ for all $r > 0$.

We now prove $H^r(\mathfrak{G}, C_S) \cong H^r(\mathfrak{G}, C^{\mathfrak{H}})$ for all $r > 0$. Let $Cl(O^S_F)$ be the ideal class group of the Dedekind domain O^F_S. We note the following (natural) exact sequence (Exercise 1):

$$1 \to C_{F,S} \to F^\times_{\mathbb{A}}/U_{F,S}F^\times \to Cl(O^S_F) \to 1. \tag{4.34}$$

By the principal ideal theorem ([CFT] XIII.4), we have

$$\varinjlim_F Cl(O^S_F) = \{1\}$$

and hence

$$\varinjlim_F C_{F,S} \cong \varinjlim_F F^\times_{\mathbb{A}}/U_{F,S}F^\times, \tag{$*$}$$

where F runs through finite extensions of K inside K^S. Let $U_S =$

$\varinjlim_F U_{F,S}$. Then by the Hilbert theorem 90 (Exercise 2–3 of the previous subsection), we get the following exact sequence:

$$0 \to (K^S)^\times \to (\varinjlim_{L\subset\overline{\mathbb{Q}}} L_{\mathbb{A}}^\times)^{\mathfrak{H}} = (\varinjlim_{F\subset K^S} F_{\mathbb{A}}^\times) \to C_K^{\mathfrak{H}} \to H^1(\mathfrak{H}, \overline{\mathbb{Q}}^\times) = 0.$$

Thus

$$C_K^{\mathfrak{H}} = \varinjlim_F F_{\mathbb{A}}^\times / F^\times.$$

After taking the injective limit of (4.34) with respect to $F \subset K^S$, we have from (*) that $C_S \cong C_K^{\mathfrak{H}}/U_S$, which shows the exactness of

$$1 \to U_S \to C_K^{\mathfrak{H}} \to C_S \to 1. \tag{4.35}$$

Thus by the long exact sequence (Corollary 4.28) attached to the above short one, we only need to prove the vanishing $H^q(\mathfrak{G}, U_S) = 0$ for all $q > 0$. Since $H^q(\mathfrak{G}, U_S) = \varinjlim_F H^q(\mathrm{Gal}(F/K), U_{F,S})$ by Corollary 4.26, we only need to prove $H^q(\mathrm{Gal}(F/K), U_{F,S}) = 0$ for all finite extensions F/K in K^S.

By projecting down cocycles to components, we easily see for finite extensions F/K that $H^q(\mathrm{Gal}(F/K), U_{F,S})$ injects into $\prod_{\mathfrak{p}\nmid p\in S} H^q(\mathrm{Gal}(F/K), \prod_{\mathfrak{P}|\mathfrak{p}} O_{F,\mathfrak{P}}^\times)$. Here the lower case $\mathfrak{p}$ indicates prime ideals of K and capital $\mathfrak{P}$ indicates those of F. Note that $\prod_{\mathfrak{P}|\mathfrak{p}} O_{F,\mathfrak{P}}^\times \cong \mathrm{Hom}_{\mathbb{Z}[D]}(\mathbb{Z}[\mathrm{Gal}(F/K)], O_{F,\mathfrak{P}}^\times)$ for the decomposition group D of $\mathfrak{P}$ in $\mathrm{Gal}(F/K)$. Thus by Shapiro's lemma (4.20),

$$H^q(\mathrm{Gal}(F/K), \prod_{\mathfrak{P}|\mathfrak{p}} O_{F,\mathfrak{P}}^\times) \cong H^q(D, O_{F,\mathfrak{P}}^\times).$$

Since $\mathfrak{P}$ is unramified over $\mathfrak{p}$, $O_{\mathfrak{P}}^\times \cong \mathbb{F}_q^\times \times L$ as D-modules, where $\mathbb{F}_q$ is the residue field of $O_{\mathfrak{P}}$, and $L = 1 + \mathfrak{P}O_{\mathfrak{P}}$. Since $D \cong \mathrm{Gal}(\mathbb{F}_q/\mathbb{F})$ for the residue field $\mathbb{F}$ of $O_{\mathfrak{p}}$, it is easy to see from Proposition 4.18 that $H^q(D, \mathbb{F}_q^\times) = 0$. By Exercise 5 of the previous subsection, $H^q(D, \mathbb{F}_q) = 0$. Since $1 + \mathfrak{P}^n O_{\mathfrak{P}}/1 + \mathfrak{P}^{n+1} O_{\mathfrak{P}}$ is isomorphic to $\mathbb{F}_q$ as D-modules and $1 + \mathfrak{P}^n O_{\mathfrak{P}} \cong O_{\mathfrak{P}}$ as D-modules for n sufficiently large (by $\mathfrak{P}$-adic logarithm), we see from Corollary 4.28 that $H^q(D, 1 + \mathfrak{P}O_{\mathfrak{P}}) = H^q(D, O_{\mathfrak{P}})$. By taking a normal base of $O_{\mathfrak{P}}$ over $O_{\mathfrak{p}}$, which lifts a normal base of $\mathbb{F}_q/\mathbb{F}$, we see that $O_{\mathfrak{P}} \cong O_{\mathfrak{p}}[D]$ as D-modules, which is $O_{\mathfrak{p}}[D]$-projective. This shows that $H^q(D, O_{\mathfrak{P}}) = 0$. Thus we conclude that $H^q(D, O_{\mathfrak{P}}^\times) = 0$ and hence $H^q(\mathrm{Gal}(F/K), U_{F,S}) = 0$ as desired. □

We have shown the following fact in the above proof of Theorem 4.47:

LEMMA 4.48 *If F/K is a finite Galois extension of number fields and if*

F/K is unramified at a prime ideal $\mathfrak{p}$ of K, then

$$H^q(\mathrm{Gal}(F/K), \prod_{\mathfrak{P}|\mathfrak{p}} O_{F,\mathfrak{P}}^\times) = 0 \quad \text{and} \quad H^q(D, O_{F,\mathfrak{P}}^\times) = 0$$

for the decomposition group $D \subset \mathrm{Gal}(F/K)$ of $\mathfrak{P}/\mathfrak{p}$.

To state the exact duality statement, we first study the reciprocity map $rec_S : C_S \to \mathfrak{G}_S^{ab}$. By global class field theory, we have the following exact sequence:

$$1 \to D_K \to K_{\mathbb{A}}^\times/K^\times \xrightarrow{rec} G^{ab} \to 1,$$

where D_K is the identity connected component of $K_{\mathbb{A}}^\times/K^\times$ and is the maximal divisible subgroup of $K_{\mathbb{A}}^\times/K^\times$ (see [CFT] VII, IX). Since U_S is totally disconnected, by (4.35), the kernel $\mathrm{Ker}(rec_{S,K})$ of the induced reciprocity map $rec_{S,K} : K_{\mathbb{A}}^\times/K^\times U_{K,S} \to \mathfrak{G}^{ab}$ is the image of D_K. Thus we see that $D_K U_{K,S}/U_{K,S} \cong \mathrm{Ker}(rec_{S,K})$, which shows that $\mathrm{Ker}(rec_{S,K})$ is divisible.

COROLLARY 4.49 *Let the notation and the assumption be as in the theorem. Let M be a discrete $\mathfrak{G}_S$-module of finite type, and let $\ell \in S$. Then*

(1) *The map $\alpha_{\mathfrak{G}}^r(M)[\ell] : \mathrm{Ext}_{\mathscr{C}}^r(M, C_S)[\ell^\infty] \to H_{ct}^{2-r}(\mathfrak{G}, M)^*[\ell^\infty]$ is bijective for all $r \geq 1$.*
(2) *Suppose that M is a finite module. Let F be a finite totally imaginary Galois extension of K inside K^S such that $\mathrm{Gal}(K^S/F)$ acts trivially on M. Then if $|M|O_S = O_S$, we have the following exact sequence:*

$$\mathrm{Hom}(M, \mathrm{Ker}(rec_{S,F})) \xrightarrow{N_{F/K}} \mathrm{Hom}_{\mathfrak{G}}(M, C_S) \xrightarrow{\alpha_{\mathfrak{G}}^0(M)} H_{ct}^2(\mathfrak{G}_S, M)^* \to 0.$$

Proof For the first assertion, by Theorems 4.46 and 4.47, we need to prove that $\alpha_{\mathfrak{G}}^1(\mathbb{Z}/\ell^m\mathbb{Z})$ is an isomorphism for all m. We get, from the long exact sequence attached to $0 \to \mathbb{Z} \xrightarrow{x \mapsto \ell^m x} \mathbb{Z} \to \mathbb{Z}/\ell^m\mathbb{Z} \to 0$, a short exact sequence:

$$\begin{aligned} 0 = H_{ct}^1(\mathfrak{G}, \mathbb{Z}) &\to H_{ct}^1(\mathfrak{G}, \mathbb{Z}/\ell^m\mathbb{Z}) \to H_{ct}^2(\mathfrak{G}, \mathbb{Z})[\ell^m] \\ &= \mathrm{Hom}_{\mathbb{Z}}(\mathfrak{G}^{ab}, \mathbb{Z}/\ell^m\mathbb{Z}) \to 0. \end{aligned}$$

This shows that $H_{ct}^1(\mathfrak{G}, \mathbb{Z}/\ell^m\mathbb{Z})^* = \mathfrak{G}^{ab}/\ell^m\mathfrak{G}^{ab}$. Similarly, we get from Proposition 4.15 the following exact sequence:

$$0 \to \mathrm{Hom}_{\mathscr{C}}(\mathbb{Z}, C_S) \otimes_{\mathbb{Z}} \mathbb{Z}/\ell^m\mathbb{Z} \to \mathrm{Ext}_{\mathscr{C}}^1(\mathbb{Z}/\ell^m\mathbb{Z}, C_S) \to \mathrm{Ext}_{\mathscr{C}}^1(\mathbb{Z}, C_S)[\ell^m] \to 0.$$

Note that $\mathrm{Hom}_{\mathscr{C}}(\mathbb{Z}, C_S) \cong C_S^{\mathfrak{G}}$ and $\mathrm{Ext}_{\mathscr{C}}^1(\mathbb{Z}, C_S) = H_{ct}^1(\mathfrak{G}, C_S) = 0$ by (cf1). This shows that $\mathrm{Ext}^1_{\mathscr{C}}(\mathbb{Z}/\ell^m\mathbb{Z}, C_S) = C_S^{\mathfrak{G}}/mC_S^{\mathfrak{G}}$. Since $\mathrm{Ker}(rec_{S,K})$ is divisible, $C_S^{\mathfrak{G}}/mC_S^{\mathfrak{G}} \cong C_K^G/mC_K^G$ for $G = \mathrm{Gal}(\overline{\mathbb{Q}}/K)$. Then $\alpha^1_{\mathfrak{G}}(\mathbb{Z}/\ell^m\mathbb{Z})$: $C_S^{\mathfrak{G}}/\ell^m C_S^{\mathfrak{G}} \to \mathrm{Hom}_{\mathbb{Z}}(\mathfrak{G}, \mathbb{Z}/\ell^m\mathbb{Z})^* = \mathfrak{G}^{ab}/m\mathfrak{G}^{ab}$ is induced by rec_G and hence is an isomorphism (Theorem 4.40).

We now prove the second assertion. When $M = \mathbb{Z}/m\mathbb{Z}$ or $\mathbb{Z}$, $\alpha^0_{\mathfrak{G}}$ is just a reciprocity map: $C_S^{\mathfrak{G}}/mC_S^{\mathfrak{G}} \to \mathfrak{G}^{ab}/m\mathfrak{G}^{ab}$, and the assertion is clear from the argument preceding the corollary. As in the proof of Theorem 4.41, we take an open normal subgroup $U = \mathrm{Gal}(K^S/F) \subset \mathfrak{G}$ fixing M element by element, and for the augmentation $\pi : \mathbb{Z}[\mathfrak{G}/U] \twoheadrightarrow \mathbb{Z}$, we consider a presentation:

$$0 \to M = \mathrm{Hom}_{\mathbb{Z}}(\mathbb{Z}, M) \xrightarrow{\pi^*} \mathrm{Hom}_{\mathbb{Z}}(\mathbb{Z}[\mathfrak{G}/U], M) = I \twoheadrightarrow N \to 0.$$

Then we have the following commutative diagram:

$$\begin{array}{ccccccc}
\mathrm{Hom}_{\mathscr{C}}(N, C_S) & \longrightarrow & \mathrm{Hom}_{\mathscr{C}_U}(M, C_S) & \longrightarrow & \mathrm{Hom}_{\mathscr{C}}(M, C_S) & \longrightarrow & \cdots \\
\downarrow{\scriptstyle \alpha^0_{\mathfrak{G}}(N)} & & \downarrow{\scriptstyle \alpha^0_U(M)} & & \downarrow{\scriptstyle \alpha^0_{\mathfrak{G}}(M)} & & \downarrow\wr \\
H^2_{ct}(\mathfrak{G}, N)^* & \longrightarrow & H^2_{ct}(U, M)^* & \longrightarrow & H^2_{ct}(\mathfrak{G}, M)^* & \longrightarrow & \cdots.
\end{array}$$

All vertical maps are surjective, and are isomorphisms after the three left terms. Thus we get the following exact sequence:

$$\mathrm{Ker}(\alpha^0_{\mathfrak{G}}(N)) \to \mathrm{Ker}(\alpha^0_U(M)) \xrightarrow{N_{F/K}} \mathrm{Ker}(\alpha^0_{\mathfrak{G}}(M)) \to 0.$$

By the argument for trivial $\mathfrak{G}$-modules, we see that

$$\mathrm{Ker}(\alpha^0_U(M)) \cong \mathrm{Hom}_{\mathbb{Z}}(M, \mathrm{Ker}(rec_{S,F})).$$

Thus the kernel $\mathrm{Ker}(\alpha^0_{\mathfrak{G}}(M))$ is the image of $\mathrm{Hom}_{\mathbb{Z}}(M, \mathrm{Ker}(rec_{S,F}))$ under the norm map. □

Exercise

(1) Show that the sequence (4.34) is exact.

4.4.3 Tate–Shafarevich groups

Let S and K be as in the previous subsection. We fix a prime $p \in S$. We suppose that M is a finite discrete $\mathfrak{G}_S$-module with p-power order. Let $\Sigma = \Sigma_K$ be the set of places of K which induce places in S of $\mathbb{Q}$. For each $v \in \Sigma$, we write K_v (resp. $\overline{K}_v$) for the completion of K at v (resp. the algebraic closure of K_v), and let $D_v \subset \mathfrak{G}$ be the decomposition group at v. When $v = \mathfrak{p}$ is a finite place, we put $H^r_{ct}(K_v, M) = H^r_{ct}(\mathrm{Gal}(\overline{K}_v/K_v), M)$;

when v is a real place, we put $H^r_{ct}(K_v, M) = H^r_T(\mathrm{Gal}(\mathbb{C}/\mathbb{R}), M)$, the Tate cohomology; and when v is a complex place, we simply put $H^r_{ct}(K_v, M) = \{0\}$. We define

$$Ш^r_S(K, M) = \mathrm{Ker}(\beta = \beta^r_S(M) : H^r_{ct}(\mathfrak{G}_S, M) \to \prod_{v \in \Sigma} H^r_{ct}(K_v, M)), \quad (Ш)$$

where the map β is the product over $v \in \Sigma$ of the composites $\beta_v : H^r_{ct}(\mathfrak{G}, M) \xrightarrow{res} H^r_{ct}(D_v, M) \xrightarrow{inf} H^r_{ct}(K_v, M)$. The terminology 'Tate–Shafarevich group' is often used to indicate $Ш^r_S(K, M)$ when M is related to abelian varieties (or elliptic curves) defined over number fields. Here we call $Ш^r_S(K, M)$, for general M the Tate–Shafarevich groups, a bit abusing the language.

We then define $M^*(1) = \mathrm{Hom}_{\mathbb{Z}}(M, \mu_{p^\infty})$ as a $\mathfrak{G}$-module, where $\mu_{p^\infty} = \bigcup_\alpha \mu_{p^\alpha}$. Then $(M^*(1))^*(1) \cong M$ canonically. Note that $M^*(1) \cong \mathrm{Hom}_{\mathbb{Z}}(M, \overline{K}^\times)$ as D_v-modules. Then by the duality theorems: Theorems 4.43, 4.45 and Corollary 4.49, we have the dual map of β

$$\gamma^r_S(M) = \beta^* : \prod_{v \in \Sigma} H^r_{ct}(K_v, M^*(1)) \to H^{2-r}_{ct}(\mathfrak{G}_S, M)^*.$$

THEOREM 4.50 (J. TATE, 1962) *Let S be a finite set of places of $\mathbb{Q}$ including the infinite place. Let K be a number field and Σ be the set of places of K above S. Fix a prime $p \in S$, and let M be a discrete finite $\mathfrak{G}_S$-module with p-power order. Then we have:*

(1) *The Tate–Shafarevich groups $Ш^1_S(K, M)$ and $Ш^2_S(K, M^*(1))$ are finite, and there exists a canonical perfect pairing:*

$$Ш^1_S(K, M) \times Ш^2_S(K, M^*(1)) \to \mathbb{Q}_p/\mathbb{Z}_p$$

inducing Pontryagin duality between the two groups;

(2) *The map $\beta^0_S(M)$ is injective, $\gamma^2_S(M)$ is surjective, and we have the identity $\mathrm{Im}(\beta^r_S(M)) = \mathrm{Ker}(\gamma^r_S(M))$ for all $r = 0, 1, 2$;*

(3) *The map $\beta^r_S(M)$ induces a surjective isomorphism for all $r \geq 3$:*

$$H^r(\mathfrak{G}_S, M) \cong \prod_{v \in \Sigma(\mathbb{R})} H^r_{ct}(K_v, M),$$

where $\Sigma(\mathbb{R})$ is the set of real places of K;

(4) *We have the following exact sequence:*

$$\begin{array}{ccccc} 0 \to H^0_{ct}(\mathfrak{G}_S, M) & \xrightarrow{\beta^0} & \prod_{v\in\Sigma} H^0_{ct}(K_v, M) & \xrightarrow{\gamma^0} & H^2_{ct}(\mathfrak{G}_S, M^*(1))^* \\ & & & & \downarrow \\ H^1_{ct}(\mathfrak{G}_S, M^*(1)) & \xleftarrow{\gamma^1} & \prod_{v\in\Sigma} H^1_{ct}(K_v, M) & \xleftarrow{\beta^1} & H^1_{ct}(\mathfrak{G}_S, M) \\ \downarrow & & & & \\ H^2_{ct}(\mathfrak{G}_S, M) & \xrightarrow[\beta^2]{} & \prod_{v\in\Sigma} H^2_{ct}(K_v, M) & \xrightarrow[\gamma^2]{} & H^0_{ct}(\mathfrak{G}_S, M^*(1))^* \to 0; \end{array}$$

(5) *Let Σ_f be the subset of finite places in Σ. Let $B_v \subset H^1(K_v, M)$ be a submodule for each finite place $v \in \Sigma_f$, and let $B_v^\perp \subset H^1_{ct}(K_v, M^*(1))$ be the orthogonal complement of B_v under the pairing of Theorems 4.43 and 4.45. We write $H^1_{B^\perp}(K, M^*(1)) = \beta^1_S(M^*(1))^{-1}(\prod_{v\in\Sigma_f} B_v^\perp)$ and $H^1_B(K, M) = \beta^1_S(M)^{-1}(\prod_{v\in\Sigma_f} B_v)$. Suppose that $p > 2$. Then we have the following exact sequence:*

$$\begin{aligned} 0 &\to H^1_B(K, M) \to H^1_{ct}(\mathfrak{G}_S, M) \to \prod_{v\in\Sigma_f} H^1_{ct}(K_v, M)/B_v \\ &\to H^1_{B^\perp}(K, M^*(1))^* \to H^2_{ct}(\mathfrak{G}_S, M) \to \prod_{v\in\Sigma_f} H^2_{ct}(K_v, M) \\ &\to H^0_{ct}(\mathfrak{G}_S, M^*(1))^* \to 0. \end{aligned}$$

Proof We again follow the treatment of Milne [ADT] I.4. We first consider the finiteness of $Ш^1_S(K, M)$. We only need to prove that $H^1_{ct}(\mathfrak{G}_S, M)$ is finite. By the inflation-restriction sequence (and finiteness of M), we may assume that $\mathfrak{G}$ acts trivially on M. Then $H^1_{ct}(\mathfrak{G}_S, M) = \mathrm{Hom}_{ct}(\mathfrak{G}, M)$, which is finite by global class field theory (because S is a finite set). After proving the duality between $Ш^1_S(K, M)$ and $Ш^2_S(K, M^*(1))$, the finiteness of $Ш^2_S(K, M^*(1))$ follows from that of $Ш^1_S(K, M)$. The duality in question follows from the exact sequence of (4), because $Ш^2_S(K, M^*(1)) = \mathrm{Coker}(\gamma^1_S(M))$.

We now prove the exact sequence of (4): Let $J_S = \varinjlim_F F_S^\times$ and $E_S = \varinjlim_F (O_F^S)^\times$, where F runs over all finite extensions of K inside K^S. Then by (4.34) and (4.35), we have the following short exact sequence of discrete $\mathfrak{G}$-modules:

$$0 \to E_S \to J_S \to C_S \to 0.$$

Then we apply Proposition 4.13 to this sequence and get the following long exact sequence:

$$\begin{aligned}\cdots &\to \mathrm{Ext}^0_{\mathscr{C}}(M^*(1),E_S)\to \mathrm{Ext}^0_{\mathscr{C}}(M^*(1),J_S)\to \mathrm{Ext}^0_{\mathscr{C}}(M^*(1),C_S)\\ &\to \mathrm{Ext}^1_{\mathscr{C}}(M^*(1),E_S)\to \mathrm{Ext}^1_{\mathscr{C}}(M^*(1),J_S)\to \mathrm{Ext}^1_{\mathscr{C}}(M^*(1),C_S)\\ &\to \mathrm{Ext}^2_{\mathscr{C}}(M^*(1),E_S)\to \mathrm{Ext}^2_{\mathscr{C}}(M^*(1),J_S)\to \mathrm{Ext}^2_{\mathscr{C}}(M^*(1),C_S)\to\cdots.\end{aligned} \tag{4.36}$$

We shall look one by one into each term of the above exact sequence and relate it to the corresponding term of the exact sequence of (4).

Since $\mathrm{Hom}_{\mathscr{C}}(X,Y)=H^0(\mathfrak{G},\mathrm{Hom}_{\mathbb{Z}}(X,Y))$ for discrete $\mathfrak{G}$-modules X and Y, we see that

$$\begin{aligned}\mathrm{Hom}_{\mathscr{C}}(M^*(1),E_S) &= \mathrm{Hom}_{\mathscr{C}}(\mathrm{Hom}_{\mathbb{Z}}(M,\mu_{p^\infty}),\mu_{p^\infty})\\ &= H^0(\mathfrak{G},\mathrm{Hom}_{\mathbb{Z}}(\mathrm{Hom}_{\mathbb{Z}}(M,\mu_{p^\infty}),\mu_{p^\infty}))=H^0(\mathfrak{G},M).\end{aligned}$$

We take a totally complex (finite) Galois extension F/K inside K^S such that $\mathrm{Gal}(K^S/F)$ fixes $M^*(1)$ element by element. Then

$$\mathrm{Hom}_{\mathscr{C}}(M^*(1),J_S)=\mathrm{Hom}_{\mathbb{Z}[\mathrm{Gal}(F/K)]}(M^*(1),F_S^\times).$$

Then $F_S^\times=\prod_{\mathfrak{p}\in\Sigma}\prod_{\mathfrak{P}|\mathfrak{p}}F_{\mathfrak{P}}^\times$ and $\prod_{\mathfrak{P}|\mathfrak{p}}F_{\mathfrak{P}}^\times\cong\mathrm{Hom}_{\mathbb{Z}[D]}(\mathbb{Z}[\mathrm{Gal}(F/K)],F_{\mathfrak{P}}^\times)$, where D is the decomposition group of $\mathfrak{P}/\mathfrak{p}$. By Shapiro's lemma (4.27), we have

$$\mathrm{Hom}_{\mathscr{C}}(M^*(1),\mathrm{Hom}_{\mathbb{Z}[D]}(\mathbb{Z}[\mathrm{Gal}(F/K)],F_{\mathfrak{P}}^\times))\cong\mathrm{Hom}_{\mathbb{Z}[D]}(M^*(1),F_{\mathfrak{P}}^\times).$$

Since M is of p-power order, we have

$$\mathrm{Hom}_{\mathbb{Z}[D]}(M^*(1),F_{\mathfrak{P}}^\times)\cong\mathrm{Hom}_{\mathbb{Z}[D]}(M^*(1),\mu_{p^\infty})\cong H^0_{ct}(\mathrm{Gal}(\overline{K}_{\mathfrak{p}}/K_{\mathfrak{p}}),M).$$

By Lemma 4.51 which follows this proof, we can replace $\mathrm{Hom}_{\mathscr{C}}$ in the above argument by $\mathrm{Ext}^1_{\mathscr{C}}$, and we get

$$\mathrm{Ext}^1_{\mathscr{C}}(M^*(1),J_S)\cong\prod_{v\in\Sigma}H^1_{ct}(K_v,M). \tag{4.37}$$

We now look into the terms of (4.36) involving E_S. Since we find in K^S any ℓ-th root of elements of $(O_F^S)^\times$ for $\ell\in S$, E_S is in particular p-divisible. Then by Proposition 4.30, we have

$$\mathrm{Ext}^q_{\mathscr{C}}(M^*(1),E_S)\cong H^q(\mathfrak{G},Hom(M^*(1),E_S))\cong H^q(\mathfrak{G},M)$$

for all $q\geq 0$.

Replacing terms of (4.36) by the cohomology groups we have identified, we get the exactness of the following sequence:

$$\begin{array}{ccccc} 0 \to H^0_{ct}(\mathfrak{G}_S, M) & \xrightarrow{\beta^0} & \prod_{v\in\Sigma} H^0_{ct}(K_v, M) & \xrightarrow{\gamma^0} & H^2_{ct}(\mathfrak{G}_S, M^*(1))^* \\ & & & & \downarrow \\ H^1_{ct}(\mathfrak{G}_S, M^*(1)) & \xleftarrow{\gamma^1} & \prod_{v\in\Sigma} H^1_{ct}(K_v, M) & \xleftarrow{\beta^1} & H^1_{ct}(\mathfrak{G}_S, M) \\ \downarrow & & & & \\ H^2_{ct}(\mathfrak{G}_S, M). & & & & \end{array}$$

Replacing M by $M^*(1)$ and taking the Pontryagin dual (Corollary 4.49) of the first four terms of the above sequence, we get another exact sequence:

$$H^1_{ct}(\mathfrak{G}_S, M^*(1)) \to H^2_{ct}(\mathfrak{G}_S, M) \xrightarrow{\beta^2} \prod_{v\in\Sigma} H^2_{ct}(K_v, M) \to H^0_{ct}(\mathfrak{G}_S, M^*(1))^* \to 0.$$

This is the last five terms of the exact sequence in (4), and hence we have completed the proof of (4).

By Lemma 4.51 for $q \geq 2$, we have

$$\begin{aligned} \cdots &\to H^1_{ct}(\mathfrak{G}, M) \to \prod_{v\in\Sigma} H^1_{ct}(K_v, M) \to H^1_{ct}(\mathfrak{G}, M^*(1))^* \\ &\to H^2_{ct}(\mathfrak{G}, M) \to \prod_{v\in\Sigma} H^2_{ct}(K_v, M) \to H^0_{ct}(\mathfrak{G}, M^*(1))^* \\ &\to H^3_{ct}(\mathfrak{G}, M) \to \prod_{v\in\Sigma(\mathbb{R})} H^3_{ct}(\mathrm{Gal}(\overline{K}_v/K_v), M) \to 0; \end{aligned} \tag{4.38}$$

and

$$H^q_{ct}(\mathfrak{G}, M) \cong \prod_{v\in\Sigma(\mathbb{R})} H^q_{ct}(K_v, M) \quad \text{for all } q > 3. \tag{4.39}$$

The map: $\prod_{v\in\Sigma} H^2_{ct}(K_v, M) \to H^0_{ct}(\mathfrak{G}, M^*(1))^*$ is surjective, because it is a dual of the injection: $H^0(\mathfrak{G}, M) \hookrightarrow \prod_{v\in\Sigma} H^0(K_v, M)$. Thus (4.39) holds even for $q = 3$. This proves (1), (2), (3) and (4).

We now prove the last exact sequence (5). Since p is odd, $H^1(K_v, M) = 0$ if v is an archimedean place. We write down the last six terms of the exact sequence in (4):

$$\begin{aligned} \cdots &\to H^1_{ct}(\mathfrak{G}, M) \to \prod_{v\in\Sigma_f} H^1_{ct}(K_v, M) \\ &\to H^1_{ct}(\mathfrak{G}, M^*(1))^* \to H^2_{ct}(\mathfrak{G}, M) \to \prod_{v\in\Sigma_f} H^2_{ct}(K_v, M) \\ &\to H^0_{ct}(\mathfrak{G}, M^*(1))^* \to 0. \end{aligned} \tag{4.40}$$

We simply write $H^1_{ct}(M)$ for $H^1_{ct}(\mathfrak{G}, M)$. From this, we get the following commutative diagram:

$$\begin{array}{ccccccc} & & \prod_{v\in\Sigma_f} B_v & \cong & \prod_{v\in\Sigma_f}\left(\frac{H^1_{ct}(K_v,M^*(1))}{B_v^\perp}\right)^* & & \\ & & \downarrow & & \downarrow & & \\ H^1_{ct}(M) & \to & \prod_{v\in\Sigma_f} H^1_{ct}(K_v,M) & \longrightarrow & H^1_{ct}(M^*(1))^* & \to & H^2_{ct}(M) \\ \| & & \downarrow & & \downarrow & & \| \\ H^1_{ct}(M) & \to & \prod_{v\in\Sigma_f}\frac{H^1_{ct}(K_v,M)}{B_v} & \longrightarrow & H^1_B(\mathfrak{G},M^*(1))^* & \to & H^2_{ct}(M), \end{array}$$

in which the middle line and all columns are exact. Then the desired exact sequence follows from a simple diagram chasing. □

We now need to prove the following lemma:

LEMMA 4.51 *Let the notation be as in Theorem 4.50. Then we have*

$$\mathrm{Ext}^q_{\mathscr{C}}(M^*(1), J_S) \cong \prod_{v\in\Sigma} H^q_{ct}(K_v, M) \quad \textit{for all } q \geq 0.$$

Proof The case $q = 0$ has been shown in the proof of the above theorem. So we may assume that $q \geq 1$. Let X be a $\mathfrak{G}$-module of finite type. For the category $\mathscr{C}$ of discrete $\mathfrak{G}$-modules, we note that

$$\begin{aligned} \mathrm{Ext}^\bullet_{\mathscr{C}}(X, J_S) &= \varinjlim_{F\subset K^S} \mathrm{Ext}^\bullet_{\mathbb{Z}[\mathrm{Gal}(F/K)]}(X, J_S^{\mathrm{Gal}(K^S/F)}) \\ &= \varinjlim_{F\subset K^S} \mathrm{Ext}^\bullet_{\mathbb{Z}[\mathrm{Gal}(F/K)]}(X, F_S^\times), \end{aligned} \tag{4.41}$$

where F runs over all finite Galois extensions of K inside K^S such that $\mathrm{Gal}(K^S/F)$ acts trivially on X. This follows from Remark 4.29. Let $v \in S$ and write $G = G_v = \mathrm{Gal}(\overline{K}_v/K_v)$ and $H = \mathrm{Gal}(\overline{F}_w/F_w)$ for a place w of F over v. We consider the extension group $Ext^r_{Gal(F/K)}(X, I)$ for $I = \prod_{w|v} F_w^\times \cong \mathrm{Ind}_D^{\mathrm{Gal}(F/K)} P^H$ with $P = \overline{K}_v^\times$, where $D = G/H$ is the decomposition group of the place w. We take a resolution $X^\bullet \twoheadrightarrow X$ of free $\mathrm{Gal}(F/K)$-modules. So $X^\bullet \twoheadrightarrow X$ is also a free resolution of D-modules. Then we have

$$\begin{aligned} \mathrm{Ext}^q_{Gal(F/K)}(X, I) &= H^q(\mathrm{Hom}_{\mathrm{Gal}(F/K)}(X^\bullet, I)) \\ &\cong H^q(\mathrm{Hom}_D(X^\bullet, P^H)) = \mathrm{Ext}^q_D(X, P^H). \end{aligned}$$

The middle isomorphism follows from the Shapiro's lemma (see (4.17) and (3.5)).

Now we suppose that $q = 1$. By Hilbert's theorem 90, we have the vanishing: $H^1(H,P)=0$. So from Theorem 4.34, we get $\mathrm{Ext}^1_D(X,P^H)\cong \mathrm{Ext}^1_G(X,P^H)$. Thus we get by (4.41)

$$\mathrm{Ext}^1_{\mathscr{C}}(X,J_S)\cong \prod_{v\in\Sigma}\mathrm{Ext}^1_{G_v}(X,\overline{K}_v^{\times}).$$

By Proposition 4.30, we have

$$\mathrm{Ext}^1_{\mathscr{C}_{G_v}}(X,\overline{K}_v^{\times})\cong H^1_{ct}(K_v,X^*(1)).$$

This proves the assertion for $q=1$.

We now consider the case $q=2$. We now change the notation and write D for the decomposition group of v in $\mathfrak{G}_S$. Write $D=G/H$. Then H is a closed normal subgroup of G and is not open. Let U be an open normal subgroup of G. Since the pair (HU,P) satisfies the axiom of class formation (see Example 4.42), by (CF2), we have

$$\begin{aligned} H^2_{ct}(H,P)[p^\infty] &= \varinjlim_U H^2(H/U\cap H,P^{UH})[p^\infty] \\ &= \varinjlim_U H^2(HU/U,P^U)[p^\infty] = \varinjlim_U \frac{1}{[HU:U]}\mathbb{Z}_p/\mathbb{Z}_p, \end{aligned}$$

where U runs over open normal subgroups of H, $[p^\infty]$ indicates p-torsion part, and the last limit is taken with respect to the map $i_{U,V}$: $\frac{1}{[HU:U]}\mathbb{Z}_p/\mathbb{Z}_p \to \frac{1}{[HV:V]}\mathbb{Z}_p/\mathbb{Z}_p$ given by

$$i_{U,V}(x)=[UH:VH]x.$$

Since $p\in S$, for any given p-power p^r, we find $V\subset U$ such that $p^r|[HU:HV]$. Thus the last injective limit in the above equation vanishes, and hence

$$H^2_{ct}(H,P)[p^\infty]=0 \ \text{ for } q=2 \text{ and } \ H^1_{ct}(H,P)=0$$

by Hilbert's theorem 90. We write $\mathscr{C}_G$ (resp. $\mathscr{C}_D$) for the category of discrete G-modules (resp. discrete D-modules).

Let $P\hookrightarrow I\twoheadrightarrow S$ be a $\mathscr{C}_G$-injective presentation chosen so that $P^H\hookrightarrow I^H\twoheadrightarrow S^H$ is a $\mathscr{C}_D$-injective presentation (as chosen in the proof of Theorem 4.34). We recall the exact sequence (4.26) in the proof of Theorem 4.34:

$$0\to \mathrm{Ext}^1_{\mathscr{C}_D}(X,S^H)\xrightarrow{inf}\mathrm{Ext}^1_{\mathscr{C}_G}(X,S)\xrightarrow{\delta}\mathrm{Hom}_{\mathscr{C}_G}(X,H^1(H,S)).$$

Since we have $H^2_{ct}(H,P)\cong H^1_{ct}(H,S)$, we have $H^1_{ct}(H,S)[p^\infty]\cong H^2_{ct}(H,P)$ $[p^\infty]=0$. Since we also have

$$\mathrm{Ext}^q_{\mathscr{C}_G}(X,P)\cong \mathrm{Ext}^{q-1}_{\mathscr{C}_G}(X,S) \ \text{ and } \ \mathrm{Ext}^q_{\mathscr{C}_D}(X,P^H)\cong \mathrm{Ext}^{q-1}_{\mathscr{C}_G}(X,S^H),$$

if X is of p-power order, we have

$$\mathrm{Hom}_{\mathscr{C}_G}(X, H^1(H,S)) = \mathrm{Hom}_{\mathscr{C}_G}(X, H^1(H,S)[p^\infty]) = 0.$$

Thus, we get

$$\mathrm{Ext}^2_{\mathscr{C}_D}(X, P^H) \cong \mathrm{Ext}^2_{\mathscr{C}_G}(X, P^H),$$

and by Proposition 4.30, we have the desired assertion.

For $q > 2$, we need to extend the proof. Let $\mu \subset \overline{K}_v^\times = P$ be the group of all roots of unity. We decompose $\mu = \mu_{p^\infty} \times \mu^{(p)}$. We put $Q = \mu^{(p)} \backslash P$. Note that

$$\mathrm{Ext}^q_{\mathscr{C}_G}(X, \mu^{(p)}) = \varinjlim_{p \nmid N} \mathrm{Ext}^q_{\mathscr{C}_G}(X, \mu_N),$$

and hence $\mathrm{Ext}^q_{\mathscr{C}_G}(X, \mu_N)$ is killed by N, because the multiplication by N on μ_N factors through the trivial group 0. Thus $\mathrm{Ext}^q_{\mathscr{C}_G}(X, \mu^{(p)})$ is a prime-to-p torsion module. If X is a p-power torsion module, $\mathrm{Ext}^q_{\mathscr{C}_G}(X, \mu^{(p)})$ is a p-power torsion, and hence we conclude that $\mathrm{Ext}^q_{\mathscr{C}_G}(X, \mu^{(p)}) = 0$. This applied to the (extension) long exact sequence shows that

$$\mathrm{Ext}^q_{\mathscr{C}_G}(X, P) \cong \mathrm{Ext}^q_{\mathscr{C}_G}(X, Q). \tag{4.42}$$

We also have an exact sequence $0 \to H^0(H, \mu^{(p)}) \to P^H \xrightarrow{\pi} Q^H \to H^1_{ct}(H, \mu^{(p)})$. Thus $Q^H/\mathrm{Im}(\pi)$ is a prime-to p torsion. This shows that $\mathrm{Ext}^q_{\mathscr{C}_D}(X, Q^H/\mathrm{Im}(\pi)) = 0$ for all q if X is a p-torsion, and hence

$$\mathrm{Ext}^q_{\mathscr{C}_D}(X, Q^H) \cong \mathrm{Ext}^q_{\mathscr{C}_D}(X, \mathrm{Im}(\pi)) \text{ and } \mathrm{Ext}^q_{\mathscr{C}_D}(X, \mathrm{Im}(\pi)) \cong \mathrm{Ext}^q_{\mathscr{C}_D}(X, P^H).$$

Thus we get, if X is a p-torsion D-module,

$$\mathrm{Ext}^q_{\mathscr{C}_D}(X, P^H) \cong \mathrm{Ext}^q_{\mathscr{C}_D}(X, Q^H). \tag{4.43}$$

From the cohomology long exact sequence, we have the following exact sequence:

$$H^q_{ct}(H, P) \to H^q_{ct}(H, Q) \to H^q_{ct}(H, \mu^{(p)}).$$

This shows that $H^q_{ct}(H, Q)$ is p-torsion-free and is a torsion module for $q = 1$ and $q > 2$ by the local duality theorem. For $q = 2$, by the above result, $H^2_{ct}(H, P)$ is p-torsion-free and is a torsion, so the same result holds.

$$H^q_{ct}(H, Q)[p^\infty] = 0 \textit{ and } H^q_{ct}(H, Q) \textit{ is a torsion module for all } q. \tag{4.44}$$

In Subsection 4.3.3, to make an injective presentation: $Y \hookrightarrow I_Y \twoheadrightarrow S_Y$ for a given module Y, we used the product of injective modules: $\mathrm{Hom}_{\mathbb{Z}}(Y, \mathbb{Q}/\mathbb{Z})$. If $y \mapsto Ny$ is an automorphism of Y for all integers N

prime to p, we can also use $\mathbb{Q}_p/\mathbb{Z}_p$ in place of $\mathbb{Q}/\mathbb{Z}$, because for any given $y \in Y$, we have a homomorphism $\phi_y : Y \to \mathbb{Q}_p/\mathbb{Z}_p$ with $\phi_y(y) \neq 0$ (this fails when y is a prime-to-p torsion element). Thus we may assume that we have an injective presentation $Q \hookrightarrow I \twoheadrightarrow S$ such that

(1) I is a p-torsion module (and hence so is S);
(2) I^H is an injective D-module.

By $H^1_{ct}(H,Q)[p^\infty] = 0$ combined with the p-torsion of S^H, we have the exact sequence:

$$0 \to Q^H \to I^H \to S^H \to 0.$$

From this, we get, for any p-torsion D-module X,

$$\mathrm{Ext}^q_{\mathscr{C}_G}(X,Q) \cong \mathrm{Ext}^{q-1}_{\mathscr{C}_G}(X,S), \quad \mathrm{Ext}^q_{\mathscr{C}_D}(X,Q^H) \cong \mathrm{Ext}^{q-1}_{\mathscr{C}_G}(X,S^H)$$

and

$$H^q_{ct}(H,Q) \cong H^{q-1}_{ct}(H,S),$$

and moreover, S is a p-torsion module. Then by an argument similar to the proof of Theorem 4.34, we get

$$\mathrm{Ext}^q_{\mathscr{C}_G}(X,Q) \cong \mathrm{Ext}^q_{\mathscr{C}_D}(X,Q^H).$$

This combined with (4.42) and (4.43) shows that

$$\mathrm{Ext}^q_{\mathscr{C}_G}(X,P) \cong \mathrm{Ext}^q_{\mathscr{C}_D}(X,P^H),$$

as long as X is a p-torsion finite module. Then Proposition 4.30 tells us that

$$\mathrm{Ext}^q_{\mathscr{C}_G}(X,P) \cong H^q_{ct}(G,X^*(1)).$$

From this, we conclude the desired assertion. □

4.4.4 Local Euler characteristic formula

Let $K/\mathbb{Q}_p$ be a finite extension for a prime p. Let $|\ |_K$ be the p-adic absolute value normalized so that $|p|_K = [O_K : pO_K]^{-1}$ for the p-adic integer ring O_K of K. We would like to prove the following theorem in this subsection:

THEOREM 4.52 (J. TATE, 1962) *Let $G = \mathrm{Gal}(\overline{\mathbb{Q}}_p/K)$ and M be a finite (discrete) G-module. Then we have*

$$\frac{|H^0_{ct}(G,M)| \cdot |H^2_{ct}(G,M)|}{|H^1_{ct}(G,M)|} = \frac{|H^0_{ct}(G,M)| \cdot |H^0_{ct}(G,M^*(1))|}{|H^1_{ct}(G,M)|} = ||M||_K.$$

Proof In the proof, we simply write $H^q(M)$ for $H^q_{ct}(G,M)$. Since $M = \oplus_\ell M[\ell^\infty]$ for primes ℓ, we may assume that M has an ℓ-power order for a prime ℓ. Then $H^q(M)$ is a $\mathbb{Z}_\ell$-module of finite length. Here length(M) is the length of the Jordan–Hölder sequence of the $\mathbb{Z}_\ell$-module M. Thus $|M| = \ell^{\text{length}(M)}$.

We define the local Euler characteristic of M by

$$\chi(M) = \chi(G,M) = \sum_{j=0}^{2}(-1)^j \operatorname{length}(H^j(M))$$

$$\chi'(M) = \chi'(G,M) = \log_p(||M||_K) = \begin{cases} -[K:\mathbb{Q}_p] \operatorname{length}_{\mathbb{Z}_p} M & \text{if } \ell = p, \\ 0 & \text{if } \ell \neq p. \end{cases}$$

Note that the left-hand side of the formula is just $p^{\chi(M)}$ and the right-hand side is $p^{\chi'(M)}$; so, we need to prove that $\chi(M) = \chi'(M)$. Let K/L be a finite extension, and write $H = \operatorname{Gal}(\overline{\mathbb{Q}}_p/L)$. By Shapiro's lemma (4.20),

$$H^q_{ct}(G,M) \cong H^q_{ct}(H, \operatorname{Ind}_G^H M),$$

where $\operatorname{Ind}_G^H M = \operatorname{Hom}_{\mathbb{Z}[G]}(\mathbb{Z}[H], M)$. Since

$$|\operatorname{Ind}_G^H M| = |M|^{[K:L]},$$

we have

$$\chi(G,M) = \chi(H, \operatorname{Ind}_G^H M). \tag{4.45}$$

It is easy to see that $\chi'(N) = \chi'(M) + \chi'(L)$ if $0 \to M \to N \to L \to 0$ is an exact sequence of finite G-modules. Thus $\chi'(M) = \chi'(M^{ss})$, where $M^{ss} = \oplus_{j=1}^{\infty} M_j/M_{j-1}$ for a Jordan–Hölder sequence $0 = M_0 \subset M_1 \subset \cdots \subset M_n = M$ of $\mathbb{Z}_\ell[G]$-modules.

We see easily that if

$$0 \to E_1 \to E_2 \to \cdots \to E_n \to 0$$

is an exact sequence of finite $\mathbb{Z}_\ell$-modules,

$$\prod_j |E_j|^{(-1)^j} = 1 \iff \sum_j (-1)^j \operatorname{length}(E_j) = 0. \tag{4.46}$$

If $0 \to M \to N \to L \to 0$ is an exact sequence of finite $\mathbb{Z}_\ell[G]$-modules, we have a long exact sequence:

$$\begin{aligned} 0 &\to H^0(M) \to H^0(N) \to H^0(L) \\ &\to H^1(M) \to H^1(N) \to H^1(L) \to H^2(M) \to H^2(N) \to H^2(L) \to 0. \end{aligned}$$

The sequence is of finite length, since $H^q(M) = 0$ if $q > 2$, and all cohomology groups are finite $\mathbb{Z}_\ell$-modules, both by Theorem 4.43. Then by (4.46), we see that $\chi(N) = \chi(M) + \chi(L)$ and therefore $\chi(M) = \chi(M^{ss})$. Since M^{ss} is a $\mathbb{F}[G]$-module for $\mathbb{F} = \mathbb{Z}/\ell\mathbb{Z}$, we may assume that M itself is a $\mathbb{F}[G]$-module. Thus, hereafter, we assume that all G-modules are $\mathbb{F}[G]$-modules, and we regard χ and χ' as functions on the Grothendieck group $R_{\mathbb{F}}(G)$ with values in $\mathbb{Z}$ (Subsection 2.1.6). Thus we need to check the formula for a set of generators of $R_{\mathbb{F}}(G) \otimes_{\mathbb{Z}} \mathbb{Q}$.

We first check the formula for the trivial $\mathbb{F} = \mathbb{Z}/\ell\mathbb{Z}$. We see that

$$\dim_{\mathbb{F}} H^0(G,\mathbb{F}) = 1 \ \text{ and } \ \dim_{\mathbb{F}} H^2(G,\mathbb{F}) = \dim_{\mathbb{F}} H^0(G,\mu_\ell) = \dim_{\mathbb{F}} \mu_\ell(K),$$

where $\mu_\ell(K) = \{z \in K | z^\ell = 1\}$. On the other hand, by local class field theory,

$$H^1_{ct}(G,\mathbb{F}) = \mathrm{Hom}_{ct}(G^{ab},\mathbb{F}) \cong K^\times/(K^\times)^\ell \text{ and } H^1_{ct}(G,\kappa) \cong \left(K^\times/(K^\times)^\ell\right) \otimes_{\mathbb{F}} \kappa \tag{4.47}$$

for a finite extension $\kappa/\mathbb{F}$. Since $K^\times \cong O^\times \times \mathbb{Z}$ for the p-adic integer ring O of K and $O^\times \cong O \times \mu_\ell(K)$ up to prime-to-ℓ torsion,

$$H^1_{ct}(G,\mathbb{F}) \cong \begin{cases} \mathbb{Z}/\ell\mathbb{Z} \oplus \mu_\ell(K) & \text{if } \ell \neq p, \\ O/pO \oplus \mathbb{Z}/p\mathbb{Z} \oplus \mu_p(K) & \text{if } \ell = p. \end{cases}$$

This shows that the formula holds for $\mathbb{Z}/\ell\mathbb{Z}$ and hence for any trivial G-module. By duality (Theorem 4.43), the formula also holds for $M = \mu_\ell = \mathbb{F}^*(1)$.

We take a finite Galois extension F/K such that $\mathrm{Gal}(\overline{\mathbb{Q}}_p/F)$ acts trivially on M. Write $\overline{G} = \mathrm{Gal}(F/K)$. Then by Corollary 2.11 $R_{\mathbb{F}}(\overline{G}) \otimes_{\mathbb{Z}} \mathbb{Q}$ is generated by $\mathrm{Ind}_{\overline{H}}^{\overline{G}} \rho$ for cyclic subgroups $\overline{H}$ of order prime to ℓ and characters $\rho : \overline{H} \to \kappa^\times$ for a finite extension $\kappa/\mathbb{F}$. Thus we may assume that $M = \mathrm{Ind}_{\overline{H}}^{\overline{G}} \rho$. Then by (4.45), we only need to check the Euler characteristic formula for ρ (or for the one-dimensional module $V = V(\rho)$ on which $\overline{H}$ acts via ρ). Thus replacing K by the fixed field $\overline{H}$, we may assume that $M = V(\rho)$ is one-dimensional over a finite extension κ of $\mathbb{F}$ and that F/K is cyclic of degree prime to ℓ. Then $H^q(\overline{G}, M) = 0$ for all $q > 0$, and for $G' = \mathrm{Gal}(\overline{\mathbb{Q}}_p/F)$, by the inflation and restriction sequence, we have

$$H^q(G,M) \cong H^0(\overline{G}, H^q(G',M)) \quad \text{for } q = 0,1,2,$$

and we note that

$$H^q(G',M) = \begin{cases} \kappa & \text{if } q = 0 \\ H^1_{ct}(G',\kappa) = \{F^\times/(F^\times)^\ell\} \otimes_{\mathbb{F}} \kappa & \text{if } q = 1 \\ H^2_{ct}(G',\kappa) = H^0(G',\kappa^*(1)) = \mu_p(F) \otimes_{\mathbb{F}} \kappa & \text{if } q = 2. \end{cases}$$

Thus writing the ρ-eigenspace of $\kappa[\overline{G}]$-module X as $X[\rho]$, we see that

$$\chi(G,M) = \dim_{\mathbb{F}} \kappa[\rho] - \dim_{\mathbb{F}}\{(F^\times/(F^\times)^\ell) \otimes_{\mathbb{F}} \kappa\}[\rho] + \dim_{\mathbb{F}}(\mu_\ell(F) \otimes_{\mathbb{F}} \kappa)[\rho],$$

because $H^0(\overline{G}, H^q(G',M)) \cong H^q(G',M)[\rho]$ by the definition of the action of $\overline{G}$ on cohomology groups of G' (one can check this by using inhomogeneous continuous cochains). Since we have already checked the result for $M = \mathbb{F}$ and $M = \mu_\ell$, we may assume that ρ is neither the trivial character nor the cyclotomic character. Therefore $\kappa[\rho] = (\mu_p(F) \otimes_{\mathbb{F}} \kappa)[\rho] = 0$, because the action of the Galois group on κ is trivial and on μ_p is via the cyclotomic character.

We consider the case where $\ell = p$, leaving the case where $\ell \neq p$ to the reader as an exercise. Then

$$\chi(G,M) = -\dim_{\mathbb{F}} \left(\{(F^\times/(F^\times)^p) \otimes_{\mathbb{F}} \kappa\}[\rho]\right).$$

Thus we need to show that

$$[K:\mathbb{Q}_p] \dim_{\mathbb{F}} M = \dim_{\mathbb{F}} \left(\{(F^\times/(F^\times)^p) \otimes_{\mathbb{F}} \kappa\}[\rho]\right),$$

because $||M||_K = p^{-[K:\mathbb{Q}_p]\dim_{\mathbb{F}} M}$. Writing the additive valuation of F as $v : F^\times \twoheadrightarrow \mathbb{Z}$, we have an exact sequence:

$$1 \to U/\mu \to F^\times/\mu \xrightarrow{v} \mathbb{Z} \to 0,$$

where $U = O_F^\times$ and μ is the maximal torsion-subgroup of $F^\times$. Then the above exact sequence is torsion-free, and after tensoring κ, we still have an exact sequence:

$$0 \to (U/\mu) \otimes_{\mathbb{Z}} \kappa \to (F^\times/\mu) \otimes_{\mathbb{Z}} \kappa \xrightarrow{v} \kappa \to 0.$$

Taking the ρ-isotypical component (after tensoring κ), we get

$$\{(U/U^p) \otimes_{\mathbb{F}} \kappa\}[\rho] \cong \{(U/\mu) \otimes_{\mathbb{Z}} \kappa\}[\rho] \cong \{(F^\times/\mu) \otimes_{\mathbb{Z}} \kappa\}[\rho].$$

Now we want to lift the representation ρ to characteristic 0 representation $\tilde{\rho}$. For that, we take a unique unramified extension L of $\mathbb{Q}_p$ of degree $\dim_{\mathbb{F}} \kappa$. Let W be the p-adic integer ring of L. Then we have $W/(p) \cong \kappa$ and hence, $W^\times \cong (1 + pW) \times \kappa^\times$. Using this isomorphism, we may think of ρ as having values in $W^\times$ (and also in L). We write this new character as $\widetilde{\rho} : \overline{G} \to W^\times$, which is called the Teichmüller lift of ρ.

Since (U/μ) is torsion-free and the order of $\overline{G}$ is prime to p,

$$\dim_{\mathbb{F}}\{(U/\mu) \otimes_{\mathbb{Z}} \kappa\}[\rho] = \mathrm{rank}_{\mathbb{Z}_p}\{(U/\mu) \otimes_{\mathbb{Z}} W\}[\widetilde{\rho}]$$

for the unique Teichmüller lift $\widetilde{\rho}$ of ρ (see Corollary 2.7 in Chapter 2). By p-adic logarithm, $(U/\mu) \otimes_{\mathbb{Z}} \mathbb{Q} \cong F$ as $\overline{G}$-modules; so, by the normal base theorem $F \cong K[\overline{G}] \cong (\mathbb{Q}_p[\overline{G}])^{[K:\mathbb{Q}_p]}$ as $\overline{G}$-modules. This shows that

$$[K : \mathbb{Q}_p] \dim_{\mathbb{F}} M = [K : \mathbb{Q}_p] \dim_{\mathbb{F}} \rho = \dim_{\mathbb{F}} \left(\{(F^\times/(F^\times)^p) \otimes_{\mathbb{F}} \kappa\}[\rho]\right)$$

as desired.

The case where $\ell \neq p$ can be treated similarly (it is much easier than the case where $\ell = p$) and is left to the reader (Exercise 1). □

Exercises

(1) Give a detailed proof of Theorem 4.4.4 when $|M|$ is prime to p.

(2) Prove (4.45).

4.4.5 Global Euler characteristic formula

We use the notation introduced in Subsections 4.4.2 and 4.4.3. Let M be a finite $\mathfrak{G}_S$-module with ℓ-power order for $\ell \in S$. We define the global Euler characteristic of M by

$$\chi(M) = \chi(\mathfrak{G}_S, M) = \sum_{q=0}^{2} (-1)^q \,\mathrm{length}_{\mathbb{Z}_\ell} H^q_{ct}(\mathfrak{G}_S, M). \tag{4.48}$$

We would like to prove

THEOREM 4.53 (J. TATE, 1965) *Let Σ_∞ be the set of archimedean places of K. We have*

$$\chi(\mathfrak{G}_S, M) = \sum_{v \in \Sigma_\infty} \left(\mathrm{length}_{\mathbb{Z}_\ell} H^0(\mathrm{Gal}(\overline{K}_v/K_v), M) - [K_v : \mathbb{R}]\,\mathrm{length}_{\mathbb{Z}_\ell} M\right).$$

Proof We follow the proof of Milne given in [ADT] I.5. Let $\varphi(M)$ be the difference of the above formula, and write $\chi(M) = \chi(\mathfrak{G}_S, M)$. Thus we need to prove that

$$\varphi(M) = 0.$$

We shall prove the theorem here assuming $\ell > 2$. We give a sketch of a proof for $\ell = 2$ later (see Remark 4.54). For simplicity, we write $H^q(M)$ (resp. $H^q_v(M)$)) for the cohomology group $H^q_{ct}(\mathfrak{G}, M)$ (resp. $H^q(\mathrm{Gal}(\overline{K}_v/K_v), M)$. By Theorem 4.50 (3),

$$H^q(M) \cong \prod_{v \in \Sigma_\infty} H^q_v(M)$$

for $q \geq 3$. Thus if $\ell > 2$, $H^q(M) = 0$ for all $q \geq 3$ (see Proposition 4.21). For each short exact sequence: $0 \to L \xrightarrow{\alpha} M \xrightarrow{\beta} N \to 0$ of finite $\mathbb{Z}_\ell[\mathfrak{G}_S]$-modules, we have the long exact sequence:

$$\begin{aligned} 0 \to H^0(L) &\xrightarrow{\alpha_1} H^0(M) \xrightarrow{\beta_1} H^0(N) \to \cdots \\ \to H^2(L) &\xrightarrow{\alpha_4} H^2(M) \xrightarrow{\beta_4} H^2(N) \to 0. \end{aligned} \tag{4.49}$$

Thus if $\ell > 2$, $\chi(L) + \chi(N) = \chi(M)$ and hence $\varphi(L) + \varphi(N) = \varphi(M)$, because $M \mapsto \mathrm{length}_{\mathbb{Z}_\ell} H^0_v(M)$ for $v \in \Sigma_\infty$ and $M \mapsto \mathrm{length}_{\mathbb{Z}_\ell} M$ are both additive (on the Grothendieck group $R_{\mathbb{F}}(\mathfrak{G})$).

Therefore χ and φ factor through the Grothendieck group $R_{\mathbb{F}}(\mathfrak{G})$ for $\mathbb{F} = \mathbb{Z}/\ell\mathbb{Z}$ and have values in $\mathbb{Q}$. By Theorem 4.50 (1) and Theorem 4.43, all the terms of the exact sequence of Theorem 4.50 (4) are finite. Then by (4.46), we get

$$\chi(M) + \chi(M^*(1)) = \sum_{v \in \Sigma} \chi_v(G_v, M),$$

where $\chi_v(G_v, M)$ is the local Euler characteristic

$$\chi(G_v, M) = \sum_{q=0}^{2} (-1)^q \,\mathrm{length}_{\mathbb{Z}_\ell} H^q_{ct}(G_v, M)$$

defined for $G_v = \mathrm{Gal}(\overline{K}_v/K_v)$. Thus by Theorem 4.45 (2) and Theorem 4.52,

$$\sum_{v \in \Sigma} \chi_v(G_v, M) = \log_\ell \left(\prod_{v \in \Sigma} ||M||_{K_v} \right).$$

Since $||M||_{K_v} = 1$ for $v \notin \Sigma$, by the product formula (cf. (3.20) in Chapter 3):

$$\prod_{v \in \Sigma} ||M||_{K_v} = \prod_{v} ||M||_{K_v} = 1,$$

we know that

$$\chi(M) + \chi(M^*(1)) = 0 \quad \text{and} \quad \varphi(M) + \varphi(M^*(1)) = 0.$$

We now prove that $\varphi(M) = \varphi(M^*(1))$, which will complete the proof. We take a finite Galois extension F/K such that $\mathfrak{H} = \mathrm{Gal}(\overline{\mathbb{Q}}/F)$ acts trivially on M and μ_ℓ. We write $\overline{G} = \mathrm{Gal}(F/K)$. By the same argument as in the proof of Theorem 4.52, we may assume that F/K is cyclic of degree prime to ℓ. The field F is totally imaginary; so, by Theorem 4.50 (3), $H^q_F(M) = H^q_{ct}(\mathfrak{H}, M) = 0$ if $q \geq 3$. The module $H^q_F(M)$ is a finite $\mathbb{Z}_\ell[\overline{G}]$-module (see the definition of the action just above Theorem 4.33). Thus $[M] \mapsto \sum_{q=0}^{2}(-1)^q[H^q_F(M)]$ defines an additive endomorphism χ' of $R_{\mathbb{F}}(\overline{G})$ for $\mathbb{F} = \mathbb{Z}/\ell\mathbb{Z}$. Similarly, $[M] \mapsto [M^*] = [\mathrm{Hom}_{\mathbb{Z}}(M, \mathbb{F})]$ defines an involution of $R_{\mathbb{F}}(\overline{G})$. We claim the following facts:

(1) $$\chi'(M^*(1)) = [M^*] \cdot \chi'(\mu_\ell);$$

(2) $$[M] \cdot [\mathbb{F}[\overline{G}]] = \dim_{\mathbb{F}}(M)[\mathbb{F}[\overline{G}]].$$

Here $[M] \cdot [N] = [M \otimes_{\mathbb{F}} N]$ and $\overline{G}$ acts diagonally on $M \otimes_{\mathbb{F}} N$. Since $\mathfrak{H} = \mathrm{Gal}(\overline{\mathbb{Q}}/F)$ acts trivially on μ_ℓ and M, we see that $M \otimes_{\mathbb{F}} \mu_\ell \cong \mu_\ell^{\dim M}$ as $\mathfrak{H}$-modules. The action of $\overline{G}$ is given by $\sigma u(h) = \sigma(u(\sigma^{-1}h\sigma))$ for cocycle $u : \mathfrak{H}^q \to M$ and $\sigma \in \mathfrak{G}$. Fixing a base of M, write the action of $\sigma \in \overline{G}$ on M as a matrix form $\rho_M(\sigma)$. We identify the two $\mathbb{F}$-vector spaces M and $\mu_\ell^{\dim M}$ via this base. Thus the matrix $\rho_M(\sigma)$ still acts on $\mu_\ell^{\dim M}$. Then $\overline{G}$ acts on $\mu_\ell^{\dim M}$ by $\sigma v = \rho_M(\sigma)\omega(\sigma)v$ for the (mod ℓ) Teichmüller character ω. From this, it is clear that

$$H^q_F(\mu_\ell \otimes_{\mathbb{F}} M) = H^q_F(\mu_\ell) \otimes_{\mathbb{F}} M$$

as $\overline{G}$-modules. This proves the first statement. The second follows from the isomorphism:

$$M^\circ \otimes_{\mathbb{F}} \mathbb{F}[\overline{G}] \cong M \otimes_{\mathbb{F}} \mathbb{F}[\overline{G}]$$

given by $m \otimes \sigma \mapsto \sigma m \otimes \sigma$, where M° is the trivial $\overline{G}$-module with $M^\circ \cong M$ as $\mathbb{F}$-vector spaces.

We define $\theta : R_{\mathbb{F}}(\overline{G}) \to \mathbb{Z}$ by $\theta([M]) = \dim_{\mathbb{F}} H^0(\overline{G}, M)$. Since $|\overline{G}|$ is prime to ℓ, $M \mapsto H^0(\overline{G}, M)$ preserves short exact sequences of $\mathbb{F}[\overline{G}]$-modules, and hence θ extends to a homomorphism of $R_{\mathbb{F}}(\overline{G})$ into $\mathbb{Z}$. Since $|\overline{G}|$ is prime to ℓ, $H^q(\overline{G}, M) = 0$ for all $q > 0$, and hence, by the inflation and restriction sequence Theorem 4.33, $H^q(M) \cong H^0(\overline{G}, H^q_F(M))$. Thus $\chi(M) = \theta(\chi'([M]))$. Then from the above two claims and $\mathbb{F}[\overline{G}] \cong \mathbb{F}[\overline{G}]^*$ as $\mathbb{F}[\overline{G}]$-modules, we find that

$$\chi'(M^*(1)) \cdot [\mathbb{F}[\overline{G}]^*] = [M^*] \cdot [\mathbb{F}[\overline{G}]] \cdot \chi'(\mu_\ell) = \dim_{\mathbb{F}}(M)[\mathbb{F}[\overline{G}]^*] \cdot \chi'(\mu_\ell).$$

Since $\dim_{\mathbb{F}} M = \dim_{\mathbb{F}} M^*(1)$, the right-hand side of the above formula is the same even if we replace M in the left-hand side by $M^*(1)$. Thus this shows that $\chi(M) = \chi(M^*(1))$, which implies that $\varphi(M) = \varphi(M^*(1))$. This completes the proof. □

REMARK 4.54 When $\ell = 2$, the above argument works well except for the fact that $\ell > 2$ is used to show the additivity of φ with respect to short exact sequences. Thus we need to modify the proof of $\varphi(M) = \varphi(N)+\varphi(L)$ for short exact sequences $0 \to L \xrightarrow{\alpha} M \xrightarrow{\beta} N \to 0$. By definition, we have

$$\varphi(M) = \chi(M) - \sum_{v\in\Sigma_\infty} \left(\mathrm{length}_{\mathbb{Z}_\ell} H^0(\mathrm{Gal}(\overline{K}_v/K_v), M) - [K_v : \mathbb{R}]\, \mathrm{length}_{\mathbb{Z}_\ell} M\right).$$

Then we truncate the long exact sequence associated with the short exact sequence as follows:

$$\begin{aligned} 0 \to H^0(L) &\xrightarrow{\alpha_1} H^0(M) \xrightarrow{\beta_1} H^0(N) \to \cdots \\ \to H^4(L) &\xrightarrow{\alpha_4} H^4(M) \xrightarrow{\beta_4} H^4(N) \xrightarrow{\delta} \mathrm{Ker}(H^5(L) \xrightarrow{\alpha_5} H^5(M)) \to 0. \end{aligned} \tag{4.50}$$

By Theorem 4.50 (3),

$$H^q(\mathfrak{G}_S, M) \cong \prod_{v\in\Sigma_\infty} H^q_v(M)$$

for $q \geq 3$. Thus we have

$$\mathrm{Ker}(H^5(L) \xrightarrow{\alpha_5} H^5(M)) \cong \prod_{v\in\Sigma_\infty} C_v$$

for $C_v = \mathrm{Ker}(H^5_v(L) \xrightarrow{\alpha_{5,v}} H^5_v(M))$. Since $\mathrm{Gal}(\overline{K}_v/K_v)$ is cyclic of order 1 or 2, it is easy to see from Proposition 4.18 that

$$\mathrm{length}(H^3(\mathfrak{G}_S, M)) = \mathrm{length}(H^4(\mathfrak{G}_S, M)).$$

Thus by (4.50) (and (4.46)), we get

$$\chi(L) + \chi(N) = \chi(M) + \sum_{v\in\Sigma_\infty} \mathrm{length}(C_v),$$

and hence,

$$\begin{aligned} &\varphi(L) + \varphi(N) - X(L) - X(N) = \varphi(M) - X(M) \\ &+ \sum_{v\in\Sigma_\infty} \left(\mathrm{length}(C_v) - \mathrm{length}(H^0_v(L)) + \mathrm{length}(H^0_v(M)) - \mathrm{length}(H^0_v(N))\right), \end{aligned}$$

where $X(M) = \sum_{v\in\Sigma_\infty} [K_v : \mathbb{R}]\, \mathrm{length}_{\mathbb{Z}_\ell} M$. Since $X(L) + X(N) = X(M)$, we need to prove the vanishing of

$$\left(\mathrm{length}(C_v) - \mathrm{length}(H^0_v(L)) + \mathrm{length}(H^0_v(M)) - \mathrm{length}(H^0_v(N))\right).$$

Again by the periodicity of cohomology groups of cyclic groups: Proposition 4.18, we see that see

$$C_v \cong \mathrm{Ker}(H_v^1(L) \xrightarrow{\alpha_{1,v}} H_v^1(M)).$$

We look at the local truncated exact sequence:

$$0 \to H_v^0(L) \to H_v^0(M) \to H_v^0(N) \to C_v \to 0,$$

which shows that

$$\mathrm{length}(C_v) = \mathrm{length}(H_v^0(L)) - \mathrm{length}(H_v^0(M)) + \mathrm{length}(H_v^0(N))$$

as desired.

5

Modular L-Values and Selmer Groups

Dirichlet's class number formula

$$\frac{G(\chi)L(1,\chi)}{2\pi i} = \frac{h(-D)}{w}$$

is the prototypical example of a number theoretic identity connecting an analytically defined L-value to a purely arithmetic invariant. Here $-D$ is a discriminant of an imaginary quadratic field $F = \mathbb{Q}(\sqrt{-D})$, $\chi = \left(\frac{-D}{\cdot}\right)$ is the Legendre symbol, $G(\chi)$ is the Gauss sum of χ, $h(D)$ is the class number of F, and w is the number of roots of unity in F. The following features stand out:

(1) The left-hand side is the value at 1 of an Euler product, associated with a Galois representation $\chi : \mathrm{Gal}(\overline{\mathbb{Q}}/\mathbb{Q}) \cong \{\pm 1\} \subset GL_1(\mathbb{Q})$ divided by a transcendental number (which is the period of the exponential function); the result is a simple number in the field generated by the values of χ (in this case, it is just a rational number);

(2) The principal part of the right-hand side is the order of a group, in this case, the class group of F, which is the order of the χ-eigenspace inside the group of everywhere unramified homomorphisms from the Galois group over F into $\mathbb{Q}_p/\mathbb{Z}_p$ (by class field theory);

(3) The proof of the formula proceeds as follows: first, one relates the class number $h(-D)$ to the residue of the Dedekind zeta function ζ_F of F, that is, the L-function of the self dual Galois representation

$$\mathbf{1} \oplus \chi = \mathrm{Ind}_F^{\mathbb{Q}} \mathbf{1};$$

second, one uses the fact that, $\zeta_F(s) = L(s,\chi)\zeta(s)$ for the Riemann zeta function $\zeta(s)$ and the residue formula $Res_{s=1}\zeta(s) = 1$ to complete the proof of the identity.

Generalization of each of the three properties has been proposed by different number theorists. But there have been many obstacles. For a general compatible system of Galois representations $\varphi = \{\varphi_{\mathfrak{p}}\}_{\mathfrak{p}}$ discussed in Chapter 1, even searching for evaluation points m at which we expect the value $L(m,\varphi)$ to have a good rationality property as in (1) has been a difficult question. Shimura and Deligne proposed to look at those integers m such that the Γ-factor is finite at m as well as at its reflection point under the functional equation. This type of integer m is called *critical.* When φ comes from a projective algebraic variety (or more generally from *a motive*) defined over a number field, there is a simple way of predicting the Γ-factor due to Serre (see [D2] and [H94b] Section 1). Thus we can definitely say which values are critical. For critical values, Deligne made a conjecture specifying the period one needs to divide the L-value to get an algebraic number. The rationality of critical values has been proved by Shimura and others in many cases for L-functions on the Hecke side (that is, L-functions associated with modular forms; see [H94b] for a list of such results). For general integers m, the value has to be divided by a further complicated (supposedly) transcendental factor, called a (generalized) *regulator* defined by Beilinson and others (see [BCL]).

As for the property (2), in different contexts, modern number theorists have attempted to create a group, often called the *Selmer group*, associated with each Galois representation $\varphi : \mathfrak{G} \to GL_n(A)$, such that its order is related to $L(1,\varphi)$ or $L(0,\varphi)$, as in (1)–(2) above. We will recall one of such definitions, due to Greenberg, in this chapter. Greenberg's definition is relatively elementary, for it uses only Galois cohomology theory, but it includes general local $\mathcal{O}$-algebras A. However, this definition requires an *ordinarity* condition on Galois representations. Another general definition is due to Bloch–Kato and will not be covered by this book, because it requires knowledge of the theory of motives and p-adic periods.

As for the third feature (3), the most simple (and natural) way to create a self dual representation containing the trivial representation $\mathbf{1}$ is to form the tensor product of a given φ with its contragredient $\widetilde{\varphi}$: $\varphi \otimes \widetilde{\varphi}$. We define, assuming that n is invertible in A, an (n^2-1)-dimensional representation $Ad(\varphi) : \mathfrak{G}_Q \to GL_N(A)$ $(N = n^2 - 1)$ so that $\varphi \otimes \widetilde{\varphi} \cong \mathbf{1} \oplus Ad(\varphi)$. When $n = 2$, $s = 1$ is critical with respect to $Ad(\varphi)$ if

$\det \varphi(c) = -1$ for complex conjugation c. Then we expect that $\frac{L(1,Ad(\varphi))}{\text{a period}}$ should be somehow related to the size of $\mathrm{Sel}(Ad(\varphi))$ in the most favorable cases, for the Selmer group $\mathrm{Sel}(Ad(\varphi))$ of $Ad(\varphi)$, in a fashion analogous to (3). Even if $s = 1$ is not critical for $Ad(\varphi)$, there seems to be a good way to define a natural transcendental factor of $L(1, Ad(\varphi))$ only using data from the Hecke side (see [H99a]).

It is not a mere coincidence that the Selmer group for $Ad(\overline{\rho})$ for a modular two-dimensional representation $\overline{\rho} : \mathfrak{G} \to GL_2(\mathbb{F})$ is the cohomology group $H^1_{\mathscr{D}}(\mathbb{Q}, Ad(\overline{\rho}))$ we scrutinized in the proof of the Taylor-Wiles theorem. The Taylor-Wiles theorem is powerful enough, if we combine it with one of my earlier works (see [H81a], [H81b] and [H88b]), to identify the adjoint L-value $L(1, Ad(\varphi))$ with the size of the Selmer group $\mathrm{Sel}(Ad(\varphi))$ for two-dimensional Galois representations φ attached to modular forms. We will give a sketch of this fact in Subsection 5.3.3.

As remarked above, Greenberg's definition requires φ to satisfy a certain p-ordinarity condition. Therefore we may vary Galois representations on the spectrum of the ordinary universal deformation ring R^{ord} over a discrete valuation ring $\mathcal{O}$ (finite flat over $\mathbb{Z}_p$). In other words, every point $P \in Spec(R^{ord})(\mathcal{O})$ (which is $\mathrm{Hom}_{\mathcal{O}\text{-}alg}(R^{ord}, \mathcal{O})$ as a set) parametrizes the maximal family of ordinary deformations by $P \mapsto \varphi_P = P \circ \varrho$, where ϱ is the universal ordinary deformation. Thus $P \mapsto \mathrm{Sel}(Ad(\varphi_P))$ may be considered to be local sections of the global (coherent) sheaf $\mathrm{Sel}(Ad(\varrho))$ on $Spec(R^{ord})$. Therefore, we expect to have an element $L_p = L_p(Ad(\varrho)) \in R^{ord}$ such that $L_p(P) = L_p \mod P$ giving rise to $\frac{L(1,Ad(\varphi_P))}{\text{a period}}$ somehow. The element L_p may be called the *p-adic L-function* of $Ad(\varrho)$. We study in Subsection 5.3.6 the function L_p and its restriction to various closed subschemes of $Spec(R^{ord})$. This is a non-abelian generalization of Iwasawa theory, because R^{ord} can be identified with an appropriate local ring of the universal p-ordinary Hecke algebra defined and studied in [H86a] and [H86b]. Thus the p-adic L-function, if it exists, is an object on the Hecke side. Then we are naturally led to study how deformation rings and Selmer groups vary if we change the base field $\mathbb{Q}$ to an extension, in particular, $\mathbb{Z}_p$-extensions.

In this context, the study of base-change of deformation rings therefore plays a crucial role. In the program of Langlands–Shimura, recalled briefly in Chapter 1, one of the most effective tools in studying arithmetic problems is to analyze their functoriality in Langlands' sense. That is, we use information from the Galois side to predict (more complicated) relations among modular forms on the Hecke side. The simplest example of such functorialities is the base-change for an extension $F/\mathbb{Q}$ [BCG].

We study intricate relations between various universal deformation rings of p-adic Galois representations under the base-change process. We start with a group theoretic description of base-change, and later we apply it to Galois representations.

Some of the material treated here requires a good knowledge of (topological) cohomology groups of differentiable manifolds with coefficients in locally constant sheaves and also some knowledge of Hilbert modular forms. A brief description of such cohomology theory is given in my book [LFE] as an appendix. This last chapter is added to give the reader an overview of the theory that is a little more advanced in technology than the one exposed in the earlier chapters. Therefore, we shall give detailed proofs for the first five sections (which are basically elementary), but in the last section, 5.6, we content ourselves with some indication of proof and references.

5.1 Selmer Groups

In this section, we shall first define the Selmer group attached to each p-ordinary Galois representation, and then describe basic properties of the Selmer groups. The definition we adopt is due to R. Greenberg [G] and is Galois-cohomological. There is another definition due to Bloch–Kato [BK] based on p-adic algebro-geometric cohomology theory (étale and de Rham cohomology) and Fontaine's theory of p-adic period rings [Fo3]. The comparison of the two Selmer groups (when they are both well defined) has been made in [Fl] and [Oc].

5.1.1 Definition

Let F/E be a **finite** Galois extension of a fixed number field E inside a fixed algebraic closure $\overline{\mathbb{Q}}$ of $\mathbb{Q}$. We fix a prime p. We write $F^{(p,\infty)}/F$ for the maximal extension of F unramified outside $\{p, \infty\}$. Thus infinite-ramification is limited to places over p in $\mathrm{Gal}(F^{(p,\infty)}/E)$. The Selmer group can be defined without such limitation, and we refer to Greenberg's original paper [G] for a more general treatment. The exposition of Selmer groups here follows the treatment in [H99b], where a slightly more general case is dealt with.

Although we mainly study two and three-dimensional cases, we start slightly more generally giving a definition for n-dimensional representations. Let $\mathbb{J}$ be a p-adic pro-artinian local ring with finite residue field $\mathbb{F}$, and take a Galois representation $\varphi : \mathfrak{G} = \mathrm{Gal}(F^{(p,\infty)}/E) \to GL_n(\mathbb{J})$.

We write $V = V(\varphi)$ for the representation space of φ, which is a $\mathbb{J}$-free module of rank n. Let L be an intermediate extension of $F^{(p,\infty)}/E$ with Galois group $\mathfrak{H}_L = \mathrm{Gal}(F^{(p,\infty)}/L)$. Let S_L be the set of all prime factors of p in L. For each $\mathfrak{p} \in S_L$, we fix a decomposition subgroup $D_\mathfrak{p}$ in $\mathfrak{H}_L$. For each $\mathfrak{p} \in S_E$, we assume that we are given a $D_\mathfrak{p}$-stable filtration:

$$0 = V(\varphi)_{0,\mathfrak{p}} \subset V(\varphi)_{1,\mathfrak{p}} \subset V(\varphi)_{2,\mathfrak{p}} \subset \cdots \subset V(\varphi)_{m_\mathfrak{p},\mathfrak{p}} = V(\varphi), \quad (\mathrm{fil}_\mathfrak{p})$$

where we assume that $V(\varphi)/V(\varphi)_{j,\mathfrak{p}}$ for all $j = 1, \ldots, m_\mathfrak{p} - 1$ are all $\mathbb{J}$-free. Although to define the Selmer group we only need to assume the existence of a two step filtration $(\mathrm{fil}_\mathfrak{p})$ for $\mathfrak{p} \in S_L$ over L, we insist on having the possibly longer filtration $(\mathrm{fil}_\mathfrak{p})$ valid over E and each subquotient being $\mathbb{J}$-free. Then the filtration $(\mathrm{fil}_\mathfrak{p})_{\mathfrak{p}\in S_E}$ over E induces the filtration $(\mathrm{fil}_\mathfrak{p})_{\mathfrak{P}\in S_L}$ over L.

The stabilizer of the filtration gives rise to a conjugacy class of a parabolic subgroup $P_\mathfrak{p}$ of $GL(n)$. We call the representation φ with $(\mathrm{fil}_\mathfrak{p})$ *nearly ordinary* of type $\mathcal{F} = \{P_\mathfrak{p}\}_\mathfrak{p}$ and call $\mathcal{F}$ the *nearly ordinarity datum* for φ. When all $P_\mathfrak{p}$ are Borel subgroups (that is, subgroups conjugate to groups made up of upper triangular matrices), we call φ (nearly ordinary) of Borel type. We write $\delta_{j,\mathfrak{p},\varphi}$ for the representation of $D_\mathfrak{p}$ on $V(\varphi)_{j,\mathfrak{p}}/V(\varphi)_{j-1,\mathfrak{p}}$. If φ is of Borel type (that is, the filtration is maximal), $\delta_{j,\mathfrak{p},\varphi}$ is a character.

If the reader is not familiar with the terminology of linear algebraic groups, he or she can hereafter assume that $n = 2$, then the filtration has only two steps: $V(\varphi)_{0,\mathfrak{p}} \subset V(\varphi)_{1,\mathfrak{p}} \subset V(\varphi)_{2,\mathfrak{p}} = V(\varphi)$ and $P_\mathfrak{p}$ is the subgroup of upper triangular matrices, that is, the (standard) Borel subgroup of $GL(2)$. For the reader who wants a reference on algebraic groups, we suggest the book by Jantzen [RAG].

We fix one step $0 \le j \le m$. We write $\delta_\mathfrak{p}^-$ for the representation of $D_\mathfrak{p}$ on $V(\varphi)_{j,\mathfrak{p}} = V(\delta_\mathfrak{p}^-)$. We call such a datum a local *Selmer datum* $\mathcal{S} = \{V(\delta_{\varphi,\mathfrak{p}}^-)\}_{\mathfrak{p}\in S_E}$. Then we consider the Pontryagin dual $\mathbb{J}$-module $\mathbb{J}^* = \mathrm{Hom}_{\mathbb{Z}_p}(\mathbb{J}, \mathbb{Q}_p/\mathbb{Z}_p)$. For each $\mathbb{J}$-module X, we define $X^* = X \otimes_\mathbb{J} \mathbb{J}^*$ and let $\mathfrak{G}$ or its subgroup act on X^* through the left factor if X has an action of the subgroup. Then, for the representation space $V(\varphi)(\cong \mathbb{J}^n)$, $V(\varphi)^*$ is a discrete $\mathfrak{G}$-module.

We consider the Galois cohomology group $H^1_{ct}(\mathfrak{G}, V(\varphi)^*)$, which is a discrete $\mathbb{J}$-torsion module. Since F/E is a finite Galois extension, ramification outside p of $F^{(p,\infty)}/E$ is limited to a finite set Σ_E of places of E with finite ramification indices. For an intermediate extension L of $F^{(p,\infty)}/E$, we write Σ_L for the set of all prime ideals of L over Σ_E. Then we define the Selmer group with respect to $\mathcal{S}$ and L by

$$\mathrm{Sel}_{\mathcal{S}}(\varphi)_{/L} = \mathrm{Ker}(H^1_{ct}(\mathfrak{H}_L, V(\varphi)^*) \to \prod_{\mathfrak{p}\in S_L} H^1_{ct}(I_{\mathfrak{p}}, V(\delta^+_{\mathfrak{p}})^*) \times \prod_{\mathfrak{q}\in\Sigma_L} H^1_{ct}(I_{\mathfrak{q}}, V(\varphi)^*)), \tag{Sel}$$

where $V(\delta^+_{\mathfrak{p}}) = V(\varphi_{\mathfrak{p}})/V(\delta^-_{\varphi,\mathfrak{p}})$, $I_{\mathfrak{q}}$ is the inertia subgroup of the decomposition group $D_{\mathfrak{q}} \subset \mathfrak{H}_L$ and the map is the restriction for $\mathfrak{q} \in \Sigma_L$ and the restriction composed with the projection: $V(\varphi)^* \to V(\delta^+_{\varphi,\mathfrak{p}})^*$ for $\mathfrak{p} \in S_L$. We can define the *strict Selmer group* $\mathrm{Sel}_{\mathcal{S},st}(\varphi)_{/L}$ replacing $I_{\mathfrak{p}}$ by $D_{\mathfrak{p}}$ in the above definition. If we allow infinite ramification outside p, like the case of an elliptic curve with multiplicative reduction at $\mathfrak{q}$ (see [MTT] and the treatment of the strict case in [W2]), we need to replace $H^1_{ct}(I_{\mathfrak{q}}, V(\varphi)^*)$ in the above formula by $H^1_{ct}(I_{\mathfrak{q}}, V(\delta^+_{\mathfrak{q}})^*)$ introducing a filtration $V(\varphi) \twoheadrightarrow V(\delta^+_{\mathfrak{q}})$ stable under $D_{\mathfrak{q}}$ in order to define a Selmer group canonically related to the L-function, further complicating the argument. This is the main reason why we have limited ramification outside p to be finite.

If either $p \nmid [F : E]$ or $L \supset F$, the inertia group at $\mathfrak{q} \in \Sigma_L$ has order prime to p, and hence $H^1(I_{\mathfrak{q}}, V(\varphi)^*) = 0$ by Proposition 4.21 in Chapter 4. Thus in this case, we have a simpler definition:

$$\mathrm{Sel}_{\mathcal{S}}(\varphi)_{/L} = \bigcap_{\mathfrak{p}\in S_L} \mathrm{Ker}(H^1_{ct}(\mathfrak{H}_L, V(\varphi)^*) \to H^1_{ct}(I_{\mathfrak{p}}, V(\delta^+_{\mathfrak{p}})^*)) \quad \text{if } p \nmid [F : E] \text{ or } L \supset F. \tag{5.1}$$

Here are some examples:

Example 5.1 Let $F/\mathbb{Q}$ be a quadratic extension (so we put $E = \mathbb{Q}$) and $\chi : \mathrm{Gal}(F/\mathbb{Q}) \cong \{\pm 1\}$ be the unique non-trivial character. We may regard χ as having values in $\mathbb{Z}_p^\times$, and assume $p > 2$. We take $V(\delta_p^+)$ to be the full space $V(\chi)$. Then $\mathrm{Sel}(\chi)_{/\mathbb{Q}}$ is isomorphic to the p-part $Cl_{F,p}$ of the strict class group of F. This can be proved as follows: By the inflation-restriction sequence, we have

$$H^1(\mathfrak{G}, V(\chi)^*) \cong H^0(\mathfrak{G}, H^1(\mathfrak{H}, V(\chi)^*)).$$

Since $\mathfrak{G}$ acts on $V(\chi)^*$ via χ, $H^0(\mathfrak{G}, H^1(\mathfrak{H}, V(\chi)^*)$ is actually the χ-eigenspace $H^1(\mathfrak{H}, V(\chi)^*)[\chi]$ in $H^1(\mathfrak{H}, V(\chi)^*)$ under the $\mathfrak{G}$-action. Thus we have

$$H^1(\mathfrak{G}, V(\chi)^*) \cong \mathrm{Hom}_{\mathfrak{G}}(\mathfrak{H}, V(\chi)^*).$$

Then the above isomorphism induces

$$\mathrm{Sel}(\chi) \overset{\alpha}{\cong} \mathrm{Hom}_{\mathfrak{G}}(Cl_{F,p}, V(\chi)^*) \cong \mathrm{Hom}(Cl_{F,p}[\chi], \mathbb{Q}_p/\mathbb{Z}_p) \cong Cl_{F,p},$$

because $\mathfrak{a}^\sigma = \mathfrak{a}^{-1}$ in the class group $Cl_{F,p}$, that is, the χ-eigenspace $Cl_{F,p}[\chi]$ is equal to the full group $Cl_{F,p}$. Thus the isomorphism α follows from class field theory, and the last isomorphism holds because $Cl_{F,p}[\mathrm{id}]$ is trivial (that is, $\mathbb{Z}$ is a PID!). So if F is imaginary (that is, $\chi(c) = -1$ for complex conjugation $\iff s = 1$ is critical for $L(s, \chi)$), we have from Dirichlet's class number formula that

$$\left| \frac{G(\chi)L(1,\chi)}{2\pi i} \right|_p = \left|| \mathrm{Sel}(\chi) |\right|_p$$

as long as $p \nmid 2w$ for the number w of roots of unity in F. More generally, taking a finite cyclic extension $F/\mathbb{Q}$ of degree n with $p \nmid n$, we let $\chi : \mathfrak{G} \cong \mu_n \subset \mathcal{O}^\times$ denote a character such that $F = L^{\mathrm{Ker}(\chi)}$ for $L = F^{(p,\infty)}$. Then, for the choice: $V(\delta_p^+) = V(\chi)$, the Selmer group $\mathrm{Sel}(\chi)_{/\mathbb{Q}}$ is isomorphic to the χ-eigenspace of $Cl_{F,p} \otimes_{\mathbb{Z}_p} \mathcal{O}$. If F is imaginary ($\iff s = 1$ is critical for $L(s, \chi) \iff \chi(c) = -1$), this fact combined with Kummer's (relative) class number formula comes close to an analog of the above formula. However, Kummer's formula (see [ICF] Theorem 4.17) expresses the (relative) class number as a product of $\prod_{j:odd} \frac{G(\chi^j)L(1,\chi^j)}{2\pi i}$ for odd j with $0 < j < ord(\chi)$. Thus we have a formula for the minus part $Cl_{F,p}^-$ of $Cl_{F,p}$ (on which complex conjugation acts by '-1') but not the one giving the order of each separate χ^j-eigenspace of $Cl_{F,p}^- \otimes_{\mathbb{Z}_p} \mathcal{O}$. Indeed, one had to wait until the proof of the Iwasawa conjecture by Mazur–Wiles [MW] and Wiles [W1] to have an exact formula for the χ-part (see also [R] and [W]).

EXAMPLE 5.2 Assume $p > 2$. Let $E = \mathbb{Q}$. Let $\mathscr{E}_{/\mathbb{Q}}$ be an elliptic curve with ordinary good reduction at $p > 2$. Here $\mathscr{E}$ has ordinary good reduction means that $\mathscr{E}$ mod p remains as an elliptic curve (defined over $\mathbb{F}_p$) and that $\mathscr{E}[p] \cong \mathbb{Z}/p\mathbb{Z}$ for the subgroup $\mathscr{E}[p]$ of p-torsion points of $\mathscr{E}$ (cf. [GMF]). We suppose that $\mathscr{E}$ acquires everywhere good reduction for a finite extension F. The Tate module $T_p(\mathscr{E})$ has a natural filtration:

$$0 \to T_p(\mathscr{E}^c) \to T_p(\mathscr{E}) \to T_p(\mathscr{E}^{et}) \to 0.$$

Here $\mathscr{E}^c$ (resp. $\mathscr{E}^{et}$) is the connected component (resp. the maximal étale quotient) of the p-divisible group of $\mathscr{E}$ (defined over $\mathbb{Z}_p$). We take $V(\varphi)$ to be $T_p(\mathscr{E})$ and $V(\delta_p^+)$ to be $T_p(\mathscr{E}^{et})$. We write $\mathrm{Sel}(\mathscr{E})_{/L}$ for $\mathrm{Sel}(\varphi)_{/L}$ for this φ. Then by Kummer theory for $\mathcal{E}$, we have an exact sequence:

$$0 \to \mathcal{E}(\mathbb{Q}) \otimes_{\mathbb{Z}} \mathbb{Q}_p/\mathbb{Z}_p \to \mathrm{Sel}(\mathcal{E})_{/\mathbb{Q}} \to Ш(\mathcal{E})_{/\mathbb{Q}}[p^\infty] \to 0.$$

Here $Ш(\mathcal{E})_{/\mathbb{Q}}[p^\infty]$ is the part killed by a power of p in the Tate-Shafarevich group $Ш(\mathcal{E})_{/\mathbb{Q}}$ for the discrete $\mathfrak{G}$-module $\mathcal{E}(F^{(p,\infty)})$ over $\mathbb{Q}$. The Tate-Shafarevich group is defined by

$$Ш(\mathcal{E})_{/\mathbb{Q}} = \mathrm{Ker}(H^1_{ct}(\mathfrak{G}, \mathcal{E}(F^{(p,\infty)})) \to \prod_{q \in \Sigma_{\mathbb{Q}} \cup \{p\}} H^1_{ct}(D_q, \mathcal{E}(F^{(p,\infty)}_{\mathfrak{q}}))),$$

where $F^{(p,\infty)}_{\mathfrak{q}}$ is the $\mathfrak{q}$-adic completion of $F^{(p,\infty)}$ for a prime $\mathfrak{q}|q$ giving D_q. For details on this type of Selmer group, we refer the reader to Greenberg's exposition [G1] Section 2 and to [ADT] I.6.

The following example will be described in more detail later:

EXAMPLE 5.3 Let $f \in S_k(\Gamma_0(C), \chi)$ be a Hecke eigenform with a primitive 'Neben' Dirichlet character χ. Write $C = C(\chi)$ for the conductor of χ, and assume $p \nmid C$. Regarding χ as a character of $\mathrm{Gal}(\overline{\mathbb{Q}}/\mathbb{Q})$, we put $E = \mathbb{Q}$ and $F = \overline{\mathbb{Q}}^{\mathrm{Ker}(\chi)}$ with $\mathrm{Gal}(F/\mathbb{Q}) \overset{\chi}{\cong} \mathrm{Im}(\chi)$. Write $f|T(n) = \lambda(T(n))f$ for a system of Hecke eigenvalues $\lambda(T(n))$, and let $\mathbb{Q}(\lambda)$ be the number field generated by $\lambda(T(n))$ for all n. Take a prime ideal $\mathfrak{p}|p$ of $\mathbb{Q}(\lambda)$, and let $\mathcal{O}$ be the $\mathfrak{p}$-adic integer ring. Then we have a Galois representation $\varphi : \mathfrak{G} \to GL_2(\mathcal{O})$ associated with λ characterized by the fact: $\mathrm{Tr}(\varphi(Frob_\ell)) = \lambda(T(\ell))$ for all primes $\ell \nmid Cp$ (by Theorem 3.26 in Chapter 3). We further suppose that $\lambda(T(p)) \in \mathcal{O}^\times$. Then $V(\varphi)$ has a natural filtration:

$$0 \to V(\varepsilon) \to V(\varphi) \to V(\delta) \to 0$$

stable under D_p (Theorem 3.26 (2) in Chapter 3). Here $\mathrm{rank}_{\mathcal{O}} V(\delta) = 1$. We can think of $\mathrm{Sel}(\varphi)_{/\mathbb{Q}}$ as in the previous example, but here we look into $\mathrm{Sel}(Ad(\varphi))_{/\mathbb{Q}}$. We let $\mathfrak{G}$ act on $M_2(\mathcal{O}) = \mathrm{End}_{\mathcal{O}}(V(\varphi))$ by conjugation. Then the subspace $V(Ad(\varphi)) \subset M_2(\mathcal{O})$ made up of trace zero matrices is stable under the action. In this way, we get a three-dimensional representation $Ad(\varphi)$. The one-dimensional subspace

$$W = \{\phi \in \mathrm{End}_{\mathcal{O}}(V(\varphi)) | \mathrm{Tr}(\phi) = \phi(V(\varepsilon)) = 0\} \cong \left\{\left(\begin{smallmatrix} 0 & * \\ 0 & 0 \end{smallmatrix}\right)\right\} \subset V(Ad(\varphi))$$

is then stable under D_p. We define the Selmer group $\mathrm{Sel}(Ad(\varphi))_{/\mathbb{Q}}$ taking $V(\delta^+_{Ad(\varphi),p})$ to be $V(Ad(\varphi))/W$. This Selmer group has been studied in depth by Wiles in [W2], and Wiles' work combined with Theorem 5.16

below ([H81a] and [H88b]) yields:

$$|\operatorname{Sel}(Ad(\varphi))_{/\mathbb{Q}}| = \left|\frac{\Gamma(1,Ad(\lambda))L(1,Ad(\lambda))}{\Omega(+,\lambda;A)\Omega(-,\lambda;A)}\right|_p^{-r} = \left|\frac{\Gamma(1,Ad(\lambda))L(1,Ad(\lambda))}{\Omega(+,\lambda;A)\Omega(-,\lambda;A)}\right|_{\mathfrak{p}}^{-1},$$

where $r = \operatorname{rank}_{\mathbb{Z}_p} \mathcal{O}$, $|p|_p = \frac{1}{p}$ and $|x|_{\mathfrak{p}} = [\mathcal{O} : x\mathcal{O}]^{-1}$ for $x \in \mathcal{O}$. Here $A = \mathbb{Q}(\lambda) \cap \mathcal{O}$ is the discrete valuation ring of $\mathbb{Q}(\lambda)$ (induced by $\mathcal{O}$), the $\pm$-periods $\Omega(\pm,\lambda;A)$ are the normalized $\pm$-periods of f with respect to A (whose definition we recall later), and $L(s, Ad(\lambda)) = L(s, Ad(\varphi))$ is the adjoint L-function of λ with Γ-factor $\Gamma(s, Ad(\lambda))$ (see 5.3.1).

5.1.2 Motivic interpretation

Here is a general heuristic on the relation between L-values and the size of the Selmer group, when the Selmer group is associated to a pure motive. What we would like to do in this subsection is to describe the basic principles behind the examples we have given; thus, we do not intend to make the following formulas very precise. In other words, for the validity of the conjectural formula relating the order of a Selmer group and the corresponding L-value, we warn that one might have to modify, appropriately, the transcendental factor and some Euler factors of the L-value. We ignore such subtlety in this subsection.

Since this section requires some (working) knowledge of the theory of motives, the reader can skip this section on first reading. Skipping this section does not seriously affect the understanding of the rest of the book.

Suppose for simplicity that $F = \mathbb{Q}$, $\mathbb{J} = \mathbb{Z}_p$ and that $V(\varphi)$ is the p-adic étale realization of a rank n pure motive $M_{/\mathbb{Q}}$. We follow Deligne's treatment of motivic L-values given in [D2] in this subsection. For a general theory of motives, we refer to [HMS] and [D2] (and also [BCL] and [SGL]).

The de Rham realization $H_{DR}(M)$ has the Hodge filtration:

$$0 \subset \cdots \subset \mathcal{F}^j(M) \subset \mathcal{F}^{j-1}(M) \subset \cdots \subset H_{DR}(M).$$

After tensoring with $\mathbb{C}$, we have a canonical isomorphism

$$I_{\mathbb{C}} : H_B(M) \otimes_{\mathbb{Q}} \mathbb{C} \cong H_{DR}(M) \otimes_{\mathbb{Q}} \mathbb{C}$$

for the Betti realization $H_B(M)$, and $H_B(M) \otimes_{\mathbb{Q}} \mathbb{C}$ has the Kähler–Hodge decomposition

$$H_B(M) \otimes_{\mathbb{Q}} \mathbb{C} = \bigoplus_{i,j} H^{i,j}(M).$$

Then we have

$$\mathscr{F}^j(M) \otimes_{\mathbb{Q}} \mathbb{C} = I_{\mathbb{C}}\left(\bigoplus_{i \geq j} H^{i,k}(M)\right).$$

Thus the existence of Hodge filtration tells us that $I_{\mathbb{C}}\left(\bigoplus_{i\geq j} H^{i,k}(M)\right) \cap H_{DR}(M)$ gives a rational structure on the $\mathbb{C}$-vector space $I_{\mathbb{C}}\left(\bigoplus_{i\geq j} H^{i,k}(M)\right)$. When $M = H^r(X)$ for a projective variety $X_{/\mathbb{Q}}$, the filtration comes from the Dolbeaux spectral sequence:

$$\mathbb{H}^p(X, \Omega^q_{X/\mathbb{Q}}) \Rightarrow H^q_{DR}(X, \mathbb{Q}),$$

where $\Omega^q_{X/\mathbb{Q}}$ is the sheaf of Kähler q-differentials on X and $\mathbb{H}^\bullet$ indicates hyper-cohomology (see [ECH] Appendix C).

We write $h(i,j)$ for $\dim_{\mathbb{C}} H^{i,j}(M)$. The action of complex conjugation interchanges the Hodge components $H^{i,j}$ and $H^{j,i}$. Thus complex conjugation acts on $H^{i,i}(M)$. Let $H^{i,i}_{\pm}(M)$ be the $\pm(-1)^i$-eigenspace of the action of complex conjugation. We write $h_{\pm}(i,i)$ for $\dim H^{i,i}_{\pm}(M)$. A motive is called *pure* if $i + j = w$ is independent of non-trivial Hodge components. The integer w is called the *weight* of M. From now on, all motives will be pure.

Each ℓ-adic étale realization $H_{et,\ell}(M)$ gives rise to a Galois module of dimension n, giving a Galois representation $\varphi_\ell : \mathrm{Gal}(\overline{\mathbb{Q}}/\mathbb{Q}) \to GL_n(\mathbb{Q}_\ell)$. The system $\{\varphi_\ell\}$ is (conjecturally) a compatible system yielding an L-function $L(s,M) = L(s,\varphi)$. The gamma factor of this L-function is conjectured to be given by

$$\Gamma(s,M) = \left(\prod_{i<j} \Gamma_{\mathbb{C}}(s-i)^{h(i,j)}\right) \times (\Gamma_{\mathbb{R}}(s-i)^{h_+(i,i)}\Gamma_{\mathbb{R}}(s-i+1)^{h_-(i,i)}), \quad (\Gamma)$$

where $\Gamma_{\mathbb{C}}(s) = 2(2\pi)^{-s}\Gamma(s)$ and $\Gamma_{\mathbb{R}}(s) = \pi^{-s/2}\Gamma(s)$ for the gamma function $\Gamma(s) = \int_0^\infty e^{-t}t^{s-1}dt$ and the (i,i) term shows up only when $w = 2i$.

The L-function $L(s,M)$ is conjectured to have a meromorphic continuation to the whole complex plane and to satisfy the following functional equation:

$$\begin{aligned}\Gamma(s,M)L(s,M) &= \varepsilon(M,s)\Gamma(1-s,M^\vee)L(1-s,M^\vee)\\ &= \varepsilon(M,s)\Gamma(w+1-s,M)\overline{L(w+1-\overline{s},M)},\end{aligned}$$

where '$\overline{X}$' is the complex conjugate of X, $M^\vee$ is the dual of M with $M \otimes M^\vee \cong \mathrm{End}(M)$, and $\varepsilon(s,M)$ is the ε-factor, which is an exponential function of the form: $W(M)C(M)^{(w+1)/2-s}$ for a positive integer $C(M)$

and a constant $W(M)$ with $|W(M)| = 1$ (see [D2] and [D1] 8.12.4 for the definition of the ε-factors).

The Tate motive $\mathbb{Q}(j)_{/\mathbb{Q}}$ for positive integers j is characterized by the following property: $L(s, \mathbb{Q}(j)) = \zeta(s+j)$ for the Riemann zeta function $\zeta(s)$. Thus $\operatorname{rank}\mathbb{Q}(j) = \dim H_?(M) = 1$ for $? = et$, DR and B, and $h(-i,-j) = \delta_{i,j}$ for the Kronecker symbol $\delta_{i,j}$. The Galois representation on $H_{et,\ell}$ is the j-th power of the ℓ-adic cyclotomic character $\nu_\ell^j : \mathrm{Gal}(\mathbb{Q}(\mu_{\ell^\infty})/\mathbb{Q}) \to \mathbb{Z}_\ell^\times$ such that $\nu(\Phi_p) = p^{-j}$ for the geometric Frobenius Φ_p. We write $M(j) = M \otimes \mathbb{Q}(j)$. Then we have $L(s, M(j)) = L(s+j, M)$.

We hereafter suppose that M is **self-dual** (as in feature (3) of Dirichlet's class number formula at the beginning of the chapter); so the weight w is equal to 0. Then the motive M is of weight 0, because $H^{i,j}(M) \neq 0 \iff H^{-i,-j}(M^\vee) \neq 0$. Thus $L(s,M) = \varepsilon(s,M)L(1-s,M)$ by the functional equation, and hence the root number $W(M)$ has a simple form: $W(M) = \pm 1$. The motive M is called *critical* if $I_\mathbb{C}$ induces an isomorphism

$$I_\mathbb{C}^+ : H_B^+(M) \otimes_\mathbb{Q} \mathbb{C} \cong H_{DR}^+(M) \otimes_\mathbb{Q} \mathbb{C},$$

where $H_B^+(M)$ is the subspace fixed (element by element) by the action of complex conjugation c and $H_{DR}^+(M) = H_{DR}(M)/\mathcal{F}^-(M)$ for the bigger middle step of the Hodge filtration $\mathcal{F}^-(M) = \mathcal{F}^0(M)$. In particular, c has to act by scalar multiplication by '-1' on $H^{0,0}(M)$ (that is, $h_+(0,0) = 0$). Thus M is critical if and only if $\Gamma(s,M)$ is finite at $s = 0$ and 1. If M is critical, $I_\mathbb{C}$ also induces an isomorphism

$$I_\mathbb{C}^- : H_B^-(M) \otimes_\mathbb{Q} \mathbb{C} \cong H_{DR}^-(M) \otimes_\mathbb{Q} \mathbb{C},$$

where $H_B^-(M)$ is the minus eigenspace of the action of complex conjugation c and $H_{DR}^-(M) = H_{DR}(M)/\mathcal{F}^+(M)$ for the smaller middle step of the Hodge filtration $\mathcal{F}^+(M) = \mathcal{F}^1(M)$.

If M is critical of weight 0 and $L(0,M) \neq 0$, we expect that $|\mathrm{Sel}(\varphi)|$ is finite and is related to $|\frac{L(0,M)c_p^+(M)}{c_\infty^+(M)}|_p^{-1}$, where $c_\infty^+(M)$ is the Deligne period normalized with respect to $V(\delta_p^-)$ and $\mathcal{F}^-(M)$ (see [D2], [SGL] Chapter 3, and (CN1–2) of Theorem 5.20 in the text). To define these periods, we assume to have a big ring of periods X yielding an isomorphism of functors $I_X : H_B(M) \otimes_\mathbb{Q} X \cong H_{DR}(M) \otimes_\mathbb{Q} X$ for all motives defined over $\mathbb{Q}$ with coefficients in $\mathbb{Q}$, where $H_B(M)$ is the Betti realization of M and $H_{DR}(M)$ is the de Rham realization. The archimedean choice of X is the field of complex numbers (which contains all archimedean periods), as we have explained already.

Since the p-adic étale realization $H_{et,p}(M)$ is canonically isomorphic to the scalar extension $H_B(M) \otimes_{\mathbb{Q}} \mathbb{Q}_p$, we need to impose the following condition on M and the p-adic choice X:

$$I_X(V(\delta_p^-) \otimes_{\mathbb{Q}} X) = \mathscr{F}^-(M) \otimes_{\mathbb{Q}} X \qquad \text{(adm)}$$

to facilitate a link between the Selmer group and the period. This is a kind of ordinarity condition imposed on M, first formulated by Panchishkin in this generality (see [SGL] 3.5). Under this Panchishkin's admissibility condition (adm), we have a homomorphism induced by I_X:

$$I_X^{\pm} : H_B^{\mp}(M) \otimes_{\mathbb{Q}} X \to H_{DR}^{\pm} \otimes_{\mathbb{Q}} X.$$

Choose a base $b^{\pm} = (b_j^{\mp})$ of $H_B^{\mp}(M)$ over $\mathbb{Q}$ and $\omega^{\pm} = (\omega_j^{\mp})_j$ of $H_{DR}^{\pm}$ over $\mathbb{Q}$, and define $c_X^{\pm}(M)$ to be the determinant of $I_X^{\pm}$ with respect to $b^{\pm}$ and $\omega^{\pm}$. When $X = \mathbb{C}$, we write $c_{\infty}^{\pm}(M)$ for $c_{\mathbb{C}}^{\pm}(M)$. Since complex conjugation reverses the Hodge filtration, we have $c_{\infty}^{+}(M)c_{\infty}^{-}(M) \neq 0$. Obviously, once we find a convenient $\overline{\mathbb{Q}}_p$-algebra X to supply p-adic periods, we may replace X by its subalgebra generated over $\overline{\mathbb{Q}}_p$ by matrix coefficients of I_X with respect to rational bases of H_B and H_{DR}. This is the reason why we called X the ring of periods, and therefore, X could be quite large. Several p-adic rings X have been proposed by Fontaine, depending on how wide one wants to vary the motive M. We do not touch Fontaine's definition here and only refer to his papers [Fo3], [Fo2] and [Fo1] (see also [SGL] Section 2), because we have not explained any algebro-geometric cohomology theory in this book. We simply assume the existence of such a ring, using variable notation 'X' for it.

Since the definition of $c_X^{\pm}(M)$ involves the action of complex conjugation (even if X is a p-adic ring), the choice of $c_X^{\pm}(M)$ actually depends on the choice of $i_p : \overline{\mathbb{Q}}_p \hookrightarrow X$ and $i_{\infty} : \overline{\mathbb{Q}} \hookrightarrow \mathbb{C}$. If we make a bad choice of embedding, the period could vanish (a result of Y. André, see [SGL] 3.3). Since this point is rather subtle and technical, we simply assume that we can make a good choice of the embedding to make $c_p^{\pm}(M) = c_X^{\pm}(M) \neq 0$ leaving it to [SGL] Section 3 to provide details of how to (conjecturally) find good embeddings and good periods. Thus what we say in this subsection is always under an assumption that we have chosen the embeddings and hence the periods $c_X^{\pm}(M)$ as well.

Since the periods $c_p^{\pm}(M)$ and $c_{\infty}^{\pm}(M)$ are computed using the same rational basis of $H_B^{\mp}(M)$ and $H_{DR}^{\pm}(M)$, the ratio $\frac{c_p^{\pm}(M)}{c_{\infty}^{\pm}(M)}$ is independent of the choice of the basis in a huge ring $\widetilde{X}$ containing X and $\mathbb{C}$ at the same time. Thus the value

$$\left|c_p^+(M)i_p \circ i_\infty^{-1}\left(\frac{L(0,M)}{c_\infty^+(M)}\right)\right|_p = \left|c_p^+(M(1))i_p \circ i_\infty^{-1}\left(\frac{L(1,M)}{c_\infty^+(M(1))}\right)\right|_p$$

makes sense independently of the choices being made and is expected to give the size $|\,\mathrm{Sel}(\varphi)|_p$ of the Selmer group associated with $V(\delta_p^-)$ satisfying (adm), as long as $\overline{\rho} = \varphi \mod p$ is absolutely irreducible. Here we have made a normalization of the periods $c_X^\pm$ (for X archimedean and p-adic) so that the above identity follows from the functional equation. When the choice of the basis $b^\pm$ and $\omega^\pm$ is normalized so that $c_p^\pm(M(j))$ is a p-adic unit, we call $c_\infty^\pm(M)$ the normalized (archimedean) period. If we choose normalized archimedean periods, we may drop $c_p^+(M(j))$ from the above value, because $|c_p^\pm(M(j))|_p = 1$.

We can generalize the above heuristic for a single motive M to a Galois representation $\varphi : \mathfrak{G} \to GL_n(\mathbb{J})$ for a large p-adic ring $\mathbb{J}$, if $Spec(\mathbb{J})(\mathbb{Z}_p) = \mathrm{Hom}_{\mathbb{Z}_p\text{-}alg}(\mathbb{J}, \mathbb{Z}_p)$ contains densely populated points P with $\varphi_P = P \circ \varphi$ associated to a motive M_P. Thus φ_P is the p-adic member of the compatible system of Galois representations attached to the motive M_P. Then under a $\mathbb{J}$-version of the admissibility condition, we expect to have a p-adic L-function $L_p \in \mathbb{J}$ such that

$$L_p(P) = c_p^+(M_P(1))i_p \circ i_\infty^{-1}\left(\frac{L^*(1, M_P)}{c_\infty^+(M_P(1))}\right)$$

for densely populated motivic points $P \in Spec(\mathbb{J})(\mathbb{Z}_p)$. We have written $L^*(1, M_P)$ instead of $L(1, M_P)$, since often we need to modify the p-Euler factor of the L-function to make p-adic interpolation (see [SGL] (E) on page 40). We refer the reader for details of such conjectures predicting the existence and characterization of the p-adic L-functions (and known examples) to [SGL] Sections 4–8.

The p-adic L-function L_p is expected somehow to describe the size of the (compact) Pontryagin dual $\mathrm{Sel}^*(\varphi)$ of the module $\mathrm{Sel}(\varphi)$, which is a discrete $\mathbb{J}$-module. An outline of a conjecture is as follows: Suppose that $\mathbb{J}$ is a noetherian local integral domain and that $\overline{\rho} = (\varphi \mod \mathfrak{m}_{\mathbb{J}})$ is absolutely irreducible. As conjectured by Greenberg [G], if $L_p \neq 0$, $\mathrm{Sel}^*(\varphi)$ is a torsion $\mathbb{J}$-module (of finite type). Further, we suppose that $\mathbb{J}$ is a regular local integral domain. Then supposing that $\mathrm{Sel}^*(\varphi)$ is a $\mathbb{J}$-torsion, we have a pseudo-isomorphism $\phi : \mathrm{Sel}^*(\varphi) \to \prod_j \mathbb{J}/(f_j)$ for finitely many $0 \neq f_j \in \mathbb{J}$. Here '*pseudo-isomorphism*' means a homomorphism whose kernel and cokernel are both supported by at least codimension 2 Zariski-closed subset of $Spec(\mathbb{J})$ (see [BCM] Chapter 7). Then $\Phi =$

$\prod_j f_j =: char_{\mathbb{J}}(\mathrm{Sel}^*(\varphi))$ is uniquely determined (up to $\mathbb{J}$-units), and is called the *characteristic power series* of $\mathrm{Sel}^*(\varphi)$. We expect that $\Phi = L_p$ up to $\mathbb{J}$-units.

Here are some examples of normalized periods in the cyclotomic case, in the case of Artin motives and in the case of elliptic curves:

EXAMPLE 5.4 Let $F = E = \mathbb{Q}$. We define again $\mathbb{Q}(-1) = H^1(\mathbf{G}_m)$ for $\mathbf{G}_m = Spec(\mathbb{Q}[t, \frac{1}{t}])$. Then $\mathbb{Q}(1) = \mathbb{Q}(-1)^\vee$, the dual. Thus $H_{DR}(\mathbb{Q}(-1)) = \mathbb{Q}\frac{dt}{t}$, $H_B(\mathbb{Q}(1)) = \mathbb{Q}S^1$ for the unit circle $S^1 \subset \mathbb{C}^\times = \mathbf{G}_m(\mathbb{C})$. $H_{et,p}(\mathbb{Q}(1)) = T_p(\mathbf{G}_m) \otimes_{\mathbb{Z}_p} \mathbb{Q}_p$ for $T_p(\mathbf{G}_m) = \varprojlim_n \mu_{p^n}$ with $\mu_{p^n} = \{\zeta \in \mathbf{G}_m | \zeta^{p^n} = 1\}$. On $T_p(\mathbf{G}_m)$, $\mathfrak{G}$ acts via the cyclotomic character ν: $\zeta^\sigma = \zeta^{\nu(\sigma)}$. By definition, $\nu(Frob_\ell) = \ell$ for the arithmetic Frobenius $Frob_\ell$ (and hence $\nu(\Phi_\ell) = \frac{1}{\ell}$ for the geometric Frobenius Φ_ℓ). Then $H_{et,p}(\mathbb{Q}(-1)) = \mathrm{Hom}_{\mathbb{Z}_p}(T_p(\mathbf{G}_m), \mathbb{Q}_p)$, on which $\mathfrak{G}$ acts via ν^{-1}. Note that $H_{DR}(\mathbb{Q}(1)) = \mathrm{Hom}_{\mathbb{Q}}(H^1(\mathbf{G}_m), \mathbb{Q}) = H_1(\mathbf{G}_m, \mathbb{Q})$. The comparison isomorphism $I_{\mathbb{C}}(1) : \mathbb{C}S^1 = H_B(\mathbb{Q}(1)) \otimes_{\mathbb{Q}} \mathbb{C} \cong H_{DR}(\mathbb{Q}(1)) \otimes_{\mathbb{Q}} \mathbb{C}$ is given by associating $S^1 \mapsto \{\frac{dt}{t} \mapsto \int_{S^1} t^{-1}dt\}$, which is exactly the de Rham isomorphism between the Betti homology and the dual of the de Rham cohomology group. Thus $c_\infty^+(\mathbb{Q}(1)) = 2\pi i = \int_{S^1} t^{-1}dt$. Since $\mathcal{F}^+(\mathbb{Q}(1)) = H_{DR}(\mathbb{Q}(1))$, we have $c_\infty^-(\mathbb{Q}(1)) = 1$. Note that $\mathbb{Q}(j) = \overbrace{\mathbb{Q}(1) \otimes \cdots \otimes \mathbb{Q}(1)}^{j}$ and $\mathbb{Q}(-j) = \mathbb{Q}(j)^\vee$ for $j > 0$. Thus the comparison isomorphism for $\mathbb{Q}(j)$ is given by $I_{\mathbb{C}}(j) = \overbrace{I_{\mathbb{C}}(1) \otimes \cdots \otimes I_{\mathbb{C}}(1)}^{j}$. Since $\mathcal{F}^{-(-1)^j}(\mathbb{Q}(j)) = 0$ and $\mathcal{F}^{(-1)^j}(\mathbb{Q}(j)) = H_{DR}(\mathbb{Q}(j))$, we have $c_\infty^{-(-1)^j}(\mathbb{Q}(j)) = (2\pi i)^j$ and $c_\infty^{(-1)^j}(\mathbb{Q}(j)) = 1$.

Let $\widehat{\overline{\mathbb{Q}}}_p$ be the p-adic completion of an algebraic closure of $\mathbb{Q}_p$. Then $\mathrm{Gal}(\overline{\mathbb{Q}}_p/\mathbb{Q}_p)$ acts continuously on $\widehat{\overline{\mathbb{Q}}}_p$. As for the p-adic periods, by Tate's p-adic Hodge theory [T], we can take $X = \widehat{\overline{\mathbb{Q}}}_p[t, t^{-1}]$ on which $\mathrm{Gal}(\overline{\mathbb{Q}}_p/\mathbb{Q}_p)$ acts on the coefficients naturally and on t by ν: $\sigma(t) = \nu(\sigma)t$. This ring is often written as B_{HT} (the Barsotti–Tate ring for p-adic Hodge–Tate theory). Then by the result of [T], we have a canonical isomorphism $I_{HT} : H_{et,p}(\mathbb{Q}(j)) \otimes_{\mathbb{Q}} B_{HT} \cong H_{DR}(\mathbb{Q}(j)) \otimes_{\mathbb{Q}} B_{HT}$. By [T], we have

$$H^0(\mathrm{Gal}(\overline{\mathbb{Q}}_p/\mathbb{Q}_p), H_{et,p}(\mathbb{Q}(j)) \otimes_{\mathbb{Q}} B_{HT}) = H_{DR}(\mathbb{Q}(j))_{\mathbb{Q}} \otimes \mathbb{Q}_p,$$

where the Galois group $\mathrm{Gal}(\overline{\mathbb{Q}}_p/\mathbb{Q}_p)$ acts diagonally (and semi-linearly) on the tensor product $H_{et,p}(\mathbb{Q}(j)) \otimes_{\mathbb{Q}} B_{HT}$. Then the p-adic comparison isomorphism keeps the rationality; so, $c_p^{\pm}(\mathbb{Q}(j)) = 1$. In general, if M is critical, admissible and self-dual, $I_X^{\pm}(M(j))$ (for $X = B_{HT}$ and $M(j)$) is the tensor product of $I_X(M)^{(-1)^j}$ (for M) and I_{HT}. Thus

$$c_p^+(M(j)) = c_p^{(-1)^j}(M). \tag{5.2}$$

Similarly we have

$$c_\infty^+(M(j)) = (2\pi i)^{d^{(-1)^j}} c_\infty^{(-1)^j}(M), \tag{5.3}$$

where $d^\pm = \dim H_B^\pm(M)$.

EXAMPLE 5.5 We return to the setting of Example 5.1; thus, $\chi = \left(\frac{-D}{}\right)$, $E = \mathbb{Q}$ and $F = \mathbb{Q}(\sqrt{-D})$. The Artin motive $M(\chi)$ attached to the Galois character $\chi : \mathfrak{G} \to \{\pm 1\}$ is given as follows: Let $V = Spec(F)$. Then $V(\overline{\mathbb{Q}})$ consists of two points represented by the two embeddings $F \hookrightarrow \overline{\mathbb{Q}}$. This is a projective algebraic variety defined over $\mathbb{Q}$ of dimension 0. Then $H^0_{et}(V \times_{\mathbb{Q}} \overline{\mathbb{Q}}, \mathbb{Q}_p) = \mathbb{Q}_p^{V(\overline{\mathbb{Q}})}$: direct sum of two copies of $\mathbb{Q}_p$ indexed by $V(\overline{\mathbb{Q}})$, on which the Galois group $\mathrm{Gal}(\overline{\mathbb{Q}}/\mathbb{Q})$ acts through $V(\overline{\mathbb{Q}})$ (interchanging two components). Here the zero-dimensional scheme $V \times_{\mathbb{Q}} \overline{\mathbb{Q}} = Spec(F \otimes_{\mathbb{Q}} \overline{\mathbb{Q}})$ is the base-change of V over $\mathbb{Q}$ to $\overline{\mathbb{Q}}$. Thus we have

$$H^0_{et}(V \times_{\mathbb{Q}} \overline{\mathbb{Q}}, \mathbb{Q}_p) = H_{et,p}(\mathbb{Q}(0)) \oplus H_{et,p}(M(\chi))$$

with $H_{et,p}(M(\chi)) = \mathbb{Q}_p(1 \oplus -1)$. Here H^0_{et} indicates the étale cohomology group of $V \times_{\mathbb{Q}} \overline{\mathbb{Q}}$ (see [ECH]), and $(1 \oplus -1) \in \mathbb{Q}_p^{V(\overline{\mathbb{Q}})}$ is the function on $V(\overline{\mathbb{Q}})$ which has value 1 on one point of $V(\overline{\mathbb{Q}})$ and -1 on the other. Since the Galois action interchanges the two points, $(1 \oplus -1)$ is a χ-eigenvector. Similarly

$$H^0_B(V, \mathbb{Q}) = \mathbb{Q}^{V(\overline{\mathbb{Q}})} = H_B(\mathbb{Q}(0)) \oplus H_B(M(\chi))$$

on which complex conjugation c acts via $V(\overline{\mathbb{Q}})$. Thus we have again

$$H_B(M(\chi)) = \mathbb{Q}(1 \oplus -1).$$

The de Rham realization $H^0_{DR}(V, \mathbb{Q})$ is given by the $\mathbb{Q}$-vector space F on which $\mathrm{Gal}(F/\mathbb{Q})$ acts. Then $F \otimes_{\mathbb{Q}} \overline{\mathbb{Q}} \cong \overline{\mathbb{Q}}^{V(\overline{\mathbb{Q}})}$ canonically; so we may take $X = \mathbb{C}$ for the place ∞ and $X = \overline{\mathbb{Q}}_p$ at p. The comparison isomorphism I_X sends the generator $(1 \oplus -1)$ to an element $g \in F \otimes_{\mathbb{Q}} \overline{\mathbb{Q}}$ such that $\sigma(g) = \chi(\sigma) g$ for $\sigma \in \mathrm{Gal}(\overline{\mathbb{Q}}/\mathbb{Q})$, which acts on the tensor product $F \otimes_{\mathbb{Q}} \overline{\mathbb{Q}}$ through the component $\overline{\mathbb{Q}}$. Such g is a rational multiple of the Gauss sum $G(\chi) = \sqrt{-D}$. Thus we may put: $c_p^+(M(\chi)) = G(\chi)^{-1}$ and $c_\infty^\pm(M(\chi)(1)) = G(\chi)^{-1}(2\pi i)$, which match well with Dirichlet's class number formula.

EXAMPLE 5.6 We return to the setting of Example 5.2. Then the motive M associated with the elliptic curve $\mathcal{E}_{/\mathbb{Q}}$ is given by $H^1(\mathcal{E})$. Thus

$$H_B(M) = H^1(\mathcal{E}(\mathbb{C}), \mathbb{Q}) = \mathrm{Hom}_{\mathbb{Q}}(H_1(\mathcal{E}, \mathbb{Q}), \mathbb{Q}),$$

$H_{et,p}(M) = H^1_{et}(\mathcal{E}_{/\overline{\mathbb{Q}}}, \mathbb{Q}_p)$ with its Galois action and $H_{DR}(M) = H^1_{DR}(\mathcal{E}, \mathbb{Q})$ for its algebraic de Rham cohomology group. Here $\mathcal{E}_{/\overline{\mathbb{Q}}}$ is the base-change of $\mathcal{E}_{/\mathbb{Q}}$ to the elliptic curve over $\overline{\mathbb{Q}}$, and $H^\bullet_{et}$ indicates the étale cohomology group of $\mathcal{E}_{/\overline{\mathbb{Q}}}$ (see [ECH]). The Hodge filtration can be interpreted as the Hodge exact sequence:

$$0 \to H^0(\mathcal{E}, \Omega_{\mathcal{E}/\mathbb{Q}}) \to H^1_{DR}(\mathcal{E}, \mathbb{Q}) \to H^1(\mathcal{E}, \mathcal{O}_{\mathcal{E}}) \to 0.$$

When $\mathcal{E}$ has ordinary good reduction modulo p, we may assume that

$$I_{HT}(H^0(\mathcal{E}_{/\mathbb{Q}_p}, \Omega_{\mathcal{E}/\mathbb{Q}_p}) \otimes_{\mathbb{Q}_p} B_{HT}) = T_p(\mathcal{E}^c) \otimes_{\mathbb{Z}_p} B_{HT}.$$

This is the admissibility condition (adm) in this case. For an elliptic curve, there is a canonical smooth group scheme $\mathcal{E}_{/\mathbb{Z}}$ called the Néron model (see [BLR]) which gives rise to $\mathcal{E}_{/\mathbb{Q}}$. Then $H^1(\mathcal{E}_{/\mathbb{Z}}, \Omega_{\mathcal{E}/\mathbb{Z}}) = \mathbb{Z}\omega$ for a unique differential ω up to sign. We can take $\pm$-eigenspace $H_1(\mathcal{E}(\mathbb{C}), \mathbb{Z})[\pm] = \mathbb{Z}b^\pm$ in $H_1(\mathcal{E}(\mathbb{C}), \mathbb{Z})$ under the action of complex conjugation and their generators $b^\pm$. Since, by the de Rham isomorphism, $H_B(M) = H^1(\mathcal{E}(\mathbb{C}), \mathbb{Q})$ is the dual of $H_1(\mathcal{E}(\mathbb{C}), \mathbb{Q})$ via integration of differentials over cycles, we find (see [SGL] pages 47–48) that:

$$c^\pm_\infty(M(1)) = \int_{b^\pm} \omega.$$

It has been shown by Faltings [Fa] Theorem 5.3 that $c^\pm_p(M(1))$ is a p-adic unit if $p > 2$ (see [SGL] 4.3, page 48). Thus one expects, if $\mathcal{E}[p]$ for $p > 2$ is absolutely irreducible and $L(1, \mathcal{E}) \neq 0$,

$$\left| i_p \left(\frac{L(1, \mathcal{E})}{c^\pm_\infty(M(1))} \right) \right|_p = || \mathrm{Sel}(\mathcal{E})_{/\mathbb{Q}} ||_p.$$

When $L(1, \mathcal{E})$ vanishes, the conjecture of Birch–Swinnerton Dyer predicts the order of the vanishing and the leading term of the Taylor expansion at $s = 1$ of $L(s, \mathcal{E})$ in terms of the rank of $\mathcal{E}(\mathbb{Q})$ and the order of the Tate-Shafarevich group Ш$(\mathcal{E})$, and for its generalization to general motives, we refer to [BK] and [G1].

We can think of $Ad(M)$ for the motive $M = H^1(\mathcal{E})$. Note that $H^2(\mathcal{E} \times \mathcal{E}) = \oplus_{i+j=2} H^i(\mathcal{E}) \otimes H^j(\mathcal{E})$ for $0 \leq i, j \leq 2$ for any one of the cohomology theories: Betti, de Rham and étale, and we have

$$H^2(\mathcal{E} \times \mathcal{E}) \supset M \otimes M \cong Ad(M)(1) \oplus \mathbb{Q}(1).$$

A geometric study of $Ad(M)$ for modular $\mathscr{E}$ was initiated in [H81a] and [H81b], and later Flach studied $Ad(M)$ in the above context of motives inside $H^2(\mathscr{E}\times\mathscr{E})$. In [Fl], under some conditions, the finiteness of $\mathrm{Sel}(Ad(M))$ is proved. Anyway, in this case, we see that $c_X^+(M(1))c_X^-(M(1)) = c_X^+(Ad(M)(1))c_X^+(\mathbb{Q}(1))^{-1}$ (see [SGL] Theorem 5.2.2).

5.1.3 Character twists

A Galois extension E_∞/E is called a $\mathbb{Z}_p$-extension if $\Gamma = \mathrm{Gal}(E_\infty/E) \cong \mathbb{Z}_p$. Classical Iwasawa theory is an in-depth study of the arithmetic of $\mathbb{Z}_p$-extensions, in particular, the cyclotomic $\mathbb{Z}_p$-extension inside $E(\mu_{p^\infty})$. Since the inertia subgroup of an abelian Galois group $\mathrm{Gal}(X/E)$ for a prime $\ell \neq p$ cannot have $\mathbb{Z}_p$ as a quotient (see Subsection 3.2.5 in Chapter 3), we have $E_\infty \subset F^{(p,\infty)}$, because a subgroup of $\mathbb{Z}_p$ is either isomorphic to $\mathbb{Z}_p$ or trivial. We have the tautological character $\kappa : \Gamma \hookrightarrow \mathbb{Z}_p[[\Gamma]]^\times$ for the Iwasawa algebra $\mathbb{Z}_p[[\Gamma]]$. Here the Iwasawa algebra is by definition $\mathbb{Z}_p[[\Gamma]] \cong \varprojlim_n \mathbb{Z}_p[\Gamma/\Gamma^{p^n}]$. We pick an intermediate finite extension L of $F^{(p,\infty)}/E$ linearly disjoint from E_∞. For simplicity, we assume that either $p \nmid [F : E]$ or $L \supset F$. Thus we can use (5.1) as the definition of $\mathrm{Sel}(\varphi)_{/L}$. We put $L_\infty = LE_\infty$.

Fix a generator $\gamma \in \Gamma$. Then we can identify $\mathbb{J}[[\Gamma]]$ with a power series ring $\mathbb{J}[[T]]$ so that $\kappa(\gamma) = 1 + T$ (see Corollary 2.23 in Chapter 2). Thus if $\mathbb{J} = \mathbb{Z}_p[[T_1, \ldots, T_r]]$, then $\mathbb{J}[[\Gamma]] \cong \mathbb{Z}_p[[T_1, \ldots, T_r, T]]$. We regard κ as a character of $\mathfrak{G}$ and then consider $\varphi \otimes \kappa : \mathfrak{G} \to GL_n(\mathbb{J}[[\Gamma]])$. We want to relate $\mathrm{Sel}(\varphi \otimes \kappa)_{/L}$ with respect to $\{V(\delta_\mathfrak{p}^- \otimes \kappa)\}_\mathfrak{p}$ and $\mathrm{Sel}(\varphi)_{/L_\infty}$ with respect to $\{V(\delta_\mathfrak{p}^-)\}_\mathfrak{p}$. For simplicity, we assume that

$$\textit{Every prime factor of } p \textit{ in } L \textit{ totally ramifies in } L_\infty/L. \qquad (\mathrm{TR}_p)$$

Let L_j be the j-th layer of L_∞. Thus $\Delta_j = \mathrm{Gal}(L_j/L) = \Gamma/\Gamma_j$ for $\Gamma_j = \mathrm{Gal}(L_\infty/L_j) = \Gamma^{p^j}$. We write φ_j for φ restricted to $\mathfrak{H}_j = \mathrm{Gal}(F^{(p,\infty)}/L_j)$ and κ_j for the Galois character κ composed with the projection: $\mathbb{Z}_p[[\Gamma]] \twoheadrightarrow \mathbb{Z}_p[[\Delta_j]]$. We then look at the induced module:

$$\begin{aligned} \mathrm{Ind}_{L_j}^L V(\varphi_j)^* &= \mathrm{Hom}_{\mathbb{Z}[\mathfrak{H}_j]}(\mathbb{Z}[\Gamma], V(\varphi)^*) \\ &= \mathrm{Hom}_{SETS}(\Delta_j, V(\varphi)^*) \cong V(\varphi \otimes \kappa_j)^*. \end{aligned}$$

Here $\mathrm{Hom}_{SETS}(X, Y)$ is the set of all maps from a set X into another set Y. By (TR_p), $I_\mathfrak{p}/I_{\mathfrak{p},j} \cong \Delta_j$, where $I_\mathfrak{p}$ is the inertia group of $\mathfrak{H} = \mathfrak{H}_L$ and $I_{\mathfrak{p},j}$ is the inertia group $I_\mathfrak{p} \bigcap \mathfrak{H}_j$ of $\mathfrak{H}_j$. Then similarly to the above formula, we get

$$\mathrm{Ind}_{L_{j,\mathfrak{p}}}^{L_{\mathfrak{p}}} V(\delta_j^-)^* = \mathrm{Hom}_{SETS}(\Delta_j, V(\delta_{\mathfrak{p}}^-)^*) \cong V(\delta_{\mathfrak{p}}^- \otimes \kappa_j)^*,$$

where δ_j^- is the restriction of $\delta_{\mathfrak{p},\varphi}^-$ to $I_{\mathfrak{p},j}$. Then by Shapiro's lemma (4.20), we have

$$H^1(\mathfrak{H}_L, V(\varphi \otimes \kappa_j)^*) \cong H^1(\mathfrak{H}_j, V(\varphi)^*) \tag{5.4}$$
$$H^1(I_{\mathfrak{p}}, V(\delta_{\mathfrak{p}}^- \otimes \kappa_j)^*) \cong H^1(I_{\mathfrak{p},j}, V(\delta_{\mathfrak{p}}^-)^*). \tag{5.5}$$

From this we conclude

Proposition 5.7 *Suppose* (TR_p). *Then*

$$\mathrm{Sel}(\varphi \otimes \kappa)_{/L} \cong \mathrm{Sel}(\varphi_\infty)_{/L_\infty}.$$

Although we proved this proposition supposing the assumption (TR_p), one can check that (TR_p) is not necessary to show the assertion of the above proposition. Since we will not use this general fact later, we leave the proof in the general case to the reader.

Example 5.8 We keep the notation of Example 5.1. Thus χ is a character inducing $\mathrm{Gal}(F/\mathbb{Q}) \cong \mu_n \subset \mathcal{O}^\times$. Let $E_\infty = \mathbb{Q}_\infty$ be the cyclotomic $\mathbb{Z}_p$-extension. We suppose that χ is an odd character, that is, $\chi(c) = -1$ for complex conjugation c. We write $\kappa : \Gamma = \mathrm{Gal}(\mathbb{Q}_\infty/\mathbb{Q}) \hookrightarrow \mathcal{O}[[\Gamma]]$ for the tautological character. We then regard $\chi\kappa$ as a character of $\mathfrak{G}$ having values in $\mathcal{O}[[\Gamma]]^\times$. Then taking $V(\delta_p^+)$ to be the full space $V(\chi\kappa)$, we have $\mathrm{Sel}(\chi\kappa)_{/\mathbb{Q}} \cong \mathrm{Sel}(\chi)_{/\mathbb{Q}_\infty}$. The Pontryagin dual $\mathrm{Sel}^*(\chi)_{/\mathbb{Q}_\infty}$ is the classical Iwasawa module studied by Iwasawa (see Introduction of [G1]).

Example 5.9 We keep the notation in Example 5.2. The study of the Selmer groups $\mathrm{Sel}(\mathcal{E})_{/\mathbb{Q}_\infty}$ (and $\mathrm{Sel}(\mathcal{E})_{/E_\infty}$) was initiated by Mazur (see [G1]). In particular, suppose that $\mathcal{E}$ is modular. The modularity is now known to be true for almost all rational elliptic curves by Wiles and others. Breuil, Conrad, Diamond and Taylor have announced recently the proof of the conjecture without any assumption (see [CDT] for the case where the conductor of $\mathcal{E}$ is not divisible by 27). Mazur constructed a p-adic L-function of $\mathcal{E}_{/\mathbb{Q}}$ (see, for example, [LFE] Chapter 6), and he conjectured that the characteristic power series of $\mathrm{Sel}(\mathcal{E})_{/\mathbb{Q}_\infty}$ is given by the p-adic L-function ([G1] Conjecture 1.13).

Proposition 5.7 frees Iwasawa-theoretic understanding of the Selmer groups from $\mathbb{Z}_p$-extensions and tells us that the Selmer groups associated with a motive over a given $\mathbb{Z}_p$-extension is simply a special case of the Selmer group over number fields for a Galois representation with coefficients in a larger ring. Thus, the study of the Selmer groups of the type $\mathrm{Sel}(\varphi)_{/L_\infty}$ appears to belong to the Iwasawa theory, but it can be regarded to be a part of the Hecke theory dealing with modular Galois representations. In any case, we may consider $\mathbb{J}[[\Gamma]]$ as the coefficient ring of $\mathrm{Sel}(\varphi \otimes \kappa)_{/F}$. The two descriptions of the Selmer group are useful as we will see later.

For a general local $\mathcal{O}$-algebra R (with common residue field) and a Galois representation $\rho : \mathfrak{G} \to GL_n(R)$, Greenberg has conjectured that $\mathrm{Sel}(\varphi \otimes \kappa)_{/E}$ is a torsion module over its coefficient ring R as long as the associated p-adic L-function exists and does not vanish ([G] Conjecture 4.1), although he does not explain how to characterize or construct an appropriate p-adic L-function for a general ρ. The conjectures from different viewpoints on p-adic L-functions are given in [SGL] and [FLP].

5.2 Adjoint Selmer Groups

In this section, we apply the generalities described in the previous section to adjoint Galois representations of a given Galois representation φ. The adjoint Selmer groups have many special features. They are closely related to the universal deformation ring of $\overline{\rho} = (\varphi \mod \mathfrak{m}_{\mathbb{J}})$ via Kähler differentials, and therefore,

(1) commutative algebra supplies us with useful tools to study the arithmetic of adjoint Selmer groups;
(2) if the deformation ring is canonically identified with a p-adic Hecke algebra, we can deepen our knowledge of the adjoint Selmer group via Hecke theory.

Therefore, the theory of adjoint Selmer groups is at the intersection of principal branches of number theory: Iwasawa theory as explained in Subsection 5.1.3, Hecke theory as pointed out in (2) above, and arithmetic algebraic geometry as described in Subsection 5.1.2 and (1) above.

Throughout this section, we keep the notation introduced in the previous section.

5.2.1 Adjoint Galois representations

We now let $\mathfrak{G}$ act on $M_n(\mathbb{J})$ by conjugation: $x \mapsto \varphi(\sigma)x\varphi(\sigma)^{-1}$. The trace zero subspace $\mathfrak{sl}$ is stable under this action. This new Galois module of dimension $n^2 - 1$ is called the adjoint representation of φ and written as $Ad(\varphi)$. Thus

$$V = V(Ad(\varphi)) = \{T \in \mathrm{End}_{\mathbb{J}}(V(\varphi)) | \operatorname{Tr}(T) = 0\}.$$

This space has a three-step filtration: $0 \subset V_{\mathfrak{p}}^{+} \subset V_{\mathfrak{p}}^{-} \subset V$ given by

$$V_{\mathfrak{p}}^{+}(Ad(\varphi)) = \{T \in V(Ad(\varphi)) | T(V(\varphi)_{j,\mathfrak{p}}) \subset V(\varphi)_{j-1,\mathfrak{p}} \text{ for all } j\}, \quad (+)$$
$$V_{\mathfrak{p}}^{-}(Ad(\varphi)) = \{T \in V(Ad(\varphi)) | T(V(\varphi)_{j,\varphi,\mathfrak{p}}) \subset V(\varphi)_{j,\varphi,\mathfrak{p}} \text{ for all } j\}. \quad (-)$$

When $n = 2$, ($\mathrm{fil}_{\mathfrak{p}}$) is a two-step filtration and we take a base of $V(\varphi)$ so that $\varphi|_{D_{\mathfrak{p}}} = \begin{pmatrix} \varepsilon_{\mathfrak{p}} & * \\ 0 & \delta_{\mathfrak{p}} \end{pmatrix}$, then we have, for example,

$$V_{\mathfrak{p}}^{+}(Ad(\varphi)) = \left\{\left(\begin{smallmatrix} 0 & * \\ 0 & 0 \end{smallmatrix}\right)\right\} \subset V_{\mathfrak{p}}^{-}(Ad(\varphi)) = \left\{\left(\begin{smallmatrix} a & * \\ 0 & b \end{smallmatrix}\right) | a + b = 0\right\} \subset M_2(\mathbb{J}).$$

Therefore, this is the filtration we have studied in the proof of the Taylor-Wiles theorem (Subsection 3.2.8 in Chapter 3). We can think of four different Selmer groups for each $\mathfrak{p}$ depending on $\mathscr{S}$, which should correspond with different p-adic L-functions. We simply write

$$\mathrm{Sel}_0(Ad(\varphi)), \quad \mathrm{Sel}(Ad(\varphi)), \quad \mathrm{Sel}_{-}(Ad(\varphi)) \quad \text{or} \quad \mathrm{Sel}_{full}(Ad(\varphi))$$

according as $\mathscr{S} = 0$, $V_{\mathfrak{p}}^{+}(Ad(\varphi))$, $V_{\mathfrak{p}}^{-}(Ad(\varphi))$ or $V(Ad(\varphi))$ for all $\mathfrak{p}|p$. Then we have the associated filtration of the Selmer groups:

$$\mathrm{Sel}_0(Ad(\varphi)) \subset \mathrm{Sel}(Ad(\varphi)) \subset \mathrm{Sel}_{-}(Ad(\varphi)) \subset \mathrm{Sel}_{full}(Ad(\varphi)).$$

Among these Selmer groups, if $E = \mathbb{Q}$ and φ is two-dimensional odd, $\mathrm{Sel}(Ad(\varphi))$ associated with $V_{\mathfrak{p}}^{+}(Ad(\varphi))$ is considered to be standard (because $Ad(\varphi)$ for two-dimensional φ often satisfies (adm) for $V_{\mathfrak{p}}^{+}(\varphi)$) and should be directly related to the normalized p-adic L-function described in [SGL] Section 4.3 (see the following example).

Example 5.10 We begin with a regular, pure motive M of rank n (with coefficients in $\mathbb{Q}$) defined over $E = \mathbb{Q}$. Here a motive is called *regular* if its Hodge component $H^{p,q}(M)$ has at most dimension 1. The numbers (p, q) with $\dim H^{p,q}(M) > 0$ are called the Hodge numbers of M. By the regularity, we may arrange the Hodge numbers (p_i, q_i) for $i = 1, 2, \ldots, n$ so that they satisfy

$$p_1 < p_2 < \cdots < p_n.$$

We consider the tensor product $M \otimes \check{M}$ and decompose it as $M \otimes \check{M} = \mathbf{1} \oplus Ad(M)$. Then $Ad(M)$ and $M \otimes \check{M}$ are self-dual motives of weight 0; so we can write down their Hodge numbers in the following matrix form:

$$\begin{pmatrix} 0 & p_2 - p_1 & \cdots & p_n - p_1 \\ p_1 - p_2 & 0 & \cdots & p_n - p_2 \\ \vdots & \vdots & \ddots & \vdots \\ p_1 - p_n & p_2 - p_n & \cdots & 0 \end{pmatrix}.$$

Thus $Ad(M)$ is critical at 0 and 1 if complex conjugation acts by the scalar multiplication -1 on $H^{0,0}(Ad(M))$, and the middle term of Hodge filtration $\mathcal{F}^-(Ad(M)) = \mathcal{F}^0(Ad(M)) \subset H_{DR}(Ad(M))$ corresponds to the upper triangular part of the above matrix, and $\mathcal{F}^+(Ad(M)) = \mathcal{F}^1(Ad(M))$ corresponds to the upper nilpotent part of the matrix. When we twist $Ad(M)$ by an Artin motive $M(\psi)$ of rank m with Galois representation ψ of $\Delta = \mathrm{Gal}(F/\mathbb{Q})$ for a totally real field F, the situation does not change. In other words, $Ad(M) \otimes \psi$ and $Ad(M)$ are critical at 0 and 1 at the same time, and $\mathcal{F}^\pm(Ad(M) \otimes \psi) = (\mathcal{F}^\pm Ad(M)) \otimes \psi$. We consider the p-adic Galois representation φ on the p-adic étale realization $H_p(M)$. Since $Ad(\varphi)$ is self-dual, $L(1, Ad(\varphi) \otimes \psi) \neq 0$ at least conjecturally, because $s = 1$ is the abscissa of convergence of $L(s, Ad(\varphi) \otimes \psi)$. Since the conjectural functional equation is of the form $s \leftrightarrow 1 - s$, we should have $L(0, Ad(\varphi) \otimes \psi) \neq 0$. Thus the Selmer group $\mathrm{Sel}(Ad(M) \otimes \psi)_{/\mathbb{Q}}$ should be finite. We now suppose that

(1) $\overline{\rho} = \varphi \mod p$ *is absolutely irreducible over* F;

(2) φ *restricted to* D_p *is isomorphic to upper triangular representation with diagonal characters* $\delta_1, \delta_2, \ldots, \delta_n$ *from the top left corner*;

(3) $\overline{\delta}_j = \delta_j \mod p$ $(j = 1, 2, \ldots, n)$ *are all distinct on the inertia subgroup at* p *over* F;

(4) $\mathbb{F}_p[S_F]$ *and* $\overline{\psi}$, *which is* ψ *modulo the maximal ideal, are disjoint as* Δ*-modules,*

where S_F is the set of primes over p in F and Δ acts on the space $\mathbb{F}_p[S_F]$ of formal linear combinations through its action on S_F. Note that $n - 1$ times the dimension of the ψ-isotypic component of $\mathbb{F}_p[S_F]$ is equal to the number of linear p-Euler factors of $L(s, Ad(\varphi) \otimes \psi)$ which vanishes at $s = 0$. These Euler factors often need to be removed when one makes p-adic interpolation of the values $L(1, Ad(\varphi) \otimes \psi)$ (see [SGL] page 40 (E) and (Γ)). Thus (4) implies that no trivial zero occurs at $s = 0$ for the p-adic L-function of $\varphi \otimes \psi\kappa$. More generally, if $\varphi : \mathfrak{G} \to GL_n(\mathbb{J})$ specializes to the Galois representation associated with a motive M as above, the p-adic L-function should specialize to the order of $\mathrm{Sel}(Ad(M) \otimes \psi)_{/\mathbb{Q}}$, and hence $\mathrm{Sel}(Ad(\varphi) \otimes \psi)_{/\mathbb{Q}}$ has to be a torsion $\mathbb{J}$-module of finite type.

In the above example, it is easy to see that $Ad(M)$ is critical if and only if $n = 2$ and $\det(\varphi(c)) = -1$ for a complex conjugation $c \in \mathfrak{G}_p$. However, if the image of φ is small, there is a way to create a smaller motive, similar to $Ad(M)$, using the Zariski closure G of $\mathrm{Im}(\varphi) \subset GL(n)$, which is an algebraic subgroup of $GL(n)$. Since this book concerns principally two-dimensional representations, we refer to [H99b] for this (slightly more general) construction.

EXAMPLE 5.11 Let $E = \mathbb{Q}$ and $F = \mathbb{Q}(\sqrt{-D})$ be an imaginary quadratic field in which $p > 2$ splits into a product of two distinct primes $\mathfrak{p}$ and $\overline{\mathfrak{p}}$. We start from a character $\overline{\varphi} : \mathfrak{H} \to \mathbb{F}^\times$ with $\overline{\varphi} \neq \overline{\varphi}_c$, where $c \in \mathfrak{G}$ is a complex conjugation and $\overline{\varphi}_c(g) = \overline{\varphi}(cgc^{-1})$. This condition is to guarantee that $\overline{\rho} = \mathrm{Ind}_F^{\mathbb{Q}} \overline{\varphi}$ is absolutely irreducible. Then we consider the universal character $\Phi : \mathfrak{H} \to \mathbb{J} = \mathscr{O}[[\mathfrak{H}_p^{ab}]]$ deforming the character $\overline{\varphi}$ (see Subsection 2.3.1 in Chapter 2). For simplicity, we assume that the class number $h(-D)$ is prime to p. Then the maximal p-profinite abelian quotient $\mathfrak{H}_p^{ab}$ of $\mathfrak{H}$ is isomorphic to $O_p \cong \mathbb{Z}_p \times \mathbb{Z}_p$ for the integer ring $O \subset F$. Thus we have $\mathbb{J} \cong \mathscr{O}[[T, S]]$. We then consider $\rho = \mathrm{Ind}_F^{\mathbb{Q}} \Phi$. Since p splits in F, we may assume that $D_p = D_{\mathfrak{p}}$, and hence F is fixed by D_p. This shows that $\rho|_{D_p} \cong \Phi|_{D_{\mathfrak{p}}} \oplus \Phi_c|_{D_{\mathfrak{p}}}$, where $\Phi_c(g) = \Phi(cgc^{-1})$. Thus we may take $V(\delta_p^-) = V(\Phi|_{D_{\mathfrak{p}}}) \subset V(\rho)$.

In this case, a (canonical) p-adic L-function $L_p = L_p(\Phi) \in \mathbb{J}$ has been constructed by N. M. Katz and Ehud de Shalit (see [ICM] and [K4]). We briefly recall the interpolation property of L_p. We choose an elliptic curve $\mathscr{E}$ with complex multiplication defined over a valuation ring $\mathscr{V}$ (extending $O_{\mathfrak{p}} \cap F$) of an algebraic extension K_0/F (unramified at $\mathfrak{p}$ over F). We assume that $\mathscr{E}(\mathbb{C}) \cong \mathbb{C}/O$. We write $\iota : F \hookrightarrow \mathbb{C}$ for an embedding given by $a^*\omega_\infty = \iota(a)\omega_\infty$ for $\omega_\infty = du$ for the variable $u \in \mathbb{C}$ under the identification $\mathscr{E}(\mathbb{C}) \cong \mathbb{C}/O$. Then $c : F \hookrightarrow \mathbb{C}$ is the complex conjugate of ι. Note that

$$H^0(\mathscr{E}_{/\mathbb{C}}, \Omega_{\mathscr{E}/\mathbb{C}}) = \mathbb{C}\omega_\infty.$$

We write $\mathscr{O}$ for the completion of $\mathscr{V}$. We may assume only here that the residue field of $\mathscr{O}$ is the algebraic closure $\overline{\mathbb{F}}_p$. Since $\overline{\mathscr{E}} = \mathscr{E} \mod p$ is an ordinary elliptic curve, $\overline{\mathscr{E}}[p^\infty] \cong \mu_{p^\infty} \times \mathbb{Q}_p/\mathbb{Z}_p$ over the algebraic closure $\overline{\mathbb{F}}_p = \mathbb{F}$, where $\overline{\mathscr{E}}[p^\infty] = \bigcup_m \mathscr{E}[p^m]$ and $\mu_{p^\infty} = \bigcup_m \mu_{p^m}$. Thus we may assume that there is an isomorphism $\phi : \mu_{p^\infty} \cong \mathscr{E}[\mathfrak{p}^\infty]$ defined over $\mathscr{O}$ of p-divisible groups, where $\mathscr{E}[\mathfrak{p}^m] = \{x \in \mathscr{E} | \mathfrak{p}^m x = 0\}$ and $\mathscr{E}[\mathfrak{p}^\infty] = \bigcup_m \mathscr{E}[\mathfrak{p}^m]$. In particular, the map ϕ induces an isomorphism from $\widehat{\mathbf{G}}_m = Spf(\mathscr{O}[[(1+t)]])$ (the formal multiplicative group) onto the formal group $\widehat{\mathscr{E}}$ of $\mathscr{E}$ along the origin. Then the canonical differential $\frac{dt}{1+t}$ on $\widehat{\mathbf{G}}_m$ induces a unique $\omega_p \in H^0(\mathscr{E}_{/\mathscr{O}}, \Omega_{\mathscr{E}/\mathscr{O}})$ such that

$$H^0(\mathscr{E}_{/\mathscr{O}}, \Omega_{\mathscr{E}/\mathscr{O}}) = \mathscr{O}\omega_p.$$

Since $\mathscr{E}$ is of genus 1, we find a 1-differential ω such that

$$H^0(\mathscr{E}_{/\mathscr{V}}, \Omega_{\mathscr{E}/\mathscr{V}}) = \mathscr{V}\omega.$$

We then define $\Omega_p \in \mathscr{O}^\times$ and $\Omega_\infty \in \mathbb{C}^\times$ by $\omega = \Omega_\infty\omega_\infty$ and $\omega = \Omega_p\omega_p$. The ratio $\frac{\Omega_p}{\Omega_\infty}$ is well defined and independent of the choice of ω.

We take an arithmetic Hecke character $\varphi \in \mathscr{M}_{m\iota+nc}(1)$ with infinity type $\infty(\varphi) = m\iota + nc$ (see Subsection 1.2.2). Then the Galois character $\widehat{\varphi}$ (associated with φ by the theorem of Weil; see Subsection 1.1.3) factors through $\mathfrak{H}^{ab}$. We define $\Omega_\infty(\varphi) = (2\pi i)^{-n}\Omega_\infty^{n-m}$ and $\Omega_p(\varphi) = \Omega_p^{n-m}$. Then it has been shown (for example, see [EEK] and [K4]) that if $m > 0 \geq n$, $s = 1$ is critical with respect to $L(s, \varphi)$ and

$$\frac{\Gamma(m)(1-\varphi^{-1}(\overline{\mathfrak{p}}))(1-\varphi(\mathfrak{p})p^{-1})L(1,\varphi)}{\Omega_\infty(\varphi)} \in \overline{\mathbb{Q}} \cap \mathscr{O}.$$

Suppose that $\widehat{\varphi}$ is a deformation of $\overline{\varphi}$. Then we have a unique $\mathscr{O}$-algebra homomorphism $\iota_\varphi : \mathbb{J} = \mathscr{O}[[\mathfrak{H}_p^{ab}]] \to \mathscr{O}$ such that $\iota_\varphi \circ \Phi = \varphi$. We define $L_p(P, \Phi) = P(L_p)$ for any $\mathscr{O}$-algebra homomorphism $P \in Spec(\mathbb{J})(\mathscr{O}) = \mathrm{Hom}_{\mathscr{O}\text{-}alg}(\mathbb{J}, \mathscr{O})$. In particular, we write $L_p(\varphi, \Phi)$ for $L_p(\iota_\varphi)$. Then the interpolation property is given by

$$\frac{L_p(\varphi, \Phi)}{\Omega_p(\varphi)} = \frac{\Gamma(m)(1-\varphi^{-1}(\overline{\mathfrak{p}}))(1-\varphi(\mathfrak{p})p^{-1})L(1,\varphi)}{(\sqrt{-D})^n\Omega_\infty(\varphi)},$$

as long as $\varphi \mod \mathfrak{m}_{\mathscr{O}} = \overline{\varphi}$ and $m > 0 \geq n$.

The Selmer group $\mathrm{Sel}(\rho)$ is studied by K. Rubin in depth, and he proved in [Rb] that the characteristic power series of $\mathrm{Sel}^*(\rho)$ is equal to $L_p(\Phi)$ up to $\mathbb{J}$-units:

$$char_{\mathbb{J}}(\mathrm{Sel}^*(\mathrm{Ind}_F^{\mathbb{Q}}\Phi)) = L_p(\Phi) \text{ up to } \mathbb{J}\text{-units.} \qquad \text{(Rb)}$$

His result is actually stronger than what we have described and covers more general cases. We refer the reader to details in his paper [Rb].

We can also look into the standard Selmer group $\mathrm{Sel}(Ad(\rho))$ for $\rho = \mathrm{Ind}_F^{\mathbb{Q}}\Phi$. Note that

$$Ad(\mathrm{Ind}_F^{\mathbb{Q}}\Phi) \cong \chi \oplus \mathrm{Ind}_F^{\mathbb{Q}}\Phi\Phi_c^{-1},$$

where $\chi = \left(\frac{-D}{\cdot}\right)$, the Legendre symbol. The character $\Phi^- = \Phi\Phi_c^{-1}$ is called the *anticyclotomic* projection of Φ, and any Galois character which is a specialization of $\Phi\Phi_c^{-1}$ is called anticyclotomic. Since the conjugation by $c:g \mapsto cgc^{-1}$ fixes a one-dimensional subspace of $\mathfrak{H}_p^{ab}$ element by element, the character $\Phi\Phi_c^{-1}$ actually has values in a subring $\mathbb{J}^- \subset \mathbb{J}$, which is isomorphic to a one-variable power series ring $\mathcal{O}[[T^-]]$, and $\Phi\Phi_c^{-1}$ induces a projection $\pi : \mathbb{J} \twoheadrightarrow \mathbb{J}^-$. We write $L_p^- = L_p(\Phi\Phi_c^{-1}) = \pi(L_p)$. We can verify that

$$\mathrm{Sel}(Ad(\mathrm{Ind}_F^{\mathbb{Q}}\Phi)) \cong (\mathrm{Sel}(\chi)\otimes_{\mathcal{O}}\mathbb{J}) \oplus \left(\mathrm{Sel}(\mathrm{Ind}_F^{\mathbb{Q}}\Phi\Phi_c^{-1})\otimes_{\mathbb{J}^-}\mathbb{J}\right).$$

Since we have assumed that $p \nmid h(-D)$, we have $\mathrm{Sel}(\chi) \cong Cl_{F,p} = \{1\}$. Thus it is natural to expect that the characteristic power series of $\mathrm{Sel}^*(Ad(\rho))_{/\mathbb{Q}}$ is given by $L_p(\Phi\Phi_c^{-1})$, which has been proved by Tilouine [Ti] and Mazur–Tilouine [MT]:

$$char_{\mathbb{J}}(\mathrm{Sel}^*(Ad(\mathrm{Ind}_F^{\mathbb{Q}}\Phi))) = h(-D)L_p(\Phi\Phi_c^{-1}) \text{ up to } \mathbb{J}\text{-units}, \qquad \text{(MT)}$$

which implies

$$char_{\mathbb{J}^-}(\mathrm{Sel}^*(\mathrm{Ind}_F^{\mathbb{Q}}\Phi\Phi_c^{-1})) = L_p(\Phi\Phi_c^{-1}) \text{ up to } \mathbb{J}^-\text{-units}.$$

We have a generalization of (MT) in the case where E is totally real and F is a totally imaginary quadratic extension of E (the CM case; see [HT] and [HT1]).

5.2.2 Universal deformation rings

Hereafter we suppose that $n = 2$ and $F = E = \mathbb{Q}$, since we have proved the identity of Hecke algebras and universal deformation rings only when $n = 2$ (and limiting ramification to the single prime p). However the formal part of the theory works well for more general algebraic groups. If the reader wants to know the general case, he or she can consult [H99b]. We suppose that $\mathbb{J} \in CNL_{\mathcal{O}}$. We put $\overline{\rho} = \varphi \mod \mathfrak{m}_{\mathbb{J}} : \mathfrak{G} \to GL_2(\mathbb{F})$. Since $n = 2$, $(\mathrm{fil}_{\mathfrak{p}})$ is just a two-step filtration (for primes $\mathfrak{p}|p$ of L). We write $\varepsilon_{\mathfrak{p},\varphi} = \delta_{1,\mathfrak{p},\varphi}$ and $\delta_{\mathfrak{p},\varphi} = \delta_{2,\mathfrak{p},\varphi}$. We also write $\overline{\delta}_{\mathfrak{p}} = \delta_{\mathfrak{p},\overline{\rho}}$.

Taking a base of $V(\varphi)$ so that $\varphi|_{D_{\mathfrak{p}}} = \begin{pmatrix} \varepsilon_{\mathfrak{p},\varphi} & * \\ 0 & \delta_{\mathfrak{p},\varphi} \end{pmatrix}$, let $\mathscr{P}_{\mathfrak{p}}(A)$ be the subalgebra of $\mathrm{End}_A(V(\varphi))$ made up of upper triangular matrices. In other words, $\mathscr{P}_{\mathfrak{p}}$ is the Lie algebra of the Borel subgroup $P_{\mathfrak{p}}$, and $\mathscr{P}_{\mathfrak{p}}$ is made up of endomorphisms of the representation space preserving the filtration $(\mathrm{fil}_{\mathfrak{p}})$. For any $\mathbb{J}$-algebra A, we simply put $\mathscr{P}_{\mathfrak{p}}(A) = \mathscr{P}_{\mathfrak{p}} \otimes_{\mathbb{J}} A$, regarding $\mathscr{P}_{\mathfrak{p}}$ as a functor from $\mathbb{J}$-ALG into the category of Lie algebras.

Let $\mathfrak{g} = \mathrm{End}(V(\varphi)) = V(\varphi) \otimes_{\mathbb{J}} \mathrm{Hom}_{\mathbb{J}}(V(\varphi), \mathbb{J})$ be the Lie algebra of $GL(n)$, where $\mathfrak{G}$ acts on $\mathrm{Hom}_{\mathbb{J}}(V(\varphi), \mathbb{J})$ by the contragredient $\widetilde{\varphi} = ({}^t\varphi)^{-1}$, that is, $\tilde{\varphi}(\sigma)\phi(v) = \phi(\varphi(\sigma)^{-1}v)$ for $\phi \in \mathrm{Hom}_{\mathbb{J}}(V(\varphi), \mathbb{J})$. Then $\mathfrak{g}$ has a natural four-step filtration:

$$0 \subset V_{\mathfrak{p}}^{+}(Ad(\varphi)) = \left\{\left(\begin{smallmatrix} 0 & * \\ 0 & 0 \end{smallmatrix}\right)\right\} \subset \left\{\left(\begin{smallmatrix} * & * \\ 0 & 0 \end{smallmatrix}\right)\right\} \subset \mathscr{P}_{\mathfrak{p}} \subset \mathfrak{g}.$$

This filtration is stable under the adjoint action of $D_{\mathfrak{p}}$ and is the double filtration induced on $\mathfrak{g}(\mathbb{J}) = V(\varphi) \otimes V(\tilde{\varphi})$ by the tensor product of the filtration $(\mathrm{fil}_{\mathfrak{p}})$ on $V(\varphi)$ and its dual on the contragredient $V(\tilde{\varphi})$. Then

$$(\mathfrak{g}/\mathscr{P}_{\mathfrak{p}})(\mathbb{J}) \cong V(\delta_{\mathfrak{p},\varphi}^{-1}) \otimes V(\varepsilon_{\mathfrak{p},\varphi}) \cong \mathrm{Hom}_{\mathbb{J}}(\delta_{\mathfrak{p},\varphi}, \varepsilon_{\mathfrak{p},\varphi})$$

as $D_{\mathfrak{p}}$-modules.

Let $D_{\mathfrak{p}}$ be the decomposition group at $\mathfrak{p}|p$ in $\mathfrak{H}_L$, and we write $I_{\mathfrak{p}}$ for the inertia subgroup of $D_{\mathfrak{p}}$. We consider the following two conditions (cf. [DGH] Chapter 6):

(AI_L) $\overline{\rho}_L$ *is absolutely irreducible as a representation of* $\mathfrak{H}_L$ *into* $GL_2(\mathbb{F})$*;*
(Rg_L) $\overline{\varepsilon}_{\mathfrak{p}} \neq \overline{\delta}_{\mathfrak{p}}$*, which is equivalent to* $H^0_{ct}(D_{\mathfrak{p}}, (\mathfrak{g}/\mathscr{P}_{\mathfrak{p}})(\mathbb{F})) = 0$*, for all* $\mathfrak{p}|p$ *in* L*.*

A deformation ρ is called *nearly ordinary* if $\rho|_{D_{\mathfrak{p}}}$ is isomorphic to an upper triangular representation; in other words, we have the following filtration of $D_{\mathfrak{p}}$-modules for all $\mathfrak{p} \in S$ whose stabilizer is in the conjugacy class of $P_{\mathfrak{p}}$ over A:

$$0 \subset V(\varepsilon_{\mathfrak{p},\rho}) \subset V(\rho|_{D_{\mathfrak{p}}}), \qquad (\mathrm{fil}_{\mathfrak{p}})$$

where $V(\rho)/V(\varepsilon_{\mathfrak{p},\rho})$ is a A-free module of rank 1 on which $D_{\mathfrak{p}}$ acts via the character $\varepsilon_{\mathfrak{p},\rho}$ with $\varepsilon_{\mathfrak{p},\rho} \mod \mathfrak{m}_A \cong \overline{\varepsilon}_{\mathfrak{p}}$ for all $\mathfrak{p}$.

We write $\phi = \det(\varphi)$. Since these characters have values in $\mathcal{O}^{\times}$, we may regard them as characters having values in any object $A \in CNL_{\mathcal{O}}$ by composing the structure homomorphism: $\mathcal{O} \to A$. Under (AI_L) and (Rg_L), the functor $\mathscr{F}_L^{\phi,n.ord}$, associating with $A \in CNL_{\mathcal{O}}$ strict equivalence classes of nearly ordinary deformations $\rho : \mathfrak{H}_L \to GL_2(A)$ of $\overline{\rho}$ with $\det(\rho) = \phi$, is representable as shown by Mazur, Boston and Tilouine (cf. [DGH] Chapter 6), where ϕ is regarded as having values in A via the structure homomorphism: $\mathcal{O} \to A$ (see Example 4.4). The representability can also be shown similarly to Proposition 3.30, so we leave its proof to the reader. Thus there exists a unique couple $(R, \varrho) = (R_L^{\phi,n.ord}, \varrho_L^{\phi,n.ord})$ made up of $R \in CNL_{\mathcal{O}}$ and a continuous deformation $\varrho = \varrho^{\phi,n.ord}$: $\mathfrak{H}_L \to GL_2(R)$ of $\overline{\rho}$ such that for each nearly ordinary deformation $\rho : \mathfrak{H}_L \to GL_2(A) \in \mathscr{F}_{G,L}^{\phi,n.ord}(A)$ of $\overline{\rho}$, there exists a unique $\mathcal{O}$-algebra homomorphism $\iota_{\rho} : R \to A$ such that ρ is strictly equivalent to $\iota_{\rho} \circ \varrho$. In particular, we have a unique $\mathcal{O}$-algebra homomorphism $\pi : R \to \mathbb{J}$ such that $\pi \circ \varrho \approx \varphi$.

Write $\overline{\xi} = \overline{\delta}_{\mathfrak{p}}$. The deformation functor for $H = I_{\mathfrak{p}}$ or $D_{\mathfrak{p}}$:

$$\Phi^H_{\mathfrak{p},\overline{\xi}}(A) = \{\xi : H \to A^\times | \xi \mod \mathfrak{m}_A = \overline{\xi}\} / \approx$$

is representable over $CNL_{\mathcal{O}}$. We write $(R^H_{\mathfrak{p},\overline{\xi}}, \xi_H)$ for the universal couple. Since this is an abelian deformation, as seen in Theorem 2.21,

$$R^H_{\mathfrak{p},\overline{\xi}} \cong \mathcal{O}[[H^{ab}_p]],$$

and for the Teichmüller lift $\xi : H \to \mathcal{O}^\times$ of $\overline{\xi}$, we have

$$\xi(h) = \xi(h)[h],$$

where $[h] \in H^{ab}_p$ is the image of h in the maximal p-profinite abelian quotient H^{ab}_p.

Here we insert the definition of *completed tensor product* of p-profinite modules M and N with continuous action over A, which is an object of $CL_{\mathcal{O}}$. Write $M = \varprojlim_j M_j$ and $N = \varprojlim_j N_j$ as projective limits of finite A-modules. Then $M \widehat{\otimes}_A N = \varprojlim_j M_j \otimes N_j$, whose transition maps are given by the tensor product of the transition maps of each projective system. This tensor product satisfies the following universal property: For a given continuous A-bilinear map $b : M \times N \to L$ for a continuous A-module L (not necessarily profinite), there exists a unique extension $\widetilde{b} : M \widehat{\otimes}_A N \to L$ such that $\widetilde{b}(a \otimes b) = b(a, b)$ for all $a \in M$ and $b \in N$. When M and N are A-algebras in $CL_{\mathcal{O}}$, $M \widehat{\otimes}_B N$ becomes naturally an A-algebra, by the multiplication given by $(a \otimes b) \cdot (c \otimes d) = ac \otimes bd$, which is an object in $CL_{\mathcal{O}}$. For example, $A[[T]] \widehat{\otimes}_A A[[S]] \cong A[[T, S]]$, where A is a profinite local ring . If A is a B-algebra for another object B in $CL_{\mathcal{O}}$, $A \widehat{\otimes}_B A$ is an A-algebra by $i : a \mapsto 1 \otimes a$, and the map i is a section of the multiplication $m : A \times A \to A$ given by $m(a, b) = ab$, that is, $m \circ i = \mathrm{id}_A$.

For the universal nearly p-ordinary deformation $\varrho \in \mathcal{F}^{\phi,n.ord}(R)$, we have $\delta_{\mathfrak{p},\varrho} : H \to R^\times$, which is a deformation of $\overline{\delta}_{\mathfrak{p}}$ over H. Thus we have a canonical $\mathcal{O}$-algebra homomorphism $i_{\mathfrak{p}} : R^H_{\mathfrak{p},\overline{\delta}_{j,\mathfrak{p}}} \to R$ inducing $\delta_{\mathfrak{p},\varrho}$ from the corresponding universal representation of H. For $H = I$ and D, we write R^H_F for

$$\widehat{\bigotimes_{\mathfrak{p}}}(R^{H_{\mathfrak{p}}}_{\mathfrak{p},\overline{\delta}_{\mathfrak{p}}}) = \mathcal{O}[[\prod_{\mathfrak{p}} (H_{\mathfrak{p}})^{ab}_p]]$$

and $\iota_H : R^H_F \to R$ for the tensor product ι_H of the morphisms $i_{\mathfrak{p}}$ over $\mathfrak{p}|p$. Again by definition, R^D_L is naturally an R^I_L-algebra. In any case, we found the following fact:

(LG) The global universal deformation ring $R_L^{\phi,n.ord}$ is an algebra over the local one R_L^H for $H = I$ and D.

5.2.3 *Kähler differentials*

We recall here the definition of Kähler 1-differentials and some of their properties for our later use. Let A be a B-algebra, and suppose that A and B are objects in $CNL_{\mathcal{O}}$. The module of 1-differentials $\Omega_{A/B}$ for a B-algebra A ($A, B \in CNL_{\mathcal{O}}$) indicates the module of **continuous** 1-differentials with respect to the profinite topology.

For a module M with continuous A-action (in short, a continuous A-module), let us define the module of B-*derivations* by

$$Der_B(A, M) = \left\{ \delta : A \to M \in \mathrm{Hom}_B(A, M) \middle| \begin{array}{c} \delta : \text{continuous} \\ \delta(ab) = a\delta(b) + b\delta(a) \\ \text{for all } a, b \in A \end{array} \right\}.$$

Here the B-linearity of a derivation δ is equivalent to $\delta(B) = 0$, because

$$\delta(1) = \delta(1 \cdot 1) = 2\delta(1) \Rightarrow \delta(1) = 0.$$

Then $\Omega_{A/B}$ represents the covariant functor $M \mapsto Der_B(A, M)$ from the category of **continuous** A-modules into $\mathbb{Z}$-MOD. Since $Der_B(A, M)$ only depends on the image of B in A under the algebra homomorphism $\iota : B \to A$ giving the B-algebra structure of A, we have

$$\Omega_{A/B} \cong \Omega_{A/\iota(B)}. \tag{5.6}$$

The construction of $\Omega_{A/B}$ is easy. The multiplication $a \otimes b \mapsto ab$ induces a B-algebra homomorphism $m : A\widehat{\otimes}_B A \to A$ taking $a \otimes b$ to ab. We put $I = \mathrm{Ker}(m)$, which is an ideal of $A\widehat{\otimes}_B A$. Then we define $\Omega_{A/B} = I/I^2$. It is easy to check that the map $d : A \to \Omega_{A/B}$ given by $d(a) = a \otimes 1 - 1 \otimes a \mod I^2$ is a continuous B-derivation (Exercise 1). Thus we have a morphism of functors: $\mathrm{Hom}_A(\Omega_{A/B}, ?) \to Der_B(A, ?)$ given by $\phi \mapsto \phi \circ d$. Since $\Omega_{A/B}$ is generated by $d(A)$ as A-modules (Exercise 2), the above map is injective. To show that $\Omega_{A/B}$ represents the functor, we need to show the surjectivity of the above map.

Proposition 5.12 *The above morphism of two functors $M \mapsto \mathrm{Hom}_A(\Omega_{A/B}, M)$ and $M \mapsto Der_B(A, M)$ is an isomorphism, where M runs over the category of continuous A-modules. Thus the functor: $M \mapsto Der_B(A, M)$ is represented by $\Omega_{A/B}$ in the category of continuous A-modules.*

Proof Define $\phi : A \times A \to M$ by $(a,b) \mapsto a\delta(b)$ for $\delta \in Der_B(A,M)$. Then $\phi(ab,c) = ab\delta(c) = a\phi(b,c)$ and $\phi(a,bc) = a\delta(bc) = ab\delta(c) = b\phi(a,c)$ for $a,c \in A$ and $b \in B$. Thus ϕ gives a continuous B-bilinear map. By the universality of the tensor product, $\phi : A \times A \to M$ extends to a B-linear map $\phi : A\widehat{\otimes}_B A \to M$. Now we see that $\phi(a \otimes 1 - 1 \otimes a) = a\delta(1) - \delta(a) = -\delta(a)$ and

$$\begin{aligned}\phi((a \otimes 1 - 1 \otimes a)(b \otimes 1 - 1 \otimes b)) &= \phi(ab \otimes 1 - a \otimes b - b \otimes a + 1 \otimes ab) \\ &= -a\delta(b) - b\delta(a) + \delta(ab) = 0.\end{aligned}$$

This shows that $\phi|_I$-factors through $I/I^2 = \Omega_{A/B}$ and $\delta = \phi \circ d$, as desired. □

Corollary 5.13 *Let the notation be as in the proposition.*

(i) *Suppose that B is a C-algebra for an object $C \in CL_{\mathcal{O}}$. Then we have the following natural exact sequence:*

$$\Omega_{B/C}\widehat{\otimes}_B A \longrightarrow \Omega_{A/C} \longrightarrow \Omega_{A/B} \to 0.$$

(ii) *Let $\pi : A \twoheadrightarrow C$ be a surjective morphism in $CL_{\mathcal{O}}$, and write $J = Ker(\pi)$. Then we have the following natural exact sequence:*

$$J/J^2 \longrightarrow \Omega_{A/B}\widehat{\otimes}_A C \longrightarrow \Omega_{C/B} \to 0.$$

Moreover, if $B = C$, then $J/J^2 \cong \Omega_{A/B}\widehat{\otimes}_A C$.

Proof By assumption, we have algebra morphisms $C \to B \to A$ in Case (i) and $B \to A \twoheadrightarrow C = A/J$ in Case (ii). By the unicity-lemma (Lemma 4.3), we only need to prove that

$$0 \to Der_B(A,M) \xrightarrow{\alpha} Der_C(A,M) \xrightarrow{\beta} Der_C(B,M)$$

is exact in Case (i) for all continuous A-modules M and that

$$0 \to Der_B(C,M) \xrightarrow{\alpha} Der_B(A,M) \xrightarrow{\beta} \mathrm{Hom}_C(J/J^2, M)$$

is exact in Case (ii) for all continuous C-modules M. The first α is just the inclusion and the second α is the pull-back map. Thus the injectivity of α is obvious in two cases. Let us prove the exactness at the mid-term of the first sequence. The map β is the restriction of derivation D on A to B. If $\beta(D) = D|_B = 0$, then D kills B and hence D is actually a B-derivation, i.e. in the image of α. The map β in the second sequence is defined as follows: For a given B-derivation $D : A \to M$, we regard D as a B-linear map of J into M. Since J kills M, $D(jj') = jD(j') + j'D(j) = 0$

for $j, j' \in J$. Thus D induces the C-linear map: $J/J^2 \to M$. Then for $b \in B$ and $x \in J$, $D(bx) = bD(x) + xD(b) = bD(x)$. Thus D is B-linear, and $\beta(D) = D|_J$. Now prove the exactness at the mid-term of the second exact sequence. The fact that $\beta \circ \alpha = 0$ is obvious. If $\beta(D) = 0$, then D kills J and hence a derivation well defined on $C = A/J$. This shows that D is in the image of α.

Now suppose that $B = C$ in the assertion (ii). We need to create a surjective C-linear map: $\gamma : \Omega_{A/B} \otimes C \twoheadrightarrow J/J^2$ such that $\gamma \circ \alpha = \mathrm{id}$. Let $\pi : A \to C$ be the projection and $\iota : B = C \hookrightarrow A$ be the structure homomorphism giving the B-algebra structure on A. To do that, we first look at the map $\delta : A \to J/J^2$ given by $\delta(a) = a - P(a) \mod J^2$ for $P = \iota \circ \pi$. Then

$$\begin{aligned} a\delta(b) + b\delta(a) - \delta(ab) &= a(b - P(b)) + b(a - P(a)) - ab - P(ab) \\ &= (a - P(a))(b - P(b)) \equiv 0 \mod J^2. \end{aligned}$$

Thus δ is a B-derivation. By the universality of $\Omega_{A/B}$, we have an A-linear map $\phi : \Omega_{A/B} \to J/J^2$ such that $\phi \circ d = \delta$. By definition, $\delta(J)$ generates J/J^2 over A, and hence ϕ is surjective. Since J kills J/J^2, the surjection ϕ factors through $\Omega_{A/B} \otimes_A C$ and induces γ as desired. □

For any continuous A-module X, we write $A[X]$ for the A-algebra with square zero ideal X. Thus $A[X] = A \oplus X$ with the multiplication given by

$$(r \oplus x)(r' \oplus x') = rr' \oplus (rx' + r'x).$$

It is easy to see that $A[X] \in CNL_{\mathcal{O}}$, if X is of finite type, and $A[X] \in CL_{\mathcal{O}}$ if X is a p-profinite A-module. By definition,

$$Der_B(A, X) \cong \{\phi \in \mathrm{Hom}_{B\text{-}alg}(A, A[X]) | \phi \mod X = \mathrm{id}\},$$

where the map is given by $\delta \mapsto (a \mapsto (a \oplus \delta(a))$. Note that $i : A \to A\widehat{\otimes}_B A$ given by $i(a) = a \otimes 1$ is a section of $m : A\widehat{\otimes}_B A \to A$. We see easily that $A\widehat{\otimes}_B A/I^2 \cong A[\Omega_{A/B}]$ by $x \mapsto m(x) \oplus (x - i(x))$. Note that $d(x) = x - i(x)$.

Exercises

(1) Show that the map $d : A \to \Omega_{A/B}$ given by $d(a) = a \otimes 1 - 1 \otimes a \mod I^2$ is a continuous B-derivation.

(2) Prove that $\Omega_{A/B}$ is generated by $d(A)$ as A-modules and that it is an A-module of finite type if A is noetherian.

5.2.4 Adjoint Selmer groups and differentials

We recall the argument of Mazur (cf. [MT]) to relate 1-differentials on $Spec(R)$ ($R = R_L^{\phi,n.ord}$) with the Selmer group $\mathrm{Sel}(Ad_S(\varphi))_{/L}$.

Write simply R for $R_L^{\phi,n.ord}$ and $\mathcal{F}$ for $\mathcal{F}_L^{\phi,n.ord}$. Let X be a profinite R-module. Then X is profinite, and $R[X]$ is an object in $CL_{\mathcal{O}}$. We consider the $\mathcal{O}$-algebra homomorphism $\xi : R \to R[X]$ with $\xi \mod X = \mathrm{id}$. Then we can write $\xi(r) = r \oplus d_\xi(r)$ with $d_\xi(r) \in X$. By the above definition of the product, we get $d_\xi(rr') = rd_\xi(r') + r'd_\xi(r)$ and $d_\xi(\mathcal{O}) = 0$. Thus d_ξ is an $\mathcal{O}$-derivation, i.e., $d_\xi \in Der_{\mathcal{O}}(R, X)$. For any derivation $d : R \to X$ over $\mathcal{O}$, $r \mapsto r \oplus d(r)$ is obviously an $\mathcal{O}$-algebra homomorphism, and we get

$$\begin{aligned}
&\{\rho \in \mathcal{F}(R[X]) | \rho \mod X = \varrho\} / \approx_X \qquad (5.7)\\
&\cong \{\rho \in \mathcal{F}(R[X]) | \rho \mod X \approx \varrho\} / \approx \\
&\cong \{\xi \in \mathrm{Hom}_{\mathcal{O}-alg}(R, R[X]) | \xi \mod X = \mathrm{id}\} \\
&\cong Der_{\mathcal{O}}(R, X) \cong \mathrm{Hom}_R(\Omega_{R/\mathcal{O}}, X),
\end{aligned}$$

where '$\approx_X$' is conjugation under $(1 \oplus M_2(X)) \cap GL_2(R[X])$.

Let ρ be the deformation in the left-hand side of (5.7). Then we may write $\rho(\sigma) = \varrho(\sigma) \oplus u'_\rho(\sigma)$. We see that

$$\varrho(\sigma\tau) \oplus u'_\rho(\sigma\tau) = (\varrho(\sigma) \oplus u'_\rho(\sigma))(\varrho(\tau) \oplus u'_\rho(\tau)) = \varrho(\sigma\tau) \oplus (\varrho(\sigma)u'_\rho(\tau) + u'_\rho(\sigma)\varrho(\tau)),$$

and we have

$$u'_\rho(\sigma\tau) = \varrho(\sigma)u'_\rho(\tau) + u'_\rho(\sigma)\varrho(\tau).$$

Define $u_\rho(\sigma) = u'_\rho(\sigma)\varrho(\sigma)^{-1}$. Then, $x(\sigma) = \rho(\sigma)\varrho(\sigma)^{-1}$ has values in $SL_2(R[X])$, and $x = 1 \oplus u \mapsto u = x - 1$ is an isomorphism from the multiplicative group of the kernel of the reduction map $SL_2(R[X]) \twoheadrightarrow SL_2(R)$: $\{x \in SL_2(R[X]) | x \equiv 1 \mod X\}$ onto the additive group

$$Ad(X) = \{x \in M_2(X) | \mathrm{Tr}(x) = 0\} = V(Ad(\varrho)) \otimes_R X.$$

Thus we may regard u as having values in $Ad(X) = V(Ad(\varrho)) \otimes_R X$.

We also have

$$\begin{aligned}
u_\rho(\sigma\tau) &= u'_\rho(\sigma\tau)\varrho(\sigma\tau)^{-1} \qquad (5.8)\\
&= \varrho(\sigma)u'_\rho(\tau)\varrho(\sigma\tau)^{-1} + u'_\rho(\sigma)\varrho(\tau)\varrho(\sigma\tau)^{-1} = Ad_S(\varrho)(\sigma)u_\rho(\tau) + u_\rho(\sigma).
\end{aligned}$$

Hence $u_\rho : \mathfrak{H}_L \to Ad_S(X)$ is a 1-cocycle. It is a straightforward computation to see the injectivity of the map:

$$\{\rho \in \mathcal{F}(R[X]) | \rho \mod X = \varrho\} / \approx_X \hookrightarrow H^1_{ct}(\mathfrak{H}_L, Ad(X))$$

given by $\rho \mapsto [u_\rho]$ (Exercise 2). We put $V_{\mathfrak{p}}^{\pm}(Ad(X)) = V_{\mathfrak{p}}^{\pm}(Ad(\varrho)) \otimes_R X$. Then we see from the fact that $\mathrm{Tr}(u_\rho) = 0$ that

$$u_\rho(I_{\mathfrak{p}}) \subset V_{\mathfrak{p}}^{+}(Ad(X)) \iff u'_\rho(I_{\mathfrak{p}}) \subset V_{\mathfrak{p}}^{+}(Ad(X)) \iff d_\xi(R^I_{\bar{\delta}_{\mathfrak{p}}}) = 0 \tag{5.9}$$

if $\xi \in \mathrm{Hom}_{\mathcal{O}-alg}(R, R[X])$ induces ρ.

Since $\mathbb{J} = \varprojlim_n \mathbb{J}/\mathfrak{m}_{\mathbb{J}}^n$ for the finite rings $\mathbb{J}/\mathfrak{m}_{\mathbb{J}}^n$, we have $\mathbb{J}^* = \varinjlim_n (\mathbb{J}/\mathfrak{m}_{\mathbb{J}}^n)^*$, which is a discrete R-module, which shows that $R[\mathbb{J}^*] = \bigcup_n R[(\mathbb{J}/\mathfrak{m}_{\mathbb{J}}^n)^*]$. We put the profinite topology on the individual $R[(\mathbb{J}/\mathfrak{m}_{\mathbb{J}}^n)^*]$. On $R[\mathbb{J}^*]$, we give a injective-limit topology. Thus a map $\phi : X \to R[\mathbb{J}^*]$ is continuous if $\phi^{-1}(R[(\mathbb{J}/\mathfrak{m}_{\mathbb{J}}^n)^*]) \to R[(\mathbb{J}/\mathfrak{m}_{\mathbb{J}}^n)^*]$ is continuous for all n with respect to the induced topology from X on the source and the profinite topology on the target. From this, any deformation having values in $GL_2(R[\mathbb{J}^*])$ gives rise to a continuous 1-cocycle (see [HT1] Chapter 2 for details about continuity) with values in the discrete $\mathfrak{G}$-module $V(Ad_S(\varphi))^*$. In this way, we get

$$(\Omega_{R/A} \otimes_R \mathbb{J})^* \cong \mathrm{Hom}_R(\Omega_{R/\mathcal{O}}, \mathbb{J}^*) \hookrightarrow H^1_{ct}(\mathfrak{H}_L, V(Ad_S(\varphi))^*) \tag{5.10}$$

if R is an A-algebra for an object A of $CNL_{\mathcal{O}}$. By definition, ρ is nearly p-ordinary if and only if u_ρ restricted to $I_{\mathfrak{p}}$ has values in $V^{-}(Ad_S(\varphi))^*$.

We can argue in the same way as above, replacing the inertia groups by the decomposition groups, and we get the corresponding results on the strict Selmer groups. Thus we get from this and (5.9) the following fact:

THEOREM 5.14 (B. MAZUR) *Assume* (AI_L) *and* (Rg_L). *Then*

$$\begin{aligned} \mathrm{Sel}^*(Ad(\varphi))_{/L} &\cong \Omega_{R_L^{\phi,n.ord}/R_L^I} \otimes_{R_L^{\phi,n.ord}} \mathbb{J}, \quad \textit{and,} \\ \mathrm{Sel}^*_{-}(Ad(\varphi))_{/L} &\cong \Omega_{R_L^{\phi,n.ord}/\mathcal{O}} \otimes_{R_G^{\phi,n.ord}} \mathbb{J}, \\ \mathrm{Sel}^*_{st}(Ad_S(\varphi))_{/L} &\cong \Omega_{R_L^{\phi,n.ord}/R_L^D} \otimes_{R_L^{\phi,n.ord}} \mathbb{J} \end{aligned}$$

as $\mathbb{J}$*-modules. Moreover, we have the following exact sequence:*

$$\Omega_{R_L^D/R_L^I} \otimes_{R_L^D} \mathbb{J} \to \mathrm{Sel}^*(Ad(\varphi))_{/L} \to \mathrm{Sel}^*_{st}(Ad_S(\varphi))_{/L} \to 0.$$

Exercises

(1) Prove (5.7).

(2) Prove the injectivity of the map:

$$\left\{\rho \in \mathscr{F}_L^{\phi,n.ord}(R[X]) | \rho \mod X = \varrho\right\} / \approx_X \hookrightarrow H^1_{ct}(\mathfrak{H}_L, Ad(X)).$$

Hint: Show that the equivalence: '$\approx_X$' in the right-hand side gives rise to the equivalence in 1-cocycles modulo coboundary.

(3) Give a detailed proof of Theorem 5.14.

5.3 Arithmetic of Modular Adjoint L-Values

In this section, we assume that $\varphi = \rho_\lambda$ for an algebra homomorphism $\lambda : h_k(p,\chi;\mathbb{Z}[\chi]) \to \overline{\mathbb{Q}}_p$, and study the arithmetic and analytic properties of $L(s, Ad(\lambda)) = L(s, Ad(\rho_\lambda))$.

5.3.1 Analyticity of adjoint L-functions

We summarize here the known facts on analyticity of the adjoint L-function $L(s, Ad(\lambda)) = L(s, Ad(\rho_\lambda))$ for a $\mathbb{Z}[\chi]$-algebra homomorphism λ of $h_k(C,\chi;\mathbb{Z}[\chi])$ into $\overline{\mathbb{Q}}$ and the compatible system ρ_λ of Galois representations attached to λ. We always assume that $k \geq 2$ in this subsection.

Note that the Galois representation ρ_λ depends only on the equivalence class of λ introduced in Subsection 3.2.1. Thus we may assume that λ is primitive of exact level C. Then writing the (reciprocal) Hecke polynomial at a prime ℓ as

$$L_\ell(X) = 1 - \lambda(T(\ell))X + \chi(\ell)\ell^{k-1}X^2 = (1-\alpha_\ell X)(1-\beta_\ell X),$$

we have the following Euler product convergent absolutely if $Re(s) > 1$:

$$L(s, Ad(\lambda)) = \prod_\ell \left\{(1 - \frac{\alpha_\ell}{\beta_\ell}\ell^{-s})(1-\ell^{-s})(1-\frac{\beta_\ell}{\alpha_\ell}\ell^{-s})\right\}^{-1}.$$

The meromorphic continuation and functional equation of this L-function was proved by Shimura in 1975 [Sh1]. Recently Shimura incorporated ideas of Garrett and Böcherer into a general setting for $GSp(2m)$ and $U(m,m)$ and has found another way of proving meromorphic continuation and the functional equation. This new method is described in his paper [Sh2] and in his new book [EPE].

The earlier method of Shimura in [Sh1] is generalized, using the language of Langlands' theory, by Gelbart and Jacquet. Taking the primitive cusp form f such that $T(n)f = \lambda(T(n))f$ for all n, let π be the automorphic representation of $GL_2(\mathbb{A})$ spanned by f and its right translations. We write $L(s, Ad(\pi))$ for the L-function of the adjoint lift $Ad(\pi)$ to $GL(3)$ [GJ]. This L-function coincides with $L(s, Ad(\lambda))$, has a meromorphic continuation to the whole complex s-plane and satisfies a functional equation of the form $1 \leftrightarrow 1-s$ whose Γ-factor is given by

$$\Gamma(s, Ad(\lambda)) = \Gamma_{\mathbb{C}}(s+k-1)\Gamma_{\mathbb{R}}(s+1),$$

where $\Gamma_{\mathbb{C}}(s) = 2(2\pi)^{-s}\Gamma(s)$ and $\Gamma_{\mathbb{R}}(s) = \pi^{-s/2}\Gamma(\frac{s}{2})$ (see (Γ) in Subsection 5.1.2).

If ρ_λ is not an induced representation of a Galois character, the L-function is known to be entire (and the adjoint lift of Gelbart–Jacquet is a cusp from). This holomorphy always holds if $C|p$, $k \geq 2$ and $\lambda(T(p))$ is a unit, simply because $\lambda(T(p))$ is a unit multiple of $\left(\sqrt{(-1)^{(p-1)/2}p}\right)^{k-1}$ if ρ_λ is an induced representation (actually from $\mathbb{Q}(\sqrt{(-1)^{(p-1)/2}p})$).

Suppose that ρ_λ is an induced representation $\mathrm{Ind}_{\mathbb{Q}(\sqrt{D})}^{\mathbb{Q}} \varphi$ for a Galois character $\varphi : \mathrm{Gal}(\overline{\mathbb{Q}}/\mathbb{Q}(\sqrt{D})) \to \overline{\mathbb{Q}}_p^\times$ (associated with a Hecke character; cf. Theorem 1.1 in Chapter 1). Then we have $Ad(\rho_\lambda) \cong \chi \oplus \mathrm{Ind}_{\mathbb{Q}(\sqrt{D})}^{\mathbb{Q}}(\varphi\varphi_\sigma^{-1})$, where $\chi = \left(\frac{D}{\cdot}\right)$ is the Legendre symbol, and $\varphi_\sigma(g) = \varphi(\sigma g \sigma^{-1})$ for $\sigma \in \mathrm{Gal}(\overline{\mathbb{Q}}/\mathbb{Q})$ inducing a non-trivial automorphism on $\mathbb{Q}(\sqrt{D})$. Since λ is cuspidal, ρ_λ is irreducible, and hence $\varphi\varphi_\sigma^{-1} \neq 1$. Thus $L(s, Ad(\lambda)) = L(s,\chi)L(s,\varphi\varphi_\sigma^{-1})$ is still an entire function, but $L(s, Ad(\lambda)\otimes\chi)$ has a simple pole at $s = 1$.

After summarizing what we have said, we shall give a sketch of a proof of the meromorphic continuation of $L(s, Ad(\lambda))$ and its analyticity around $s = 1$ following [LFE] Chapter 9.

THEOREM 5.15 *Let χ be a primitive character modulo C. Let $\lambda : h_k(C, \chi; \mathbb{Z}[\chi]) \to \mathbb{C}$ be a $\mathbb{Z}[\chi]$-algebra homomorphism for $k \geq 2$. Then $\Gamma(s, Ad(\lambda))L(s, Ad(\lambda))$ has an analytic continuation to the whole complex s-plane and*

$$\Gamma(1, Ad(\lambda))L(1, Ad(\lambda)) = 2^k C^{-1} \int_{\Gamma_0(C)\backslash\mathfrak{H}} |f|^2 y^{k-2} dxdy,$$

where $f = \sum_{n=1}^{\infty} \lambda(T(n))q^n \in S_k(\Gamma_0(C), \chi)$ and $z = x + iy \in \mathfrak{H}$. If $C = 1$, we have the following functional equation:

$$\Gamma(s, Ad(\lambda))L(s, Ad(\lambda)) = \Gamma(1-s, Ad(\lambda))L(1-s, Ad(\lambda)).$$

Proof We consider $L(s-k+1, \rho_\lambda \otimes \widetilde{\rho}_\lambda)$ for the Galois representation associated with λ (Theorem 3.26). Since $\rho_\lambda \otimes \widetilde{\rho}_\lambda = \mathbf{1} \oplus Ad(\rho_\lambda)$, we have

$$L(s, \rho_\lambda \otimes \widetilde{\rho}_\lambda) = L(s, Ad(\lambda))\zeta(s) \tag{5.11}$$

for the Riemann zeta function $\zeta(s)$. Then, the Rankin convolution method tells us (cf. [LFE] Theorem 9.4.1) that

$$\left(2^{2-s}\prod_{p|C}(1-\frac{1}{p^{s-k+1}})\right)\Gamma_{\mathbb{C}}(s)L(s-k+1,\rho_\lambda\otimes\widetilde{\rho}_\lambda)$$
$$=\int_{\Gamma_0(C)\backslash\mathfrak{H}}|f|^2E'_{0,C}(s-k+1,\mathbf{1})y^{-2}dxdy,$$

where $E'_0(s,\mathbf{1})$ is the Eisenstein series of level C for the trivial character $\mathbf{1}$ defined in [LFE] page 297. Since the Eisenstein series is slowly increasing and f is rapidly decreasing, the integral converges absolutely on the whole complex s-plane outside the singularity of the Eisenstein series. The Eisenstein series has a simple pole at $s=1$ with constant residue: $\pi\prod_{p|C}(1-\frac{1}{p})$, which yields

$$Res_{s=k}\left(\left(2^{2-s}\prod_{p|C}(1-\frac{1}{p^{s-k+1}})\right)\Gamma_{\mathbb{C}}(s)L(s-k+1,\rho_\lambda\otimes\widetilde{\rho}_\lambda)\right)$$
$$=\pi\prod_{p|C}(1-\frac{1}{p})\int_{\Gamma_0(C)\backslash\mathfrak{H}}|f|^2y^{-2}dxdy.$$

This combined with (5.11) yields the residue formula and analytic continuation of $L(s,Ad(\lambda))$ over the region of $Re(s)\geq 1$. Since $\Gamma_{\mathbb{C}}(s)E'_{0,C}(s,\mathbf{1})$ satisfies a functional equation of the form $s\mapsto 1-s$ (see [LFE] Theorem 9.3.1), we have the meromorphic continuation of $\Gamma_{\mathbb{C}}(s)\Gamma_{\mathbb{C}}(s-k+1)L(s-k+1,\rho_\lambda\otimes\widetilde{\rho}_\lambda)$. Dividing the above zeta function by $\Gamma_{\mathbb{R}}(s-k+1)\zeta(s-k+1)$, we get the L-function $\Gamma(s-k+1,Ad(\lambda))L(s-k+1,Ad(\lambda))$, and hence meromorphic continuation of $\Gamma(s,Ad(\lambda))L(s,Ad(\lambda))$ to the whole s-plane and its holomorphy around $s=1$.

When $C=1$, the functional equation of the Eisenstein series is particularly simple:

$$\Gamma_{\mathbb{C}}(s)E_{0,1}(s,\mathbf{1})=2^{1-2s}\Gamma_{\mathbb{C}}(1-s)E_{0,1}(1-s,\mathbf{1}),$$

which combined with the functional equation of the Riemann zeta function (e.g. [LFE] Theorem 2.3.2 and Corollary 8.6.1) yields the functional equation of the adjoint L-function $L(s,Ad(\lambda))$. □

5.3.2 Rationality of adjoint L-values

By the explicit form of the Gamma factor, $\Gamma(s,Ad(\lambda))$ is finite at $s=0,1$, and hence $L(1,Ad(\lambda))$ is a critical value in the sense of Deligne and

Shimura, as long as $L(s, Ad(\lambda))$ is finite at these points. Thus we expect the L-value divided by a period of the λ-eigenform to be algebraic. This fact was first shown by Sturm (see [St1] and [St2]) by using Shimura's integral expression (in [Sh1]). Here we shall describe the integrality of the value, following [H81a] and [H88b], whose approach is different from Sturm. Then we shall relate in the following subsection, as an application of the Taylor-Wiles theorem, the size of the module $\mathrm{Sel}(Ad(\rho_\lambda))$ and the p-primary part of the critical value $\frac{\Gamma(1,Ad(\lambda))L(1,Ad(\lambda))}{\Omega(+,\lambda;A)\Omega(-,\lambda;A)}$.

To define the periods $\Omega(\pm,\lambda;A)$, we need to look into modular (topological) cohomology groups. Let A be a $\mathbb{Z}[\chi]$-algebra. Let $L(n,\chi;A)$ be the space of homogeneous polynomials in (X, Y) of degree n with coefficients in A. We let $\gamma = \left(\begin{smallmatrix} a & b \\ c & d \end{smallmatrix}\right) \in M_2(\mathbb{Z}) \cap GL_2(\mathbb{Q})$ act on $P(X, Y) \in L(n,\chi;A)$ by

$$(\gamma P)(X, Y) = \chi(d)P((X, Y)^t\gamma^\iota),$$

where $\gamma^\iota = (\det\gamma)\gamma^{-1}$.

Then we can define the *cuspidal cohomology group* $H^1_{cusp}(\Gamma_0(C), L(n,\chi;A))$ as in [IAT] Chapter 8 and [LFE] Chapter 6, whose definition we recall later in this subsection.

Let us prepare preliminary facts for the definition of cuspidal cohomology groups. Let $\Gamma = \Gamma_C = \Gamma(3) \cap \Gamma_0(C)$ for $\Gamma(3) = \{\gamma \in SL_2(\mathbb{Z}) | \gamma \equiv 1 \mod 3\}$. The good point of Γ_C is that it acts on $\mathfrak{H}$ freely without fixed point. To see this, let Γ_z be the stabilizer of $z \in \mathfrak{H}$ in Γ. Since the stabilizer of z in $SL_2(\mathbb{R})$ is a maximal compact subgroup C_z of $SL_2(\mathbb{R})$ (see Subsection 3.1.3), $\Gamma_z = \Gamma \cap C_z$ is compact-discrete and hence is finite. Thus if Γ is torsion-free, it acts freely on $\mathfrak{H}$. Pick a torsion-element $\gamma \in \Gamma$. Then two eigenvalues ζ and $\overline{\zeta}$ of γ are roots of unity complex conjugate each other. Since Γ cannot contain -1, $\zeta \notin \mathbb{R}$. Thus if $\gamma \neq 1$, we have $-2 < \mathrm{Tr}(\gamma) = \zeta + \overline{\zeta} < 2$. Since $\gamma \equiv 1 \mod 3$, $\mathrm{Tr}(\gamma) \equiv 2 \mod 3$, which implies $\mathrm{Tr}(\gamma) = -1$. Thus γ satisfies $\gamma^2 + \gamma + 1 = 0$ and hence $\gamma^3 = 1$. Thus $\mathbb{Z}[\gamma] \cong \mathbb{Z}[\omega]$ for a primitive cubic root ω. Since 3 ramifies in $\mathbb{Z}[\omega]$, $\mathbb{Z}[\omega]/3\mathbb{Z}[\omega]$ has a unique maximal ideal $\mathfrak{m}$ with $\mathfrak{m}^2 = 0$. The ideal $\mathfrak{m}$ is principal and is generated by ω. Thus the matrix ($\gamma - 1 \mod 3$) corresponds to ($\omega - 1 \mod 3$), which is non-zero nilpotent. This $\gamma - 1 \mod 3$ is non-zero nilpotent, showing that $\gamma \notin \Gamma(3)$, a contradiction.

By the above argument, the fundamental group of $Y = \Gamma_C\backslash\mathfrak{H}$ is isomorphic to Γ_C. Then we may consider the locally constant sheaf $\mathscr{L}(n,\chi;A)$ of sections associated with the following covering:

$$\mathscr{X} = \Gamma_C\backslash(\mathfrak{H} \times L(n,\chi;A)) \twoheadrightarrow Y \quad \text{via } (z, P) \mapsto z.$$

Since Γ_C acts on $\mathfrak{H}$ without fixed point, the space $\mathscr{X}$ is locally isomorphic to Y, and hence $\mathcal{L}(n,\chi;A)$ is a well-defined locally constant sheaf. In this setting, there is a canonical isomorphism (see [LFE] Appendix Theorem 1 and Proposition 4):

$$H^1(\Gamma_C, L(n,\chi;A)) \cong H^1(Y, \mathcal{L}(n,\chi;A)).$$

Note that $\Gamma_0(C)/\Gamma_C$ is a finite group whose order is a factor of 24. Thus as long as 6 is invertible in A, we have

$$H^0(\Gamma_0(C)/\Gamma_C, H^1(\Gamma_C, L(n,\chi;A))) = H^1(\Gamma_0(C), L(n,\chi;A)). \quad (5.12)$$

Hereafter, we always assume that 6 is invertible in A.

Let $\mathfrak{S} = \Gamma_C\backslash\mathbf{P}^1(\mathbb{Q}) \cong \Gamma_C\backslash SL_2(\mathbb{Z})/\Gamma_\infty$ for $\Gamma_\infty = \{\gamma \in SL_2(\mathbb{Z}) | \gamma(\infty) = \infty\}$. Thus $\mathfrak{S}$ is the set of cusps of $Y = \Gamma_C\backslash\mathfrak{H}$. We can take a neighborhood of ∞ in Y isomorphic to the cylinder $\mathbb{C}/\mathbb{Z}$. Since we have a neighborhood of each cusp isomorphic to a given neighborhood of ∞, we can take an open neighborhood of each cusp of Y isomorphic to the cylinder. We then compactify Y adding a circle at every cusp. We write $\overline{Y}$ for the compactified space. Then

$$\partial\overline{Y} = \bigsqcup_{\mathfrak{S}} S^1,$$

and

$$H^q(\partial\overline{Y}, \mathcal{L}(n,\chi;A)) \cong \bigoplus_{s\in\mathfrak{S}} H^q(\Gamma_s, L(n,\chi;A)),$$

where Γ_s is the stabilizer in Γ_C of a cusp $s \in \mathbf{P}^1(\mathbb{Q})$ representing an element in $\mathfrak{S}$. Since $\Gamma_s \cong \mathbb{Z}$, $H^q(\partial\overline{Y}, \mathcal{L}(n,\chi;A)) = 0$ if $q > 1$.

We have a commutative diagram whose horizontal arrows are given by the restriction maps:

$$\begin{array}{ccc} H^1(Y, \mathcal{L}(n,\chi;A)) & \xrightarrow{res} & H^1(\partial\overline{Y}, \mathcal{L}(n,\chi;A)) \\ \wr\downarrow & & \wr\downarrow \\ H^1(\Gamma_C, L(n,\chi;A)) & \xrightarrow{res} & \bigoplus_{s\in\mathfrak{S}} H^1(\Gamma_s, L(n,\chi;A)). \end{array}$$

We then define H^1_{cusp} by the kernel of the restriction map.

We have the boundary exact sequence (cf. [LFE] Appendix Corollary 2):

$$\begin{aligned} 0 &\to H^0(Y, \mathcal{L}(n,\chi;A)) \to H^0(\partial\overline{Y}, \mathcal{L}(n,\chi;A)) \to H^1_c(Y, \mathcal{L}(n,\chi;A)) \\ &\xrightarrow{\pi} H^1(Y, \mathcal{L}(n,\chi;A)) \to H^1(\partial\overline{Y}, \mathcal{L}(n,\chi;A)) \to H^2_c(Y, \mathcal{L}(n,\chi;A)) \to 0. \end{aligned}$$

Here H^1_c is the sheaf cohomology group with compact support, and the map π sends each compactly supported cohomology class to its usual cohomology class. Thus H^1_{cusp} is equal to the image of π, made up of cohomology classes rapidly decreasing towards cusps (when $A = \mathbb{C}$). We also have (cf. [LFE] Chapter 6 and Appendix):

$$H^2_c(Y, \mathscr{L}(n,\chi;A)) \cong L(n,\chi;A)/\sum_{\gamma\in\Gamma_C}(\gamma-1)L(n,\chi;A),$$

$$H^0_c(Y, \mathscr{L}(n,\chi;A)) = 0 \quad\text{and}\quad H^0(Y, \mathscr{L}(n,\chi;A)) = H^0(\Gamma_C, L(n,\chi;A)). \tag{5.13}$$

When $A = \mathbb{C}$, the isomorphism $H^2_c(Y, \mathscr{L}(n,\chi;A)) \cong A$ is given by $[\omega] \mapsto \int_Y \omega$, where ω is a compactly supported 1-form representing the cohomology class $[\omega]$ (de Rham theory; cf. [LFE] Appendix Proposition 6).

Suppose that $n!$ is invertible in A. Then the $\binom{n}{j}^{-1} \in A$ for binomial symbols $\binom{n}{j}$. We can then define a pairing $[\,,\,] : L(n,\chi;A)\times L(n,\chi^{-1};A) \to A$ by

$$[\sum_j a_j X^{n-j}Y^j, \sum_j b_j X^{n-j}Y^j] = \sum_{j=0}^{n}(-1)^j \binom{n}{j}^{-1} a_j b_{n-j}. \tag{5.14}$$

By definition, $[(X - zY)^n, (X - \overline{z}Y)^n] = (z - \overline{z})^n$. It is easy to check that $[\gamma P, \gamma Q] = \det\gamma^n [P, Q]$ for $\gamma \in GL_2(A)$ (Exercise 1). Thus we have a Γ_C-homomorphism $L(n,\chi;A) \otimes_A L(n,\chi^{-1};A) \to A$, and we get the cup product pairing

$$[\,,\,] : H^1_c(Y, \mathscr{L}(n,\chi;A)) \times H^1(Y, \mathscr{L}(n,\chi^{-1};A)) \longrightarrow H^2_c(Y, A) \cong A.$$

This pairing induces the cuspidal pairing

$$[\,,\,] : H^1_{cusp}(Y, \mathscr{L}(n,\chi;A)) \times H^1_{cusp}(Y, \mathscr{L}(n,\chi^{-1};A)) \longrightarrow A. \tag{5.15}$$

By (5.12), we identify $H^1_{cusp}(\Gamma_0(C), L(n,\chi;A))$ as a subspace of $H^1_{cusp}(Y, \mathscr{L}(n,\chi;A))$ and write $[\,,\,]$ for the pairing induced on $H^1_{cusp}(\Gamma_0(C), \mathscr{L}(n,\chi;A))$ by the above pairing of $H^1_{cusp}(Y, \mathscr{L}(n,\chi;A))$.

The (quasi) involution τ induced by the action of $\tau = \begin{pmatrix} 0 & -1 \\ C & 0 \end{pmatrix}$ defines a quasi-involution on the cohomology

$$\tau : H^1_{cusp}(\Gamma_0(C), L(n,\chi;A)) \to H^1_{cusp}(\Gamma_0(C), L(n,\chi^{-1};A)),$$

which is given by $u \mapsto \{\gamma \mapsto \tau u(\tau\gamma\tau^{-1})\}$ for 1-cocycle u. The cocycle $u|\tau$ has values in $L(n,\chi^{-1};A)$ because conjugation by τ interchanges the diagonal entries of γ. We have $\tau^2 = (-C)^n$ and $[x|\tau, y] = [x, y|\tau]$. Then we modify the above pairing $[\,,\,]$ by τ and define $\langle x, y\rangle = [x, y|\tau]$ ([LFE]

6.3 (6)). As described in [IAT] Chapter 8 and [LFE] Chapter 6, we have a natural action of Hecke operators $T(n)$ on $H^1_{cusp}(\Gamma_0(C), L(n,\chi;A))$. The operator $T(n)$ is symmetric with respect to this pairing:

$$\langle x|T(n), y\rangle = \langle x, y|T(n)\rangle. \tag{5.16}$$

We have an isomorphism, called the Eichler–Shimura isomorphism,

$$\delta : S_k(\Gamma_0(C), \chi) \oplus \overline{S}_k(\Gamma_0(C), \chi) \cong H^1_{cusp}(\Gamma_0(C), L(n,\chi;\mathbb{C})), \tag{5.17}$$

where $k = n+2$, $\overline{S}_k(\Gamma_0(C), \chi)$ is the space of anti-holomorphic cusp forms of weight k of 'Neben' type character χ, and

$$H^1_{cusp}(\Gamma_0(C), L(n,\chi;\mathbb{C})) \subset H^1(\Gamma_0(C), L(n,\chi;\mathbb{C}))$$

is the *cuspidal* cohomology groups defined in [IAT] Chapter 8 (see also [LFE] Chapter 6 under the formulation close to this chapter; in these books H^1_{cusp} is actually written as H^1_P and is called the *parabolic* cohomology group).

The Eichler–Shimura map δ is specified in [LFE] as follows: We put

$$\omega(f) = \begin{cases} f(z)(X - zY)^n dz & \text{if } f \in S_k(\Gamma_0(C), \chi), \\ f(z)(X - \overline{z}Y)^n d\overline{z} & \text{if } f \in \overline{S}_k(\Gamma_0(C), \chi). \end{cases}$$

Then we associate to f the cohomology class of the 1-cocycle $\gamma \mapsto \int_z^{\gamma(z)} \omega(f)$ of $\Gamma_0(C)$ for a fixed point z on the upper half complex plane. The map δ does not depend on the choice of z.

There are two natural operations on the cohomology group (cf. [LFE] 6.3): One is the action of Hecke operators $T(n)$ on $H^1_{cusp}(\Gamma_0(C), L(n,\chi;A))$, and the other is an action of complex conjugation c given by $c\omega(z) = \varepsilon\omega(-\overline{z})$ for $\varepsilon = \left(\begin{smallmatrix}-1 & 0\\ 0 & 1\end{smallmatrix}\right)$ and a differential form ω. In particular, δ and c commute with $T(n)$. We write $H^1_{cusp}(\Gamma_0(C), L(n,\chi;A))[\pm]$ for the $\pm$-eigenspace of c. Then it is known ([IAT] or [LFE] (11) in Section 6.3) that $H^1_{cusp}(\Gamma_0(C), L(n,\chi;\mathbb{Q}(\lambda)))[\pm]$ is $h_\kappa(C,\chi;\mathbb{Q}(\lambda))$-free of rank 1. Supposing that A contains the eigenvalues $\lambda(T(n))$ for all n, we write $H^1_{cusp}(\Gamma_0(C), L(n,\chi;A))[\lambda,\pm]$ for the λ-eigenspace under $T(n)$.

We now regard λ as actually having values in $\mathcal{O} \cap \overline{\mathbb{Q}}$ (fixing an embedding of $\overline{\mathbb{Q}} \hookrightarrow \overline{\mathbb{Q}}_p$). Put $A = \mathcal{O} \cap \mathbb{Q}(\lambda)$. Then A is a valuation ring of $\mathbb{Q}(\lambda)$ of residual characteristic p. Thus for the image L of $H^1_{cusp}(\Gamma_0(C), L(n,\chi;A))$ in $H^1_{cusp}(\Gamma_0(C), L(n,\chi;\mathbb{Q}(\lambda)))$,

$$H^1_{cusp}(\Gamma_0(C), L(n,\chi;\mathbb{Q}(\lambda)))[\lambda,\pm] \cap L = A\xi_\pm$$

for a generator $\xi_\pm$. Then for the normalized eigenform $f \in S_\kappa(\Gamma_0(C), \chi)$

with $T(n)f = \lambda(T(n))f$, we define $\Omega(\pm,\lambda;A) \in \mathbb{C}^\times$ by

$$\delta(f) \pm c(\delta(f)) = \Omega(\pm,\lambda;A)\xi_\pm.$$

The above definition of the period $\Omega(\pm,\lambda;A)$ can be generalized to the Hilbert modular case [H94b].

We now compute

$$\langle \Omega(+,\lambda;A)\xi_+, \Omega(-,\lambda;A)\xi_- \rangle = \Omega(+,\lambda;A)\Omega(-,\lambda;A)\langle \xi_+,\xi_- \rangle.$$

Note that $\delta(f)|\tau = W(\lambda)(-1)^n C^{(n/2)}\delta(f_c)$, where $f_c = \sum_{m=1}^{\infty} \overline{\lambda(T(m))q^m}$ and $f|\tau = W(\lambda)f_c$ and $W(\lambda) \in \mathbb{C}$ with $|W(\lambda)| = 1$. By definition, we have

$$2\Omega(+,\lambda;A)\Omega(-,\lambda;A)\langle \xi_+,\xi_- \rangle = [\delta(f) + c\delta(f), (\delta(f) - c\delta(f))|\tau],$$

which is equal to, up to sign,

$$\begin{aligned} 4i \int_{Y_0(C)} [\delta(f)|\tau, c\delta(f)]dx \wedge dy &= 2^k i^{k+1} W(\lambda) C^{(k/2)-1} \int_{Y_0(C)} |f_c|^2 y^{k-2} dxdy \\ &= 2^k i^{k+1} W(\lambda) C^{(k/2)-1} \int_{Y_0(C)} |f|^2 y^{k-2} dxdy \\ &= i^{k+1} W(\lambda) C^{k/2} \Gamma(1, Ad(\lambda)) L(1, Ad(\lambda)), \end{aligned} \tag{5.18}$$

where $Y_0(C) = \Gamma_0(C)\backslash\mathfrak{H}$. This shows

THEOREM 5.16 *Let χ be a character of conductor C. Let $\lambda : h_k(C,\chi;\mathbb{Z}[\chi]) \to \overline{\mathbb{Q}}$ $(k \geq 2)$ be a $\mathbb{Z}[\chi]$-algebra homomorphism. Then for a valuation ring A of $\mathbb{Q}(\lambda)$, we have, up to sign,*

$$\frac{i^{k+1} W(\lambda) C^{k/2} \Gamma(1, Ad(\lambda)) L(1, Ad(\lambda))}{\Omega(+,\lambda;A)\Omega(-,\lambda;A)} = \langle \xi_+,\xi_- \rangle \in \mathbb{Q}(\lambda).$$

If the residual characteristic of A is prime to $n!$ ($\Leftrightarrow$ $p > n$), then $\langle \xi_+,\xi_- \rangle \in A$.

The proof of rationality of the adjoint L-values as above can be generalized to even non-critical values $L(1, Ad(\lambda) \otimes \alpha)$ for quadratic Dirichlet characters α (see [H99a]). If one insists on p-ordinarity: $\lambda(T(p)) \in A^\times$ for the residual characteristic $p \geq 5$ of A, we can show that $\langle \xi_+,\xi_- \rangle \in A$. This follows from the perfectness of the duality pairing $\langle\ ,\ \rangle$ on the p-ordinary cohomology groups defined below even if $n!$ is not invertible in A (see [H88b] for details).

Let $\mathcal{O}$ be the completion of the valuation ring A. Let h be the local ring of $h_k(C,\chi;\mathcal{O})$ through which λ factors. Let 1_h be the idempotent of h in the Hecke algebra. Since the conductor of χ coincides with C, $h_k(C,\chi;\mathcal{O})$ is reduced. Thus for the quotient field K of $\mathcal{O}$, the unique local ring h_K of $h_k(C,\chi;K)$ through which λ factors is isomorphic to K. Let 1_λ be the idempotent of h_K in $h_k(C,\chi;K)$. Then we have the following important corollary.

Corollary 5.17 *Let the assumption be as in Theorem 5.16. Let A be a valuation ring of residual characteristic $p > 3$. Suppose that $\langle\ ,\ \rangle$ induces a perfect duality on $1_h H^1_{cusp}(\Gamma_0(C), L(n,\chi;\mathcal{O}))$. Then*

$$\left|\frac{i^{k+1}W(\lambda)C^{k/2}\Gamma(1,Ad(\lambda))L(1,Ad(\lambda))}{\Omega(+,\lambda;A)\Omega(-,\lambda;A)}\right|_p^{-r} = |L^\lambda/L_\lambda|,$$

where $r = \mathrm{rank}_{\mathbb{Z}_p}\mathcal{O}$, $L^\lambda = 1_\lambda L$ for the image L of $H^1_{cusp}(\Gamma_0(C), L(n,\chi;\mathcal{O}))[+]$ in the cohomology $H^1_{cusp}(\Gamma_0(C), L(n,\chi;K))[+]$, and L_λ is given by the intersection $L^\lambda \cap L$ in $H^1_{cusp}(\Gamma_0(C), L(n,\chi;K))[+]$.

Proof By our choice, ξ_+ is the generator of L_λ. Similarly we define $M^\lambda = 1_\lambda M$ for the image M of $H^1_{cusp}(\Gamma_0(C), L(n,\chi;\mathcal{O}))[-]$ in $H^1_{cusp}(\Gamma_0(C), L(n,\chi;K))[-]$, and $M_\lambda = M^\lambda \cap M$ in $H^1_{cusp}(\Gamma_0(C), L(n,\chi;K))[-]$. Then ξ_- is a generator of M_λ. Since the pairing is perfect, $L_\lambda \cong \mathrm{Hom}_{\mathcal{O}}(M^\lambda,\mathcal{O})$ and $L^\lambda \cong \mathrm{Hom}_{\mathcal{O}}(M_\lambda,\mathcal{O})$ under $\langle\ ,\ \rangle$. Then it is easy to see that $|\langle\xi_+,\xi_-\rangle|_p^{-1} = |L^\lambda/L_\lambda|$ (Exercise 2). □

As for the assumption of the perfect duality, we quote the following slightly technical result from [H81a] and [H88b] Section 3:

Theorem 5.18 *Let the notation and assumption be as in Theorem 5.16. Suppose that $C|p$ and $p > 3$. If either $\lambda(T(p)) \in A^\times$ or $\frac{1}{n!} \in A$, then*

1. *$1_h H^1_{cusp}(\Gamma_0(C), L(n,\chi;\mathcal{O}))$ is $\mathcal{O}$-free;*
2. *the pairing $\langle\ ,\ \rangle$ induces a perfect duality on $1_h H^1_{cusp}(\Gamma_0(C), L(n,\chi;\mathcal{O}))$.*

What is really proved in [H88b] Section 3 is the $\mathcal{O}$-freeness and the perfect self-duality of $H^1_{cusp,ord}(\Gamma_1(C), L(n;\mathcal{O}))$, where $H^1_{cusp,ord} = eH^1_{cusp}$ for $e = \lim_{n\to\infty} T(p)^{n!}$ (as in Subsection 3.1.8). Thus if $C = 1$, the theorem follows from this result. If $C = p$, we have an orthogonal decomposition (from the fact that $p \nmid |(\mathbb{Z}/p\mathbb{Z})^\times|$ and the inflation-restriction sequence):

$$H^1_{cusp,ord}(\Gamma_1(p), L(n;\mathcal{O})) \cong \bigoplus_\chi H^1_{cusp,ord}(\Gamma_0(C), L(n,\chi;\mathcal{O})),$$

and thus the theorem follows from [H88b] Theorem 3.1. If the reader scrutinizes the argument in [H88b] Section 3, replacing $(\Gamma_1(Np^r), L(n;\mathcal{O}))$ there by $(\Gamma_0(C), L(n,\chi;\mathcal{O}))$ here, he or she will find that the above theorem holds without assuming $C|p$, but we do not use this general fact later.

Exercises

(1) Prove $[\gamma P, \gamma Q] = \det \gamma^n [P, Q]$ for the pairing given in (5.14).
(2) Give a detailed proof of Corollary 5.17.

5.3.3 Congruences and adjoint L-values

Here we study a non-abelian adjoint version of the analytic class number formula, which follows from the theorem of Taylor–Wiles (Theorem 3.31 in Chapter 3) and some earlier work of the author (presented in the previous subsection). It relates the size of the Selmer group $\mathrm{Sel}(Ad(\varphi))$ to the L-value $L(1, Ad(\varphi))$, where $\varphi = \rho_\lambda : \mathfrak{G}_p \to GL_2(\mathcal{O})$ is the modular Galois representation attached to an $\mathcal{O}$-algebra homomorphism $\lambda : h_k(p,\chi;\mathcal{O}) \to \mathcal{O}$ $(k \geq 2)$ with $\lambda(U(p)) \in \mathcal{O}^\times$ (see Theorem 3.26). We fix an embedding $i_p : \overline{\mathbb{Q}} \hookrightarrow \overline{\mathbb{Q}}_p$ and write $\lambda^\circ : h_k(C,\chi;\mathbb{Z}) \to \overline{\mathbb{Q}}$ for the primitive algebra homomorphism associated with λ. Thus the conductor C of λ satisfies $C|p$. This condition is not necessary to obtain the formula, but we assume it to make our presentation simple.

The idea is to relate the size of the Selmer group to congruences of Hecke eigensystems λ. To describe such congruence among cusp forms in terms of Hecke algebras and deformation rings of Galois representations, we introduce here a general notion of congruence modules and differential modules: Let R be an algebra over a normal noetherian domain A. We assume that R is an A-flat module of finite type. Let $\phi : R \to A$ be an A-algebra homomorphism. We define

$$C_1(\phi; A) = \Omega_{R/A} \otimes_{R,\phi} \mathrm{Im}(\phi),$$

which we call the *differential* module of ϕ , and as we have observed in Theorem 5.14, if R is a deformation ring, this module is the Pontryagin dual of a certain Selmer group. We suppose that R is reduced (that is, the nilradical of R vanishes). Then the total quotient ring $Frac(R)$ can be decomposed uniquely into $Frac(R) = Frac(\mathrm{Im}(\phi)) \times X$ as an algebra direct product. Write 1_ϕ for the idempotent of $Frac(\mathrm{Im}(\phi))$ in $Frac(R)$. Let $\mathfrak{a} = \mathrm{Ker}(R \to X) = (1_\phi R \cap R)$, $S = \mathrm{Im}(R \to X)$ and $\mathfrak{b} = \mathrm{Ker}(\phi)$. Here the intersection $1_\phi R \cap R$ is taken in $Frac(R) = Frac(\mathrm{Im}(\phi)) \times X$.

Then we put

$$C_0(\phi;A) = (R/\mathfrak{a}) \otimes_{R,\phi} \mathrm{Im}(\phi) \cong \mathrm{Im}(\phi)/(\phi(\mathfrak{a})) \cong 1_\phi R/\mathfrak{a} \cong S/\mathfrak{b},$$

which is called the *congruence* module of ϕ but is actually a ring (cf. [H88b] Section 6). We can split the isomorphism $1_\phi R/\mathfrak{a} \cong S/\mathfrak{b}$ as follows: First note that $\mathfrak{a} = (R \cap (1_\phi R \times 0))$ in $Frac(\mathrm{Im}(\phi)) \times X$. Then $\mathfrak{b} = (0 \times X) \cap R$, and we have

$$1_\phi R/\mathfrak{a} \cong R/(\mathfrak{a} \oplus \mathfrak{b}) \cong S/\mathfrak{b},$$

where the maps $R/(\mathfrak{a} \oplus \mathfrak{b}) \to 1_\phi R/\mathfrak{a}$ and $R/(\mathfrak{a} \oplus \mathfrak{b}) \to S/\mathfrak{b}$ are induced by two projections from R to $1_\phi R$ and S.

Suppose now that A is the integer ring of a number field in $\overline{\mathbb{Q}}$. Since the spectrum $Spec(C_0(\phi;A))$ of the congruence ring $C_0(\phi;A)$ is the scheme theoretic intersection of $Spec(Im(\phi))$ and $Spec(R/\mathfrak{a})$ in $Spec(R)$:

$$Spec(C_0(\lambda;A)) = Spec(\mathrm{Im}(\phi)) \times_{Spec(R)} Spec(R/\mathfrak{a}),$$

we conclude that

> *a prime $\mathfrak{p}$ is in the support of $C_0(\phi;A)$ if and only if there exists an A-algebra homomorphism $\phi' : R \to \overline{\mathbb{Q}}$ factoring through $R/\mathfrak{a}$ such that*
>
> $$\phi(a) \equiv \phi'(a) \mod \mathfrak{p}$$
>
> *for all $a \in R$.*

In other words, ϕ mod $\mathfrak{p}$ factors through $R/\mathfrak{a}$ and can be lifted to ϕ'. Therefore, if A is sufficiently large, $\bigcup_\phi Supp(C_0(\phi;A))$ consists of primes dividing the absolute different $\mathfrak{d}(R/\mathbb{Z})$ of R over $\mathbb{Z}$, and each prime appearing in the absolute discriminant of $R/\mathbb{Z}$ divides the order of the congruence module for some ϕ.

By Corollary 5.13 applied to $0 \to \mathfrak{b} \to R \xrightarrow{\phi} A \to 0$, we know that $C_1(\phi;A) \cong \mathfrak{b}/\mathfrak{b}^2$. Since $C_0(\phi;A) \cong S/\mathfrak{b}$, we may further define *higher congruence modules* by $C_n(\phi;A) = \mathfrak{b}^n/\mathfrak{b}^{n+1}$. The knowledge of all $C_n(\phi;A)$ is almost equivalent to the knowledge of the entire ring R. Therefore the study of $C_n(\phi;A)$ and the graded algebra

$$\bigoplus_n C_n(\phi;A)$$

is important and interesting, when R is a Galois deformation ring. As we will see, even in the most favorable cases, so far, we only theoretically know the cardinality of modules C_0 and C_1 for universal deformation rings R.

As already described, primes appearing in the discriminant of the Hecke algebra gives congruences among algebra homomorphisms of the Hecke algebra into $\overline{\mathbb{Q}}$, which are points in $Spec(h_k)(\overline{\mathbb{Q}})$. In order to show that we have a rather large amount of non-trivial congruence among λ, we quote here from [HM] a table of the discriminant of $h_k(1, \mathbf{1}; \mathbb{Z})$ over $\mathbb{Z}$ (the computation was done by Y. Maeda). Here '**1**' denotes the trivial character. In the table below, the first column gives weight k, the second gives $\mathrm{rank}_{\mathbb{Z}}\, h_k(1, \mathbf{1}; \mathbb{Z})$ and the third gives the discriminant. As already explained, the primes appearing in the discriminant are divisible by a prime ideal $\mathfrak{p}$ of the Hecke algebra sitting in $Supp(C_0(\lambda; \mathbb{Z}))$ for some λ. If the Hecke algebra $h_k(1, \mathbf{1}; \mathbb{Z})$ is an integral domain (which is the case in the limit of the following table and has been conjectured by Maeda to be the case for all k; [HM]), these primes appear as congruences among Galois-conjugate Hecke eigenforms. If the Hecke algebra $h_k(1, \mathbf{1}; \mathbb{Q})$ splits into a product of at least two fields, some of these primes appear as congruences among non-conjugate Hecke eigenforms. When $N > 1$, splitting of $h_k(N, \chi; \mathbb{Q})$ does occur (see tables of Hecke eigenvalues at the end of [MFM]).

We return to the original setting of $\lambda : h_k(p, \chi; \mathbb{Z}) \to \overline{\mathbb{Q}}$. Let K be the closure of $\mathbb{Q}(\lambda^\circ)$ in $\overline{\mathbb{Q}}_p$ (as we have fixed the embedding $\overline{\mathbb{Q}} \hookrightarrow \overline{\mathbb{Q}}_p$), and let $\mathcal{O}$ be the p-adic integer ring of K. We put $n = k - 2 \geq 0$. We assume that $\lambda(U(p)) \in \mathcal{O}^\times$. Then φ satisfies (ord) in Subsection 3.2.4. We define the residual representation $\overline{\rho} : \mathrm{Gal}(\overline{\mathbb{Q}}/\mathbb{Q}) \to GL_2(\mathbb{F})$ by ρ_λ mod $\mathfrak{m}$ for $\mathfrak{m} = \mathfrak{m}_{\mathcal{O}}$.

Let E be a number field in $\mathbb{Q}^{(p,\infty)}$, and put $\mathfrak{H}_E = \mathrm{Gal}(\mathbb{Q}^{(p,\infty)}/E)$. For any representation ρ of $\mathfrak{G}$, we write ρ_E for the restriction of ρ to $\mathfrak{H}_E$. We assume (ai_κ): absolute-irreducibility of $\overline{\rho}_\kappa$ for $\kappa = \mathbb{Q}(\sqrt{(-1)^{(p-1)/2}p})$ and (rg_p) (= ($\mathrm{Rg}_{\mathbb{Q}}$)) for $\overline{\rho}$. Since $\det(\overline{\rho}) = \chi\omega^{k-1}$ mod $\mathfrak{m}$, if $\chi\omega^{k-1} \not\equiv 1$ mod $\mathfrak{m}$, the condition (rg_p) holds.

Under (ai_κ) and (rg_p), for $\phi = \chi\nu^{k-1}$ with the p-adic cyclotomic character $\nu : \mathfrak{G}_p \to \mathcal{O}^\times$, the functor $\mathcal{F}_{\mathbb{Q}}^{\phi,ord} : CNL_{\mathcal{O}} \to SETS$ is representable by a universal couple $(R, \varrho) = (R^{\phi,ord}, \varrho^{\phi,ord})$. Hereafter, we simply write Φ for $\mathcal{F}_{\mathbb{Q}}^{\phi,ord}$. When χ is trivial and $k = 2$, formally, we need to consider also the flat case as described in Subsection 3.2.7 in Chapter 3. However, as explained there, the flat-ordinary case is empty when the auxiliary level is 1; so, we forget about it.

For an element a in a $\mathbb{Z}_p$-algebra A, which is of finite type as a $\mathbb{Z}_p$-module, $\varepsilon = \lim_{n\to\infty} a^{n!}$ always exists in A and gives an idempotent of A (see [LFE] Lemma 7.2.1). The direct summand εA is the product of all local rings of A in which the projected image of a is a unit.

Discriminant of Hecke algebras

weight	dim	Discriminant
24	2	$2^6 \cdot 3^2 \cdot 144169$
28	2	$2^6 \cdot 3^6 \cdot 131 \cdot 139$
30	2	$2^{12} \cdot 3^2 \cdot 51349$
32	2	$2^6 \cdot 3^2 \cdot 67 \cdot 273067$
34	2	$2^8 \cdot 3^4 \cdot 479 \cdot 4919$
36	3	$2^{24} \cdot 3^6 \cdot 5^2 \cdot 7^2 \cdot 23 \cdot 1259 \cdot 269\,461\,929\,553$
38	2	$2^{10} \cdot 3^2 \cdot 181 \cdot 349 \cdot 1009$
40	3	$2^{20} \cdot 3^{10} \cdot 5^2 \cdot 13^2 \cdot 73 \cdot 59\,077 \cdot 92\,419\,245\,301$
42	3	$2^{22} \cdot 3^6 \cdot 5^2 \cdot 7^2 \cdot 1465\,869\,841 \cdot 578\,879\,197\,969$
44	3	$2^{22} \cdot 3^8 \cdot 5^2 \cdot 7^2 \cdot 37 \cdot 92\,013\,596\,772\,457\,847\,677$
46	3	$2^{31} \cdot 3^{12} \cdot 5^2 \cdot 227 \cdot 454\,287\,770\,269\,681\,529$
48	4	$2^{40} \cdot 3^{14} \cdot 5^6 \cdot 7^4 \cdot 31 \cdot 6093\,733 \cdot 1675\,615\,524\,399\,270\,726\,046\,829\,566\,281\,283$
50	3	$2^{22} \cdot 3^{10} \cdot 5^4 \cdot 7^4 \cdot 12\,284\,628\,694\,131\,742\,619\,401$

Let

$$e_0 = \lim_{n\to\infty} T(p)^{n!} \in h_k(1;\mathcal{O}) \quad \text{and} \quad e = \lim_{n\to\infty} U(p)^{n!} \in h_k(p,\chi;\mathcal{O}).$$

Since $T(p)$ of level 1 and that of level p are different, we have used the symbol $U(p)$ for $T(p)$ of level p. We know that $e_0 \neq e$ and we have a surjective $\mathcal{O}$-algebra homomorphism of $h^{ord}(p,\mathrm{id};\mathcal{O}) = eh_k(p,\mathrm{id};\mathcal{O})$ onto $h_k^{ord}(1,\mathcal{O}) = e_0 h_k(1;\mathcal{O})$ taking $T(n)$ to $T(n)$ for all n prime to p [MT]. If $k > 2$, $h_k^{ord}(1;\mathcal{O}) \cong h_k^{ord}(p,\mathrm{id};\mathcal{O})$ (see (ism) in Subsection 3.2.4), which is also a consequence of [H86b] Proposition 4.7.

By our assumption that $\lambda(U(p)) \in \mathcal{O}^\times$, λ factors through $h_k^{ord}(p,\chi;\mathcal{O})$, which implies that ρ_λ is p-ordinary (see Theorem 3.26). Let h be the unique local ring of $h_k^{ord}(p,\chi;\mathcal{O})$ through which λ factors. Let $\rho_h : \mathfrak{G}_p \to GL_2(h)$ be the Galois representations constructed in Theorem 3.29.

We have already seen in Theorem 3.31 that Φ is representable by the pair (h, ρ_h), under (ai_κ) and (rg_p). Thus in this case, the natural morphisms $\pi : R \to h$ for $R = R^{\phi,ord}$ with $\pi \circ \varrho \approx \rho_h$ is a surjective isomorphism in $CNL_{\mathcal{O}}$.

By (red) in Subsection 3.2.4, h is reduced. In the course of the proof of the above result: $R \cong h$, it is shown that R is a reduced local complete intersection (see Theorem 3.35) over $\mathcal{O}$. This fact implies, under reducedness of R,

$$|C_1(\lambda;\mathcal{O})| = |C_0(\lambda;\mathcal{O})|. \qquad \text{(Cg1)}$$

The above assertion is a purely ring theoretic fact due to Tate ([MRo] A.3 Conclusion 4) and is almost equivalent to the complete intersection property (see [W2] Appendix and Lenstra [Lt] page 103). We will recall Tate's proof of the above in the following subsection (see Corollary 5.26).

Let 1_h be the idempotent of h in $h_k(p,\chi;\mathcal{O})$ and write $X[h]$ for 1_hX for each module X over $h_k(p,\chi;\mathcal{O})$. A local complete intersection ring is a Gorenstein ring ([CRT] Section 21); so, $h \cong \mathrm{Hom}_{\mathcal{O}}(h,\mathcal{O}) = h^*$ as h-modules. We have proved in [H81b] and [H86b] the existence of the following exact sequence of h-modules:

$$0 \to h \longrightarrow H^1_{cusp}(\Gamma_0(p), L(n,\chi;\mathcal{O}))[h] \longrightarrow \mathrm{Hom}_{\mathcal{O}}(h,\mathcal{O}) \to 0.$$

Thus the Gorenstein-ness of h implies, for any given parity $\varepsilon \in \{\pm 1\}$:

THEOREM 5.19 *Let h be a local ring of $h_k^{ord}(p,\chi;\mathcal{O})$ for $k \geq 2$ and $p > 3$. Suppose that h is a Gorenstein ring. Then*

$$H^1_{cusp}(\Gamma_0(C), L(n,\chi;\mathcal{O}))[h,\varepsilon] \cong h \cong R \quad \textit{as } h\textit{-modules.} \qquad (\mathrm{mlt}_{\mathbb{Q}})$$

Here we identify h with the local ring of $h_k(C,\chi;\mathcal{O})$ for the conductor C of λ (thus C is either p or 1). Under (ai_κ) and (rg_p), we know by the Taylor-Wiles theorem that h is a local complete intersection and hence is Gorenstein. Thus the assertion of the theorem holds in this case.

The Gorenstein-ness of the p-adic Hecke algebras was first noticed by Mazur [Ma] for weight 2 Hecke algebras of prime level and is a necessary and sufficient condition for $\overline{\rho} = (\rho_\lambda \mod \mathfrak{m}_{\mathcal{O}})$ to appear with multiplicity one in the p-torsion points of the jacobian of the modular curve. After Mazur's work, many mathematicians have generalized the result in many other cases. For a list of results valid for weight 2 Hecke algebras, see [W2] Section 2.1.

Supposing that the residual Galois representation $\overline{\rho}$ as above attached to a local ring h of $h_k(C,\chi;\mathcal{O})$ is absolutely irreducible over $\kappa = \mathbb{Q}[\sqrt{(-1)^{(p-1)/2}p}]$, we shall give here a sketch of a proof of the above theorem using a Taylor-Wiles system:

Proof For a set of primes $Q = \{q_1,\dots,q_r\}$ as in Theorem 3.35 in Chapter 3, we consider $\Gamma_Q = \Gamma_1(Q) \cap \Gamma_0(p) \cap \Gamma_C$. We shall use the notation introduced in Theorem 3.35 in Chapter 3. Since the argument is independent of the parity ε, we may assume that $\varepsilon = +$. Since $p > 3$, we may identify the following two cohomology groups

$$H^1_{cusp}(\Gamma_1(Q)\cap\Gamma_0(p), L(n,\chi;A))[+] \cong H^0(SL_2(\mathbb{F}_3), H^1_{cusp}(\Gamma_Q, L(n,\chi;A)))[+],$$

where $A = \mathcal{O}$, $\mathcal{O}/p^m\mathcal{O}$ and $K/\mathcal{O}$ for the field of fractions K of $\mathcal{O}$ ((5.12)). The Hecke algebra $h_k(\Gamma_1(Q)\cap\Gamma_0(p),\chi;\mathcal{O})$ and the group algebra $\mathcal{O}[\Delta_Q]$ naturally act on $H^1_{cusp}(\Gamma_1(Q)\cap\Gamma_0(p), L(n,\chi;A))[+]$. Then we take M_Q to be the Pontryagin dual of the h_Q-eigenspace of $H^1_{ord,cusp}(\Gamma_1(Q)\cap\Gamma_0(p), L(n,\chi;K/\mathcal{O}))[+]$. The module M_Q is free of finite rank over $\mathcal{O}$ (cf. [H86b] Lemma 4.6 and [H88b] (2.4)). The arguments in these papers show that $H^1_{cusp,ord}(\Phi, L(n,\chi;\mathcal{O}))$ (resp. $H^1_{cusp,ord}(\Phi, L(n,\chi;K/\mathcal{O}))$) is $\mathcal{O}$-free (resp. p-divisible) for any Φ with $\Gamma_1(Np^r)\subset\Phi\subset SL_2(\mathbb{Z})$.

In the above definition of M_Q, we can drop the subscript '*cusp*', because on the boundary cohomology, the Hecke operator acts through the Eisenstein quotient of the Hecke algebra (cf. [H88b] Proposition 2.3) contradicting the absolute irreducibility of $\overline{\rho}$.

We now prove that M_Q is $\mathcal{O}[\Delta_Q]$-free. For any character $\psi : \Delta_Q \to \mathcal{O}^\times$, regarding ψ as a character of $\chi_\Phi : \Phi/\Gamma_1(Q)\cap\Gamma_0(p) \to \mathcal{O}$, we have an exact sequence from the inflation-restriction sequence for $\Phi = \Gamma_0(Q)\cap\Gamma_0(p)$:

$$\begin{aligned} 0 &\to H^1(\mathrm{Im}(\chi_\Phi), H^0(\mathrm{Ker}(\chi_\Phi), L(n,\chi\psi;A))) \to H^1(\Phi, L(n,\chi\psi;A)) \qquad (5.19)\\ &\to H^1(\mathrm{Ker}(\chi_\Phi), L(n,\chi;A))[\psi] \to H^2(\mathrm{Im}(\chi_\Phi), H^0(\mathrm{Ker}(\chi_\Phi), L(n,\chi\psi;A))),\end{aligned}$$

where '$[\psi]$' indicates ψ-eigenspace.

We claim that a high power of $T(p)$ annihilates the cohomology group of finite quotient $\mathrm{Im}(\chi_\Phi)$ (cf. [H86b] Lemmas 6.1–2):

$$T(p^m)\{H^q(\mathrm{Im}(\chi_\Phi), H^0(\mathrm{Ker}(\chi_\Phi), L(n,\chi\psi;A)))\} = 0$$

for $q = 1, 2$ and $A = \mathcal{O}/p^r\mathcal{O}$. The key point of the proof is that we can choose a decomposition:

$$\Phi\left(\begin{smallmatrix}1&0\\0&p^m\end{smallmatrix}\right)\Phi = \bigsqcup_\alpha \Phi\alpha$$

so that $\alpha \equiv \left(\begin{smallmatrix}1&*\\0&1\end{smallmatrix}\right) \mod N_Q^s$ if $m = \prod_i q_i^{s-1}(q_i-1)$. By definition (see [LFE] 6.3), for each inhomogeneous q-cocycle $u : \mathrm{Im}(\chi_\Phi) \to H^0(\mathrm{Ker}(\chi_\Phi), L(n,\chi\psi;A))$, $[u]|T(p^m) = [u|T(p^m)]$ with $u|T(p^m)(g) = \sum_\alpha \alpha^\iota u(\alpha g_1\alpha^{-1},\dots,\alpha g_q\alpha^{-1})$, where $\alpha^\iota = (\det\alpha)\alpha^{-1}$. Then if $s > 1$, for each q-cocycle $\alpha^\iota u(\alpha g_1\alpha^{-1},\dots,\alpha g_q\alpha^{-1}) = u(g_1\dots,g_q)$, and hence $[u]|T(p^m) = p^m[u]$. This proves the claim.

By the claim, after taking $+$-eigenspace for complex conjugation and h_Q-eigenspace, we know that $M_{Q,\psi} = M_Q/\sum_{\delta\in\Delta_Q}(\delta - \psi(\delta))M_Q$ is the Pontryagin dual of the eigenspace $H^1(\Phi, L(n, \chi\psi; K/\mathcal{O}))[h_Q, +]$, which is p-divisible. Thus $M_{Q,\psi}$ is $\mathcal{O}$-free of finite rank. This is enough to conclude that M_Q is $\mathcal{O}[\Delta_Q]$-free by the argument in the proof of Corollary 3.20.

Again by (5.19), we conclude that the Δ_Q-coinvariant M_{Q,Δ_Q} is a factor of the Pontryagin dual module of $H^1_{cusp,ord}(\Gamma_0(N_Q p), L(n,\chi;K/\mathcal{O}))$. Then by our choice of $q \in Q$ so that $\overline{\rho}(Frob_q)$ has two distinct eigenvalues, M_{Q,Δ_Q} has to be sitting inside the Pontryagin dual of $H^1_{cusp,ord}(\Gamma_0(p), L(n,\chi;K/\mathcal{O}))$. This shows that $M_{Q,\Delta_Q} \cong M_\emptyset = M$ satisfying (tw4–5). The verification of (tw1,2,3) is independent of the choice of M_Q and has been done in Subsection 3.2.8. Thus (R, R_Q, M_Q) satisfies the axioms of the Taylor-Wiles system, and therefore by the control criterion Theorem 3.35, we conclude that M is free of finite rank over $h \cong R$. By the self-duality of $H^1_{cusp,ord}(\Gamma_0(C), L(n,\chi;\mathcal{O}))$ (see Theorem 5.18), we have $M \cong H^1_{cusp,ord}(\Gamma_0(C), L(n,\chi;\mathcal{O}))[h, +]$. After extending the scalar to K, $H^1_{cusp,ord}(\Gamma_0(C), L(n,\chi;K))[+]$ is free of rank 1 over the Hecke algebra with coefficients in K. This shows $M \cong h$. □

By virtue of ($\mathrm{mlt}_{\mathbb{Q}}$), we can compute $C_0(\lambda^\circ;\mathcal{O})$ using cohomology groups. Let $A = \mathbb{Q}(\lambda^\circ) \cap \mathcal{O}$, which is a valuation ring of $\mathbb{Q}(\lambda^\circ)$ with residual characteristic p. By our choice of $\mathcal{O}$, we have $\mathcal{O} = \varprojlim_n A/\mathfrak{m}_A^n$. For any integral domain B over A with quotient field M, we write $L(B)$ for the image of $H^1_{cusp}(\Gamma_0(C), L(n,\chi;B))[\varepsilon]$ in $H^1_{cusp}(\Gamma_0(C), L(n,\chi;M))[\varepsilon]$. Then $L(\mathcal{O}) = L(A) \otimes_A \mathcal{O}$, because $L(A)$ is a free A-module of finite rank. Decomposing $h_k(C,\chi;\mathbb{Q}(\lambda^\circ)) = \mathbb{Q}(\lambda^\circ) \times X$, we define $L^\lambda(A)$ to be the image of $L(A)$ in $L(A) \otimes_{h_k(C,\chi;A),\lambda^\circ} \mathbb{Q}(\lambda^\circ)$ and a cohomological congruence module by

$$C_0^H(\lambda^\circ;A) = L^\lambda(A)/(L(A) \cap L^\lambda(A)).$$

Then ($\mathrm{mlt}_{\mathbb{Q}}$) shows that

$$C_0(\lambda^\circ;\mathcal{O}) \cong C_0^H(\lambda^\circ;A) \otimes_A \mathcal{O} \cong C_0^H(\lambda^\circ;\mathcal{O}). \tag{5.20}$$

The last identity follows from our assumption $\mathcal{O} = \varprojlim_n A/\mathfrak{m}_A^n$.

As shown in Corollary 5.17, the p-adic absolute value of

$$\frac{C^{k/2}W(\lambda)\Gamma(1, Ad(\lambda))L(1, Ad(\lambda))}{\Omega_1(+,\lambda^\circ;A)\Omega_1(-,\lambda^\circ;A)}$$

is the inverse of the order of the right-hand side module of the above equation (5.20). That is, under (ai_κ), (rg_p) and (ord) for $\overline{\rho}$, if $p \geq 5$,

$$\left|\frac{C^{k/2}W(\lambda)\Gamma(1,Ad(\lambda))L(1,Ad(\lambda))}{\Omega_1(+,\lambda^\circ;A)\Omega_1(-,\lambda^\circ;A)}\right|_p^{-r} = |C_0^H(\lambda^\circ;A)| = |C_0(\lambda^\circ;A)|$$
$$= |C_1(\lambda^\circ;A)|, \qquad \text{(Cg2)}$$

where $r = r(\mathcal{O}) = \operatorname{rank}_{\mathbb{Z}_p}\mathcal{O}$, and $|\ |_p$ is the p-adic absolute value of A normalized so that $|p|_p = p^{-1}$. It is easy to see from the Gorenstein property: $h^* \cong h$ as h-modules that for a non-zero element $\eta(\lambda) \in A$, $C_0(\lambda^\circ;A) \cong A/\eta(\lambda)A$ (see Lemma 5.21), and (Cg2) is equivalent to saying that the L-value is equal to $\eta(\lambda)$ up to A-units. In (Cg2), if $C = 1$, we can drop the factor $C^{k/2}W(\lambda)$, since $C = W(\lambda) = 1$ in this case.

As we have seen in Theorem 5.14, it is a general fact [MT] that

$$\operatorname{Sel}(Ad(\rho_\lambda))_{/\mathbb{Q}} \cong \operatorname{Hom}_{\mathbb{Z}_p}(C_1(\lambda;\mathcal{O}),\mathbb{Q}_p/\mathbb{Z}_p) \ \text{ if } n > 0. \qquad \text{(Cg3)}$$

Here we have used the fact, in applying Theorem 5.14, that the image of $R^I_{\mathbb{Q}} = \mathcal{O}[[(I_p)^{ab}_p]]$ is just $\mathcal{O}$ because $\delta_{\rho_\lambda,p}$ is unramified. When $\chi = \mathrm{id}$, h is isomorphic to a unique local ring in $h_k(1;\mathcal{O})$; so, we can replace λ in the above formula by λ°.

Combining all that we have said, we get the following order formula of the Selmer group:

THEOREM 5.20 *Let $\lambda^\circ : h_k(C,\chi;\mathbb{Z}[\chi]) \to \overline{\mathbb{Q}}$ be a primitive $\mathbb{Z}[\chi]$-algebra homomorphism of conductor $C|p$. Suppose $k \geq 2$, $p \geq 5$, (ai_κ), (ord), (rg_p) for $\overline{\rho}$ and that $\lambda^\circ(T(p))$ is a unit in $\mathcal{O}$. Let $\lambda : h_k(p,\chi;\mathcal{O}) \to \mathcal{O}$ be the algebra homomorphism equivalent to λ° with $\lambda(U(p)) \in \mathcal{O}^\times$. We put $A = \mathbb{Q}(\lambda^\circ)\cap\mathcal{O}$. Take an element $\eta(\lambda) \in A$ such that $A/\eta(\lambda) \cong C_0(\lambda^\circ;A)$. Then we have*

$$\frac{C^{k/2}W(\lambda)\Gamma(1,Ad(\lambda))L(1,Ad(\lambda))}{\Omega_1(+,\lambda;A)\Omega_1(-,\lambda;A)} = \eta(\lambda) \quad \textit{up to A-units, and} \qquad \text{(CN1)}$$

$$\left|\frac{C^{k/2}W(\lambda)\Gamma(1,Ad(\lambda))L(1,Ad(\lambda))}{\Omega_1(+,\lambda^\circ;A)\Omega_1(-,\lambda^\circ;A)}\right|_p^{-[\mathcal{O}:\mathbb{Z}_p]} = \#(\operatorname{Sel}(Ad(\rho_\lambda))_{/\mathbb{Q}}). \qquad \text{(CN2)}$$

The definition of the Selmer group can also be obtained through Fontaine's theory as was done by Bloch–Kato, and the above formula can be viewed as an example of the Tamagawa number formula of Bloch and Kato for the motive $M(Ad(\rho_\lambda))$ (see [W2] page 466, [BK] Section 5, [Fl] and [Oc]). The finiteness of the Bloch–Kato Selmer groups $\operatorname{Sel}(Ad(\rho_\lambda))$ for λ of weight 2 associated with an elliptic curve (under some additional assumptions) was first proved by M. Flach [Fl], and then relating Bloch–Kato Selmer groups to Greenberg Selmer groups, he also showed the finiteness of Greenberg Selmer groups. By adopting the definition of

Bloch–Kato, we can define the Selmer group $\mathrm{Sel}_{crys}(Ad(\rho_\lambda))$ in the flat case, and the formula (CN2) is valid even for the flat case (if we allow auxiliary level $N > 1$) for this Selmer group of the crystalline motive $M(Ad(\lambda^\circ))$.

When $M(\lambda^\circ)$ is ordinary and crystalline of weight 1 (so $M(\lambda)$ corresponds to a weight 2 cusp form), we have discrepancy of the two Selmer groups. The difference is described by the p-Euler factor $(1-\alpha_p^2)$, where $\alpha_p = \lambda(U(p))$ is a p-adic unit. In other words, using p-ordinary deformation, the L-value $L(1, Ad(\lambda^\circ))$ multiplied by $(1-\alpha_p^2)$ gives rise to the size of $\mathrm{Sel}(Ad(\rho_\lambda))$, while the single $L(1, Ad(\rho_\lambda))$ describes the size of $\mathrm{Sel}_{crys}(Ad(\rho_\lambda))$ in the flat case (therefore, level prime to p). The factor $(1-\alpha_p^2)$ is in turn the size of the congruence module between old forms of level Np and new forms of level $N > 1$. Study of congruence between primitive forms (of the same weight) were started by Doi (and Hida) [DHI] in 1976, and the study of congruence between primitive and old Hecke eigenforms was initiated by K. Ribet in the early 80's [R1] (and [R2] and [MR]).

The formula (CN2) in the above theorem can be called a non-abelian generalization of a classical analytic class number formula (see [W2] Chapter 4). For a bigger modular Galois representation φ into $GL(2)$ with coefficients in power series rings $\mathcal{O}[[T]]$ and $\mathcal{O}[[T,S]]$, there is an Iwasawa theoretic version of the formula which describes the characteristic power series of $\mathrm{Sel}(Ad(\varphi))$ in terms of the p-adic L-functions of $Ad(\varphi)$. Interested readers can consult [SGL] Section 2.10 and [HTU]. We shall give some examples in the following subsections after proving (Cg1).

An obvious reason for calling (CN2) a non-abelian class number formula is that the p-part of the strict class group of a number field K is isomorphic to $\mathrm{Sel}_0(\varphi)$ for the factor φ of $\mathrm{Ind}_K^{\mathbb{Q}}\,\mathrm{id}$ given by $\mathrm{Ind}_K^{\mathbb{Q}}\,\mathrm{id} = \varphi \oplus \mathrm{id}$. Here the subscript 0 indicates that we have taken the trivial filtration $\{0\} \subset V(\varphi)$ at p to define the Selmer group. The classical class number formula can be written as

$$L(1,\varphi) = \frac{2^r(2\pi)^t R_\infty h(K)}{w\sqrt{|D|}},$$

where D is the (absolute) discriminant of K, R_∞ is the regulator of K, r (resp. t) is the number of real (resp. complex) places of K and $h(K)$ is the class number of K. Thus for $p > 2$, $|h(K)|_p = \|\mathrm{Sel}_0(\varphi)\|_p$. Note here that $\Omega = \frac{2^r(2\pi)^t R_\infty}{w\sqrt{|D|}}$ is the period of a full degree differential ω on $X/O_{K+}^\times$ for $X = \{x \in K \otimes_{\mathbb{Q}} \mathbb{R} | N(x) = 1\}$ normalized so that the integration of

Hecke character $\lambda : K_{\mathbb{A}}^{\times}/K^{\times} \to \mathbb{C}^{\times}$:

$$\int_{K_{\mathbb{A}}^{\times}} \Phi(x)\lambda(x)|x|_{\mathbb{A}}^{s} d\mu^{(\infty)} d\mu_{\infty}$$

gives rise to the Hecke L-function with appropriate Γ-factor, for a suitable choice of Shwartz–Bruhat function Φ on $K_{\mathbb{A}}^{\times}$. Here $d\mu_{\infty}$ is the measure attached to ω (cf. [LFE] Chapter 8). Our period $\Omega(\pm, \lambda; A)$ for the Hecke eigenvalue system λ for $GL(2)$ is again a period of differential form $\omega(f)$ whose Mellin transform gives rise to the modular L-function $L(s, \rho_{\lambda})$, or if we write $f^{*} : GL_2(\mathbb{Q})\backslash GL_2(\mathbb{A}) \to \mathbb{C}$ for the adelic cusp form associated with f, we have

$$\int_{\mathbb{A}^{\times}} f(x)|x|_{\mathbb{A}}^{s} dx = \Gamma\text{-factor} \times L(s, \rho_{\lambda}).$$

Thus the formula in the theorem is an exact analogy of the classical class number formula, and in this sense, we call the formula a non-abelian class number formula.

There is one more technical reason to support this naming. In the classical case, the formula is proven by first showing $Res_{s=1}\zeta_K(s)$ is the quantity we want, for the Dedekind zeta function $\zeta_K(s)$. Note that $\zeta_K(s) = \zeta(s)L(s, \varphi)$, and the fact: $Res_{s=1}\zeta(s) = 1$ yields the desired formula. In the $GL(2)$ case, we have $\varphi \otimes \widetilde{\varphi} = \mathrm{id} \oplus Ad(\varphi)$ for $\varphi = \rho_{\lambda}$, and thus

$$\left|\frac{Res_{s=1}(\Gamma\text{-factor} \times L(s, \varphi \otimes \widetilde{\varphi}))}{\Omega_1(+, \lambda^{\circ}; A)\Omega_1(-, \lambda^{\circ}; A)}\right|_p^{-r(\mathcal{O})} = |C_0^H(\lambda; \mathcal{O})|_p$$

by (CN1). However, the above formula was first proven in [H81a], and then, it is related to $L(1, Ad(\varphi))$ using the fact: $L(s, \varphi \otimes \widetilde{\varphi}) = \zeta(s)L(s, Ad(\varphi))$. Thus the technique here is again similar to the classical case. There is a generalization of this fact (or (CN1)) to cohomological cusp forms over $GL(2)_{/K}$ for an arbitrary number field, and there is a good possibility that this generalizes to modular forms on classical groups. Interested readers should consult [U], [H99a], [H99b] and [H94b].

5.3.4 Gorenstein and complete intersection rings

We would like to prove the identity

$$|C_0(\lambda; \mathcal{O})| = |C_1(\lambda; \mathcal{O})|$$

that we claimed in the previous subsection for a local complete intersection R over $\mathcal{O}$ and an $\mathcal{O}$-algebra homomorphism $\lambda : R \to \mathcal{O}$.

We start slightly more generally and follow Tate's treatment in the Appendix of [MRo]. Let A be a normal noetherian integral domain of characteristic 0 and R be a reduced A-algebra free of finite rank r over A. The algebra R is called a *Gorenstein* algebra over A if $\mathrm{Hom}_A(R, A) \cong R$ as R-modules. Since R is free of rank r over A, we choose a base $(x_1, \ldots, x_r)$ of R over A. Then for each $y \in R$, we have an $r \times r$-matrix $\rho(y)$ with entries in A defined by $(yx_1, \ldots, yx_r) = (x_1, \ldots, x_r)\rho(y)$. Define $\mathrm{Tr}(y) = \mathrm{Tr}(\rho(y))$. Then $\mathrm{Tr} : R \to A$ is an A-linear map, well defined independently of the choice of the base. Suppose that $\mathrm{Tr}(xR) = 0$. Then in particular, $\mathrm{Tr}(x^n) = 0$ for all n. Therefore all eigenvalues of $\rho(x)$ are 0, and hence $\rho(x)$ and x is nilpotent. By the reducedness of R, $x = 0$ and hence the pairing $(x, y) = \mathrm{Tr}(xy)$ on R is non-degenerate.

LEMMA 5.21 *Let A be a normal noetherian integral domain of characteristic 0 and R be an A-algebra. Suppose the following three conditions:*

(1) *R is free of finite rank over A;*

(2) *R is Gorenstein: $i : \mathrm{Hom}_A(R, A) \cong R$ as R-modules;*

(3) *R is reduced.*

Then for an A-algebra homomorphism $\lambda : R \to A$, we have

$$C_0(\lambda; A) \cong A/\lambda(i(\mathrm{Tr}_{R/A}))A.$$

In particular, $\mathrm{length}_A\, C_0(\lambda; A)$ *is equal to the valuation of $d = \lambda(i(\mathrm{Tr}_{R/A}))$ if A is a discrete valuation ring.*

Proof Let $\phi = i^{-1}(1)$. Then $\mathrm{Tr}_{R/A} = \delta\phi$. The element $\delta = \delta_{R/A}$ is called the different of R/A. Then the pairing $(x, y) \mapsto \mathrm{Tr}_{R/A}(\delta^{-1}xy) \in A$ is a perfect pairing over A, where $\delta^{-1} \in S = Frac(R)$ and we have extended $\mathrm{Tr}_{R/A}$ to $S \to K = Frac(A)$. Since R is commutative, $(xy, z) = (y, xz)$. Decomposing $S = K \oplus X$ so that the projection onto K coincides with λ on A, we have

$$C_0(\lambda; A) = \mathrm{Im}(\lambda)/\lambda(\mathfrak{a}) \cong A/R \cap (K \oplus 0).$$

From this, it is easy to conclude that the pairing (,) induces a perfect A-duality between $R \cap (K \oplus 0)$ and $A \oplus 0$. Thus $R \cap (K \oplus 0)$ is generated by $\lambda(\delta) = \lambda(i(\mathrm{Tr}_{R/A}))$. □

We now introduce two A-free resolutions of R, in order to compute $\delta_{R/A}$. We start slightly more generally. Let X be an algebra. A sequence $f = (f_1, \ldots, f_n) \in X^n$ is called *regular* if $x \mapsto f_j x$ is injective on $X/(f_1, \ldots, f_{j-1})$ for all $j = 1, \ldots, n$. We now define a complex $K_X^{\bullet}(f)$

(called *the Koszul complex*) out of a regular sequence f. Let $V = X^n$ with a standard base $e_1, \dots, e_n$ (see [CRT] Section 16). Then we consider the exterior algebra

$$\bigwedge^{\bullet} V = \bigoplus_{j=0}^{n} (\wedge^j V).$$

The graded piece $\wedge^j V$ has a base $e_{i_1,\dots,i_j} = e_{i_1} \wedge e_{i_2} \wedge \cdots \wedge e_{i_j}$ indexed by sequences $(i_1, \dots, i_j)$ satisfying $0 < i_1 < i_2 < \cdots < i_j \le n$. We agree to put $\bigwedge^0 V = X$ and $\bigwedge^j V = 0$ if $j > n$. Then we define an X-linear differential $d : \bigwedge^j X \to \bigwedge^{j-1} X$ by

$$d(e_{i_1} \wedge e_{i_2} \wedge \cdots \wedge e_{i_j}) = \sum_{r=1}^{j} (-1)^{r-1} f_{i_r} e_{i_1} \wedge \cdots \wedge e_{i_{r-1}} \wedge e_{i_{r+1}} \wedge \cdots \wedge e_{i_j}.$$

In particular, $d(e_j) = f_j$ and hence,

$$\bigwedge^0 V / d(\bigwedge^1 V) = X/(f).$$

Thus, $(K_X^{\bullet}(f), d)$ is a complex and X-free resolution of $X/(f_1, \dots, f_n)$. We also have

$$d_n(e_1 \wedge e_2 \wedge \cdots \wedge e_n) = \sum_{j=1}^{n} (-1)^{j-1} f_j e_1 \wedge \cdots \wedge e_{j-1} \wedge e_{j+1} \wedge \cdots \wedge e_n.$$

Suppose now that X is a B-algebra. Identifying $\bigwedge^{n-1} V$ with V by

$$e_1 \wedge \cdots \wedge e_{j-1} \wedge e_{j+1} \wedge \cdots \wedge e_n \mapsto e_j$$

and $\bigwedge^n V$ with X by $e_1 \wedge e_2 \wedge \cdots \wedge e_n \mapsto 1$, we have

$$\mathrm{Im}(d_n^* : \mathrm{Hom}_B(\bigwedge^{n-1} V, Y) \to \mathrm{Hom}_B(\bigwedge^n V, Y)) \cong (f)\,\mathrm{Hom}_B(X, Y),$$

where $(f)\,\mathrm{Hom}_B(X, Y) = \sum_j f_j \,\mathrm{Hom}_B(X, Y)$, regarding $\mathrm{Hom}_B(X, Y)$ as an X-module by $y\phi(x) = \phi(xy)$. This shows that if X is a B-algebra free of finite rank over B, $K_X^{\bullet}(f)$ is a B-free resolution of $X/(f)$, and

$$\mathrm{Ext}_B^n(X/(f), Y) = H^n(\mathrm{Hom}_B(K_X^{\bullet}(f), Y)) \cong \frac{\mathrm{Hom}_B(X, Y)}{(f)\,\mathrm{Hom}_B(X, Y)} \tag{5.21}$$

for any B-module Y.

We now suppose that R is a local complete intersection over A. Thus R is free of finite rank over A and $R \cong B/(f_1, \dots, f_n)$ for $B = A[[T_1, \dots, T_n]]$. Write t_j for T_j mod $(f_1, \dots, f_n)$ in R. Since R is local, t_j are contained in the maximal ideal $\mathfrak{m}_R$ of R. We consider $C = B \otimes_A R \cong R[[T_1, \ldots, T_n]]$. Then

$$R = R[[T_1, \ldots, T_n]]/(T_1 - t_1, \ldots, T_n - t_n),$$

and $g = (T_1 - t_1, \ldots, T_n - t_n)$ is a regular sequence in $C = R[[T_1 \ldots, T_n]]$. Since C is B-free of finite rank, the two complexes $K_B^\bullet(f) \twoheadrightarrow R$ and $K_C^\bullet(g) \twoheadrightarrow R$ are B-free resolutions of R.

We have a Λ-algebra homomorphism $\Phi : B \hookrightarrow C$ given by $\Phi(x) = x \otimes 1$. We extend Φ to $\Phi^\bullet : K_B^\bullet(f) \to K_C^\bullet(g)$ in the following way. Write $f_j = \sum_{j=1}^n b_{ij} g_j$. Then we define $\Phi^1 : K_B^1(f) \to K_C^1(g)$ by $\Phi^1(e_i) = \sum_{j=1}^n b_{ij} e_j$. So $\Phi^j = \bigwedge^j \Phi^1$. One can check that this map $\Phi^\bullet$ is a morphism of complexes. In particular,

$$\Phi_n(e_1 \wedge \cdots \wedge e_n) = \det(b_{ij}) e_1 \wedge \cdots \wedge e_n. \tag{5.22}$$

Since $\Phi^\bullet$ is the lift of the identity map of R to the B-projective resolutions $K_B^\bullet(f)$ and $K_C^\bullet(g)$, it induces an isomorphism of extension groups computed by $K_C^\bullet(g)$ and $K_B^\bullet(f)$:

$$\Phi^* : H^\bullet(\mathrm{Hom}_B(K_C^\bullet(g), B)) \cong \mathrm{Ext}_B^j(R, B) \cong H^\bullet(\mathrm{Hom}_B(K_B^\bullet(f), B)).$$

In particular, identifying $\bigwedge^n B^n = B$, we have from (5.21) that

$$H^n(\mathrm{Hom}_B(K_B^\bullet(f), B)) = \mathrm{Hom}_B(B, B)/(f)\,\mathrm{Hom}_B(B, B) = B/(f) = R$$

and similarly

$$H^n(\mathrm{Hom}_B(K_C^\bullet(g), B)) = \frac{\mathrm{Hom}_B(C, B)}{(g)\,\mathrm{Hom}_B(C, B)}.$$

The isomorphism between R and $\frac{\mathrm{Hom}_B(C,B)}{(g)\,\mathrm{Hom}_B(C,B)}$ is induced by Φ_n which is a multiplication by $d = \det(b_{ij})$ (see (5.22)). Thus we have

Lemma 5.22 *Let $\pi : B = A[[T_1, \ldots, T_n]] \twoheadrightarrow R$ be the projection. We have an isomorphism:*

$$h : \frac{\mathrm{Hom}_B(C, B)}{(T_1 - t_1, \ldots, T_n - t_n)\,\mathrm{Hom}_B(C, B)} \cong R$$

given by $h(\phi) = \pi(\phi(d))$ for $d = \det(b_{ij}) \in C$.

We have a base-change map:

$$\iota : \mathrm{Hom}_A(R, A) \longrightarrow \mathrm{Hom}_B(C, B) = \mathrm{Hom}_B(B \otimes_A R, B \otimes_A A),$$

taking ϕ to $\mathrm{id} \otimes \phi$. Identifying C and B with power series rings, $\iota(\phi)$ is simply applying the original ϕ to coefficients of power series in $R[[T_1, \ldots, T_n]]$. We define $I = h \circ \iota : \mathrm{Hom}_A(R, A) \to R$.

LEMMA 5.23 *The above map I is an R-linear isomorphism, satisfying $I(\phi) = \pi(\imath(\phi(d)))$. Thus the ring R is Gorenstein.*

Proof We first check that I is an R-linear map. Since $I(\phi) = \pi(\imath(\phi(d)))$, we compute $I(\phi \circ b)$ and $rI(\phi)$ for $b \in B$ and $r = \pi(b)$. By definition, we see that

$$I(\pi(bx)) = \pi(\imath(\phi(r \otimes 1)d)) \quad \text{and} \quad rI(\phi) = \pi(b\imath(\phi(d))).$$

Thus we need to check that $\pi(\imath(\phi)((r \otimes 1 - 1 \otimes b)d)) = 0$. This follows from:

$$r \otimes 1 - 1 \otimes b \in (g) \quad \text{and} \quad \det(b_{ij})g_i = \sum_i b'_{ij} f_i,$$

where b'_{ij} are the (i,j)-cofactors of the matrix (b_{ij}). Thus I is R-linear. Since $\imath \mod \mathfrak{m}_B$ for the maximal ideal $\mathfrak{m}_B$ of B is a surjective isomorphism from

$$\mathrm{Hom}_A((A/\mathfrak{m}_A)^r, A/\mathfrak{m}_A) = \mathrm{Hom}_A(R,A) \otimes_A A/\mathfrak{m}_A$$

onto

$$\mathrm{Hom}_B((B/\mathfrak{m}_B)^r, B/\mathfrak{m}_B) = \mathrm{Hom}_B(C,B) \otimes_B B/\mathfrak{m}_B,$$

the map $\imath$ is non-trivial modulo $\mathfrak{m}_C$. Thus $I \mod \mathfrak{m}_R$ is non-trivial. Since h is an isomorphism, $\mathrm{Hom}_B(C,B) \otimes_C C/\mathfrak{m}_C$ is one-dimensional, and hence $I \mod \mathfrak{m}_R$ is surjective. By Nakayama's lemma, I itself is surjective. Since the target and the source of I are A-free of equal rank, the surjectivity of I implies its injectivity. This completes the proof. □

COROLLARY 5.24 *We have $I(\mathrm{Tr}_{R/A}) = \pi(d)$ for $d = \det(b_{ij})$, and hence the different $\delta_{R/A}$ is equal to $\pi(d)$.*

Proof The last assertion follows from the first by $I(\phi) = \pi(\imath(\phi(d)))$. To show the first, we choose dual basis $x_1, \ldots, x_r$ of R/A and $\phi_1, \ldots, \phi_r$ of $\mathrm{Hom}_A(R,A)$. Thus for $x \in R$, writing $xx_i = \sum_i a_{ij}x_j$, we have $\mathrm{Tr}(x) = \sum_i a_{ii} = \sum_i \phi_i(xx_i) = \sum_i x_i\phi_i(x)$. Thus $\mathrm{Tr} = \sum_i x_i\phi_i$.

Since x_i is also a base of C over B, we can write $d = \sum_j b_j x_i$ with $\imath(\phi_i)(d) = b_i$. Then we have

$$\begin{aligned} I(\mathrm{Tr}_{R/A}) &= \sum_i x_i I(\phi_i) = \sum_i x_i \pi(\imath(\phi_i)(d)) = \sum_i x_i \pi(b_i) \\ &= \pi(\sum_i b_i x_i) = \pi(d). \end{aligned}$$

This proves the desired assertion. □

Proposition 5.25 *Let A be a discrete valuation ring, and let R be a reduced complete intersection over A. Then for an A-algebra homomorphism $R \to A$, we have*

$$\mathrm{length}_A\, C_0(\lambda, A) = \mathrm{length}_A\, C_1(\lambda, A).$$

In fact the assertion of the proposition is equivalent to R being a complete intersection over A (see [Lt] for a proof).

Proof Let X be a torsion A-module, and suppose that we have an exact sequence:

$$A^r \xrightarrow{L} A^r \to X \to 0$$

of A-modules. Then we claim that $\mathrm{length}_A X = \mathrm{length}_A(A/\det(L)A)$. By elementary divisor theory applied to L, we may assume that L is a diagonal matrix with diagonal entry $d_1, \ldots, d_r$. Then the assertion is clear, because $X = \bigoplus_j A/d_jA$ and the length A/dA is equal to the valuation of d.

Since R is reduced, $\Omega_{R/A}$ is a torsion R-module, and hence $\Omega_{R/A} \otimes_R A = C_1(\lambda; A)$ is a torsion A-module. Since R is a complete intersection over A, we can write

$$R \cong A[[T_1, \ldots, T_r]]/(f_1, \ldots, f_r).$$

Then by Corollary 5.13 (ii), we have the following exact sequence for $J = (f_1, \ldots, f_r)$:

$$J/J^2 \otimes_{A[[T_1,\ldots,T_r]]} A \longrightarrow \Omega_{A[[T_1,\ldots,T_r]]/A} \otimes_{A[[T_1,\ldots,T_r]]} A \longrightarrow \Omega_{R/A} \otimes_R A \to 0.$$

This gives rise to the following exact sequence:

$$\bigoplus_j A df_j \xrightarrow{L} \bigoplus_j A dT_j \longrightarrow C_1(\lambda; A) \to 0,$$

where $df_j = f_j \mod J^2$. Since $C_1(\lambda; A)$ is a torsion A-module, we see that $\mathrm{length}_A(A/\det(L)A) = \mathrm{length}_A\, C_1(\lambda; A)$. Since $g = (T_1 - t_1, \ldots, T_n - t_n)$, we see easily that $\det(L) = \pi(\lambda(d))$. This combined with Corollary 5.24 and Lemma 5.21 proves the desired assertion. □

If A is a discrete valuation ring with finite residue field, we have

$$|X| = [A : \mathfrak{m}_A]^{\mathrm{length}_A X}$$

for an A-torsion module X of finite type. Thus we have

COROLLARY 5.26 *Suppose that A is a discrete valuation ring with finite residue field. If R is a reduced complete intersection over A, then we have*

$$|C_0(\lambda; A)| = |C_1(\lambda; A)|$$

for each A-algebra homomorphism $\lambda : R \to A$.

Let A be a normal noetherian domain. Then we see easily that $C_j(\lambda_P; A_P) = C_j(\lambda; A)_P$ for primes $P \subset A$, where $\lambda_P : R_P \to A_P$ is the localization of λ at P. Since A is a normal noetherian integral domain, for a height 1 prime P, A_P is a discrete valuation ring (see [CRT] Section 11). For any torsion A-module X of finite type, $\ell_P(X) = \mathrm{length}_{A_P} X_P$ is finite for all height 1 primes P and $\ell_P(X) = 0$ except for finitely many primes P. Thus $char_A(X) = \prod_P P^{\ell_P(X)}$ is a well-defined ideal of A, where P runs over all height 1-primes of A. The ideal $char_A(X)$ is called the *characteristic ideal* of X. If $A = \mathcal{O}[[T_1, \ldots, T_r]]$ for a discrete valuation ring $\mathcal{O}$, then A is a unique factorization domain ([CRT] Theorem 20.8). Thus all height 1 prime ideals are principal, and hence $char_A(X)$ is a principal ideal, whose generator is called the *characteristic power series* of X. This definition of $char_A(X)$ is equivalent to the one already given (see [BCM] Chapter 7). Thus we have

COROLLARY 5.27 *Let A be a normal noetherian integral local domain of characteristic* 0 *and R be an A-algebra. Suppose the following three conditions:*

(1) *R is free of finite rank over A;*
(2) *R is a local complete intersection over A;*
(3) *R is reduced.*

Then for any A-algebra homomorphism $\lambda : R \to A$, *we have*

$$char_A(C_0(\lambda; A)) = char_A(C_1(\lambda; A)).$$

5.3.5 Universal p-ordinary Hecke algebras

We keep our assumption that $\overline{\rho}$ is associated to a Hecke eigenform f in $S_k(\Gamma_0(p), \chi\omega^{-k}; \mathcal{O})$ for the Teichmüller character ω; in other words, $\overline{\rho} \equiv \rho_\lambda$ mod $\mathfrak{m}_{\mathcal{O}}$ for the $\mathcal{O}$-algebra homomorphism $\lambda : h_k(p, \chi\omega^{-k}; \mathcal{O}) \to \mathcal{O}$ given by $T(n)f = \lambda(T(n))f$ for all n (thus, $\phi = \det(\overline{\rho}) = \chi\omega^{-1}$). We suppose that $\lambda(T(p)) \in \mathcal{O}^\times$. Then automatically, as we have seen, $p \geq 11$. In any case, $\overline{\rho}$ is p-ordinary and unramified outside p (Theorem 3.26).

We are going to create a Hecke algebra, called the universal p-ordinary

Hecke algebra, from the Hecke side, which is canonically isomorphic to the p-ordinary universal couple (R^{ord}, ϱ^{ord}).

Let $\Lambda = \mathcal{O}[[T]]$ be the one variable power series ring. Write $\mathbf{h}^{ord}(\chi;\Lambda)_{/\mathbb{Q}}$ for the universal p-ordinary Hecke algebra of level p^∞ of character χ with coefficients in Λ defined in [LFE] page 218 in Chapter 7, which is naturally a Λ-algebra (by definition). Let P_ℓ be the ideal of Λ generated by $(1+T)-(1+p)^\ell$. Then P_ℓ is the kernel of the $\mathcal{O}$-algebra homomorphism: $\pi_\ell : \Lambda \to \mathcal{O}$ given by $\Phi(T) \mapsto \Phi((1+p)^\ell - 1) \in \mathcal{O}$. The algebra $\mathbf{h}^{ord}(\chi;\Lambda)$ is characterized by the following property:

$$\mathbf{h}^{ord}(\chi;\Lambda) \otimes_{\Lambda,\pi_\ell} \mathcal{O} \cong h_\ell^{ord}(p, \chi\omega^{-\ell}; \mathcal{O}) \text{ for all } \ell \geq 2, \tag{5.23}$$

where the isomorphism sends the Hecke operator $T(n) \in \mathbf{h}^{ord}(\chi;\Lambda)$ to $T(n) \in h_\ell^{ord}(p, \chi\omega^{-\ell}; \mathcal{O})$ for all n. The statement, like (5.23) relating the universal p-ordinary Hecke algebra with the corresponding algebra of each weight ℓ, is often called a vertical control theorem. By (5.23), as we will see, $\mathbf{h}^{ord}(\chi;\Lambda)$ is free of finite rank over Λ.

By the theorem of Taylor-Wiles, assuming the two conditions (ai$_\kappa$) and (rg$_p$) ($\kappa = \mathbb{Q}(\sqrt{(-1)^{(p-1)/2}p})$), we can construct directly from R^{ord} the local ring $\mathbf{h} = \mathbf{h}^{ord}_{\mathbb{Q}}$ of $\mathbf{h}^{ord}(\chi,\Lambda)$, through which λ factors. We describe this fact here. Let $\phi_\ell = \chi\omega^{-\ell}\nu^{\ell-1}$ for the p-adic cyclotomic character ν. Thus $\phi_k = \det(\rho_\lambda) = \phi$. Identify $\mathcal{O}[[T]]$ with $\mathcal{O}[[\Gamma]]$ for $\Gamma = 1 + p\mathbb{Z}_p \cong \mathfrak{G}_p^{ab}$ (Corollary 2.23) by $1+p \mapsto 1+T$. Since P_ℓ is generated by $\gamma - \phi_\ell(\gamma)$ for $\gamma \in \Gamma$, by construction (see Proposition 3.30), the $\mathcal{O}$-algebra homomorphism $\pi_\ell : R^{ord} \to R^{ord,\phi_\ell}$ such that $\pi_\ell \circ \varrho^{ord} \approx \varrho^{ord,\phi_\ell}$ induces an isomorphism $\pi_\ell : R^{ord}/P_\ell R^{ord} \cong R^{ord,\phi_\ell}$. But by the theorem of Taylor-Wiles, $R^{ord,\phi_\ell} \cong h_\ell$ for the local ring $h_\ell^{ord}(p, \chi\omega^{-\ell}; \mathcal{O})$ whose residual representation modulo $\mathfrak{m}_{h_\ell}$ is isomorphic to $\overline{\rho}$, if $\ell \geq 2$. The existence of such h_ℓ follows from the existence of $\lambda_\ell : h_\ell(p, \chi\omega^{-k}; \mathcal{O}) \to \mathcal{O}$ with $\lambda_\ell(T(n)) \equiv \lambda(T(n)) \mod \mathfrak{m}_{\mathcal{O}}$ for all n, which follows from the construction of $\mathbf{h}^{ord}(\chi;\Lambda)$ in [LFE] Chapter 7 (this and $\mathcal{O}$-freeness of h_ℓ are the points where we need automorphic information from [LFE] after knowing the Taylor-Wiles theorem; so these two pieces of information are not the consequence of the Taylor-Wiles theorem). Thus R^{ord} satisfies a version of the property (5.23), so to speak, restricted to $\mathbf{h}$.

We now prove that R^{ord} is free of finite rank over Λ. Write $r = \mathrm{rank}_{\mathcal{O}} h_k$, and choose a base $\overline{x}_1, \ldots, \overline{x}_r$ of h_k over $\mathcal{O}$. We choose $x_1, \ldots, x_r \in R^{ord}$ so that $\pi_k(x_j) = \overline{x}_j$ for all j. Define a Λ-linear map $\pi : \Lambda^r \to R^{ord}$ by $\pi(a_1, \ldots, a_r) = \sum_j a_j x_j$. Then by Nakayama's lemma, π is a surjection (because $\pi \mod P_k$ is a surjection by definition). Since r is a minimal number of generators over Λ of R^{ord}, h_ℓ is free of rank r over $\mathcal{O}$ again by Nakayama's lemma. Since $\pi \mod P_\ell : (\Lambda/P_\ell)^r \to h_\ell$ is a surjective $\mathcal{O}$-linear map from an $\mathcal{O}$-free module of rank r onto another $\mathcal{O}$-free module of rank r, it has to be an isomorphism. Thus we see that $\mathrm{Ker}(\pi \mod P_\ell) \subset (P_\ell\Lambda)^r$ and that $\mathrm{Ker}(\pi) \subset \bigcap_{\ell \geq 2}(P_\ell\Lambda)^r = \{0\}$. Therefore, π is an isomorphism. The same argument, replacing R^{ord} and h_ℓ by $\mathbf{h}^{ord}(\chi;\Lambda)$ and $h_\ell^{ord}(p,\chi\omega^{-\ell};\mathcal{O})$, respectively, shows the freeness of $\mathbf{h}^{ord}(\chi;\Lambda)$. We have proved

THEOREM 5.28 *Suppose* (ai$_\kappa$) *and* (rg$_p$) *for* $\overline{\rho} = \rho_{h_k} \mod \mathfrak{m}_{h_k}$ *for a local ring* h_k *of* $h_k^{ord}(p,\chi\omega^{-k};\mathcal{O})$, *where* $k \geq 2$ *is an integer. Then we have:*

(1) *For each integer* $\ell \geq 2$, *there exists a unique local ring* h_ℓ *of the weight* ℓ *Hecke algebra* $h_\ell^{ord}(p,\chi\omega^{-\ell};\mathcal{O})$ *such that* $\overline{\rho} = \rho_{h_\ell} \mod \mathfrak{m}_{h_\ell}$. *Moreover,* $r = \mathrm{rank}_{\mathcal{O}}\, h_\ell = \mathrm{rank}_{\mathcal{O}}\, h_k$ *is independent of* ℓ;
(2) *The universal p-ordinary deformation ring* R^{ord} *deforming* $\overline{\rho}$ *is free of rank* r *over* Λ *and satisfies* $R^{ord}/P_\ell R^{ord} \cong h_\ell$ *canonically for all* $\ell \geq 2$.

We have a unique local ring $\mathbf{h}^{ord} = \mathbf{h}_{\mathbb{Q}}^{ord}$ of $\mathbf{h}^{ord}(\chi;\Lambda)_{/\mathbb{Q}}$ such that λ factors through $\mathbf{h}^{ord}$. By Definition (5.23), $\mathbf{h}^{ord} = \varprojlim_k \mathbf{h}^{ord}/P_2 \cap \cdots \cap P_k$. In the same manner as in the proof of Theorem 3.29, we have a degree 2 pseudo-representation $\pi_k : \mathfrak{G}_p \to \mathbf{h}^{ord}/P_2 \cap \cdots \cap P_k$ such that

$$\mathrm{Tr}(\pi_k(Frob_\ell)) = T(\ell) \mod P_2 \cap \cdots \cap P_k$$

for all $\ell \nmid p$. Taking the projective limit of π_k, we get the pseudo-representation $\pi : \mathfrak{G}_p \to \mathbf{h}^{ord}$. Out of π, we can construct a Galois representation $\rho^{ord} : \mathfrak{G}_p \to GL_2(\mathbf{h}^{ord})$, by Proposition 3.27, having the same trace with π. The representation constructed is a p-ordinary deformation of $\overline{\rho}$, since all specialization of $\rho^{ord} \mod P_k$ ($k \geq 2$) is p-ordinary (Theorem 3.26 (2)). Thus, we have a surjective morphism $\iota : R_{\mathbb{Q}}^{ord} \to \mathbf{h}^{ord}$.

THEOREM 5.29 *Suppose that* $p > 3$. *Then under* (ai$_\kappa$) *and* (rg$_p$), *we have*

$$R_{\mathbb{Q}}^{ord} \cong \mathbf{h}_{\mathbb{Q}}^{ord}.$$

Proof We identify Λ with $\mathcal{O}[[\mathfrak{G}_p^{ab}]]$ by class field theory. Let

$$\phi_\ell = \lambda_\ell(\det(\rho^{ord})) = \chi\omega^{-\ell+1}.$$

By the Taylor-Wiles theorem and (5.23),

$$R^{ord,\phi_\ell} \cong \mathbf{h}^{ord} \otimes_{\Lambda,\phi_\ell} \mathcal{O}$$

for all $\ell \geq 2$. By our construction of the universal deformation ring R^{ord,ϕ_ℓ} for $\phi = \chi\omega^{-\ell+1}$ in Proposition 3.30 of Chapter 3, we have

$$\mathbf{h}^{ord} \otimes_{\Lambda,\phi_\ell} \mathcal{O} \cong R^{ord,\phi_\ell} \cong R^{ord} \otimes_{\Lambda,\phi_\ell} \mathcal{O}$$

for all $\ell \geq 2$. This identity provides the result, because R^{ord} is noetherian. □

To date, we have four different proofs of the freeness of $\mathbf{h}^{ord}$ over Λ. The fourth one is the one presented above, using Galois deformation theory. The first one, which I found in 1982, is exposed in [H86a]. Its key ingredient is the solution (studied by N. Katz) to a p-ordinary moduli problem of elliptic curves with a p-power level structure and a nowhere vanishing differential (the moduli problem of Weierstrass type). This proof seems amenable to further generalization to modular forms on Shimura varieties (see [H99c]). The second one in [H86b] uses the arithmetic of jacobians of modular curves in conjunction with cohomology groups of such curves. This technique is now generalized to $GL(n)$ ([H95] and [H98]) and $GSp(2g)$ ([TU]). The third one in [LFE] Chapter 7 is the most elementary and is based on the technique of multiplying cusp forms by Λ-adic Eisenstein series.

The second approach to the freeness of $\mathbf{h}^{ord}$ gives a seemingly different definition of $\mathbf{h}^{ord}(\chi;\Lambda)$ [H86b]: The algebra $\mathbf{h}^{ord}(\chi;\Lambda)_{/\mathbb{Q}}$ is the χ-part of

$$\mathbf{h}^{ord}(p^\infty;\Lambda)_{/\mathbb{Q}} = \varprojlim_\alpha h_\ell^{ord}(\Gamma_1(p^\alpha);\mathcal{O})_{/\mathbb{Q}} \quad \text{for any } \ell \geq 2.$$

Some explanation is due on the meaning of the χ-part. The inclusion: $i : S_k(\Gamma_1(p^\alpha)) \hookrightarrow S_k(\Gamma_1(p^\beta))$ $(\alpha \leq \beta)$ commutes with Hecke operators: $T(n) \circ i = i \circ T(n)$ for all n prime to p and $U(p) \circ i = i \circ U(p)$. Thus the restriction of Hecke operators h of level p^β to the subspace of level p^α is a Hecke operator of level p^α. This map therefore induces the projection morphism: $h_k(p^\beta;\mathcal{O}) \twoheadrightarrow h_k(p^\alpha;\mathcal{O})$ taking $T(n)$ to $T(n)$ as long as $\alpha > 0$. By a theorem of Katz (Theorem 3.13), $(\mathbb{Z}/p^\alpha\mathbb{Z})^\times$ acts on $h_\ell^{ord}(p^\alpha;\mathcal{O})_{/\mathbb{Q}}$. Thus $\mathbb{Z}_p^\times = (1+p\mathbb{Z}_p) \times \mu_{p-1}$ acts on the limit $\mathbf{h}^{ord}(p^\infty;\Lambda)$. Then we take the subspace of $\mathbf{h}^{ord}(p^\infty;\Lambda)$ on which μ_{p-1} acts by $\chi\omega^{-\ell}$ (not χ). This algebra is by definition $\mathbf{h}^{ord}(\chi;\Lambda)$ and is independent of the choice of $\ell \geq 2$.

5.3.6 *p-adic adjoint L-functions*

Let (R^{ord}, ϱ^{ord}) be the universal p-ordinary couple. Under the assumption of Theorem 5.29 (which we keep in this subsection), we have shown that

$\mathbf{h}^{ord} \cong R^{ord}$. Let $\lambda : \mathbf{h}^{ord} \to \Lambda$ be a Λ-algebra homomorphism. Write $\rho_\lambda = \lambda \circ \varrho^{ord}$. By definition, we have an exact sequence:

$$0 \to V(\varepsilon_\lambda) \to V(\rho_\lambda) \to V(\delta_\lambda) \to 0$$

giving the p-ordinarity to ρ_λ. We thus have $\mathrm{Sel}(Ad(\rho_\lambda))$ with respect to $\mathscr{S} = \{V(\varepsilon_\lambda)\}$. By Theorem 5.14 (and Corollary 5.13 (ii)), we have, as Λ-modules,

$$\mathrm{Sel}^*(Ad(\rho_\lambda)) \cong \mathrm{Ker}(\lambda)/\mathrm{Ker}(\lambda)^2 = C_1(\lambda;\Lambda). \tag{5.24}$$

Thus $\mathrm{Sel}^*(Ad(\rho_\lambda))$ is a torsion Λ-module, and hence we have a characteristic power series $\Phi(T) \in \Lambda$. We would like to construct a p-adic L-function $L_p(T) \in \Lambda$ from the Hecke side such that $L_p(T) = \Phi(T)$ up to units in Λ.

After tensoring $\mathcal{O}$ via ϕ_ℓ, we get an $\mathcal{O}$-algebra homomorphism

$$\lambda_\ell : h_\ell(p,\chi\omega^{-\ell};\mathcal{O}) \twoheadrightarrow \mathbf{h}^{ord}/P_\ell\mathbf{h}^{ord} \to \mathcal{O}.$$

Thus λ_ℓ is associated to a Hecke eigenform f_ℓ. We then have $\eta_\ell \in \mathcal{O}$ such that $C_0(\lambda_\ell;\mathcal{O}) \cong \mathcal{O}/\eta_\ell\mathcal{O}$ and

$$\frac{W(\lambda_\ell^\circ)C(\lambda_\ell^\circ)^{\ell/2}\Gamma(1,Ad(\lambda_\ell))L(1,Ad(\lambda_\ell))}{\Omega(+,\lambda_\ell;A_\ell)\Omega(-,\lambda_\ell;A_\ell)} = \eta_\ell$$

up to units in $\mathcal{O}$, where $A_\ell = \mathbb{Q}(\lambda_\ell) \cap \mathcal{O}$. We require to have

$$L_p(P_\ell) = (L_p \mod P_\ell) = L_p((1+p)^\ell - 1) = \eta_\ell$$

up to $\mathcal{O}$-units for all $\ell \geq 2$.

Since $\mathbf{h}^{ord}$ can be embedded into $\prod_\ell h_\ell^{ord}(p,\chi\omega^{-\ell};\mathcal{O})$, the reducedness of the Hecke algebras $h_\ell^{ord}(p,\chi\omega^{-\ell};\mathcal{O})$ shows that $\mathbf{h}^{n.ord}$ is reduced. Thus for the field of fractions $\mathscr{L}$ of Λ,

$$\mathbf{h}^{ord} \otimes_\Lambda \mathscr{L} \cong \mathscr{L} \oplus X,$$

where the projection to $\mathscr{L}$ is $\lambda \otimes \mathrm{id}$. We then define $C_j(\lambda;\Lambda)$ as in Subsection 5.3.3.

We now claim

THEOREM 5.30 *Assume* (ai_K) *and* (rg_p) *for* $\bar{\rho}$. *Then* $\mathbf{h}^{ord}$ *is a complete intersection over* Λ. *Thus* $\mathbf{h}^{ord}$ *is free of finite rank over* Λ *and*

$$\mathbf{h}^{ord} \cong \Lambda[[T_1,\dots,T_r]]/(f_1,\dots,f_r).$$

Proof We already know that $\mathbf{h}^{ord}$ is free of finite rank over Λ (see Theorem 5.28 (2)). By the Taylor-Wiles theorem (Theorem 3.31), we have an isomorphism

$$h_k \cong \mathcal{O}[[T_1,\dots,T_r]]/(\overline{f}_1,\dots,\overline{f}_r).$$

We write $\overline{t}_j$ for the image of T_j in h_k. Then we lift it to $t_j \in \mathbf{h}^{ord}$ so that $t_j \otimes 1 = \overline{t}_j$ under $\mathbf{h}^{ord} \otimes_{\Lambda,\phi_k} \mathcal{O} \cong h_k$. Then we define a surjective Λ-linear map $\pi : \Lambda[[T_1,\dots,T_r]] \to \mathbf{h}^{ord}$ by $T_j \mapsto t_j$. So $\mathrm{Ker}(\pi) \otimes_\Lambda \mathcal{O} \cong (\overline{f}_1,\dots,\overline{f}_r)$ because $\mathbf{h}^{ord}$ is Λ-free. Then by the Nakayama's lemma, taking $f_i \in \mathrm{Ker}(\pi)$ so that $f_i \otimes 1 = \overline{f}_i$ under $\mathrm{Ker}(\pi) \otimes_\Lambda \mathcal{O} \cong (\overline{f}_1,\dots,\overline{f}_r)$, we have $\mathrm{Ker}(\pi) = (f_1,\dots,f_r)$. Thus we see that

$$\mathbf{h}^{ord} \cong \Lambda[[T_1,\dots,T_r]]/(f_1,\dots,f_r)$$

as desired. □

Again by Corollary 5.27, Theorem 5.30 tells us that the characteristic power series of $C_1(\lambda;\Lambda)$ and $C_0(\lambda;\Lambda)$ coincide. By Theorem 5.30, $\mathbf{h}^{ord}$ is Gorenstein, and hence $\mathrm{Hom}_\Lambda(\mathbf{h}^{ord},\Lambda) \cong \mathbf{h}^{ord}$ as $\mathbf{h}^{ord}$-modules, which tells us that

$$C_0(\lambda;\Lambda) \cong \Lambda/L_p\Lambda$$

for a non-zero element $L_p \in \Lambda$. Since $h_\ell \cong \mathbf{h}^{ord} \otimes_{\Lambda,\phi_\ell} \mathcal{O}$, we easily see from a diagram chasing that

$$C_0(\lambda,\Lambda) \otimes_{\Lambda,\phi_\ell} \mathcal{O} \cong C_0(\lambda_\ell;\mathcal{O}).$$

This assures us that $L_p(P_\ell) = \eta_\ell$ up to $\mathcal{O}$-units. The Iwasawa module $C_0(\lambda;\Lambda)$ was first introduced in [H86a] to study behavior of congruence between modular forms as one varies Hecke eigenforms f_ℓ associated with λ_ℓ. The fact that the characteristic power series of $C_0(\lambda;\Lambda)$ interpolates p-adically the adjoint L-values was pointed out in [H86a] (see also [H88b]). We record here what we have proved:

COROLLARY 5.31 *Let the notation and the assumption be as in Theorem 5.30. Then there exists $0 \neq L_p(T) \in \Lambda$ such that:*

(1) *$L_p(T)$ gives a characteristic power series of* $\mathrm{Sel}^*(Ad(\rho_\lambda))$;
(2) *We have, for all $\ell \geq 2$,*

$$L_p(P_\ell) = \frac{W(\lambda_\ell^\circ)C(\lambda_\ell)^{\ell/2}\Gamma(1,Ad(\lambda_\ell))L(1,Ad(\lambda_\ell))}{\Omega(+,\lambda_\ell;A_\ell)\Omega(-,\lambda_\ell;A_\ell)}$$

up to units in $\mathcal{O}$, where $C(\lambda_\ell)$ is the conductor of λ_ℓ.

5.4 Control of Universal Deformation Rings

In this section, we describe a general theory (given in [H96]) of controlling the deformation rings of representations of a normal subgroup under the action of the quotient finite group. The general result here will be used in the following section to get a control theorem of the adjoint Selmer group over the cyclotomic $\mathbb{Z}_p$-extension.

Throughout the section, we fix a profinite group G and an open normal subgroup H. We write the quotient $\Delta = G/H$. Our deformation functor can be defined over the category $CL_{\mathcal{O}}$, but if the following finite p-Frattini condition is satisfied by G, all the functors introduced here, if representable in $CL_{\mathcal{O}}$, are actually representable over the smaller full subcategory $CNL_{\mathcal{O}}$ of noetherian pro-artinian rings (see Proposition 2.30):

$$\text{Each open subgroup of } G \text{ has finite } p\text{-Frattini quotient.} \qquad (\Phi)$$

The p-Frattini quotient of a profinite group G is $G/\overline{G^p(G:G)}$ for the commutator subgroup $(G:G)$. As already remarked in Proposition 2.30, by class field theory, this condition is satisfied by $\mathrm{Gal}(F^S/F)$ for a number field F, where F^S/F is the maximal extension unramified outside a finite set S of places of F. Thus, assuming (Φ) does not cause any harm to our later application, we will assume (Φ) throughout this section for simplicity.

5.4.1 Deformation functors of group representations

We fix a representation $\overline{\rho} : G \to GL_n(\mathbb{F})$ and consider the following condition

$$\overline{\rho}_H = \overline{\rho}|_H \text{ is absolutely irreducible.} \qquad (\mathrm{AI}_H)$$

In this subsection, we study various deformation problems of $\overline{\rho}$ and relations among the universal rings.

We consider a deformation functor $\mathcal{F}_H : CNL_{\mathcal{O}} \to SETS$ given by

$$\mathcal{F}_H(A) = \{\rho : H \to GL_n(A) \mid \rho \equiv \overline{\rho} \mod \mathfrak{m}_A\}/\approx$$

where '$\approx$' is the strict equivalence in $GL_n(A)$. The functor $\mathcal{F}_H$ is representable under (AI_H) by Theorem 2.26 (see [DGH] Theorem 3.3 for another proof). We write (R_H, ρ_H) for the universal couple. Since ρ_G restricted to H is an element in $\mathcal{F}_H(R_H)$, we have an $\mathcal{O}$-algebra homomorphism (called the base-change map) $\alpha : R_H \to R_G$ such that $\alpha\rho_H = \rho_G|_H$.

We would like to determine $\mathrm{Ker}(\alpha)$ and $\mathrm{Im}(\alpha)$ in terms of Δ. We briefly recall the theory of extending representation described in Subsection 4.3.5. By choosing a lift $c_0(\sigma) \in GL_n(\mathcal{O})$ for $\sigma \in G$ such that $c_0(\sigma) \equiv \overline{\rho}(\sigma) \mod \mathfrak{m}_{\mathcal{O}}$, we can define for any $\rho \in \mathcal{F}_G(A)$, $\rho^\sigma(g) = \rho(\sigma g \sigma^{-1})$ and $\rho^{[\sigma]}(g) = c_0(\sigma)^{-1}\rho^\sigma(g)c_0(\sigma)$ in $\mathcal{F}_H(A)$. In this way, Δ acts via $\sigma \mapsto [\sigma]$ on $\mathcal{F}_H$ and R_H. Then as seen in Subsection 4.3.5, we can attach a 2-cocycle b of Δ with values in $\widehat{\mathbf{G}}_m(A)$ to any representation $\rho \in \mathcal{F}_H(A)$ with $\rho^{[\sigma]} \approx \rho$ in the following way. Let us recall the construction of b briefly: First choose a lift $c(\sigma)$ of $\overline{\rho}(\sigma)$ in $GL_n(A)$ for each $\sigma \in G$ such that $c(1) = 1$, $\rho = c(\sigma)^{-1}\rho^\sigma c(\sigma)$ and $c(h\tau) = \rho(h)c(\tau)$ for $h \in H$ and $\tau \in G$. Then we have that $c(\sigma)c(\tau) = b(\sigma,\tau)c(\sigma\tau)$ for a 2-cocycle b of Δ with values in $\widehat{\mathbf{G}}_m(A)$. The cohomology class $[\rho]$ is uniquely determined by ρ independently of the choice of c and is called the *obstruction* class to extending ρ to G. If $[\rho] = 0$, then $b(\sigma,\tau) = \zeta(\sigma)^{-1}\zeta(\tau)^{-1}\zeta(\sigma\tau)$ for a 1-cochain ζ. We then modify c by $c\zeta$. Then c extends the representation ρ to a representation $\pi = c$ of G (Theorem 4.35).

LEMMA 5.32 *Let $\rho \in \mathcal{F}_H(A)$. Assume* ($\mathrm{AI}_H$) *and that n is prime to p and $\rho^{[\sigma]} \approx \rho$ for all $\sigma \in \Delta$. If* $\det(\rho)$ *can be extended to a deformation of* $\det\overline{\rho}$ *(over G) having values in an A-algebra B containing A, then ρ can be extended uniquely to a deformation $\pi : G \to GL_n(B)$ of $\overline{\rho}$ whose determinant coincides with the extension to G of* $\det(\rho)$.

Proof By applying 'det' to c and b, we know that $[\det(\rho)] = [\det(b)] = n[\rho]$. If n is prime to p, the vanishing of $n[\rho]$ in $H^2(\Delta, \widehat{\mathbf{G}}_m(B))$ is equivalent to the vanishing of the obstruction class $[\rho]$. Thus if $\det(\rho)$ extends to G (that is $n[\rho] = 0$), then ρ extends to a representation π of G which has determinant equal to the extension of $\det(\rho)$ prearranged. Since $[\overline{\rho}_H] = 0$, we may assume that π is a deformation of $\overline{\rho}$. We now show the uniqueness of π. We get, out of π, other extensions $\pi \otimes \chi \in \mathcal{F}_G(B)$ for $\chi \in H^1(\Delta, \widehat{\mathbf{G}}_m(B)) = \mathrm{Hom}(\Delta, \widehat{\mathbf{G}}_m(B))$. Conversely, if π and π' are two extensions of ρ in $\mathcal{F}_G(B)$, then for $h \in H$, $\pi'(\sigma)\rho(h)\pi'(\sigma)^{-1} = \pi(\sigma)\rho(h)\pi(\sigma)^{-1}$ and hence $\pi(\sigma)^{-1}\pi'(\sigma)$ commutes with ρ. Then by Lemma 2.12 in Chapter 2, $\chi(\sigma) = \pi(\sigma)^{-1}\pi'(\sigma)$ is a scalar in $\widehat{\mathbf{G}}_m(B)$.

$$\begin{aligned}\chi(\sigma\tau) = \pi(\sigma\tau)^{-1}\pi'(\sigma\tau) &= \pi(\tau)^{-1}\pi(\sigma)^{-1}\pi'(\sigma)\pi'(\tau)\\ &= \pi(\tau)^{-1}\chi(\sigma)\pi'(\tau) = \chi(\sigma)\chi(\tau).\end{aligned}$$

Thus χ is an element in $H^1(\Delta, \widehat{\mathbf{G}}_m(B))$ and $\pi' = \pi \otimes \chi$, which shows that $\det(\pi')$ is equal to $\det(\pi)\chi^n$. If $\det(\pi') = \det(\pi)$, then $\chi^n = 1$. Since χ is of p-power order, if n is prime to p, $\chi = 1$. □

Here is a consequence of the proof of the lemma:

COROLLARY 5.33 *Let $\pi_0 \in \mathcal{F}_G(B)$ be an extension of $\rho \in \mathcal{F}_H(A)$ for an A-algebra B containing A. Then we have*

$$\{\pi_0 \otimes \chi \mid \chi \in \mathrm{Hom}(\Delta, \widehat{\mathbf{G}}_m(B))\} = \{\pi \in \mathcal{F}_G(B) \mid \pi_{|H} = \rho\}.$$

It is easy to see that if $H^2(\Delta, \mathbb{F}) = 0$, then $H^2(\Delta, \widehat{\mathbf{G}}_m(A)) = 0$ for all A in CNL (Exercise 1). Therefore we see that, if $H^2(\Delta, \mathbb{F}) = 0$,

$$\mathcal{F}_H^{\Delta}(A) = H^0(\Delta, \mathcal{F}_H(A)) \cong \mathcal{F}_G(A)/\widehat{\Delta}(A) \text{ for } \widehat{\Delta}(A) = \mathrm{Hom}(\Delta, \widehat{\mathbf{G}}_m(A)). \quad (*)$$

Here we let $\chi \in \widehat{\Delta}(A)$ act on $\mathcal{F}_G(A)$ via $\pi \mapsto \pi \otimes \chi$. Suppose that $\mathcal{F}_H^{\Delta}$ is represented by a universal couple $(R_{H,\Delta}, \rho_{H,\Delta})$ and $[\rho_{H,\Delta}] = 0$ in $H^2(\Delta, \widehat{\mathbf{G}}_m(R_{H,\Delta}))$. Then for each $\rho \in \mathcal{F}_H^{\Delta}(A)$, we have $\varphi : R_{H,\Delta} \to A$ such that $\varphi\rho_{H,\Delta} \approx \rho$. Then $\varphi_*[\rho_{H,\Delta}] = [\rho]$ and therefore, $[\rho] = 0$ in $H^2(\Delta, \widehat{\mathbf{G}}_m(A))$. This again shows (*).

Under (AI_H), by Proposition 2.13, $\mathcal{F}_H^{\Delta}(A) \ni \rho \mapsto \mathrm{Tr}(\rho)$ sends representations ρ to Δ-invariant pseudo-representations which are deformations of $\mathrm{Tr}(\overline{\rho})$, bijectively. In the same way as in the proof of Theorem 2.26, it is easy to check that this deformation functor of pseudo-representations is representable (Exercise 2). Then by the unicity-lemma Subsection 4.1.3, the subfunctor $\mathcal{F}_H^{\Delta}$ is represented by a residue ring $R_H/\mathfrak{a}$ for an ideal $\mathfrak{a}$. Again by the unicity-lemma, $\mathcal{F}_H^{\Delta}$ is represented by $R_{H,\Delta} = R_H/\Sigma_{\sigma\in\Delta}R_H([\sigma]-1)R_H$ (Exercise 3).

PROPOSITION 5.34 *Suppose (AI_H). Then $\mathcal{F}_H^{\Delta}$ is represented by $(R_{H,\Delta}, \rho_{H,\Delta})$ for $R_{H,\Delta} = R_H/\mathfrak{a}$ with $\mathfrak{a} = \Sigma_{\sigma\in\Delta}R_H([\sigma]-1)R_H$ and $\rho_{H,\Delta} = \rho_H \mod \mathfrak{a}$. If either $[\rho_{H,\Delta}] = 0$ in $H^2(\Delta, \widehat{\mathbf{G}}_m(R_{H,\Delta}))$ or $H^2(\Delta, \mathbb{F}) = 0$, then we have $\mathcal{F}_G/\widehat{\Delta} \cong \mathcal{F}_H^{\Delta}$ via $\pi \mapsto \pi|_H$.*

We now consider the following subfunctor $\mathcal{F}_{G,H}$ of $\mathcal{F}_H$ given by

$$\mathcal{F}_{G,H}(A) = \{\rho|_H \in \mathcal{F}_H(A) | \rho \in \mathcal{F}_G(B) \text{ for a flat } A\text{-algebra } B \text{ in } CNL_{\mathcal{O}}\}.$$

algebra B may not be unique and depends on A. Let us check that $\mathcal{F}_{G,H}$ is really a functor. If $\varphi : A \to A'$ is a morphism in CNL and $\rho|_H \in \mathcal{F}_{G,H}(A)$ with $\rho \in \mathcal{F}_G(B)$, B being flat over A, then $A'\widehat{\otimes}_A B$ is a flat A'-algebra in CNL. Then $(\varphi \otimes id)\rho \in \mathcal{F}_G(A'\widehat{\otimes}_A B)$ such that $\varphi(\rho|_H) = ((\varphi \otimes id)\rho)|_H$. Thus $\mathcal{F}_H(\varphi)$ takes $\mathcal{F}_{G,H}(A)$ into $\mathcal{F}_{G,H}(A')$, which shows that $\mathcal{F}_{G,H}$ is a well-defined functor. For each $\rho \in \mathcal{F}_{G,H}(A)$, we have an extension $\rho \in \mathcal{F}_G(B)$. By the universality of (R_G, ρ_G), we have $\varphi : R_G \to B$ such that $\varphi\rho_G = \rho$. Then $\rho|_H = (\varphi\rho_G)|_H = \varphi(\rho_G|_H) = \varphi\alpha\rho_H$. This shows that

$\varphi\alpha$ is uniquely determined by $\rho|_H \in \mathcal{F}_{G,H}(A)$. Therefore φ restricted to $\mathrm{Im}(\alpha)$ has values in A and is uniquely determined by $\rho|_H \in \mathcal{F}_{G,H}(A)$. Conversely, supposing that $[\alpha\rho_H] = 0$ in $H^2(\Delta, \widehat{\mathbf{G}}_m(B))$ for a flat extension B of $\mathrm{Im}(\alpha)$ in CNL, for a given $\varphi : \mathrm{Im}(\alpha) \to A$ which is a morphism in CNL, we shall show that $\rho = \varphi\alpha\rho_H$ is an element of $\mathcal{F}_{G,H}(A)$. In any case, $\alpha\rho_H$ can be extended to G as an element in $\mathcal{F}_G(B)$, and hence $\alpha\rho_H \in \mathcal{F}_{G,H}(\mathrm{Im}(\alpha))$. We note that ρ can be extended to G because $[\varphi\alpha\rho_H] = \varphi_*[\alpha\rho_H]$ which vanishes in $H^2(\Delta, \widehat{\mathbf{G}}_m(B'))$ for $B' = B\widehat{\otimes}_{\mathrm{Im}(\alpha),\varphi}A$. Thus $\rho \in \mathcal{F}_{G,H}(A)$, and $\mathcal{F}_{G,H}$ is represented by $(\mathrm{Im}(\alpha), \alpha\rho_H)$ as long as $[\alpha\rho_H] = 0$ in $H^2(\Delta, \widehat{\mathbf{G}}_m(B))$ for a flat extension B of $\mathrm{Im}(\alpha)$ in CNL.

We have the following inclusions of functors: $\mathcal{F}_G/\widehat{\Delta} \hookrightarrow \mathcal{F}_{G,H} \subset \mathcal{F}_H^{\Delta} \subset \mathcal{F}_H$, the first map being given by $\rho \mapsto \rho|_H$, which is injective by Corollary 5.33. The functor $\mathcal{F}_H^{\Delta}$ is represented by $R_H/\mathfrak{a}$ for $\mathfrak{a} = \Sigma_{\sigma\in\Delta}R_H([\sigma]-1)R_H$. Because of the above inclusion, if $[\alpha\rho_H] = 0$ in $H^2(\Delta, \widehat{\mathbf{G}}_m(B))$ for a flat extension B of $\mathrm{Im}(\alpha)$ in CNL, the ring $\mathrm{Im}(\alpha)$ is a surjective image of $R_H/\mathfrak{a} = R_{H,\Delta}$. If $[\rho_{H,\Delta}] = 0$ (for $\rho_{H,\Delta} = \rho_H$ mod $\mathfrak{a}$) in $H^2(\Delta, \widehat{\mathbf{G}}_m(B'))$ for a flat extension B' of $R_{H,\Delta}$ in CLN, then $\rho_{H,\Delta} \in \mathcal{F}_{G,H}(R_{H,\Delta})$ and thus $\mathcal{F}_H^{\Delta} = \mathcal{F}_{G,H}$.

PROPOSITION 5.35 *Assume* (AI_H) *and that* $[\alpha\rho_H] = 0$ *in* $H^2(\Delta, \widehat{\mathbf{G}}_m(B))$ *for a flat extension* B *of* $\mathrm{Im}(\alpha)$ *in* CNL. *Then* $\mathcal{F}_{G,H}$ *is represented by* $(\mathrm{Im}(\alpha), \alpha\rho_H)$. *If further* $[\rho_{H,\Delta}] = 0$ *in* $H^2(\Delta, \widehat{\mathbf{G}}_m(B'))$ *for a flat extension* B' *of* $R_{H,\Delta}$, *then we have* $\mathcal{F}_{G,H} = \mathcal{F}_H^{\Delta}$.

The character $\det(\rho_H)$ induces an $\mathcal{O}$-algebra homomorphism: $\mathcal{O}[[H^{ab}]] \to R_H$ for the maximal continuous abelian quotient H^{ab} of H. We write its image as Λ_H and simply write Λ for Λ_G. Since the map $\mathcal{O}[[H^{ab}]] \to R_H$ factors through the local ring $\mathcal{O}[[H_p^{ab}]]$ in $CNL_{\mathcal{O}}$ for the maximal p-profinite quotient H_p^{ab} of H^{ab}, Λ_H is an object in $CNL_{\mathcal{O}}$. Thus we have a character $\det(\rho_H) : H \to \Lambda_H^{\times}$. We consider the category CNL_{Λ_H} of complete noetherian local Λ_H-algebras with residue field $\mathbb{F}$. We consider the functor $\mathcal{F}_{\Lambda_H,H} : CNL_{\Lambda_H} \to SETS$ given by

$$\mathcal{F}_{\Lambda_H,H}(A) = \{\rho : H \to GL_n(A) \mid \rho \equiv \overline{\rho} \mod \mathfrak{m}_A \text{ and } \det(\rho) = \det(\rho_H)\}/\approx.$$

Pick $\rho : H \to GL_n(A) \in \mathcal{F}_{\Lambda_H,H}(A)$. Then regarding A as an $\mathcal{O}$-algebra naturally, we know that $\rho \in \mathcal{F}_H(A)$. Thus there is a unique morphism $\varphi : R_H \to A$ such that $\varphi\rho_H \approx \rho$. Then $\varphi(\det(\rho_H)) = \det(\rho)$, and φ is a morphism in CNL_{Λ_H}. Therefore (R_H, ρ_H) represents $\mathcal{F}_{\Lambda_H,H}$. We consider another functor on CNL_Λ similar to $\mathscr{F}_{G,H}$:

$$\mathscr{F}_{\Lambda,G,H}(A) = \{\rho|_H \in \mathscr{F}_H(A) | \rho \in \mathscr{F}_{\Lambda,G}(B) \text{ for a flat } A\text{-algebra } B \text{ in } CNL_\Lambda\}.$$

Take $\rho \in \mathscr{F}_{\Lambda,G,H}(A)$ such that $\rho = \rho'|_H$ for $\rho' \in \mathscr{F}_{\Lambda,G}(B)$. Then there exists a unique $\varphi : R_G \to B$ with $\det(\rho') = \varphi(\det(\rho_G))$. Since the Λ-algebra structure of B is given by $\det(\rho')$, φ induces a Λ-algebra homomorphism of $\mathrm{Im}(\alpha)\Lambda$ into B for the algebra $\mathrm{Im}(\alpha)\Lambda$ generated by $\mathrm{Im}(\alpha)$ and Λ. From $\rho = (\varphi\rho_G)|_H = \varphi(\rho_G|_H) = \varphi\alpha\rho_H$, we see that the Λ-algebra homomorphism φ restricted to $\mathrm{Im}(\alpha)\Lambda$ is uniquely determined by ρ. Supposing that $[\alpha\rho_H]$ vanishes in $H^2(\Delta, \widehat{\mathbf{G}}_m(B))$ for a flat extension B of $\mathrm{Im}(\alpha)$, we know that $[\alpha\rho_H]$ vanishes in the cohomology group $H^2(\Delta, \widehat{\mathbf{G}}_m(\mathrm{Im}(\alpha)\Lambda \otimes_{\mathrm{Im}(\alpha)} B))$. For any morphism $\varphi : \mathrm{Im}(\alpha)\Lambda \to A$ in CNL_Λ, $[\varphi\alpha\rho_H] = \varphi_*[\alpha\rho_H]$ vanishes in $H^2(\Delta, \widehat{\mathbf{G}}_m(B'))$ for $B' = A \otimes_{\mathrm{Im}(\alpha)} B$ which is flat over A. Thus we have an extension π of ρ to G having values in B'. Suppose further that n is prime to p. In this case, as already remarked, we can always extend ρ without extending A and without assuming the vanishing of $[\alpha\rho_H]$, because $\det(\rho)$ can be extended to G by $\varphi \circ \det(\rho_G)$. Thus we know that:

$$\mathscr{F}_{\Lambda,G,H}(A) = \{\rho|_H \in \mathscr{F}_H(A) | \rho \in \mathscr{F}_{\Lambda,G}(A)\}.$$

Since $\det(\rho)$ can be extended to G without changing A, there is a unique extension of π with values in $GL_n(A)$ such that $\det(\pi) = \iota \circ (\det(\rho_G))$, which implies that $\pi \in \mathscr{F}_{\Lambda,G}(A)$ and hence $\pi|_H \in \mathscr{F}_{\Lambda,G,H}(A)$. Thus $\mathscr{F}_{\Lambda,G,H}$ is represented by $(\mathrm{Im}(\alpha)\Lambda, \alpha\rho_H)$ if n is prime to p. We consider the morphism of functors: $\mathscr{F}_{\Lambda,G} \to \mathscr{F}_{\Lambda,G,H}$ sending π to $\pi|_H$. As we have already remarked, the extension of $\rho \in \mathscr{F}_{\Lambda,G,H}(A)$ to $\pi \in \mathscr{F}_\Lambda(A)$ is unique if n is prime to p. Thus in this case, the morphism of functors is an isomorphism of functors. Therefore $(R_G, \rho_G) \cong (\mathrm{Im}(\alpha)\Lambda, \alpha\rho_H)$. Thus we get

THEOREM 5.36 *Suppose* (AI_H) *and that either n is prime to p or $[\alpha\rho_H]$ vanishes in $H^2(\Delta, \widehat{\mathbf{G}}_m(B))$ for a flat extension B of* $\mathrm{Im}(\alpha)$. *Then $\mathscr{F}_{\Lambda,G,H}$ is representable by* $(\mathrm{Im}(\alpha)\Lambda_G, \alpha\rho_H)$. *Moreover, if n is prime to p, we have the equality $R_G = \mathrm{Im}(\alpha)\Lambda_G$.*

Since α restricted to Λ_H coincides with the algebra homomorphism induced by the inclusion $H \subset G$, $\alpha(\Lambda_H) \subset \Lambda$. We put $R' = \mathrm{Im}(\alpha) \otimes_{\Lambda_H} \Lambda$. By definition, the character $1 \otimes \det(\rho_G)$ of G coincides on H with $(\alpha \circ \det(\rho_H)) \otimes 1$ in R'. Thus $\alpha\rho_H$ can be extended uniquely to $\rho'_G : G \to GL_n(R')$ such that $\det(\rho'_G) = 1 \otimes \det(\rho_G)$ if n is prime to p. Thus we have a natural map $\imath : R_G \to R'$ such that $\imath\rho_G = \rho'_G$. Since R_G is an algebra over Λ and $\mathrm{Im}(\alpha)$, it is an algebra over R'. Thus we have the structural morphism $\imath' : R' \to R_G$. By Theorem 5.36, $\imath'$ is surjective. By definition, $\imath\alpha\rho_H = \imath\rho_H|_H = \imath\rho'_G|_H = \alpha\rho_H \otimes 1$ and $\imath\det(\rho_G) = \det(\rho'_G) = 1 \otimes \det(\rho_G)$. Thus $\imath'\imath\alpha\rho_H = \imath'(\alpha\rho_H \otimes 1) = \alpha\rho_H$ and $\imath'\imath\det(\rho_G) = \imath'(1 \otimes \det(\rho_G)) = \det(\rho_G)$. Thus $\imath'\imath$ is the identity on Λ and $\mathrm{Im}(\alpha)$, and hence $\imath'\imath = id$. Similarly, $\imath\imath'\rho'_g = \imath\rho_G = \rho'_G$. This shows that

$$\imath\imath'(\alpha\rho_H \otimes 1) = \imath(\alpha\rho_H) = (\alpha\rho_H \otimes 1) \text{ and}$$
$$\imath\imath'(1 \otimes \det(\rho_G)) = \imath(\det(\rho_G)) = 1 \otimes \det(\rho_G).$$

Thus $\imath\imath'$ is again the identity on $\mathrm{Im}(\alpha) \otimes 1$ and $1 \otimes \Lambda$, and $\imath\imath' = id$. Let X_p (resp. $X^{(p)}$) indicate the maximal p-profinite (resp. prime-to-p profinite) quotient of a profinite group X. Write ω for the restriction of $\det(\rho_G)$ to $(G^{ab})^{(p)}$. Define $\kappa : G^{ab} \to \mathcal{O}[[G_p^{ab}]]^\times$ by $\kappa(g) = \omega(g)[g_p]$ for the projection g_p of g into G_p^{ab}, where $[x]$ denotes the group element of $x \in G_p^{ab}$ in the group algebra. Assuming that $\mathbb{F}$ is big enough to contain all g-th roots of unity for the order g of $\mathrm{Im}(\omega)$, we can perform the same argument replacing $(\Lambda_H, \Lambda_G, \det(\rho_G))$ by $(\mathcal{O}[[H_p^{ab}]], \mathcal{O}[[G_p^{ab}]], 1 \otimes \kappa)$. Thus we get

Corollary 5.37 *Assume* (AI_H) *and that n is prime to p. Then we have*

$$(R_G, \rho_G) \cong (\mathrm{Im}(\alpha) \otimes_{\Lambda_H} \Lambda_G, \alpha\rho_H \otimes \det(\rho_G))$$
$$\cong (\mathrm{Im}(\alpha) \otimes_{\mathcal{O}[[H_p^{ab}]]} \mathcal{O}[[G_p^{ab}]], \alpha\rho_H \otimes \kappa).$$

In particular, R_G is flat over $\mathrm{Im}(\alpha)$.

Exercises

(1) Show that $H^2(\Delta, \widehat{\mathbb{G}}_m(A)) = 0$ for all A in $CNL_{\mathcal{O}}$ if $H^2(\Delta, \mathbb{F}) = 0$. Hint: $\widehat{\mathbb{G}}_m(A)$ has a Δ-invariant filtration whose subquotients are isomorphic to $\mathbb{F}$.
(2) Show that $\mathcal{F}_H^\Delta$ is representable in $CNL_{\mathcal{O}}$.
(3) Show that $\mathcal{F}_H^\Delta$ is represented by $R_{H,\Delta}$.

5.4.2 Nearly ordinary deformations

Hereafter we assume that $n = 2$. We would like to describe nearly p-ordinary Galois deformations. Let us first introduce some notation: Let $S = S_G$ be a finite set of closed subgroups of G. For each $D \in S$, let $S(D)$ be a complete representative set for H-conjugacy classes of

$\{gDg^{-1} \cap H \mid g \in G\}$. In application, $G = \mathfrak{G}_S$ and D is given by decomposition subgroups of primes in S for a finite set of primes S. For simplicity, we assume that $D \cap H \in S(D)$ always. Then the disjoint union $S_H = \bigsqcup_{D \in S} S(D)$ is a finite set, because $|S(D)| = |H\backslash G/D|$.

Let $V = \mathcal{O}^2$ be rank 2-free $\mathcal{O}$-modules consisting of column vectors. We identify $GL_2(\mathcal{O})$ with the group of $\mathcal{O}$-linear automorphisms $Aut_{\mathcal{O}}(V)$. Then the algebraic group $GL(2)$ defined over $\mathcal{O}$ can be regarded as a covariant functor from $CL_{\mathcal{O}}$ into the category of groups given by $GL_2(A) = Aut_A(V \otimes_{\mathcal{O}} A)$. An algebraic subgroup $B \subset GL(2)$ is called the *Borel subgroup* defined over $\mathcal{O}$ if there exists an $\mathcal{O}$-submodule $W \subset V$ with $V/W \cong \mathcal{O}$ such that

$$B(A) = \{x \in GL_2(A) | x(W(A)) \subset W(A)\},$$

where $W(A) = W \otimes_{\mathcal{O}} A \subset V \otimes_{\mathcal{O}} A = V(A)$. Thus any two Borel subgroups defined over $\mathcal{O}$ are conjugate each other by an element in $GL_2(\mathcal{O})$.

Let $\{B_D\}_{D \in S}$ be a set of Borel subgroup of $GL(2)_{/\mathcal{O}}$ defined over $\mathcal{O}$ indexed by $D \in S$. For each $D' \in S(D)$ such that $D' = H \cap gDg^{-1}$, we define $B_{D'} = c(g)B_D c(g)^{-1}$ for a lift $c(g) \in GL_n(\mathcal{O})$ of $\overline{\rho}(g)$. Now we impose the following additional condition to our deformation problem: We assume that

$$\overline{\rho}(D) \subset B_D(\mathbb{F}) \text{ for each } D \in S_G. \tag{NO}$$

Then we consider the following condition:

(no_H) There exists $g_D \in \widehat{GL}_2(A)$ for each $D \in S_H$ such that

$$g_D \rho(D) g_D^{-1} \subset B_D(A),$$

where $\widehat{GL}_n(A) = 1 + \mathfrak{m}_A M_n(A)$.

We define a subfunctor $\mathcal{F}_X^{n.ord}$ of the functor $\mathcal{F}_X$ by

$$\mathcal{F}_X^{n.ord}(A) = \{\rho \in \mathcal{F}_X(A) \mid \rho \text{ satisfies } (\mathrm{no}_X)\},$$

where X denotes either G or H depending on the group concerned. Then by (NO), (no_X) and our choice of B_D, $\mathcal{F}_X^{n.ord}(\mathbb{F}) = \{\overline{\rho}|_X\} \neq \emptyset$.

For each $D \in S_H$, we have $B_D \subset GL(2)$ fixing rank 1 $\mathcal{O}$-free module $W_D \subset V$. Suppose (no_H) for $\rho \in \mathcal{F}_X^{n.ord}(A)$. Then $\rho(D)$ leaves $g_D^{-1}W_D(A)$ stable. Thus $\rho(d)$ for $d \in D$ induces a scalar multiplication on $g_D^{-1}W(A) \cong W(A)$ and $g_D^{-1}V(A)/g_D^{-1}W(A) \cong V(A)/W(A)$. In other words, $\rho(d)w = \varepsilon_{D,\rho}(d)w$ for $w \in g_D^{-1}W(A)$ and $\rho(d)v = \delta_{D,\rho}(d)v$ for $v \in g_D^{-1}V(A)/g_D^{-1}W(A)$. The maps $\varepsilon_\rho, \rho_\rho : D \to A^\times$ are continuous characters and are, respectively, deformations of $\overline{\varepsilon}_D = \varepsilon_{D,\overline{\rho}}$ and $\overline{\delta}_D = \delta_{D,\overline{\rho}}$. We consider the regularity condition:

$$\overline{\varepsilon}_D \neq \overline{\delta}_D \quad \text{on } D \in S_H. \tag{Rg$_D$}$$

We can prove in exactly the same manner as in the proof of Proposition 3.30 the following fact:

PROPOSITION 5.38 *Assume* (AI$_H$), (NO) *and* (Rg$_H$) *for* $\overline{\rho}$. *Then the functor* $\mathcal{F}_X^{n.ord}$ *is representable by a universal couple* $(R_X^{n.ord}, \varrho_X^{n.ord})$ *in* $CNL_{\mathcal{O}}$.

In the same manner as in the previous subsection, we can check that Δ acts on $\mathcal{F}_H^{n.ord}$ via $\rho \mapsto \rho^{[\sigma]}$. Take $D \in S$ and put $D' = D \cap H \in S(D)$. Since $\overline{\rho}$ is invariant under Δ and $\overline{\rho} \in \mathcal{F}_G^{n.ord}(\mathbb{F})$,

$$\overline{\varepsilon}_{D'}^{[\sigma]} = \overline{\varepsilon}_{D'} \text{ and } \overline{\delta}_{D'}^{[\sigma]} = \overline{\delta}_{D'} \quad \text{for all } \sigma \in D. \tag{Inv}$$

Now suppose that $\rho \in \mathcal{F}_H^{\Delta,n.ord}(A)$ and $[\rho] = 0$ in $H^2(\Delta, \widehat{\mathbb{G}}_m(B))$ for a flat A-algebra B. Then we find an extension $\pi : G \to GL_n(B)$ of ρ. Let $\sigma \in D$ and $D' = H \cap D$. Thus $\pi(\sigma)\rho(d')\pi(\sigma)^{-1} = \rho(\sigma d' \sigma^{-1}) \in g_{D'}^{-1} B_D(A) g_{D'}$ for all $d' \in D'$ and hence

$$\varepsilon_{D',\rho}(d') = \varepsilon_{D',\rho}(\sigma d' \sigma^{-1}) \text{ and } \delta_{D',\rho}(d') = \delta_{D',\rho}(\sigma d' \sigma^{-1}).$$

By taking $d' \in D'$ with $\overline{\varepsilon}_{D'}(d') \neq \overline{\delta}_{D'}(d')$, the above equalities imply that $\pi(\sigma)$ has to be upper triangular (if we take a base of $V(\rho) \otimes_A B$ so that $g_{D'}^{-1} B_D(B) g_{D'}$ is upper triangular). Thus $\pi(D) \subset g_{D'}^{-1} B_D(B) g_{D'}$, and, taking $g_D = g_{D'}$, we confirm that $\pi \in \mathcal{F}_G^{n.ord}(A)$. Since $\mathcal{F}_G^{n.ord}$ is stable under the action of $\widehat{\Delta}$, all the arguments given for $\mathcal{F}_X$ in the previous paragraph are valid for $\mathcal{F}_X^{n.ord}$ for $X = G$ and H. Writing $(R_X^{n.ord}, \rho_X^{n.ord})$ for the universal couple representing $\mathcal{F}_X^{n.ord}$, we obtain

THEOREM 5.39 *Suppose* (AI$_H$), (Rg$_D$) *for all* $D \in S_H$ *and that* n *is prime to* p. *Then we have the equality* $R_G^{n.ord} = \mathrm{Im}(\alpha^{n.ord})\Lambda_G^{n.ord}$, *where* $\alpha^{n.ord} : R_H^{n.ord} \to R_G^{n.ord}$ *is the base-change map given by* $\alpha^{n.ord}\rho_H^{n.ord} \approx \rho_G^{n.ord}|_H$ *and* $\Lambda_G^{n.ord}$ *is the image of* $\mathcal{O}[[G_p^{ab}]]$ *in* $R_G^{n.ord}$. *Moreover, we have*

$$(R_G^{n.ord}, \rho_G^{n.ord}) \cong (\mathrm{Im}(\alpha^{n.ord}) \otimes_{\mathcal{O}[[H_p^{ab}]]} \mathcal{O}[[G_p^{ab}]], \alpha^{n.ord}\rho_H^{n.ord} \otimes \kappa).$$

One can generalize the notion of nearly ordinary representation to $GL(n)$-representations, requiring to have $\rho(D) \subset g_D^{-1} B_D(A) g_D$ for a proper parabolic subgroup $P_D \subset GL(n)$ defined over $\mathcal{O}$. We refer to [DGH] and [H96] Appendix for a general treatment of such deformations.

Exercises

(1) Show that $\mathcal{F}_H^{n.ord}$ is representable under (AI_H) and (Rg_D).
(2) Show that $\pi(\sigma) \in g_{D'}^{-1} P_D(B) g_{D'}$ under $(\mathrm{Rg}_{D'})$.

5.4.3 Ordinary deformations

In this subsection, we continue to assume that $n = 2$ and all B_D are conjugate to the subgroup made up of upper triangular matrices. Fix a normal closed subgroup $I = I_D$ of each $D \in S$. For $D' = gDg^{-1} \cap H \in S(D)$, we put $I_{D'} = gI_D g^{-1} \cap H$. We call $\rho \in \mathcal{F}_X^{n.ord}(A)$ ordinary if ρ satisfies the following conditions:

$$I \subset \mathrm{Ker}(\delta_{D,\rho}) \text{ for every } D \in S_X. \qquad (\mathrm{Ord}_X)$$

We then consider the following subfunctor $\mathcal{F}_X^{ord}$ of $\mathcal{F}_X^{n.ord}$:

$$\mathcal{F}_X^{ord}(A) = \{\rho \in \mathcal{F}_X^{n.ord}(A) \mid \rho \text{ is ordinary}\}.$$

It is easy to see that the functor $\mathcal{F}_X^{ord}$ is representable by $(R_X^{ord}, \rho_X^{ord})$ under (Rg_D) for every $D \in S_X$ (see Proposition 3.30).

Let $\rho \in \mathcal{F}_H^{ord}(A)$. Suppose $[\rho] = 0$ in $H^2(\Delta, \widehat{\mathbf{G}}_m(B))$ for a flat A-algebra B. Then we have at least one extension π of ρ in $\mathcal{F}_G^{n.ord}(B)$. We consider $\delta_{D,\pi} : D \to A^\times$ for $D \in S$. We suppose one of the following two conditions for each $D \in S$:

(Tr_D) $|I_D/I_D \cap H|$ is prime to p;
(Ex_D) Every p-power order character of $I_D/I_D \cap H$ can be extended to a character of Δ having values in a flat extension B' of B so that it is trivial on $I_{D'}$ for all $D' \in S$ different from D.

Under (Tr_D), as a homomorphism of groups, $\delta_{D,\pi}$ restricted to I_D factors through $\overline{\delta}_{D,\rho}$ which is trivial on I. Thus $\delta_{D,\pi}$ is trivial on I_D. We note that $\delta_{D,\pi}$ is of p-power order on $I_D/H \cap I_D$ because $\overline{\delta}_{D,\rho}$ is trivial on I_D and $\delta_{D,\rho}$ is trivial on $I_D \cap H$. Thus we may extend $\delta_{D,\pi}$ to a character η of Δ congruent to 1 modulo $\mathfrak{m}_{B'}$. Then we twists π by η^{-1}, getting an extension $\pi' = \pi \otimes \eta^{-1}$ such that $\delta'_{D,\pi}$ is trivial on I_D. Repeating this process for the D's satisfying (Ex_D), we find an extension $\pi \in \mathcal{F}_G^{ord}(B)$ for a flat extension B of A. We now consider

$$\mathcal{F}_{G,H}^{ord}(A) = \{\rho|_H \in \mathcal{F}_H^{ord}(A) | \rho \in \mathcal{F}_G^{ord}(B) \text{ for a flat extension } B \text{ of } A\}.$$

In the same manner as in Subsection 5.4.1, if either $p > 2 = n$ or $[\alpha^{ord} \rho_H^{ord}] = 0$ in $H^2(\Delta, \widehat{\mathbf{G}}_m(B))$ for a flat extension B of $\mathrm{Im}(\alpha^{ord})$ in

$CNL_{\mathcal{O}}$, we know that $\mathcal{F}^{ord}_{G,H}$ is represented by $(\mathrm{Im}(\alpha^{ord}), \alpha^{ord}\rho^{ord}_H)$, where $\alpha^{ord} : R^{ord}_H \to R^{ord}_G$ is an $\mathcal{O}$-algebra homomorphism given by $\alpha^{ord}\rho^{ord}_H \approx \rho^{ord}_G|_H$.

Let $\rho \in \mathcal{F}^{ord}_{G,H}(A)$ and π be its extension in $\mathcal{F}^{ord}_G(B)$ for a flat A-algebra B in $CNL_{\mathcal{O}}$. The character $\det(\pi)$ is uniquely determined by ρ on the subgroup of G^{ab}_p generated by all $I_{D,p}$, because another choice is $\pi \otimes \chi$ for a character χ of Δ and $(\delta)_{D,\pi\otimes\chi} = \chi$ on $I_{D,p}$. If G^{ab}_p is generated by the $I_{D,p}$'s and H_p, $\det(\pi)$ is uniquely determined by ρ. Thus assuming that $p > 2$, π itself is uniquely determined by ρ. Therefore the morphism of functors: $\mathcal{F}^{ord}_G \to \mathcal{F}^{ord}_{G,H}$ given by $\rho \mapsto \rho|_H$ identifies $\mathcal{F}^{ord}_G$ with a subfunctor of $\mathcal{F}^{ord}_{G,H}$, inducing a surjective $\mathcal{O}$-algebra homomorphism $\beta : \mathrm{Im}(\alpha^{ord}) \to R^{ord}_G$ such that $\rho^{ord}_G|_H = \beta\alpha\rho^{ord}_H$. Since $\rho^{ord}_G|_H = \alpha\rho^{ord}_H$, β is the identity on $\mathrm{Im}(\alpha^{ord})$, and we conclude that $\mathrm{Im}(\alpha^{ord}) = R^{ord}_G$. This implies

Theorem 5.40 *Suppose that $n = 2$ and $p > 2$. Suppose (AI_H), (Rg_D) for $D \in S_H$ and either (Tr_D) or (Ex_D) for each $D \in S$. Suppose further that the $I_{D,p}$'s for all $D \in S$ and H_p generate G^{ab}_p. Then we have $\mathrm{Im}(\alpha^{ord}) = R^{ord}_G$. In particular, for any deformation $\rho \in \mathcal{F}^{ord}_{G,H}(A)$, there is a unique extension $\pi \in \mathcal{F}^{ord}_G(A)$ such that $\pi|_H = \rho$. If further $[\rho^{\Delta,ord}_H] = 0$ in $H^2(\Delta, \widehat{\mathbf{G}}_m(B))$ for a flat extension B of $R^{ord}_{H,\Delta}$, then*

$$R^{ord}_{H,\Delta} \cong \mathrm{Im}(\alpha^{ord}) = R^{ord}_G,$$

where $R^{ord}_{H,\Delta} = R^{ord}_H / \Sigma_{\sigma\in\Delta} R^{ord}_H([\sigma]-1)R^{ord}_H$.

5.4.4 Deformations with fixed determinant

We take a character $\chi : G \to \mathcal{O}^\times$ such that $\chi \equiv \det(\overline{\rho}) \mod \mathfrak{m}_{\mathcal{O}}$. We then define

$$\mathcal{F}^{\chi,?}_X(A) = \left\{\rho \in \mathcal{F}^?_X(A) | \det(\rho) = \chi|_X\right\}.$$

Supposing the representability of $\mathcal{F}^?_X$, it is easy to check that $\mathcal{F}^{\chi,?}_X$ is representable. Since the determinant is already fixed and can be extended to G, by the argument in the previous subsections shows that if n is prime to p,

$$\mathcal{F}^{\chi,?,\Delta}_H = \mathcal{F}^{\chi,?}_{G,H} = \mathcal{F}^\chi_G.$$

Write $(R^{\chi,?}_X, \rho^{\chi,?}_X)$ for the universal couple representing $\mathcal{F}^{\chi,?}_X$ and define $\alpha^{\chi,?} : R^{\chi,?}_H \to R^{\chi,?}_G$ so that $\alpha^{\chi,?}\rho^{\chi,}_H \approx \rho^{\chi,?}_G$. Then we have

PROPOSITION 5.41 *Assume* (AI_H), (Rg_D) *for* $D \in S_H$ *and that n is prime to p. Then we have*

$$R_H^{\chi,?}/\Sigma_{\sigma\in\Delta}R_H^{\chi,?}([\sigma]-1)R_H^{\chi,?} = R_{G,H}^{\chi,?} \cong \mathrm{Im}(\alpha^{\chi,?}) = R_G^{\chi,?},$$

where $R_G^{\chi,?}$ *is either* R_G^{χ} *or* $R_G^{\chi,n.ord}$.

The above proposition is actually valid without assuming $n = 2$ when we do not impose near-ordinarity or ordinarity. One can generalize the notion of near-ordinarity to general $n \geq 2$, and then the analogous fact also holds (see [H96] Appendix).

5.5 Base Change of Deformation Rings

We now apply the results obtained in the previous section to Galois deformations in the following setting: Fix an odd prime p. We take a continuous Galois representation $\overline{\rho}$ of $\mathrm{Gal}(\overline{\mathbb{Q}}/\mathbb{Q})$ into $GL_2(\mathbb{F})$ for a finite field $\mathbb{F}$ of characteristic p. Since $\overline{\rho}$ is continuous, it factors through the Galois group $\mathfrak{G} = \mathfrak{G}_S$ of the maximal extension of $\mathbb{Q}$ unramified outside a finite set of primes S. In this book, for simplicity, we take $S = \{p, \infty\}$, although our ideas certainly work well in a more general setting. Let $\mathfrak{H}$ be a closed normal subgroup of $\mathfrak{G}$, and $\Delta = \mathfrak{G}/\mathfrak{H} = \mathrm{Gal}(F/\mathbb{Q})$. We fix a valuation ring $\mathcal{O}$ finite flat over $\mathbb{Z}_p$ with residue field $\mathbb{F}$ and consider the category $CNL = CNL_{\mathcal{O}}$ of complete noetherian local $\mathcal{O}$-algebras with residue field $\mathbb{F}$.

5.5.1 *Various deformation rings*

A deformation of $\overline{\rho}|_{\mathfrak{H}}$ is a continuous representation $\rho : \mathfrak{H} \to GL_2(A)$ for an object A of CNL such that $\rho \bmod \mathfrak{m}_A = \overline{\rho}$. We call a deformation ρ nearly p-ordinary if for a decomposition subgroup $D_{\mathfrak{p}}$ of $\mathfrak{H}$ at each p-adic place $\mathfrak{p}$, ρ restricted to $D_{\mathfrak{p}}$ is isomorphic to an upper triangular representation. Thus we have two characters $\varepsilon_{D_{\mathfrak{p}},\rho}$ and $\delta_{D_{\mathfrak{p}},\rho}$ of $D_{\mathfrak{p}}$ realized as diagonal entries. We then consider the following two deformation functors $\mathcal{F} = \mathcal{F}_F : CNL \to SETS$ given by

$$\mathcal{F}_F(A) = \{\rho : \mathfrak{H} \to GL_2(A) \ \text{ is a deformation of } \overline{\rho}\}/\approx,$$
$$\mathcal{F}_F^{n.ord}(A) = \{\rho \in \mathcal{F}_F(A) \text{ is nearly } p\text{-ordinary}\}.$$

It has been shown in Theorem 2.26 that $\mathcal{F}_F$ is representable in $CL_{\mathcal{O}}$ under the following condition:

$$\overline{\rho} \text{ restricted to } \mathfrak{H} \text{ is absolutely irreducible.} \qquad (\mathrm{AI}_F)$$

If, further, $[F:\mathbb{Q}]$ is finite (that is, $\mathfrak{H}$ is open in $\mathfrak{G}$), the group satisfies (Φ) and hence, the functor is representable in $CNL_{\mathcal{O}}$ (see Proposition 2.30). In addition to the above condition, to assure the representability of $\mathcal{F}_F^{n.ord}$, we need to assume

$$(\mathrm{Rg}_F) \qquad \overline{\varepsilon}_{D_{\mathfrak{p}}} = \varepsilon_{D_{\mathfrak{p}},\overline{\rho}} \text{ and } \overline{\delta}_{D_{\mathfrak{p}}} = \delta_{D_{\mathfrak{p}},\overline{\rho}} \text{ are distinct for each } \mathfrak{p}.$$

When representable, write (R_F, ϱ_F) (resp. $(R_F^{n.ord}, \varrho_F^{n.ord})$) for the universal couple representing $\mathcal{F}_F$ (resp. $\mathcal{F}_F^{n.ord}$). When we consider a deformation problem with restriction '?', (for example ? = *n.ord*), we write $(R_F^?, \varrho_F^?)$ for the universal couple with the condition '?'. We list here two more restrictions we would like to study: We call a nearly p-ordinary deformation ρ p-ordinary if $\delta_{D,\rho}$ is unramified for every decomposition subgroup D of $\mathfrak{H}$ over p. For a given character $\chi : \mathfrak{G} \to \mathcal{O}^\times$, we say that a deformation ρ has fixed determinant χ if $\det \rho = \chi$ in $A^\times$. Then we define the following subfunctors of $\mathcal{F}_F$:

$$\mathcal{F}_F^{ord}(A) = \{\rho \in \mathcal{F}_F^{n.ord}(A) | \rho \text{ is } p\text{-ordinary}\}$$
$$\mathcal{F}_F^{\chi}(A) = \{\rho \in \mathcal{F}_F(A) | \det(\rho) = \chi\}$$
$$\mathcal{F}_F^{\chi,n.ord}(A) = \mathcal{F}_F^{\chi}(A) \cap \mathcal{F}_F^{n.ord}(A), \quad \mathcal{F}_F^{\chi,ord}(A) = \mathcal{F}_F^{\chi}(A) \cap \mathcal{F}_F^{ord}(A).$$

It is easy to check that the above subfunctors of $\mathcal{F}_F^{n.ord}$ are representable under (AI_F) and (Rg_F), and $\mathcal{F}_F^{\chi}$ is representable under (AI_F) (cf. Proposition 3.30).

For the moment, we assume that $[F:\mathbb{Q}] < \infty$. Let $\mathfrak{H}^{ab} = \mathfrak{H}/\overline{(\mathfrak{H},\mathfrak{H})}$ be the maximal (*continuous*) abelian quotient. We write $\mathfrak{H}_p^{ab}$ for the maximal p-profinite quotient of $\mathfrak{H}^{ab}$. Thus $\mathfrak{H}^{ab} = \mathfrak{H}_p^{ab} \times \mathfrak{H}_{ab}^{(p)}$, and by class field theory, $\mathfrak{H}_p^{ab} \cong \mathbb{Z}_p^d \times \mu$ for a finite p-group μ, where d is an integer with $1 \leq d < [F:\mathbb{Q}]$. Then as seen in Theorem 2.21, the functor $\mathcal{F}_{F,\det(\overline{\rho})}$ obtained by replacing $\overline{\rho}$ by $\det(\overline{\rho})$ is represented by the continuous group algebra $(\mathcal{O}[[\mathfrak{H}_p^{ab}]], \kappa)$ for a suitable character κ with $\kappa(h) = h$ for $h \in \mathfrak{H}_p^{ab}$. Since

$$\det(\varrho_F^?) \in \mathcal{F}_{F,\det(\overline{\rho})}(R_F^?) \cong \mathrm{Hom}_{CNL}(\mathcal{O}[[\mathfrak{H}_p^{ab}]], R_F^?),$$

there is a unique $\mathcal{O}$-algebra homomorphism $\iota^? : \mathcal{O}[[\mathfrak{H}_p^{ab}]] \to R_F^?$ such that $\iota^?\kappa = \det(\varrho_F^?)$. Thus R_F is an $\mathcal{O}[[\mathfrak{H}_p^{ab}]]$-algebra. Similarly, since $\varrho_{\mathbb{Q}}^? \in \mathcal{F}_{\mathbb{Q}}^?(R_{\mathbb{Q}}^?)$, we see that $\varrho_{\mathbb{Q}}^?|_{\mathfrak{H}} \in \mathcal{F}_F^?(R_{\mathbb{Q}}^?)$. Thus there exists a unique $\mathcal{O}$-algebra homomorphism $\alpha^? : R_F^? \to R_{\mathbb{Q}}^?$ such that

$$\alpha^? \circ \varrho_F^? = \varrho_{\mathbb{Q}}^?|_{\mathfrak{H}}.$$

We call $\alpha^?$ the base-change map (of Galois side). We now describe $\mathrm{Im}(\alpha^?)$

and $\mathrm{Ker}(\alpha^?)$ using the result in the previous section. For that, we take a complete representative set Δ' in $\mathfrak{G}$ for $\Delta = \mathfrak{G}/\mathfrak{H}$. We lift $\overline{\rho}(\sigma)$ ($\sigma \in \Delta'$) to an element $c(\sigma) \in GL_n(\mathcal{O})$ so that $c(\sigma) \bmod \mathfrak{m}_{\mathcal{O}} = \overline{\rho}(\sigma)$. Then we let Δ act on $\mathcal{F}_F$ by $\rho^\sigma(g) = c(\sigma)^{-1}\rho(\sigma g \sigma^{-1})c(\sigma)$. As seen in Section 5.4, this is a well-defined functorial action on $\mathcal{F}_F^?$. By universality, Δ acts on $R_F^?$ via $\mathcal{O}$-algebra automorphisms. We consider the following condition:

$$p \text{ totally ramifies in } F/\mathbb{Q}. \qquad \text{(TR)}$$

Thus we have from the results in Section 5.4 the following fact:

THEOREM 5.42 (CONTROL THEOREM) *Let F be a finite Galois extension of $\mathbb{Q}$ (with $\Delta = \mathrm{Gal}(F/\mathbb{Q})$) unramified outside $\{p, \infty\}$. We assume* (AI_F) *and* (Rg_F) *for $\overline{\rho}$.*

(i) *If $? = \emptyset$ or n.ord, suppose either that $H^2(\Delta, \mathbb{F}) = 0$ or that Δ is cyclic. Then we have*

$$R_{\mathbb{Q}}^? \cong \mathrm{Im}(\alpha^?) \otimes_{\mathcal{O}[[\mathfrak{H}_p^{ab}]]} \mathcal{O}[[\mathfrak{G}_p^{ab}]] \quad \textit{and} \quad \mathrm{Ker}(\alpha^?) = \sum_{\sigma \in \Delta} R_F^?(\sigma - 1)R_F^?.$$

(ii) *If $? = \chi$, suppose that p is odd. Then we have*

$$R_{\mathbb{Q}}^\chi \cong R_F^\chi / \sum_{\sigma \in \Delta} R_F^\chi(\sigma - 1)R_F^\chi.$$

(iii) *If $? = ord$, suppose* (TR), *$p > 2$ and either that $H^2(\Delta, \mathbb{F}) = 0$ or Δ is cyclic. Then we have*

$$R_{\mathbb{Q}}^{ord} \cong R_F^{ord} / \sum_{\sigma \in \Delta} R_F^{ord}(\sigma - 1)R_F^{ord}.$$

In all the above cases, $Spec(\mathrm{Im}(\alpha^?))$ is isomorphic to the maximal closed subscheme of $Spec(R_F^?)$ fixed under Δ.

We study the relation among the various subfunctors of $\mathcal{F}_F$. Suppose that p is odd and that $\chi \bmod \mathfrak{m}_{\mathcal{O}} = \det(\overline{\rho})$. Then we have a natural transformation for $? = \emptyset$ or *n.ord*: $\mathcal{F}_{F,\overline{\rho}}^? \to \mathcal{F}_F^{\chi,?} \times \mathcal{F}_{F,\det(\overline{\rho})}$ given by $\rho \mapsto (\rho^\chi, \det(\rho))$, where

$$\rho^\chi = \rho \otimes (\det(\rho)^{-1}\chi)^{1/2}.$$

Note here that $\det(\rho)^{-1}\chi$ is of p-power order with p odd, and hence its square root is uniquely determined. By this remark, we can recover ρ from $(\rho^\chi, \det(\rho))$. Thus we have $\mathcal{F}_{F,\overline{\rho}}^? \cong \mathcal{F}_F^{\chi,?} \times \mathcal{F}_{F,\det(\overline{\rho})}$ and hence

$$R_F^? \cong R_F^{\chi,?} \widehat{\otimes}_{\mathcal{O}} \mathcal{O}[[\mathfrak{H}_p^{ab}]] \cong R_F^{\chi,?}[[\mathfrak{H}_p^{ab}]].$$

When $F = \mathbb{Q}$, the restriction of a character ξ of D to the inertia subgroup I has a unique extension $\xi^{\mathfrak{G}}$ to $\mathfrak{G}$, because the image of I in D^{ab} is naturally isomorphic to $\mathfrak{G}^{ab}$. Then, assuming that $\overline{\rho}$ is p-ordinary, we see that $\rho \mapsto (\rho \otimes (\delta_{D,\rho}^{-1})^{\mathfrak{G}}, (\delta_{D,\rho})^{\mathfrak{G}})$ induces a natural transformation: $F_{\mathbb{Q}}^{n.ord} \cong \mathcal{F}_{\mathbb{Q}}^{ord} \times \mathcal{F}_{\mathbb{Q},(\delta_{D,\overline{\rho}})^{\mathfrak{G}}}$. Thus we get

$$R_{\mathbb{Q}}^{n.ord} \cong R_{\mathbb{Q}}^{ord} \widehat{\otimes}_{\mathcal{O}} \mathcal{O}[[\Gamma]] \cong R_{\mathbb{Q}}^{ord}[[\Gamma]],$$

where we have written Γ for $\mathfrak{G}_p^{ab}$ ($\cong 1 + p\mathbb{Z}_p$ if p is odd) following the tradition in the Iwasawa theory. We summarize the above argument into the following

Proposition 5.43 *Suppose the assumption of Theorem 5.42 depending on the restriction '?'. Suppose that χ mod $\mathfrak{m}_{\mathcal{O}} = \det(\overline{\rho})$. Then we have the following canonical isomorphisms:*

(i) *For $? = \emptyset$ or n.ord,*

$$R_F^? \cong R_F^{\chi,?} \widehat{\otimes}_{\mathcal{O}} \mathcal{O}[[\mathfrak{H}_p^{ab}]] \cong R_F^{\chi,?}[[\mathfrak{H}_p^{ab}]].$$

(ii) *Suppose that $\overline{\rho}$ is p-ordinary. Then*

$$R_{\mathbb{Q}}^{n.ord} \cong R_{\mathbb{Q}}^{ord} \widehat{\otimes}_{\mathcal{O}} \mathcal{O}[[\Gamma]] \cong R_{\mathbb{Q}}^{ord}[[\Gamma]].$$

In particular, we have a canonical isomorphism (if $F = \mathbb{Q}$):

$$R_{\mathbb{Q}}^{ord} \cong R_{\mathbb{Q}}^{\chi}$$

under the assumptions of (i) *and* (ii).

5.6 Hilbert Modular Hecke Algebras

In this section, after a brief description of the theory of Hecke algebras for GL(2) over general number fields, we are going to prove the torsion property of adjoint Selmer group over (infinite) cyclotomic extensions.

5.6.1 *Various Hecke algebras for* $GL(2)$

In this section, we briefly describe how to generalize the definition of Hecke algebras to cohomological cusp forms on $GL(2)$ over number fields. The author hopes to come back to this topic in more detail in a future book. We require a good knowledge of sheaf and group cohomology (see, for example, Appendix of [LFE], [H88a] and [H94b]).

Let Σ be the set of all infinite places of F. We regard Σ as the set of all field embeddings of F into $\mathbb{C}$ modulo complex conjugation on the left.

Let $G = Res_{F/\mathbb{Q}}GL(2)$. Thus for each $\mathbb{Q}$-algebra A, $G(A) = GL_2(A \otimes_{\mathbb{Q}} F)$. In particular, $G(\mathbb{R}) = \prod_{\sigma\in\Sigma} GL_2(F_\sigma)$, where F_σ is the closure of $\sigma(F)$ in $\mathbb{C}$. We write Z (resp. $C_{\infty+}$) for the center of G (resp. the connected component of the standard maximal compact subgroup of $G(\mathbb{R})$). Here the words: 'standard maximal compact' mean that we take $O_2(\mathbb{R})$ for real places and $U_2(\mathbb{C})$ for complex places. Thus $C_{\infty+}$ is the product of copies of $SO_2(\mathbb{R})$ and $U_2(\mathbb{C})$. Then for the identity connected component $G(\mathbb{R})_+$ of $G(\mathbb{R})$, $\mathfrak{Z} = G(\mathbb{R})_+/C_{\infty+}$ is the symmetric space of $G(\mathbb{R})_+$ with the left action of $G(\mathbb{R})_+$. We consider for each open compact subgroup U of $G(\mathbb{A}^{(\infty)})$ the following modular variety

$$X(U) = G(\mathbb{Q})\backslash G(\mathbb{A})_+/UZ(\mathbb{R})C_{\infty+},$$

where $G(\mathbb{A})_+ = G(\mathbb{A}^{(\infty)})G(\mathbb{R})_+$. The space $X(U)$ is a disjoint union of finitely many quotients of the form $\Gamma\backslash\mathfrak{Z}$ for discrete congruence subgroups $\Gamma \subset G(\mathbb{Q})_+ = G(\mathbb{Q})\cap G(\mathbb{R})_+$. In particular, if U is sufficiently small, $X(U)$ is a smooth differential manifold (possibly with several connected components).

Let K be a sufficiently large finite extension of $\mathbb{Q}_p$ containing all conjugates of F over $\mathbb{Q}$. Write $\mathcal{O}$ for the p-adic integer ring of K. Let I be the set of all field embeddings of F into $\overline{\mathbb{Q}}$. We consider $v = \sum_{\sigma\in I} v_\sigma\sigma$ and $n = \sum_{\sigma\in I} n_\sigma\sigma$ in $\mathbb{Z}[I]$. Here $\mathbb{Z}[I]$ is the module of formal linear combinations of elements in I. For each embedding $\sigma : F \hookrightarrow K$, we have the projection $\sigma : G(\mathbb{Q}_p) \to GL_2(K)$. Then we look into the space of the following polynomial representation

$$L(n, v; K) \cong \bigotimes_\sigma (\det(\sigma)^{v_\sigma} \otimes Sym^{\otimes n_\sigma}(\sigma))$$

for the symmetric m-th tensor representation $Sym^{\otimes m}(\sigma)$ of $\sigma : G(\mathbb{Q}_p) \to GL_2(K)$.

More specifically, we take $L(n, v; A)$ to be the space of polynomials in the variables $\{(X_\sigma, Y_\sigma)\}_{\sigma\in I}$, with coefficients in a general $\mathcal{O}$-algebra A, homogeneous of degree n_σ for each pair (X_σ, Y_σ). We let γ act on $P \in L(n, v; A)$ by

$$\gamma P((X_\sigma, Y_\sigma)) = \det(\gamma)^v P((X_\sigma, Y_\sigma)^t\sigma(\gamma)^\iota),$$

where $a^v = \prod_\sigma \sigma(a)^{v_\sigma}$ and $\gamma^\iota = \det(\gamma)\gamma^{-1}$. Then we define a covering space $\mathcal{X}(U)$ of $X(U)$ by

$$G(\mathbb{Q})\backslash(G(\mathbb{A}) \times L(n, v; A))/UZ(\mathbb{R})C_{\infty+},$$

where the action is given by $\gamma(x, P)u = (\gamma x u, u_p^\iota P)$ for $\gamma \in G(\mathbb{Q})$ and $u \in UZ(\mathbb{R})C_{\infty+}$ with $u^\iota = \det(u)u^{-1}$.

We consider the sheaf of locally constant sections of $\mathscr{X}(U)$ over $X(U)$, which we write again as $L(n, v; A)$. For $q = r_1 + r_2$, we define

$$\mathscr{S}(U; A) = H^q_{cusp}(X(U), L(n, v; A)).$$

Note here, as is well known (see [LFE] Appendix), that there is a canonical isomorphism:

$$H^q(X(U), L(n, v; A)) \cong \bigoplus_{\Gamma} H^q(\Gamma, L(n, v; A)),$$

if $X(U) = \bigsqcup_\Gamma \Gamma\backslash\mathscr{Z}$ for discrete subgroups $\Gamma \subset G(\mathbb{Q})$. The *cuspidal* cohomology group is defined as follows: We define

$$H^q_{cusp}(X(U), L(n, v; A)) = H^q_{cusp}(X(U), L(n, v; \mathcal{O})) \otimes_{\mathcal{O}} A.$$

Thus we only need to define $H^q_{cusp}(X(U), L(n, v; \mathcal{O}))$. When $n \neq 0$, we define $H^q_{cusp}(X(U), L(n, v; \mathcal{O}))$ by the image under the natural map of the compactly supported cohomology group $H^q_c(X(U), L(n, v; \mathcal{O}))$ in $H^q(X(U), L(n, v; K))$. When $n = 0$, there is a possibility of having cohomology classes invariant under the action of $G(\mathbb{Q})$ not just Γ (cf. [H88a] Theorem 6.2, [H94b] Section 10 and [Ha]). We write the totality of such classes as $Inv(U) \subset H^q(X(U), L(0, v; K))$. Then $H^q_{cusp}(X(U), L(n, v; \mathcal{O}))$ is defined by the image of the compactly supported cohomology group $H^q_c(X(U), L(n, v; \mathcal{O}))$ in $H^q(X(U), L(n, v; K))/Inv(U)$. By our definition, $H^q_{cusp}(X(U), L(n, v; \mathcal{O}))$ is $\mathcal{O}$-free, and if A is a finite $\mathcal{O}$-algebra, $H^q_{cusp}(X(U), L(n, v; A))$ is A-free of finite rank.

We suppose that $\mathscr{S}(U; K) \neq 0$ for sufficiently small U. Under this assumption, we have $n + 2v = [n + 2v]t$ for $t = \sum_{\sigma \in I} \sigma \in \mathbb{Z}[I]$ and an integer $[n + 2v] \in \mathbb{Z}$. On this space, Hecke operators $T(y) = \left[U \left(\begin{smallmatrix} y & 0 \\ 0 & 1 \end{smallmatrix}\right) U\right]$ naturally act ([H88a] Section 7); further, $\mathbb{T}(y) = y_p^{-v} T(y)$ preserves the image of $H^q_{cusp}(X(U), L(n, v; \mathcal{O}))$. Thus, when $F = \mathbb{Q}$, the weight $(n, 0)$ corresponds to the space of cusp forms of weight $k = n + 2$ (cf. (5.17)). In particular, when $F = \mathbb{Q}$ and if we are dealing with classical space of cusp forms, there is no need to modify Hecke operators.

Now we take

$$U = U_\alpha = \left\{ x \in GL_2(\widehat{O}_F) | x \equiv \left(\begin{smallmatrix} 1 & * \\ 0 & 1 \end{smallmatrix}\right) \mod p^\alpha \right\},$$

where $\widehat{O}_F = \prod_{\mathfrak{q}} O_{F,\mathfrak{q}}$ over all prime ideals $\mathfrak{q} \subset O_F$ for the $\mathfrak{q}$-adic completion $O_{F,\mathfrak{q}}$ of O_F. We consider the $\mathcal{O}$-subalgebra $h_{n,v}(U)$ of $End_K(\mathscr{S}(U; K))$ generated by $\mathbb{T}(y)$ for all integral ideles y, which is an algebra free of finite rank over $\mathcal{O}$. Then we can show that the central action $\langle z\rangle = \nu^{-[n+2v]-1}(z)[UzU]$ of $z \in Z(\mathbb{A}^{(\infty)})$ is naturally contained in $h_{n,v}(U)$ (cf. (3.27) and (3.31) both in Chapter 3), where ν is the p-adic cyclotomic character. The central action factors through the ray class group $Cl_F(p^\infty)$ modulo p^∞: $Cl_F(p^\infty) = (\mathfrak{H}_F)^{ab} \cong F_{\mathbb{A}}^\times/\overline{F^\times \det(U^{(p)})F_{\infty+}^\times}$, where $U^{(p)} = \{u \in U | u_p = 1\}$ and $\mathfrak{H}_F = \mathrm{Gal}(F^{(p,\infty)}/F)$. Thus, the algebra $h_{n,v}(U)$ is an algebra over $\mathcal{O}[[\mathbf{G}]]$ via the character

$$z\begin{pmatrix} y & 0 \\ 0 & 1\end{pmatrix} \longmapsto \langle z\rangle\mathbb{T}(y),$$

where $\mathbf{G} = Cl_F(p^\infty) \times O_{F,p}^\times$ for $O_{F,p} = O_F \otimes_{\mathbb{Z}} \mathbb{Z}_p = \varprojlim_n O_F/p^n O_F$.

We have a commutative diagram for $0 < \alpha < \beta$

$$\begin{array}{ccc} \mathcal{S}(U_\alpha;K) & \xrightarrow{\subset} & \mathcal{S}(U_\beta;K) \\ T\big\downarrow & & \big\downarrow T \\ \mathcal{S}(U_\alpha;K) & \xrightarrow[\subset]{} & \mathcal{S}(U_\beta;K) \end{array}$$

where $T = \mathbb{T}(y)$ and $\langle z\rangle$, respectively.

Thus restriction of Hecke operators gives surjective $\mathcal{O}$-algebra homomorphisms: $h(U_\beta) \to h(U_\alpha)$. We then define

$$\mathbf{h}(p^\infty;\mathcal{O})_{/F} = \varprojlim_\alpha h(U_\alpha).$$

The Hecke algebra $\mathbf{h}(p^\infty;\mathcal{O})_{/F}$ is independent of (n, v) ([H89a]) if F is totally real (see [H94a] for the situation in the general case) and is a compact ring. Then each member $h(U_\alpha)$ of the projective system is $\mathcal{O}$-free of finite rank and hence is compact semi-local. Since the projection maps take $\mathbb{T}(y)$ to $\mathbb{T}(y)$, we have well-defined $\mathbb{T}(y)$ in $\mathbf{h}(p^\infty;\mathcal{O})$. In particular, for a prime element $\varpi_{\mathfrak{q}}$ of $F_{\mathfrak{q}}$ ($\mathfrak{q} \nmid p$), $\mathbb{T}(\varpi_{\mathfrak{q}})$ and $\langle\varpi_{\mathfrak{q}}\rangle$ are independent of the choice of $\varpi_{\mathfrak{q}}$, which we therefore write as $\mathbb{T}(\mathfrak{q})$ and $\langle\mathfrak{q}\rangle$, respectively.

In any case, hereafter we assume that

$$F \text{ is totally real.} \qquad (\mathbb{R})$$

Writing $h(U_\alpha) = \prod_h h$ as a product of local rings h, we write $\mathbb{T}_h(y)$ for the projection of $\mathbb{T}(y)$ to h. Then we put

$$h^{n.ord}(U_\alpha) = \prod_{\mathbb{T}_h(p)\in h^\times} h,$$

and we define the nearly p-ordinary part of $\mathbf{h}(p^\infty;\mathcal{O})_{/F}$ by

$$\mathbf{h}^{n.ord}(p^\infty;\mathcal{O})_{/F} = \varprojlim_\alpha h^{n.ord}(U_\alpha).$$

The algebra $\mathbf{h}^{n.ord}(p^\infty;\mathcal{O})_{/\mathbb{Q}}$ is actually bigger than $\mathbf{h}^{ord}(p^\infty;\mathcal{O})_{/\mathbb{Q}}$ which we have already defined, and we will later relate the two definitions.

We write W for the torsion-free part of $\mathbf{G}$. Then, as seen in [H89a],

(TF) $\mathbf{h}^{n.ord}(p^\infty;\mathcal{O})_{/F}$ *is reduced and is a torsion-free* $\mathcal{O}[[W]]$*-module of finite type.*

Actually, we can prove the freeness of $\mathbf{h}^{n.ord}(p^\infty;\mathcal{O})_{/F}$ over $\mathcal{O}[[W]]$, generalizing the argument in [H86a] (see [H99c]). However, since we do not need the freeness later, we just content ourselves with the weaker result as above.

For each local ring $h = h_{/F}$ of $\mathbf{h}^{n.ord}(p^\infty;\mathcal{O})_{/F}$ and prime ideal P of h, we have a unique semi-simple Galois representation $\rho_P : \mathrm{Gal}(\overline{\mathbb{Q}}/F) \to GL_2(k(P))$ for the residue field $k(P) = h_P/P$ such that

(cnt) ρ_P preserves an h-submodule L of finite type generating $k(P)^2$ over $k(P)$ and is $\mathfrak{m}_h$-adically continuous as a map of $\mathrm{Gal}(\overline{\mathbb{Q}}/F)$ into $Aut_h(L)$;

(unr) ρ_P is unramified outside p;

(n.o) ρ_P is nearly p-ordinary and satisfies $\delta_{D_{\mathfrak{p}},\rho}([y;F_{\mathfrak{p}}]) = \mathbb{T}_h(y)$ for $y \in F_{\mathfrak{p}}$ and the decomposition subgroup $D_{\mathfrak{p}}$ at $\mathfrak{p}$,

where $[y, F_{\mathfrak{p}}]$ is the Artin symbol of local class field theory;

(ch) $$\det(1_2 - \rho_P(Frob_{\mathfrak{q}})X) = 1 - \mathbb{T}(\mathfrak{q})X + \langle\mathfrak{q}\rangle X^2 \in k(P)[X]$$

for all prime $\mathfrak{q}$ outside p. When $\overline{\rho}$ $(= \rho_m$ for the maximal ideal m of $h)$ satisfies (AI_F), we have a continuous $\rho_h : \mathrm{Gal}(\overline{\mathbb{Q}}/F) \to GL_2(h)$ satisfying the above conditions [H89b]. Let $\chi : \mathrm{Gal}(\overline{\mathbb{Q}}/F) \to \mathcal{O}$ be a Galois character such that $\chi \bmod \mathfrak{m}_{\mathcal{O}} = \det(\overline{\rho})$. We consider the Galois representation $\rho_h^\chi = \rho_h \otimes (\det(\rho_h)^{-1}\chi)^{1/2}$ which satisfies $\det(\rho_h) = \chi$. We call $\rho_{\mathbf{h}}^\chi$ the χ-projection of ρ_h. Write $\mathbf{h}^\chi = \mathbf{h}_F^\chi$ for the closed $\mathcal{O}$-subalgebra of h generated by $\mathrm{Tr}(\rho_{\mathbf{h}}^\chi)$.

We can construct the Hecke algebra $\mathbf{h}^\chi$ directly. First we consider the maximal subspace $\mathcal{S}(U_\alpha;K)_\chi$ of $\mathcal{S}(U_\alpha;K)$ on which $\langle z\rangle$ for $z \in Z(\mathbb{A}^{(\infty)})$ acts via χ. Let $Cl_F(p^\infty)$ be the ray class group modulo p^∞. Then we see that

$$Cl_F(p^\infty) \cong Z(\mathbb{A})/\overline{Z(\mathbb{Q})Z(\widehat{\mathbb{Z}}^{(p)})Z(\mathbb{R})_+}$$

for the identity connected component $Z(\mathbb{R})_+$ of $Z(\mathbb{R})$. This shows that $Cl_F(p^\infty)$ acts on $\mathcal{S}(U_\alpha;K)$ via $\langle z\rangle$ and on $\mathcal{S}(U_\alpha;K)_\chi$ via χ. Then writing

$h^{\chi}(U_\alpha)$ for the Hecke algebra generated by $\mathbb{T}(y)$ and $\langle z\rangle$ over $\mathcal{O}$ in $End_K(\mathcal{S}(U_\alpha;K)_\chi)$, we get

$$h^{n.ord,\chi}(p^\infty;\mathcal{O}) = \varprojlim_\alpha h^{n.ord,\chi}(U_\alpha),$$

where $h^{n.ord,\chi}(U_\alpha)$ is the maximal quotient (and hence the maximal direct summand) of $h^{\chi}(U_\alpha)$ in which $\mathbb{T}(p)$ is invertible. When $\chi\nu^{-1}$ is a square $\chi\nu^{-1} = \xi^2$, the above Hecke algebra is basically the Hecke algebra of $Res_{F/\mathbb{Q}}PGL(2)$ twisted by ξ. By duality between the space of modular forms and Hecke algebras ([H91] Section 3 and [SGL] Section 2.6; see also Theorem 3.17 for a version over $\mathbb{Q}$), $h^{n.ord,\chi}(p^\infty;\mathcal{O})$ is a residue algebra of $\mathbf{h}^{n.ord}(p^\infty;\mathcal{O})$. In fact we have a surjective $\mathcal{O}$-algebra homomorphism

$$\mathbf{h}^{n.ord}/\sum_{z\in Cl_F(p^\infty)}(\langle z\rangle - \chi(z))\mathbf{h}^{n.ord} \to h^{n.ord,\chi}.$$

We decompose $Cl_F(p^\infty) = \mu\times\mathbb{W}^+(p)$ so that $\mathbb{W}^+(p)$ is the maximal p-profinite subgroup (and hence also the maximal p-profinite quotient). Thus μ is a finite group of order prime to p. We then write $\mathbf{h}^{n.ord}(p^\infty;\mathcal{O})[\xi]$ for the ξ-eigenspace under the action of μ for a character $\xi:\mu\to\mathcal{O}$. We note here that

$$\mathcal{O}[[Cl_F(p^\infty)]] \cong \mathcal{O}[[\mathbb{W}^+(p)]][\mu] \cong \prod_\xi \mathcal{O}[[\mathbb{W}^+(p)]],$$

where ξ runs over all characters of μ. We now state an analog of Proposition 5.43 in the Hecke-side:

THEOREM 5.44 *Suppose that $p>2$. Then we have the following canonical isomorphism of $\mathcal{O}[[\mathbf{G}]]$-algebras:*

$$\imath:\mathbf{h}^{n.ord}(p^\infty;\mathcal{O})[\chi|_\mu] \cong h^{n.ord,\chi}(p^\infty;\mathcal{O})\widehat{\otimes}_{\mathcal{O}}\mathcal{O}[[\mathbb{W}^+(p)]].$$

In particular, for each local ring $\mathbf{h}$ of $\mathbf{h}^{n.ord}(p^\infty;\mathcal{O})$ with absolutely irreducible residual representation $\overline{\rho}$, $\mathbf{h}^{\chi}$ is a local ring of $h^{n.ord,\chi}(p^\infty;\mathcal{O})$,

$$\mathbf{h}\cong\mathbf{h}^{\chi}\widehat{\otimes}_{\mathcal{O}}\mathcal{O}[[\mathbb{W}^+(p)]],$$

and $\mathcal{O}[[\mathbb{W}^-(p)]]$ is contained in $\mathbf{h}^{\chi}$, where the group $\mathbb{W}^-(p)$ is the maximal p-profinite quotient of $O_{F,p}^{\times}\subset\mathbf{G}$.

We shall prove this theorem at the end of this chapter. Although the theorem is intuitive, its proof is actually long and requires considerable knowledge of Hilbert modular forms. For the moment, the reader is asked to take the theorem on faith.

Note that $Cl_{\mathbb{Q}}(p^\infty) \cong \mathbb{Z}_p^\times$. Then Theorem 5.44 combined with the fact that

$$\mathbf{h}^{n.ord}(p^\infty;\mathcal{O})_{/\mathbb{Q}} \cong \mathbf{h}^{ord}(p^\infty;\mathcal{O})\widehat{\otimes}_{\mathcal{O}}\mathcal{O}[[\mathbb{Z}_p^\times]]$$

given in remark (iv) in the introduction of [H89b] shows the following result, which also follows from Theorem 5.44 and Proposition 5.43 under some additional assumptions:

COROLLARY 5.45 *Let* $\mathbf{h}_{\mathbb{Q}}^{ord}$ *be the local ring of* $\mathbf{h}^{ord}(p^\infty;\mathcal{O})_{/\mathbb{Q}}$ *which is the image of* $\mathbf{h}$. *Let* χ *be a character of* $\mathfrak{G}$ *with values in* $\mathcal{O}^\times$ *such that* $\det \rho_h \equiv \chi$ *mod* $\mathfrak{m}_h$. *Suppose p-ordinarity for* ρ_h *mod* $\mathfrak{m}_h$ *and* $p > 2$. *Then we have a canonical ring isomorphism*

$$\mathbf{h}_{\mathbb{Q}}^{\chi} \cong \mathbf{h}_{\mathbb{Q}}^{ord}.$$

Then Proposition 5.43 and Theorem 5.44 show

COROLLARY 5.46 *Let* $\mathbf{h}_{\mathbb{Q}}^{n.ord}$ *be a local ring of* $\mathbf{h}^{n.ord}(p^\infty;\mathcal{O})_{/\mathbb{Q}}$ *with* $\overline{\rho} = \rho_h$ *mod* $\mathfrak{m}_h$ *satisfying* (AI_κ) *for* $\kappa = \mathbb{Q}(\sqrt{(-1)^{(p-1)/2}p})$ *and* $(\mathrm{Rg}_{\mathbb{Q}})$. *Then we have a canonical ring isomorphism*

$$R_{\mathbb{Q}}^{n.ord} \cong \mathbf{h}_{\mathbb{Q}}^{n.ord}.$$

5.6.2 Automorphic base change

We fix a local component $\mathbf{h} = \mathbf{h}_{\mathbb{Q}}^{n.ord}$ of the Hecke algebra $\mathbf{h}^{n.ord}(p^\infty;\mathcal{O})_{/\mathbb{Q}}$. Let $\overline{\rho}$ be the residual Galois representation associated with the maximal ideal $\mathfrak{m}_h$. We suppose $(\mathrm{AI}_{\mathbb{Q}})$ for $\overline{\rho}$. Then there is a unique continuous Galois representation $\rho_{\mathbf{h}/\mathbb{Q}} : \mathfrak{G} \to GL_2(\mathbf{h})$ satisfying (cnt), (n.o), (unr) and (ch). We now assume that $\Delta = \mathfrak{G}/\mathfrak{H}$ is a solvable group. Let F be the fixed field of $\mathfrak{H}$. We assume $(\mathbb{R})$ and (AI_F) for $\overline{\rho}$.

By the existence of automorphic base change proved by Langlands [BCG], there is a unique local ring $\mathbf{h}_F$ of $\mathbf{h}^{n.ord}(p^\infty;\mathcal{O})_{/F}$ such that the attached Galois representation $\rho_{\mathbf{h}/F} : \mathfrak{H} \to GL_2(\mathbf{h}_F)$ is a deformation of $\overline{\rho}|_{\mathfrak{H}}$. Note that, by (n.o), $\mathbf{h}_F$ is generated by $\mathbb{T}(\mathfrak{q}) = \mathrm{Tr}(\rho_{h/F}(Frob_{\mathfrak{q}}))$ for primes $\mathfrak{q}$ outside any finite set of primes including p. Thus the base change map $\alpha_{F/\mathbb{Q}}^{n.ord} : R_F^{n.ord} \to R_{\mathbb{Q}}^{n.ord}$ induces a unique $\mathcal{O}$-algebra homomorphism $\beta = \beta_{F/\mathbb{Q}}$ of $\mathbf{h}_F$ into $\mathbf{h}_{\mathbb{Q}}$ such that $\beta(\mathbb{T}(\mathfrak{q})) = \mathrm{Tr}(\rho_{\mathbf{h}/\mathbb{Q}}(Frob_{\mathfrak{q}})) \in \mathbf{h}_{\mathbb{Q}}$. By (n.o), we conclude that $\beta(\mathbb{T}(y)) = \mathbb{T}(N_{F/\mathbb{Q}}(y))$ for $y \in F_p$. Then $\beta\rho_{h/F} \approx \rho_{h/\mathbb{Q}}|_{\mathfrak{H}}$ and $\beta(\det(\rho_{h/F})) = \det(\rho_{h/\mathbb{Q}})|_{\mathfrak{H}}$. Now we fix a character $\chi : \mathfrak{G} \to \mathcal{O}^\times$ such that $\det(\overline{\rho}) = \chi$ mod $\mathfrak{m}_{\mathcal{O}}$. Then β induces $\beta^\chi : \mathbf{h}_F^\chi \to \mathbf{h}_{\mathbb{Q}}^\chi$. We now assume that p ramifies totally in $F/\mathbb{Q}$. Write $\delta_F = \delta_{D_{\mathfrak{p}},\rho_{\mathbf{h}/F}}$ for the character of $D_{\mathfrak{p}}$ given by the lower-right-corner diagonal character of $\rho_{h/F}|_{D_{\mathfrak{p}}}$. By definition, for the local Artin symbol $[y, F_{\mathfrak{p}}]$ for $y \in F_{\mathfrak{p}}$, we have $\delta_F([y, F_{\mathfrak{p}}]) = \mathbb{T}_h(y)$. Thus, by local class field theory, we have a commutative diagram

$$\begin{array}{ccc} N : \mathcal{O}[[\mathbb{W}^-(p)]] & \longrightarrow & \mathcal{O}[[1+p\mathbb{Z}_p]] \\ \downarrow & & \downarrow \\ \beta : \mathbf{h}_F^{\chi} & \longrightarrow & \mathbf{h}_{\mathbb{Q}}^{\chi}, \end{array}$$

where N is induced by the norm map: $O_{F,p}^{\times} \to \mathbb{Z}_p^{\times}$. We may extend a little bit the above diagram. Let $D_{\mathfrak{p}}^{ab}$ be the maximal (continuous) abelian quotient of $D_{\mathfrak{p}}$. Then δ_F is a character of $D_{\mathfrak{p}}^{ab}$ and we have an extended commutative diagram:

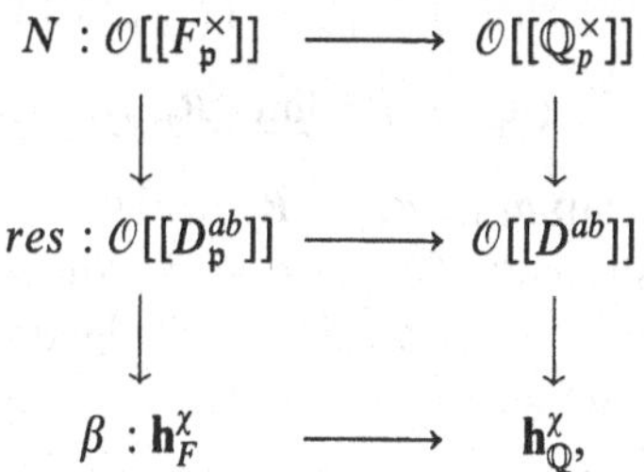

where we have written the decomposition group of $\mathfrak{G}_p$ at p simply as D (which is the unique one containing $D_{\mathfrak{p}}$). Note that the image of δ is generated by $\mathbb{T}(y)$ for $y \in O_{\mathfrak{p}} - \{0\}$ for the integer ring O of F. We take a uniformizer ϖ of $O_{\mathfrak{p}}$ such that $N(\varpi) = p$. Write

$$U = Ker(N : \mathbb{W}^-(p) \to \mathbb{Z}_p^{\times}).$$

Since β takes $\mathcal{O}[[U]]$ onto $\mathcal{O}$ and $\mathbb{T}(p)$ is transcendental in $\mathbf{h}_{\mathbb{Q}}$ over $\mathcal{O}$, we conclude from the above diagram that $\mathcal{O}[[U]][[t]] \hookrightarrow \mathbf{h}_F$, where t goes to $\mathbb{T}(\varpi) - a$ for a unique root of unity $a \in \mathcal{O}$ with $a \equiv \mathbb{T}_{\mathfrak{h}}(\varpi) \mod \mathfrak{m}_{\mathfrak{h}}$. Since the specialization of $\mathbb{T}(p) \in \mathbf{h}_{\mathbb{Q}}^{ord}$ to weight k has complex absolute value $\sqrt{p^{k-1}}$ for all $k > 2$ by Deligne's proof of the Ramanujan conjecture ([D]), $\mathbb{T}(p)$ assumes infinitely many distinct specializations to $\mathcal{O}$ over $Spec(\Lambda)$. Thus the transcendence of $\mathbb{T}(p) \in \mathbf{h}_{\mathbb{Q}}^{\chi}$ over $\mathcal{O}$ follows from the natural isomorphism $\mathbf{h}_{\mathbb{Q}}^{\chi} \cong \mathbf{h}_{\mathbb{Q}}^{ord}$.

THEOREM 5.47 *Suppose* ($\mathbb{R}$) *and that p ramifies totally in $F/\mathbb{Q}$. Then δ_F embeds $\mathcal{O}[[U]][[t]]$ into $\mathbf{h}_F^{\chi}$, where the variable t is sent to $\mathbb{T}(\varpi) - a$ for a unique root of unity $a \in \mathcal{O}$ with $a \equiv \mathbb{T}(\varpi) \mod \mathfrak{m}_{\mathfrak{h}}$.*

5.6.3 An Iwasawa theory for Hecke algebras

Under some simple conditions, we shall give an explicit description of local rings of Hecke algebras (and Galois deformation rings) over cyclotomic p-ramified p-extensions over $\mathbb{Q}$ as an application of the theorems we have proved.

We fix here the restriction '?' among $\{\emptyset, n.ord\}$. We fix a local component $\mathbf{h}^?_{\mathbb{Q}}$ of $\mathbf{h}^?(p^\infty;\mathcal{O})_{/\mathbb{Q}}$ associated with $\overline{\rho}$, and choose a character $\chi : \mathfrak{G} \to \mathcal{O}^\times$ with χ mod $\mathfrak{m}_{h^?} = \det(\overline{\rho})$ for $\overline{\rho} = \rho_{h^?}$ mod $\mathfrak{m}_{h^?}$. We assume $(\mathrm{AI}_{\mathbb{Q}})$ for $\overline{\rho}$. We take the unique $\mathbb{Z}_p$-extension $\mathbb{Q}_\infty/\mathbb{Q}$ and apply the above theory to $\Delta_j = \mathrm{Gal}(\mathbb{Q}_j/\mathbb{Q}) \cong \mathbb{Z}/p^j\mathbb{Z}$ for each layer $\mathbb{Q}_j$. We write Γ for $\Delta_\infty = \mathrm{Gal}(\mathbb{Q}_\infty/\mathbb{Q})$ and define Γ_j by $\Gamma/\Gamma_j = \Delta_j$. We note here that $(\mathrm{AI}_{\mathbb{Q}})$ is equivalent to $(\mathrm{AI}_{\mathbb{Q}_\infty})$ as seen in (AI) in Subsection 4.3.5. Thus $\mathcal{F}^{\chi,?}_F$ is representable for $F = \mathbb{Q}_j$ for any finite j, because $(Rg_{\mathbb{Q}_\infty})$ is obviously equivalent to $(Rg_{\mathbb{Q}})$. We write $(R_j, \varrho_j) = (R^{\chi,?}_F, \varrho^{\chi,?}_F)$ for $F = \mathbb{Q}_j$. We define

$$(R_\infty, \varrho_\infty) = \varprojlim{}_j (R_j, \varrho_j)$$

for the base-change map $\alpha_{k,j} : R_k \to R_j$ in the relative case. If R_∞ is an object in $CNL_{\mathcal{O}}$, it represents $\mathcal{F}^{\chi,?}_F$ for $F = \mathbb{Q}_\infty$, otherwise it pro-represents the functor. Note that Γ acts on R_j via $\mathcal{O}$-algebra automorphisms for $j = 0, 1, 2, \ldots, \infty$, and under the assumptions of Theorem 5.14, we have, for a generator γ_j of Γ_j,

THEOREM 5.48 *Assume either* $(\mathrm{AI}_{\mathbb{Q}})$ *or* $(\mathrm{AI}_{\mathbb{Q}})$ *and* $(\mathrm{Rg}_{\mathbb{Q}})$ *according as* $? = \emptyset$ *or n.ord. Then we have*

$$R_\infty/R_\infty(\gamma_j - 1)R_\infty \cong R_j$$

for $j = 0, 1, \ldots, \infty$.

Out of this control theorem, we can create a lot of well-controlled Iwasawa modules. We suppose that we have an algebra A in $CNL_{\mathcal{O}}$ so that

(1) R_∞ is an A-algebra;
(2) Γ acts trivially on A.

We can obviously take A to be $\mathcal{O}$. There seems no other canonical choice of A, but in any case, we state our result in this generality. Take an A-algebra B in $CNL_{\mathcal{O}}$, and suppose we have an A-algebra homomorphism $\pi : R_0 \to B$. Then we have a unique Galois representation $\varphi = \pi\varrho_{\mathbb{Q}}$. We then consider the module of *continuous* 1-differentials $\Omega_{R_j/A}$ and its B-part: $\Omega_{R_j/A} \otimes_{R_j} B$. Then from the above control theorem, we have

PROPOSITION 5.49 *Let the notation and the assumption be as in Theorem 5.48 and as above. Let A be a closed $\mathcal{O}$-subalgebra of R_∞ (in $CNL_{\mathcal{O}}$) on which Γ acts trivially. Let B be an A-algebra in $CNL_{\mathcal{O}}$ and $\pi : R_0 \to B$ be an A-algebra homomorphism. Then we have for $0 \le j \le k \le \infty$,*

$$\frac{\Omega_{R_k/A}\widehat{\otimes}_{R_k}B}{(\gamma_j - 1)\Omega_{R_k/A}\widehat{\otimes}_{R_k}B} \cong \Omega_{R_j/A}\widehat{\otimes}_{R_j}B. \tag{Ct}$$

Although this result is intuitive, we shall give a proof since this is fundamental in the sequel:

Proof We write R for $R_k\widehat{\otimes}_A B$ and R' for $R_j\widehat{\otimes}_A B$. Then $R/R(\gamma_j - 1)R \cong R'$. Write α for the projection: $R \to R'$ and π' for

$$m \circ ((\pi \circ \alpha_j) \otimes \text{id}) : R' = R_k\widehat{\otimes}_A B \to B\widehat{\otimes}_A B \xrightarrow{m} B$$

for multiplication $m : B\widehat{\otimes}_A B \to B$. Let $\lambda = \pi' \circ \alpha$. We have from Corollary 5.13 that

$$\begin{aligned}\mathrm{Ker}(\lambda) \otimes_R B = \mathrm{Ker}(\lambda)/\mathrm{Ker}(\lambda)^2 &= \Omega_{R/B} \otimes_R B = \Omega_{R_k\widehat{\otimes}_A B/A\otimes_A B} \otimes_R B \\ &\cong (\Omega_{R_k/A}\widehat{\otimes}_A B) \otimes_R B \cong \Omega_{R_k/A}\widehat{\otimes}_{R_k} B.\end{aligned}$$

Similarly, we have $\mathrm{Ker}(\pi') \otimes_R B \cong \Omega_{R_j/A}\widehat{\otimes}_{R_j} B$. We have the following exact sequence:

$$0 \to R(\gamma_j - 1)R \to \mathrm{Ker}(\lambda) \xrightarrow{\alpha} \mathrm{Ker}(\pi') \to 0. \tag{5.26}$$

Tensoring B over R to (5.26) and writing $J = R(\gamma_j - 1)R$, we get another exact sequence:

$$(J/J^2) \otimes_R B = J \otimes_R B \xrightarrow{i} \Omega_{R_k/A}\widehat{\otimes}_{R_k} B \longrightarrow \Omega_{R_j/A}\widehat{\otimes}_{R_j} B \longrightarrow 0.$$

We look into the B-linear map $\gamma_j - 1 : \mathrm{Ker}(\lambda) \to R$. Write B' for the image of B in R. Then $B' \subset \mathrm{Ker}(\gamma_j - 1)$ and $R = \mathrm{Ker}(\lambda) + B'$. Thus $(\gamma_j - 1)R = (\gamma_j - 1)\,\mathrm{Ker}(\lambda)$. Since γ_j is a B'-algebra automorphism of R and J/J^2 is a B'-module, we have

$$r(\gamma_j - 1)r' \equiv (\gamma_j - 1)rr' \mod J^2 \quad (r, r' \in J).$$

This shows that $\gamma_j - 1 : \mathrm{Ker}(\lambda) \to R$ induces a surjective morphism of B'-modules: $\mathrm{Ker}(\lambda)/\mathrm{Ker}(\lambda)^2 \to J/J^2$; thus, $\mathrm{Im}(i) = (\gamma_j - 1)(\Omega_{R_k/A}\widehat{\otimes}_{R_k} B)$, which proves the result. □

In particular, taking $A = \mathcal{O}$ and $B = \mathbb{F}$, we see that $\Omega_{R_\infty/\mathcal{O}} \otimes_{R_\infty} \mathbb{F}$ is an $\mathbb{F}[[\Gamma]]$-module of finite type, and it is of torsion $\iff$ $R_\infty \in CNL_{\mathcal{O}}$ (see Lemma 2.28). Let us now assume that $? = n.ord$ and that $\overline{\rho}$ is nearly

p-ordinary (satisfying ($\mathrm{AI}_{\mathbb{Q}}$) and ($\mathrm{Rg}_{\mathbb{Q}}$)). We write Λ for the subalgebra of R_∞ topologically generated by the image of $\delta_{D\cap\mathfrak{G}_\infty,\rho}(D\cap\mathfrak{G}_\infty)$ over $\mathcal{O}$ for $\rho=\varrho_\infty$, where $\mathfrak{G}_\infty$ is the stabilizer of $\mathbb{Q}_\infty$ in $\mathfrak{G}$ and D is a decomposition subgroup of $\mathfrak{G}$ at p. Writing K_j for the p-adic completion of $\mathbb{Q}_j$, we put

$$\mathscr{W}_j=\varprojlim_\alpha K_j^\times/(K_j^\times)^{p^\alpha}=(1+\varpi_jO_j)\times\varpi_j\mathbb{Z}_p,$$

where ϖ_j is a uniformizer of the p-adic integer ring O_j of K_j such that $\varpi_j=N_{K_{j+1}/K_j}\varpi_{j+1}$ (for all j). The norm map $N_{K_k/K_j}:K_k^\times\to K_j^\times$ for $k>j$ induces a homomorphism $N_{k,j}:\mathscr{W}_k\to\mathscr{W}_j$. We consider the p-profinite part W_j of the universal norm subgroup $\bigcap_{k>j}\mathrm{Im}(N_{k,j})\subset\mathscr{W}_j$. Then

$$U_j^{(1)}=\mathrm{Ker}(N_{K_j/\mathbb{Q}_p}:1+\varpi_jO_j\to\mathbb{Z}_p)\cong(\mathbb{Z}_p)^{p^j-1}\subset W_j\cong(\mathbb{Z}_p)^{p^j},$$

and by local class field theory, we have a natural surjection

$$\mathcal{O}[[W_j]]\to\Lambda_j=\alpha_j(\Lambda),$$

where we write $\alpha_j:R_\infty\to R_j$ for the base change map and Λ is the image of $\mathcal{O}[[W_\infty]]$ in R_∞. Note that

$$\mathcal{O}[[W_\infty]]=\varprojlim_j\mathcal{O}[[W_j]]$$

is a huge non-noetherian ring.

Let M_j be the maximal p-abelian extension of K_j. Since K_∞/K_0 is a $\mathbb{Z}_p$-extension, we see that

$$\mathrm{Gal}(M_\infty/K_j)\cong\mathrm{Gal}(K_\infty/K_j)\ltimes\mathrm{Gal}(M_\infty/K_\infty).$$

Thus

$$W_j=W_\infty/(\mathrm{Gal}(M_\infty/K_j),\mathrm{Gal}(M_\infty/K_j))\cong W_\infty/(\gamma_j-1)W_\infty.$$

This shows that

$$\mathcal{O}[[W_j]]/\mathcal{O}[[W_j]](\gamma-1)\mathcal{O}[[W_j]]\cong\mathcal{O}[[W_0]]\qquad(*)$$

for $\gamma=\gamma_0$. Note here that

$$\Omega_{\mathcal{O}[[W_j]]/\mathcal{O}}\cong W_j\otimes_{\mathbb{Z}_p}\mathcal{O}[[W_j]],\qquad(**)$$

where $\sigma\in\Gamma$ acts on $\Omega_{\mathcal{O}[[W_j]]/\mathcal{O}}$ by $\sigma(w\otimes\lambda)=\sigma(w)\otimes\sigma(\lambda)$ for $w\in W_j$ and $\lambda\in\mathcal{O}[[W_j]]$. This follows from the following fact (see Corollary 2.23):

$$\Omega_{\mathcal{O}[[\mathbb{Z}_p^r]]/\mathcal{O}}\cong\Omega_{\mathcal{O}[[T_1,\dots,T_r]]/\mathcal{O}}\cong\bigoplus_{j=1}^r\mathcal{O}[[T_1,\dots,T_r]]dT_j\cong\mathbb{Z}_p^r\otimes_{\mathbb{Z}_p}\mathcal{O}[[T_1,\dots,T_r]].$$

As seen in Subsection 5.6.2, by the existence of automorphic base-change lift, we have a unique local ring $\mathbf{h}_j$ of $\mathbf{h}^{n.ord}(p^\infty;\mathcal{O})_{/\mathbb{Q}_j}$ such that the attached Galois representation $\rho_j : \mathfrak{G}_j \to GL_2(\mathbf{h}_j)$ is a nearly p-ordinary deformation of $\overline{\rho}|_{\mathfrak{G}_j}$, where $\mathfrak{G}_j$ is the stabilizer of $\mathbb{Q}_j$ in $\mathfrak{G}$. Since $\rho_j^\chi \in \mathscr{F}_{\mathbb{Q}_j}^\chi(\mathbf{h}_j^\chi)$, we have a natural morphism $\iota_j : R_j \to \mathbf{h}_j^\chi$ for the closed subalgebra $\mathbf{h}_j^\chi$ of $\mathbf{h}_j$ generated by $\mathrm{Tr}(\rho_j^\chi)$ for the χ-projection ρ_j^χ of ρ_j. Let $U_j = 1+\varpi_j O_j$ be the maximal p-profinite (actually, torsion-free) quotient of the p-adic unit group of K_j. As shown in Theorem 5.44, $\mathbf{h}_j^\chi$ is finite and torsion-free over $\mathcal{O}[[U_j]]$ and $\mathbf{h}_j \cong \mathbf{h}_j^\chi\widehat{\otimes}_{\mathcal{O}}\mathcal{O}[[\Gamma_j]]$, because as is well known, $\Gamma_j = (\mathfrak{G}_j^{ab})_p$ for the stabilizer $\mathfrak{G}_j$ of $\mathbb{Q}_j$ in $\mathfrak{G}$. As we have deduced in Theorem 5.47, the natural image of $\mathcal{O}[[W_j]] \cong \mathcal{O}[[U_j^{(1)}]][[\delta_j([\varpi_j,K_j])-a]]$ in $\mathbf{h}_j$ is isomorphic to $\mathcal{O}[[U_j^{(1)}]][[\mathbb{T}(\varpi_j)-a]] \cong \mathcal{O}[[W_j]]$ for the root of unity $a \in \mathcal{O}$ congruent to $\delta_j([\varpi_j,K_j])$ modulo the maximal ideal. Thus we have

$$\Lambda_j \cong \mathcal{O}[[W_j]].$$

This combined with (∗) shows that

$$(\Omega_{\Lambda_j/\mathcal{O}} \otimes_{\Lambda_j} \mathbb{F})/(\gamma-1)(\Omega_{\Lambda_j/\mathcal{O}} \otimes_{\Lambda_j} \mathbb{F}) \cong \Omega_{\Lambda_0/\mathcal{O}} \otimes_{\Lambda_0} \mathbb{F} \tag{Ct1}$$

for a generator $\gamma \in \Gamma$.

Thus we have the following commutative diagram with exact rows:

$$\begin{array}{ccccccc}
\frac{(\Omega_{\Lambda_j/\mathcal{O}}\otimes_{\Lambda_j}\mathbb{F})}{(\gamma-1)(\Omega_{\Lambda_j/\mathcal{O}}\otimes_{\Lambda_j}\mathbb{F})} & \longrightarrow & \frac{(\Omega_{R_j/\mathcal{O}}\otimes_{R_j}\mathbb{F})}{(\gamma-1)(\Omega_{R_j/\mathcal{O}}\otimes_{R_j}\mathbb{F})} & \longrightarrow & \frac{(\Omega_{R_j/\Lambda_j}\otimes_{R_j}\mathbb{F})}{(\gamma-1)(\Omega_{R_j/\Lambda_j}\otimes_{R_j}\mathbb{F})} & \to & 0 \\
{\scriptstyle (Ct1)}\downarrow & & {\scriptstyle (Ct)}\downarrow & & \downarrow & & \\
(\Omega_{\Lambda_0/\mathcal{O}} \otimes_{\Lambda_0} \mathbb{F}) & \longrightarrow & (\Omega_{R_0/\mathcal{O}} \otimes_{R_0} \mathbb{F}) & \longrightarrow & (\Omega_{R_0/\Lambda_0} \otimes_{R_0} \mathbb{F}) & \to & 0.
\end{array}$$

We therefore conclude from (Ct) and (Ct1):

$$\Omega_{R_j/\Lambda_j} \otimes_{R_j} \mathbb{F}/(\gamma-1)\Omega_{R_j/\Lambda_j} \otimes_{R_j} \mathbb{F} \cong \Omega_{R_0/\Lambda_0} \otimes_{R_0} \mathbb{F}. \tag{Ct2}$$

This shows that $\Omega_{R_\infty/\Lambda_\infty} \otimes_{R_\infty} \mathbb{F}$ is an $\mathbb{F}[[\Gamma]]$-module of finite type. Moreover, for any surjective $\mathcal{O}$-algebra homomorphism $\pi : R_0 \to \mathbb{I}$ in $CNL_{\mathcal{O}}$, $\Omega_{R_\infty/\Lambda_\infty} \otimes_{R_\infty} \mathbb{I}$ is an $\mathbb{I}[[\Gamma]]$-module of finite type, because

$$\Omega_{R_\infty/\Lambda_\infty} \otimes_{R_\infty} \mathbb{I} \otimes_{\mathbb{I}[[\Gamma]]} \mathbb{F} \cong \Omega_{R_j/\Lambda_j} \otimes_{R_j} \mathbb{F}/(\gamma-1)\Omega_{R_j/\Lambda_j} \otimes_{R_j} \mathbb{F}.$$

If $\Omega_{R_0/\Lambda_0} = 0$ (that is, $R_0 \cong \mathbf{h}_0^\chi \cong \Lambda_0$), by Nakayama's lemma, $\Omega_{R_j/\Lambda_j} = 0$. This shows that

$$R_j \cong \mathbf{h}_j^\chi \cong \Lambda_j.$$

The theorem of Taylor-Wiles combined with the fact: $R_0 \cong R_{\mathbb{Q}}^{ord}$ shows that ι_0 is an isomorphism under the assumption of Theorem 5.29. Thus $\mathbf{h}_0^\chi \cong R_0 \cong R_{\mathbb{Q}}^{ord}$ is free of finite rank over $\mathcal{O}[[\Gamma]] = \mathcal{O}[[T]]$ as already seen in Theorem 5.28. Supposing that $\mathbf{h}_0^\chi = \mathcal{O}[[\Gamma]]$ and writing $a(T) = \delta_{D,\rho}(Frob_p)$ for $\rho = \rho_h^{ord}$ (which is the image of $T(p)$ in the ordinary part of $\mathbf{h}_0$), $\Lambda_0 = \mathbf{h}_0^\chi \cong R_0$ if and only if $\frac{\partial a}{\partial T}(0)$ is an $\mathcal{O}$-unit, which is in turn equivalent to the assertion that $(a(\gamma^k - 1) - a(\gamma^{k+(p-1)} - 1))/\varpi$ $(k \in \mathbb{Z})$ is an $\mathcal{O}$-unit for a prime element ϖ of $\mathcal{O}$. Thus we have

COROLLARY 5.50 *Suppose that $\overline{\rho}$ is p-ordinary and the assumptions of Theorem 5.29 hold. Let $\mathbf{h}_{\mathbb{Q}_j}^{ord}$ be the local ring of the universal ordinary Hecke algebra associated with $\overline{\rho}|_{\mathfrak{G}_j}$. If $\mathbf{h}_{\mathbb{Q}}^{ord} = \Lambda_0$ (in particular, if $\mathbf{h}_{\mathbb{Q}}^{ord} = \mathcal{O}[[\Gamma]] = \mathcal{O}[[T]]$ and $\frac{\partial a}{\partial T}(0)$ is an $\mathcal{O}$-unit), then*

$$R_{\mathbb{Q}_j}^\chi \cong \mathbf{h}_{\mathbb{Q}_j}^\chi \cong \mathcal{O}[[W_j]], \quad R_{\mathbb{Q}_j}^{ord} \cong \mathbf{h}_{\mathbb{Q}_j}^{ord} \cong \mathcal{O}[[\Gamma_j]]$$

and

$$R_{\mathbb{Q}_j}^{n.ord} \cong \mathbf{h}_{\mathbb{Q}_j}^{n.ord} \cong \mathcal{O}[[W_j \times \Gamma_j]].$$

Here are some remarks: (a) The identification of the universal deformation ring as a Hecke algebra is first conjectured by Mazur (see [MT]) and then refined by A. Wiles [W2] Conjecture 2.16, which is proved in [W] in almost all cases (see [FM] for a list of such conjectures). In the Hilbert modular case, one can make an analogous conjecture, and again in many cases, this Hilbert modular version has been proved by Fujiwara [Fu]. On the other hand, the identification of the Hecke algebra with the group algebra as in Corollary 5.50 makes the ring a little more explicit.

(b) Suppose that $\overline{\rho} = \rho_f \bmod \mathfrak{m}_{\mathcal{O}}$ for a p-ordinary modular form $f \in S_k^{n.ord}(\Gamma_0(p))$. The condition $\mathbf{h}_{\mathbb{Q}}^{ord} = \mathcal{O}[[\Gamma]]$ is equivalent to the fact that there is no congruence between f and any other Hecke eigenforms in $S_k^{n.ord}(\Gamma_0(p))(\cong S_k^{n.ord}(SL_2(\mathbb{Z}))$ if $k > 2)$. If f is Ramanujan's function Δ and $p = 11$, then

$$a(\gamma - 1) = 1 \text{ and } a(\gamma^{11} - 1) \equiv a(p; \Delta) = 534612 \bmod 11^{11},$$

and $a(p;\Delta) - 1 = 534611 = 7 \cdot 11 \cdot 53 \cdot 131$. Here we note that by our normalization of $\mathcal{O}[[\Gamma]]$-algebra structure, $a(\gamma^{k-1} - 1)$ is the p-Fourier coefficient of the weight k cusp form in the family. Thus Δ satisfies the assumption of the above corollary. Y. Maeda† has checked this non-divisibility by p^2 also for each weight 2 eigenform in $S_2(\Gamma_0(p))$ for $p = 17, 19, 23, 29, 31, 37, 41, 43, 59, 61$ and 67. As for $p = 53$, there are two families of modular forms, and one of them with $a \equiv -1 \bmod \mathfrak{m}_{\mathcal{O}[[T]]}$ has

† The author is grateful to Y. Maeda for allowing him to include his computation here.

non-unit $a'(T) = \frac{\partial a}{\partial T}(T)$. In this case, for the cusp form $f \in S_{54}(SL_2(\mathbb{Z}))$ of weight 54 in the family, $a(p;f)+1$ is divisible by a square of a prime factor of 53 in the Hecke field of f.

5.6.4 Adjoint Selmer groups over cyclotomic extensions

We now prove the torsionness of the adjoint Selmer groups over $\mathbb{Q}_\infty$. We look at $\mathcal{O}[[W_\infty]]^\Gamma = \mathcal{O}[[W_\infty^\Gamma]]$. For ϖ_j, we define $\zeta_j(\sigma) \in O_j^\times$ by $\varpi_j^\sigma = \zeta_j(\sigma)\varpi_j$. By our choice: $N_{K_{j+1}/K_j}(\varpi_{j+1}) = \varpi_j$, the sequence $\{\zeta_j(\sigma)\}_j$ is norm coherent, defining a 1-cocycle

$$\zeta(\sigma) = \varprojlim{}_j \zeta_j(\sigma)$$

on Γ with values in $U_\infty^{(1)} = \varprojlim{}_j U_j^{(1)}$. Since for any prime element ϖ of O_j, it generates K_j over $\mathbb{Q}_p$, $\varpi^\sigma \neq \varpi$ for any non-trivial automorphism σ of K_j. From this, it is easy to see that the class of ζ gives a non-trivial element in $H^1(\Gamma, U_\infty^{(1)})$ which is not a torsion element. Applying this to the long exact sequence coming from:

$$1 \to U_\infty^{(1)} \ (= U_\infty) \to W_\infty \to W_0 \to 1,$$

we see that $W_\infty^\Gamma = 1$. Thus the maximal Γ-invariant subring of Λ_∞ is $\mathcal{O}$.

Now we take $A = \mathcal{O} = B$ in Proposition 5.49. Then we can show similarly to (Ct2) that

$$\frac{\Omega_{R_\infty/\Lambda_\infty} \otimes_{R_\infty} \mathcal{O}}{(\gamma-1)\Omega_{R_\infty/\Lambda_\infty} \otimes_{R_\infty} \mathcal{O}} \cong \Omega_{R_0/\Lambda_0} \otimes_{R_0} \mathcal{O}. \tag{Ct3}$$

Let $\lambda : \mathbf{h}^{ord}(p^\infty;\mathcal{O})_{/\mathbb{Q}} \to \mathbb{I}$ be an $\mathcal{O}[[T]]$-algebra homomorphism for an integral domain $\mathbb{I}$ finite flat over $\mathcal{O}[[T]]$. We take $\overline{\rho}$ to be the residual Galois representation attached to λ. Since R_0 and $\Lambda_0 = \mathcal{O}[[t]]$ for $t = \mathbb{T}(p) - a$ have the same Krull dimension by Theorem 5.29, there exists a height one prime $P \subset \mathbb{I}$ with $\mathbb{I}/P \cong \mathcal{O}$ such that $\frac{dt}{dT} \mod P \neq 0$. Then

$$\Omega_{R_0/\Lambda_0} \otimes_{R_0} \mathbb{I} \otimes_{\mathbb{I}} \mathbb{I}/P = \Omega_{R_0/\Lambda_0} \otimes_{R_0} \mathbb{I}/P$$

is a finite module and thus is an $\mathcal{O}$-torsion module of finite type. Then $\Omega_{R_0/\Lambda_0} \otimes_{R_0} \mathbb{I}$ has to be an $\mathbb{I}$-torsion module of finite type.

Since all the P's in the dense open subset $Spec(\mathbb{I}\left[\delta, \frac{1}{\delta}\right]) \subset Spec(\mathbb{I})$ for $\delta = \frac{dt}{dT}$ in the quotient field of $\mathbb{I}$ satisfy the above non-vanishing property,

$$\Omega_{R_\infty/\Lambda_\infty} \otimes_{R_\infty, P\circ\lambda} \mathcal{O}$$

is a torsion $\mathcal{O}[[\Gamma]]$-module for all height one primes P of $\mathbb{I}$ except for finitely many. Thus $\Omega_{R_\infty/\Lambda_\infty} \otimes_{R_\infty} \mathbb{I}$ is a torsion $\mathbb{I}[[\Gamma]]$-module of finite type (as conjectured by Greenberg).

Note that

$$\Omega_{\Lambda_j/\mathcal{O}[[U_j^{(1)}]]} \cong \Lambda_j dt_j,$$

where we have written as t_j the variable $\delta_j([\varpi_j, K_j])$. We now have an exact sequence:

$$\mathbb{I} \cong \Omega_{\Lambda_j/\mathcal{O}[[U_j^{(1)}]]} \otimes_{\Lambda_j} \mathbb{I} \longrightarrow \Omega_{R_j/\mathcal{O}[[U_j^{(1)}]]} \otimes_{R_j} \mathbb{I} \longrightarrow \Omega_{R_j/\Lambda_j} \otimes_{R_j} \mathbb{I} \longrightarrow 0$$

for $j = 0, 1, \ldots, \infty$. This shows that $\Omega_{R_\infty/\mathcal{O}[[U_\infty]]} \otimes_{R_\infty} \mathbb{I}$ is a torsion $\mathbb{I}[[\Gamma]]$-module of finite type.

Let $R_j^{ord} = R_{\mathbb{Q}_j}^{ord}$ and $R_j^{n.ord} = R_{\mathbb{Q}_j}^{n.ord}$. We have a natural projection $pr : R_j^{n.ord} \to R_j^{ord}$ associated with the inclusion $\mathcal{F}^{ord} \hookrightarrow \mathcal{F}^{n.ord}$. Since $R_0 \cong R_0^{ord}$ canonically (Proposition 1.1), we may regard $\mathbb{I}$ as an $R_j^{n.ord}$-algebra as well as an R_j-algebra. If we write π^{ord} for the projection of R_0^{ord} onto $\mathbb{I}$ and $\varphi^{ord} = \pi^{ord}\varrho_{\mathbb{Q}}^{ord}$, we have a relation that $\varphi = (\varphi^{ord})^\chi$, and therefore $Ad(\varphi) = Ad(\varphi^{ord})$ for the adjoint representation $Ad(\varphi^{ord})$. We anyway have an exact sequence for $J = \mathrm{Ker}(pr)$:

$$J/J^2 \otimes_{R_j^{n.ord}} \mathbb{I} \longrightarrow \Omega_{R_j^{n.ord}/\mathcal{O}[[U_j \times \Gamma_j]]} \otimes_{R_j^{n.ord}} \mathbb{I} \longrightarrow \Omega_{R_j^{ord}/\mathcal{O}[[\Gamma_j]]} \otimes_{R_j^{ord}} \mathbb{I} \longrightarrow 0.$$

Let $W'_\infty = W_\infty/U_\infty^{(1)}$, and write J' for the kernel of the projection of $\Lambda_\infty = \mathcal{O}[[W_\infty]]$ to $\mathcal{O}[[W'_\infty]]$. Since $J'/J'^2 \otimes_{R_j^{n.ord}} \mathbb{I} \cong U_j^{(1)} \otimes_{\mathbb{Z}_p} \mathbb{I}$ canonically and $U_\infty = U_\infty^{(1)}$, for $j = \infty$, we have the following commutative diagram with exact rows:

$$\begin{array}{ccccc}
U_\infty \bigotimes_{\mathbb{Z}_p} \mathbb{I} & \longrightarrow & \Omega_{\Lambda_\infty/\mathcal{O}[[U_\infty]]} \bigotimes_{\Lambda_\infty} \mathbb{I} & \stackrel{a}{\longrightarrow} & \Omega_{\mathcal{O}[[W'_\infty]]/\mathcal{O}} \bigotimes_{\mathcal{O}[[W'_\infty]]} \mathbb{I} \\
\downarrow & & \downarrow & & \downarrow \\
J/J^2 \bigotimes_{\Lambda_\infty} \mathbb{I} & \longrightarrow & \Omega_{R_j^{n.ord}/\mathcal{O}[[U_j \times \Gamma_j]]} \bigotimes_{R_j^{n.ord}} \mathbb{I} & \underset{b}{\longrightarrow} & \Omega_{R_j^{ord}/\mathcal{O}[[\Gamma_j]]} \bigotimes_{R_j^{ord}} \mathbb{I}.
\end{array}$$

Note that ι is surjective, because by universality, $J = \sum_{u \in U_\infty} (u-1) R_j^{n.ord}$. Since a is a surjective isomorphism, b has to be an isomorphism. Thus we get

$$\Omega_{R_\infty^{n.ord}/\mathcal{O}[[U_\infty^{(1)}]]} \otimes_{R_\infty^{n.ord}} \mathbb{I} \cong \Omega_{R_\infty^{ord}/\mathcal{O}} \otimes_{R_\infty^{ord}} \mathbb{I}.$$

Note that $R_\infty^{n.ord} \cong R_\infty$ because there is no non-trivial abelian p-extension over $\mathbb{Q}_\infty$ unramified outside p and ∞. We get

$$\Omega_{R_\infty^{n.ord}/\mathcal{O}[[U_\infty^{(1)}]]} \otimes_{R_\infty^{n.ord}} \mathbb{I} \cong \Omega_{R_\infty/\mathcal{O}[[U_\infty^{(1)}]]} \otimes_{R_\infty} \mathbb{I}.$$

By Theorem 5.14,

$$\Omega_{R^{ord}_{\infty}/\mathcal{O}} \otimes_{R^{ord}_{\infty}} \mathbb{I} \cong \mathrm{Sel}(Ad(\varphi^{ord}))^{*}_{/\mathbb{Q}_{\infty}},$$

where '$*$' indicates the Pontryagin duality. Thus we get a proof of the co-torsionness of the Selmer group. We record what we have proven as

THEOREM 5.51 *Assume the assumption of Theorem 5.29 for $\overline{\rho} = \varphi \mod \mathfrak{m}_{\mathbb{I}}$. Let $\varphi = \pi\varrho_0$. Then:*

(1) $\Omega_{R_{\infty}/\mathcal{O}[[U^{(1)}_{\infty}]]} \otimes_{R_{\infty}} \mathbb{I}$ *is a torsion $\mathbb{I}[[\Gamma]]$-module of finite type.*
(2) *We have the following isomorphisms:*

$$\Omega_{R_{\infty}/\mathcal{O}[[U^{(1)}_{\infty}]]} \otimes_{R_{\infty}} \mathbb{I} \cong \Omega_{R^{n.ord}_{\infty}/\mathcal{O}[[U^{(1)}_{\infty}]]} \otimes_{R^{n.ord}_{\infty}} \mathbb{I}$$
$$\cong \Omega_{R^{ord}_{j}/\mathcal{O}} \otimes_{R^{ord}_{j}} \mathbb{I} \cong \mathrm{Sel}(Ad(\varphi))^{*}_{/\mathbb{Q}_{\infty}},$$

where '$*$' indicates the Pontryagin dual module.
(3) *Let $\varphi_P = \varphi \mod P$ for $P \in Spec(\mathbb{I})$. Then for almost all height one prime $P \in Spec(\mathbb{I})(\overline{\mathbb{Q}}_p) = \mathrm{Hom}_{\mathcal{O}-alg}(\mathbb{I}, \overline{\mathbb{Q}}_p)$, $\mathrm{Sel}(Ad(\varphi_P))^{*}_{/\mathbb{Q}_{\infty}}$ is a torsion $A[[\Gamma]]$-module of finite type, where $A = \mathrm{Im}(P) = \mathbb{I}/P$. In particular, if $\frac{dt}{dT} \not\equiv 0 \mod P$, this assertion holds.*

This shows the Selmer group $\mathrm{Sel}(Ad(\varphi))_{/\mathbb{Q}_{\infty}} = \mathrm{Sel}(Ad(\varphi) \otimes \kappa)_{/\mathbb{Q}}$ for the universal character $\kappa : \Gamma \to \mathbb{I}[[\Gamma]]^{\times}$ controls the growth of three types of universal deformation rings $R^{n.ord}_{j}$, R^{χ}_{j} and R^{ord}_{j} as j grows.

If $\mathbb{I} = \Lambda$, $P = P_2$ and φ_P is a Galois representation of an elliptic curve, combining the results of [BDGP] and [GS], it is known that $\frac{dt}{dT} \not\equiv 0 \mod P$, which gives a concrete example of torsionness of the Selmer group associated to an elliptic cusp form.

We do not formulate the main conjecture in this context, because there is no space left to give a formulation of the p-adic analytic L-functions associated with Galois representations. For that, we refer readers to [SGL] and [HTU].

5.6.5 *Proof of Theorem 5.44*

We now prove Theorem 5.44. We keep the notation introduced in the theorem.

We recall from [H88a] and [SGL] Section 2.2 the definition of the space of adelic cusp forms with respect to U_{α} briefly. Let $(k, w) \in \mathbb{Z}[I]^2$ be a double digit weight (with $k_{\sigma} > 0$ for all σ). The space $\mathrm{S}_{k,w}(U_{\alpha})$ is the collection of all functions $f : GL_2(F_{\mathbb{A}}) \to \mathbb{C}$ which satisfies an F-version of (A′1), (A2–3) in Subsection 3.1.5 for $S = U_{\alpha}$. We refer for a precise meaning of (A2–3) to [SGL] Section 2.2. As for the version of (A′1), to get the exact statement under the present notation, we just need to replace the automorphic factor there by:

$$J_{k,w}(c,i) = \prod_{\sigma\in I} \det(c_\sigma)^{-w_\sigma}(\gamma_\sigma i + \delta_\sigma)^{k_\sigma}$$

for $c = (c_\sigma) \in SO_2(F_{\mathbb{R}}) = \prod_{\sigma\in I} SO_2(\mathbb{R})$ with $c_\sigma = \left(\begin{smallmatrix} * & * \\ \gamma_\sigma & \delta_\sigma \end{smallmatrix}\right)$. We then define $\mathbf{S}_{k,w}(U_\alpha; A)$ by the collection of all functions in $\mathbf{S}_{k,w}(U_\alpha)$ with Fourier coefficients in a subring A in $\mathbb{C}$. Here the Fourier coefficients are defined similarly to Theorem 3.10 (see [H91] Section 1 and [SGL] Sections 2.3–4).

We consider the space

$$\overline{\mathbf{S}} = \overline{\mathbf{S}}(U_\infty; \mathcal{O})$$

of p-adic cusp forms defined in [H91] Section 3 and [SGL] Section 2.6 for $U_\infty = \bigcap_\alpha U_\alpha$. This space is the p-adic completion of the union $\bigcup_\alpha \mathbf{S}_{k,w}(U_\alpha; \mathcal{O})$, where $\mathbf{S}_{k,w}(U_\alpha; \mathcal{O}) = \mathbf{S}_{k,w}(U_\alpha; \mathcal{O}\cap\overline{\mathbb{Q}}) \otimes_{\mathcal{O}\cap\overline{\mathbb{Q}}} \mathcal{O}$. On $\overline{\mathbf{S}}$, we have a natural (continuous) action of the Hecke algebra $\mathbf{h}(p^\infty; \mathcal{O})$. In [H91], this space is first defined relative to a weight $(k, w) \in \mathbb{Z}[I]^2$, but later it is shown that the space is independent of weight ([H91] (3.2)). We can think of the maximal subspace $\overline{\mathbf{S}}[\chi]$ of $\overline{\mathbf{S}}$ on which $Cl_F(p^\infty)$ acts via χ. Each $f \in \overline{\mathbf{S}}$ has its q-expansion coefficients $a_p(y, f)$ which is a function of finite ideles y, vanishing outside integral ones. Then for any continuous function $\phi : Cl_F(p^\infty) \to \mathcal{O}$, we can define $f \otimes \phi \in \overline{\mathbf{S}}$ by

$$a_p(y, f \otimes \phi) = \phi(y) a_p(y, f). \tag{tw}$$

This operation is defined in [H91] page 369 for characters ϕ, which extends to the operation for general functions ϕ. The important point here is that in (tw), the function ϕ is pulled back to a function on the idele group

$$Z(\mathbb{A}^{(\infty)})/\overline{Z(\mathbb{Q})_+ Z(\widehat{\mathbb{Z}}^{(p)})} \cong Cl_F(p^\infty)$$

for $Z(\mathbb{Q})_+ = Z(\mathbb{Q}) \cap Z(\mathbb{R})_+$ and then made a product with $a_p(y, f)$. Hence, $f \mapsto f \otimes \phi$ preserves nearly p-ordinarity. This can be checked restricting ϕ to a Hecke character ξ of type A_0 in the sense of Weil with infinity type $w \in \mathbb{Z}[I]$. Then we see from the definition that

$$(f \otimes \xi)|\mathbb{T}(y) = \xi(y) y_p^w (f|\mathbb{T}(y) \otimes \xi). \tag{ctw}$$

Here $\xi(y)y_p^w$ is the value of the p-adic avatar of ξ which is a p-adic unit. In this way, we have a linear map:

$$\overline{\mathbf{S}}[\chi] \otimes_{\mathcal{O}} C(Cl_F(p^\infty); \mathcal{O}) \to \overline{\mathbf{S}}$$

given by $(f,\phi)\mapsto f\otimes\phi$. Here $C(T;T')$, for two topological spaces T and T', is the space of all continuous maps from T to T'. For the maximal p-profinite quotient $\mathbb{W}^+(p)$ of $Cl_F(p^\infty)$, we thus get a linear map:

$$m:\overline{\mathbb{S}}[\chi]\otimes_{\mathcal{O}}C(\mathbb{W}^+(p);\mathcal{O})\to\overline{\mathbb{S}}.$$

Let μ be the maximal prime-to-p torsion subgroup of $Cl_F(p^\infty)$. Then $\mathbb{W}^+(p)=Cl_F(p^\infty)/\mu$. Since we are twisting by the character of $\mathbb{W}^+(p)$, the action of μ is unaffected; in other words, m is $\mathcal{O}[\mu]$-linear. Therefore, m has values in the $\chi|_\mu$-eigenspace $\overline{\mathbb{S}}[\chi|_\mu]$. We can show by the argument in [H91] Section 7.F or (ctw) that

$$(f\otimes\phi)|\mathbb{T}(y)=(f|\mathbb{T}(y))\otimes(\phi|y)\text{ and }(f\otimes\phi)|\langle z\rangle=(f|\langle z\rangle)\otimes(\phi|z^2),$$

where $\phi|z(y)=\phi(yz)$. Thus the image of $\overline{\mathbb{S}}\otimes C(\mathbb{W}^+(p);\mathcal{O})$ is stable under $\mathbb{T}(y)$ and $\langle z\rangle$, and $\overline{\mathbb{S}}^{n.ord}\otimes C(\mathbb{W}^+(p);\mathcal{O})$ is mapped by m to $\overline{\mathbb{S}}^{n.ord}[\chi|_\mu]$. This has finite level analog. We consider the space $\mathbf{S}_{k,w}(U_\alpha;\mathcal{O})$ of p-adic modular forms introduced in [H91] Section 2 for $(k,w)=(n+2t,t-v)$ (see also [SGL] Section 2.1). Then m takes $\mathbf{S}^{n.ord}_{k,w}(U_\alpha;\mathcal{O})\otimes C(\mathbb{W}^+_\beta(p);\mathcal{O})$ into $\mathbf{S}^{n.ord}_{k,w}(U_{\alpha+2\beta};\mathcal{O})$, where $\mathbb{W}^+_\beta(p)$ is the maximal p-primary quotient of $Cl_F(p^\beta)$. Note that the Hecke algebra $h^{n.ord}(U_\alpha)$ is reduced. Thus $\mathbf{S}^{n.ord}_{k,w}(U_\alpha;K)$ is the direct sum of one-dimensional eigenspaces under Hecke operators. So to show the injectivity of m on $\mathbf{S}^{n.ord}_{k,w}(U_\alpha;\mathcal{O})_\chi\otimes C(\mathbb{W}^+_\beta(p);\mathcal{O})$, what we need to prove is that for any automorphic representation π, the isomorphism classes of $\pi\otimes\xi$ are all distinct for all characters ξ of $\mathbb{W}^+_\beta(p)$. It is well known that $\pi\otimes\xi\cong\pi$ for $\xi\neq\mathrm{id}$ if and only if ξ is a quadratic character associated with a quadratic extension M/K and π is a base-change lift from $GL(1)_{/M}$ (that is, π is spanned by the theta series of the norm form of M; see [BCG] Lemma 11.7, for example). Thus assuming that $p>2$, the map m is injective on $\mathbf{S}^{n.ord}_{k,w}(U_\alpha;\mathcal{O})\otimes C(\mathbb{W}^+_\beta(p);\mathcal{O})$. Then we have a surjection

$$h^{n.ord}(U_{\alpha+2\beta})[\chi|_\mu]\to h^{n.ord,\chi}(U_\alpha)\otimes\mathcal{O}[\mathbb{W}^+_\beta(p)].$$

Here '$X[\chi|_\mu]$' indicates the direct summand of a p-profinite $\mathcal{O}[\mu]$-module X on which μ acts by $\chi|_\mu$. The surjectivity follows from the fact that the Hecke algebra on

$$\mathbf{S}^{n.ord}_{k,w}(U_\alpha;\mathcal{O})\otimes C(\mathbb{W}^+_\beta(p);\mathcal{O})$$

is obviously a tensor product, because $\chi^{-1}(z)\langle z\rangle=\mathrm{id}\otimes z$ in $1\otimes\mathcal{O}[\mathbb{W}^+_\beta(p)]$ and $(\mathrm{id}\otimes y)^{-1}\mathbb{T}(y)=\mathbb{T}(y)\otimes\mathrm{id}$ on $\mathbf{S}^{n.ord}_{k,w}(U_\alpha;\mathcal{O})\otimes 1$. Taking the limit with respect to α and β, we get a surjective algebra homomorphism:

$$\mathbf{h}^{n.ord}(p^\infty;\mathcal{O})[\chi|_\mu]\to h^{n.ord,\chi}(p^\infty;\mathcal{O})\widehat{\otimes}\mathcal{O}[[\mathbb{W}^+(p)]].$$

Now let us go back to the cohomological situation, but changing the group G to its inner form. We take a quaternion algebra $B_{/F}$ everywhere unramified at finite places and maximally ramified at infinite places. Let q for the number of unramified places of B at infinity. Then $q = 0$ or 1 with $q \equiv [F : \mathbb{Q}] \bmod 2$. We consider the linear algebraic group $G_{B/\mathbb{Q}}$ defined by $G_B(A) = (B \otimes_{\mathbb{Q}} A)^\times$. We identify by an isomorphism $G_B(\mathbb{A}^{(\infty)})$ with $G(\mathbb{A}^{(\infty)})$; hence $U \subset G(\mathbb{A}^{(\infty)}) = G_B(\mathbb{A}^{(\infty)})$. We then consider the Shimura variety over $\mathbb{C}$

$$X(U) = X_B(U) = G_B(\mathbb{Q})\backslash G_B(\mathbb{A})_+/UZ_B(\mathbb{R})C_B,$$

where Z_B is the center of G_B, $G_B(\mathbb{A})_+ = G_B(\mathbb{A}^{(\infty)})G_B(\mathbb{R})_+$ for the connected component $G_B(\mathbb{R})_+$ of $G_B(\mathbb{R})$ with identity, and C_B is a fixed maximal connected compact subgroup of $G_B(\mathbb{R})$. We take K sufficiently large so that $L(n, v; A)$ is a well-defined $G_B(A)$-module for any $\mathcal{O}$-module A (see [H88a] Section 1). However, we normalize the action in the following way: for $P \in L(n, v; A)$

$$\gamma P((X_\sigma, Y_\sigma)) = P((X_\sigma, Y_\sigma)\sigma({}^t\gamma^\iota)),$$

while in [H88a] and [H89a], we have taken a slightly different action

$$\gamma P((X_\sigma, Y_\sigma)) = P((X_\sigma, Y_\sigma)\sigma(\gamma)).$$

This change is made in order to get the Hecke algebra $h_{n,v}(U)$ as a subalgebra of the endomorphism algebra of $H^q(X_B(U), L(n, v; K))$ by the Jacquet–Langlands–Shimizu correspondence. We fix (n, v) and consider

$$S_B(U; A) = H^q(X_B(U), L(n, v; A)).$$

We therefore have $S_B^{n.ord}(U; A)$. We write

$$S_B^{n.ord}(U_\infty; A) = \varinjlim_\alpha S_B^{n.ord}(U_\alpha; A),$$

and we obtain another expression of the Hecke algebra:

$$\mathbf{h}^{n.ord}(U_\infty) = \mathbf{h}^{n.ord}(p^\infty; \mathcal{O}) = \varprojlim_\alpha \mathbf{h}_B^{n.ord}(U_\alpha),$$

where $h_B^{n.ord}(U_\alpha)$ is the $\mathcal{O}$-subalgebra of $\mathrm{End}_K(S_B(U_\alpha; K))$ generated by $\mathbb{T}(y)$ and $\langle z \rangle$. Then it is proved in [H88a] and [H89a] that

(1) $S_B(U_\alpha; K/\mathcal{O})$ is p-divisible;
(2) $S_B(U_\alpha; \mathcal{O})$ is $\mathcal{O}$-free;
(3) We have canonical isomorphisms for finite α

$$\mathrm{End}_{\mathcal{O}}(S_B(U_\alpha; K/\mathcal{O})) \cong \mathrm{End}_{\mathcal{O}}(S_B(U_\alpha; \mathcal{O})) \cong \mathrm{End}_K(S_B(U_\alpha; K));$$

(4) $h_B^{n.ord}(U_\alpha)$ acts faithfully on $S_B^{n.ord}(U_\alpha;A)$ for $A = K$, $K/\mathcal{O}$ and $\mathcal{O}$ $(\alpha = 1, 2, \dots, \infty)$;

(5) The Pontryagin dual module of $S_B^{n.ord}(U_\infty;K/\mathcal{O})$ is $\mathcal{O}[[\mathbb{W}]]$-free of finite rank, where $\mathbb{W}$ is the p-profinite part of $\mathbf{G}$.

In particular, writing $\mathbb{W} = \mathbb{W}^+(p) \times \mathbb{W}^-(p)$ for the p-profinite part $\mathbb{W}^-(p)$ of $O_{F,p}^\times$, we see from (5) that $h^{n.ord,\chi}(p^\infty;\mathcal{O})$ acts faithfully on $\mathcal{O}[[\mathbb{W}^-(p)]]$-cofree module

$$S_B^{n.ord}(U_\infty;K/\mathcal{O})_\chi = \left\{x \in S_B^{n.ord}(U_\infty;K/\mathcal{O}) \middle| \, x|\langle z\rangle = \chi(z)x\right\}.$$

This shows that on each irreducible component, the morphism of schemes we get from the surjection:

$$\iota : \mathbf{h}^{n.ord}(p^\infty;\mathcal{O})[\chi|_\mu] \to h^{n.ord,\chi}(p^\infty;\mathcal{O})\widehat{\otimes}_{\mathcal{O}}\mathcal{O}[[\mathbb{W}^+(p)]]$$

is an isomorphism since it is injective ($\Longleftrightarrow$ 'surjective' on the side of rings), and the source and target have the same Krull dimension. Since the two schemes are reduced, it tells us that ι is an isomorphism.

We now need to prove the last assertion. Originally, $\mathbf{h}^\chi$ is defined to be a subalgebra of $\mathbf{h}^{n.ord}$ generated by $\mathrm{Tr}(\rho_{\mathbf{h}}^\chi)$ (not the quotient of $\mathbf{h}^{n.ord}$). From the first assertion, we see that $\mathbf{h} = \mathbf{h}'\widehat{\otimes}\mathcal{O}[[\mathbb{W}^+(p)]]$ for a local ring $\mathbf{h}'$ of $\mathbf{h}^{n.ord,\chi}(p^\infty;\mathcal{O})$. Then it is obvious, from the definition of $\mathbf{h}^\chi$, that $\mathbf{h}^\chi$ coincides with $\mathbf{h}'$.

Bibliography

Books

[ADT] J. S. Milne, *Arithmetic Duality Theorem*, Perspectives in Math. **1**, Academic Press, 1986

[AME] N. M. Katz and B. Mazur, *Arithmetic Moduli of Elliptic Curves*, Ann. of Math. Studies **108**, Princeton University Press, 1985

[ARL] A. Borel and W. Casselman, *Automorphic Forms, Representations and L-functions*, Proc. Symp. Pure Math. **33**, Part 1 and 2, 1979, American Math. Society

[BCG] R. P. Langlands, *Base Change for GL*(2), Annals of Math. Studies **96**, Princeton University Press, 1980

[BCL] M. Rapoport, N. Schappacher and P. Schneider, (eds.), *Beilinson's Conjectures on Special Values of L-functions*, Perspectives in Math. **4**, Academic Press, 1988

[BCM] N. Bourbaki, *Commutative Algebra*, Hermann, Paris, 1961-83

[BCT] J. Arthur and L. Clozel, *Simple Algebras, Base Change, and the Advanced Theory of the Trace Formula*, Annals of Mathematics Studies **120**, Princeton University Press, 1989

[BLR] S. Bosch, W. Lütkebohmert and M. Raynaud, *Néron Models*, Ergebnisse der Mathematik, 3 Folge, **21**, Springer, Berlin-Heidelberg-New York-Tokyo, 1990

[BTP] N. Bourbaki, *Topology*, Hermann, Paris, 1961-65

[CAL] H. Matsumura, *Commutative Algebra*, Benjamin, 1980

[CFN] J. Neukirch, *Class Field Theory*, Springer, Berlin-Heidelberg-New York-Tokyo, 1986

[CFT] E. Artin and J. Tate, *Class Field Theory*, Benjamin, 1968

[CLC] J.-P. Serre, *Corps Locaux*, Hermann, Paris, 1962

[CRT] H. Matsumura, *Commutative Ring Theory*, Cambridge Studies in Advanced Mathematics **8**, Cambridge Univ. Press, 1986

[DGH] J. Tilouine, *Deformation of Galois Representations and Hecke Algebras*, Publ. Mehta Res. Inst., Narosa Publ., New Delhi, 1996

[ECH] J. S. Milne, *Étale Cohomology*, Princeton University Press, Princeton, 1980

[EEK] A. Weil, *Elliptic Functions according to Eisenstein and Kronecker*, Springer, 1976

[EGA] A. Grothendieck and J. Dieudonné, *Eléments de Géométrie Algébrique*, Publ. IHES **4** (1960), **8** (1961), **11** (1961), **17** (1963), **20** (1964), **24** (1965), **28** (1966), **32** (1967)

[EPE] G. Shimura, *Euler Products and Eisenstein Series*, CBMS Regional Conference Series **93**, American Mathematical Society, Providence, 1997

[FLP] B. Perrin-Riou, *Fonction L p-adiques des Représentations p-adiques*, Astérisque **229**, 1995

[GAL] Y. Ihara, K. Ribet and J.-P. Serre, (eds.), *Galois Groups over* $\mathbb{Q}$, MSRI publications **16**, 1989

[GMF] H. Hida, *Geometric Modular Forms and Elliptic Curves*, to be published by World Scientific Publishing Company, Singapore

[HAL] P. J. Hilton and U. Stammback, *A Course in Homological Algebra*, Graduate Text in Math. **4**, Springer, Berlin-Heidelberg-New York-Tokyo, 1970

[HKC] J. Coates and S.-T. Yau, (eds.), *Elliptic Curves, Modular Forms, & Fermat's Last Theorem*, Series in Number Theory I, International Press, Boston, 1995

[HMS] P. Deligne, J.S. Milne, A. Ogus and K.-Y. Shih, *Hodge Cycles, Motives, and Shimura Varieties*, Lecture Notes in Math. **900** Springer, Berlin-Heidelberg-New York-Tokyo, 1982

[HMW] E. Hecke, *Mathematische Werke*, Vandenhoeck and Ruprecht, Göttingen, 1970

[IAT] G. Shimura, *Introduction to the Arithmetic Theory of Automorphic Functions*, Iwanami-Shoten and Princeton Univ. Press, 1971

[ICF] L. C. Washington, *Introduction to Cyclotomic Fields*, Graduate Text in Math. **83**, Springer, Berlin-Heidelberg-New York-Tokyo, 1980

[ICM] E. de Shalit, *Iwasawa Theory of Elliptic Curves with Complex Multiplication*, Perspective Math. **3**, Academic Press, 1987

[LFE] H. Hida, *Elementary Theory of L-functions and Eisenstein Series*, LMSST **26**, Cambridge University Press, Cambridge, 1993

[LGF] L. E. Dickson, *Linear Groups with an Exposition of the Galois Field Theory*, Teubner, Leipzig, 1901

[LRF] J.-P. Serre, *Linear Representations of Finite Groups*, GTM 42, Springer, Berlin-Heidelberg-New York-Tokyo, 1977

[MFM] T. Miyake, *Modular Forms*, Springer, Berlin-Heidelberg-New York-Tokyo, 1989

[MRT] C.W. Curtis and I. Reiner, *Methods of Representation Theory*, John Wiley and Sons, New York, 1981.

[RAG] J. C. Jantzen, *Representations of Algebraic Groups*, Academic Press, 1987

[SGL] H. Hida, *On the Search of Genuine p-adic Modular L-functions for GL(n)*, Mémoires SMF **67**, 1996

[ZSA] R. Godement and H. Jacquet, *Zeta Functions of Simple Algebras*, Lecture Notes in Math. **260**, Springer, Berlin-Heidelberg-New York-Tokyo, 1972

Articles

[BDGP] K. Barré-Sirieix, G. Diaz, F. Gramain and G. Philibert, Une preuve de la conjecture de Mahler–Manin, *Inventiones Math.* **124** (1996), 1–9

[BR] D. Blasius and J. D. Rogawski, Motives for Hilbert modular forms, *Inventiones Math.* **114** (1993), 55–87

[BK] S. Bloch and K. Kato, *L*-functions and Tamagawa numbers of motives, *Progress in Math.* (Grothendieck Festschrift 1) **86** (1990), 333–400

[C] H. Carayol, Formes modulaires et représentations galoisiennes à valeurs dans un anneau local compact, *Contemporary Math.* **165** (1994), 213–237

[C1] H. Carayol: Sur la mauvaise réduction des courbes de Shimura, *Compositio Math.* **59** (1986), 151–230

[Cl] L. Clozel, Motifs et formes automorphes: Applications du principe de fonctorialité, *Perspective Math.* **10** (1990), 77–159

[CDT] B. Conrad, F. Diamond and R. Taylor, Modularity of certain potentially Barsotti–Tate Galois representations, *J. AMS* **12** (1999), 521–567

[D] P. Deligne, Formes modulaires et représentations *l*-adiques, *Sém. Bourbaki*, exp. 335, 1969

[D1] P. Deligne, Les constantes des équations finctionnelles des fonctions *L*, *Lecture Notes in Math.* **349** (1973), 501–595

[D2] P. Deligne, Valeurs des fonctions *L* et périodes d'intégrales, *Proc. Symp. Pure Math.* **33** (1979), part 2, 313–346

[Dd] F. Diamond, The Taylor–Wiles construction and multiplicity one, *Inventiones Math.* **128** (1997), 379–391

[Dd1] F. Diamond, On deformation rings and Hecke rings, *Ann. of Math.* **144** (1996), 137–166

[DHI] K. Doi, H. Hida and H. Ishii, Discriminant of Hecke fields and the twisted adjoint *L*-values for *GL*(2), *Inventiones Math.* **134** (1998), 547–577

[DN] K. Doi and H. Naganuma, On the functional equation of certain Dirichlet series, *Inventiones Math.* **9** (1969), 1–14

[DR] P. Deligne and M. Rapoport, Les schémas de modules des courbes ellitpiques, *LNM* **349** (1973), 143–174

[DS] P. Deligne and J.-P. Serre, Formes modulaires de poids 1, *Ann. Sci. Ec. Norm. Sup.* 4th series **7** (1974), 507–530

[DT] F. Diamond and R. Taylor, Lifting modular mod ℓ representations, *Duke Math. J.* **74** (1994), 253–269

[E] B. Edixhoven, The weight in Serre's conjectures on modular forms, *Inventiones Math.* **109** (1992), 563–594

[Fa] G. Faltings, Crystalline cohomology and *p*-adic Galois representations, Proc. JAMI inaugural Conference, supplement to *Amer. J. Math.* (1988), 25–80

[Fk] Y. Flicker, On twisted liftings, *Transactions AMS* **290** (1985), 161–178

[Fl] M. Flach, A finiteness theorem for the symmetric square of an elliptic curve, *Inventiones Math.* **109** (1992), 307–327

[Fo1] J.-M. Fontaine, Modules galoisiens, modules filtrés et anneaux de Barsotti–Tate, *Astérisque* **65** (1979), 3–80

[Fo2] J.-M. Fontaine, Sur certains types de représentations p-adiques du group de Galois d'un corps local; construction d'un anneau de Barsotti–Tate, *Ann. of Math.* **115** (1982), 529–577

[Fo3] J.-M. Fontaine, Le corps des périodes p-adiques, *Astérisque* **223** (1994), 59–111

[Fu] K. Fujiwara, Deformation rings and Hecke algebras in totally real case, preprint, 1996

[FL] J.-M. Fontaine and G. Laffaille, Construction de représentaions p-adiques, *Ann. Sci. Ec. Norm. Sup.* 4th series **15** (1982), 547–608

[FM] J.-M. Fontaine and B. Mazur, Geometric Galois representations, in [HKC] (1995), 41–78

[G] R. Greenberg, Iwasawa theory and p-adic deformation of motives, *Proc. Symp. Pure Math.* **55** Part 2 (1994), 193–223

[G1] R. Greenberg, Arithmetic theory of elliptic curves, to appear in *Lecture Notes in Math.*, Springer, 1999

[GJ] S. Gelbart and H. Jacquet, A relation between automorphic representations of $GL(2)$ and $GL(3)$, *Ann. Sci. Ec. Norm. Sup.* 4th series **11** (1978), 471–542

[GS] R. Greenberg and G. Stevens, p-adic L-functions and p-adic periods of modular forms, *Inventiones Math.* **111** (1993), 407–447

[H81a] H. Hida, Congruences of cusp forms and special values of their zeta functions, *Inventiones Math.* **63** (1981), 225–261

[H81b] H. Hida, On congruence divisors of cusp forms as factors of the special values of their zeta functions, *Inventiones Math.* **64** (1981), 221–262

[H85] H. Hida, A p-adic measure attached to the zeta functions associated with two elliptic modular forms I, *Inventiones Math.* **79** (1985), 159–195

[H86a] H. Hida, Iwasawa modules attached to congruences of cusp forms, *Ann. Sci. Ec. Norm. Sup.* 4th series **19** (1986), 231–273

[H86b] H. Hida, Galois representations into $GL_2(\mathbb{Z}_p[[X]])$ attached to ordinary cusp forms, *Inventiones Math.* **85** (1986), 545–613

[H88a] H. Hida, On p-adic Hecke algebras for GL_2 over totally real fields, *Ann. of Math.* **128** (1988), 295–384

[H88b] H. Hida, Modules of congruence of Hecke algebras and L-functions associated with cusp forms, *Amer. J. Math.* **110** (1988), 323–382

[H89a] H. Hida, On nearly ordinary Hecke algebras for $GL(2)$ over totally real fields, *Adv. Studies in Pure Math.* **17** (1989), 139–169

[H89b] H. Hida, Nearly ordinary Hecke algebras and Galois representations of several variables, *Supplement to Amer. J. Math.* (Proc. JAMI Inaugural Conference), 1989, 115–134.

[H89c] H. Hida, Theory of p-adic Hecke algebras and Galois representations, *Sugaku Exposition* **2–3** (1989), 75–102

[H91] H. Hida, On p-adic L-functions of $GL(2) \times GL(2)$ over totally real fields, *Ann. l'institut Fourier*, **41** (1991), 311–391.

[H94a] H. Hida, p-adic ordinary Hecke algebras for $GL(2)$, *Ann. l'institut Fourier* **44** (1994), 1289–1322.

[H94b] H. Hida, On the critical values of L-functions of $GL(2)$ and $GL(2) \times GL(2)$, *Duke Math. J.* **74** (1994), 431–529

[H95] H. Hida, Control theorems of p-nearly ordinary cohomology groups for $SL(n)$, *Bull. Soc. Math. Fr.* **123** (1995), 425–475

[H96] H. Hida, On Selmer groups of adjoint modular Galois representations, Number Theory, Paris, 1993–94, *LMS Lecture Notes Series*, **235** (1996) 89–132.

[H98] H. Hida, Automorphic induction and Leopoldt-type conjectures for $GL(n)$, *Asian J. Math.* **2** (1998), 667–710

[H99a] H. Hida, Non-critical values of adjoint L-functions for $SL(2)$, *Proc. Symp. Pure Math.* **66** (1999), Part I, 123–175

[H99b] H. Hida, Adjoint Selmer groups as Iwasawa modules, *Israel Journal of Math.* 2000

[H99c] H. Hida, Control theorems for coherent sheaves on Shimura varieties of PEL-type, preprint, 1999

[Ha] G. Harder, Eisenstein cohomology of arithmetic groups. The case GL_2, *Inventiones Math.* **89** (1987), 37–118

[HM] H. Hida and Y. Maeda, Non-abelian base change for totally real fields, Special issue of *Pacific J. Math.* in memory of Olga Taussky Todd (1997), 189–217

[HS] G. Hochschild and J. P. Serre, Cohomology of group extensions, *Trans. Amer. Math. Soc.* **74** (1953), 110–134

[HT] H. Hida and J. Tilouine, Anti-cyclotomic Katz p-adic L-functions and congruence modules, *Ann. Sci. Ec. Norm. Sup.* 4th series **26** (1993), 189–259

[HT1] H. Hida and J. Tilouine, On the anticyclotomic main conjecture for CM fields, *Inventiones Math.* **117** (1994), 89–147

[HTU] H. Hida, J. Tilouine and E. Urban, Adjoint modular Galois representations and their Selmer groups, *Proc. Natl. Acad. Sci. U.S.A.* **94** (1997), 11121–11124

[K1] N. M. Katz, Higher congruences between modular forms, *Annals of Math.* **101** (1975), 332–367

[K2] N. M. Katz, p-adic properties of modular schemes and modular forms, *Lecture Notes in Math.* **350** (1973), 69–190

[K3] N. M. Katz, Serre–Tate local moduli, in "Surfaces algébriques", *Lecture Notes in Math.* **868** (1978), 138–202

[K4] N. M. Katz, p-adic L-functions for CM fields, *Inventiones Math.* **49** (1978), 199–297

[Kz] D. Kazhdan, On liftings, in "Lie group representations", *Lecture Notes in Math.* **1041** (1983), Springer

[L] R. P. Langlands, Modular forms and l-adic representation, in "Modular functions of one variable II", *Springer Lecture Notes* **349** (1973), 362–499

[Lg] S. Lang, Some history of the Shimura–Taniyama conjecture, *Notice AMS* **42** (1995), 1301–1307

[Lt] H. W. Lenstra, Complete intersections and Gorenstein rings, in [HKC] (1995), 99–109

[Ma] B. Mazur, Modular curves and the Eisenstein ideal, *Publ. IHES* **47** (1977), 33–186

[Ma1] B. Mazur, Deforming Galois representations, in "Galois group over $\mathbb{Q}$", *MSRI publications* **16**, (1989), 385–437

[MR] B. Mazur and K. A. Ribet, Two dimensional representations in the arithmetic of modular curves, *Astérique* **196–197** (1991), 215–255

[MRo] B. Mazur and L. Roberts, Local Euler characteristics, *Inventiones Math.* **9** (1970), 201–234

[MT] B. Mazur and J. Tilouine, Représentations Galoisiennes, différentielles de Kähler et "conjectures principales", *Publ. IHES* **71** (1990), 65–103

[MTT] B. Mazur, J. Tate and J. Teitelbaum, On p-adic analogies of the conjectures of Birch and Swimerton-Dyer, *Inventiones Math.* **84** (1986), 1–48

[MW] B. Mazur and A. Wiles, Class fields of abelian extensions of $\mathbb{Q}$, *Inventiones Math.* **76** (1984), 179–330

[MW1] B. Mazur and A. Wiles, On p-adic analytic families of Galois representations, *Compositio Math.* **59** (1986), 231–264

[MoW] C. Mœglin and J.-L. Waldspurger, Le spectre résiduel de $GL(n)$, *Ann. Sci. Ec. Norm. Sup.* 4th series **22** (1989), 605–674

[Ny] L. Nyssen, Pseudo-représentations, *Math. Ann.* **306** (1996), 257–283

[O1] M. Ohta, On l-adic representations attached to automorphic forms, *Japan J. Math.* **8** (1982), 1–47

[O2] M. Ohta, On the zeta function of an abelian scheme over the Shimura curve, *Japan J. Math.* **9** (1983), 1–25

[O3] M. Ohta, On the zeta function of an abelian scheme over the Shimura curve II, *Advanced Studies in Pure Math.* **2** (1983), 37–54

[Oc] T. Ochiai, Control theorem for Selmer groups of p-adic representations, Master Thesis, University of Tokyo, 1998

[P] P. Procesi, The invariant theory of $n \times n$ matrices, *Advances in Math.* **19** (1976), 306–381

[R] K. A. Ribet, A modular construction of unramified p-extensions of $\mathbb{Q}(\mu_p)$, *Inventiones Math.* **34** (1976), 151–162

[R1] K. A. Ribet, Congruence relations between modular forms, *Proc. Int. Cong. of Math.* (1983), 503–514

[R2] K. A. Ribet, On modular representations of $\mathrm{Gal}(\overline{\mathbb{Q}}/\mathbb{Q})$ arising from modular forms, *Inventiones Math.* **100** (1990), 431–476

[Ra] R. Ramakrishna, On a variation of Mazur's deformation functor, *Compositio Math.* **87** (1993), 269–286

[Rb] K. Rubin, The "main conjecture" in Iwasawa theory for imaginary quadratic fields, *Inventiones Math.* **103** (1991), 25–68

[Ro] J. D. Rogawski, Functoriality and the Artin conjecture, *Proc. Symp. Pure Math.* **61** (1997), 331–353

[Se] J.-P, Serre, Propriétés galoisiennes des points d'ordre fini des courbes elliptiques, *Inventiones Math.* **15** (1972), 259–331

[Se1] J.-P, Serre, Sur les représentations modulaires de degré 2 de $\mathrm{Gal}(\overline{\mathbf{Q}}/\mathbf{Q})$, *Duke Math. J.* **54** (1987), 179–230

[Sh1] G. Shimura, On the holomorphy of certain Dirichlet series, *Proc. London Math. Soc.* **31** (1975), 79–98

[Sh2] G. Shimura, Eisenstein series and zeta functions on symplectic groups, *Inventiones Math.* **119** (1995), 539–584

[Sh3] G. Shimura, Yutaka Taniyama and his time, Very personal recollections, *Bull. London Math. Soc.* **21** (1989), 186–196

[Sh4] G. Shimura, Response to Steele Prize for Lifetime Achievement, *Notice AMS* **43** (1996), 1343–1347

[St1] J. Sturm, Special values of zeta functions, and Eisenstein series of half integral weight, *Amer. J. Math.* **102** (1980), 219–240

[St2] J. Sturm, Evaluation of the symmetric square at the near center point, *Amer. J. Math.* **111** (1989), 585–598

[T] J. Tate, p-divisible groups, *Proc. of Conference on Local Fields*, Driebergen (1966), 158–183

[T1] J. Tate, Global class field theory, in *Algebraic Number Theory* (Cassels and Fröhlich (eds.)), Academic Press, 1967, 163–203

[T2] J. Tate, A review of non-archimedean elliptic functions, in [HKC] (1995), 162–184

[Ti] J. Tilouine, Sur la conjecture principale anticyclotomique, *Duke Math. J.* **59** (1989), 629–673

[Ty1] R. Taylor, On Galois representations associated to Hilbert modular forms, *Inventiones Math.* **98** (1989), 265–280

[Ty2] R. Taylor, Galois representations associated to Siegel modular forms of low weight, *Duke Math. J.* **63** (1991), 281–332

[Ty3] R. Taylor, ℓ-adic representations associated to modular forms over imaginary quadratic fields. II, *Inventiones Math.* **116** (1994), 619–643

[TU] J. Tilouine and E. Urban, Several variable p-adic families of Siegel–Hilbert cusp eigensystems and their Galois representations, to appear, *Ann. Sci. Ec. Norm. Sup.*

[TW] R. Taylor and A. Wiles, Ring theoretic properties of certain Hecke modules, *Ann. of Math.* **141** (1995), 553–572

[U] E. Urban, Formes automorphes cuspidales pour GL_2 sur un corps quadratiques imaginaire. Valeurs spéciales de fonctions L et congruences, *Compositio Math.* **99** (1995), 283–324

[W] A. Wiles, Modular curves and the class group of $\mathbb{Q}(\mu_p)$, *Inventiones Math.* **58** (1980), 1–35

[W1] A. Wiles, The Iwasawa conjecture for totally real fields, *Ann. of Math.* **131** (1990), 493–540

[W2] A. Wiles, Modular elliptic curves and Fermat's last theorem, *Ann. of Math.* **141** (1995), 443–551

[We] A. Weil, Über Bestimmung Dirichletcher Reihen durch Funktionalgleichungen, *Math. Ann.* **168** (1967), 149–156

Subject Index

List of Statements

List of Symbols